Fractals in Engineering
From Theory to Industrial Applications

Springer
London
Berlin
Heidelberg
New York
Barcelona
Budapest
Hong Kong
Milan
Paris
Santa Clara
Singapore
Tokyo

Jacques Lévy Véhel, Evelyne Lutton and
Claude Tricot (Eds)

Fractals in Engineering

From Theory to Industrial Applications

With 152 Figures
Including 6 Colour Plates

Springer

Dr J. Lévy Véhel
Dr E. Lutton
INRIA - Rocquencourt,
Domaine de Voluceau, BP 105, 78153 Le Chesnay Cedex, France

Professor C. Tricot
Ecole Polytechnique de Montréal,
CP 6079, Succ. Centre-Ville, Montréal, Québec H3C 3A7, Canada

ISBN-13: 978-1-4471-1253-2 Springer-Verlag Berlin Heidelberg New York

British Library Cataloguing in Publication Data
Fractals in engineering : from theory to industrial applications
1.Engineering mathematics - Congresses 2.Fractals - Congresses
I.Levy Vehel, Jacques II.Lutton, Evelyne III.Tricot, Claude
620'.001514742
ISBN-13: 978-1-4471-1253-2 e-ISBN-13: 978-1-4471-0995-2
DOI: 10.1007/978-1-4471-0995-2

Library of Congress Cataloging-in-Publication Data
A catalog record for this book is available from the Library of Congress

Softcover reprint of the hardcover lst edition 1997

Typesetting: Camera ready by editors

69/3830-543210 Printed on acid-free paper

Foreword

It is no longer necessary to present Fractal Analysis. So many colloquiums have been organized, so many books have been written on this subject that we do not feel obliged to explain its importance in experimental science. After about 15 years of pioneering work more and more engineers show their interest in the modeling of irregular data with the help of fractal sets or processes. Many have tried to find their way through theoretical books, essentially based upon the Geometric Measure Theory, with the risk of finding themselves eventually very far from the original starting point. After this hard experience some of them may have been tempted to decide that Fractal Analysis was just one more of those mathematical games. On the other hand, some applied books are focussed on the experimental side to such an extent that it is hard to understand how things really work and which tools are the most useful for a given purpose.

As the title points out, **Fractals in Engineering** is written for all engineers, and for experimental scientists in general. It is not a pure, theoretical treatise. It shows mathematics at work, so to speak, through important applications originating in a deep theory. Numerous subjects are addressed, since Fractal Science is a self-ramified field of research, as everyone knows. We wished to keep a balance between the classical and the new; between solid, well-referenced survey articles and bold, original developments. Contributors have been chosen with care, needless to say; the intrinsic quality of the product is theirs. But we tried to give it direction and self-consistency, with a multidisciplinary flavor which makes it a book of interest to most researchers.

In all sections the reader will find theoretical material which supports the construction; one does not make science without a good mathematical basis. This volume contains the state-of-the-art on different subjects such as Multifractal Analysis, IFS Theory, Fractal Random Processes. Strong, basic articles can also be found on Fractals and Waves, or on Chemical Engineering. But the sections also include practicable solutions to current engineering problems such as Image Segmentation, Network Traffic Modeling or Multiband Antennas Design. These applications give life and justification to the subject. New fields of research are presented as well : Financial Engineering, Environmental or Biomedical applications.

Without trying to make predictions, the future of Fractal Analysis will certainly benefit from this interaction between mathematics and industrial research. Industrial applications without their theoretical support tend to confine themselves to repetitive, computerized recipes (we wish all research heads in industry can understand that; we know that some of them do). On the other hand, isolated mathematicians tend to close themselves inside apparently sterile problems, the private property of a very few specialists. The experimental research has recently given rise to deep and exciting subjects of thought for the mathematicians. Our best hope is that this book constitutes one more evidence of these fruitful exchanges, and a major step in this direction.

It is important to mention at this point all the people who took part in the elaboration of this volume. The contributors first, who did the most important job. Our publishers of Springer-Verlag, for their assistance and comprehension. Our institution INRIA, for the technical assistance. We will not forget Nathalie Gaudechoux, for the patient typing, re-typing, and reconstruction work... Finally, the entire team of the "Groupe Fractales" at INRIA, who participated at all stages and proved, once more, their dedication and solidarity. We address a warm thank to all of them.

Jacques Lévy Véhėl,
Evelyne Lutton,
Claude Tricot.

Table of Contents

RANDOM PROCESSES

NETWORK TRAFFIC ANALYSIS

FRACTALS AND WAVES

MULTIFRACTAL ANALYSIS

Some open problems about multifractal functions

S. Jaffard[1,2]

[1] Département de Mathématiques, Université Paris XII,
Faculté des Sciences et Technologie,
61 Av. du Gal. de Gaulle, 94010 Créteil Cedex, FRANCE

[2] C.M.L.A., E.N.S. de Cachan, FRANCE

Summary. We review some results concerning multifractal functions and emphasize the limitations of these results and the open problems that are related.

1. Introduction

First, let me state that this paper reflects my personal views about multifractal functions. In particular, the open problems I will mention are basically the ones I have met, or the ones that are naturally 'just behind' the theorems that I like or that I have worked on.

The notion of multifractal function is based on *pointwise regularity*. Let me recall this notion. Let α be a positive real number and $x_0 \in I\!R^m$; a function $F : I\!R^m \to I\!R$ is $C^\alpha(x_0)$ if there exists a polynomial P of degree less than α such that

$$|F(x) - P(x - x_0)| \leq C|x - x_0|^{\alpha}. \tag{1.1}$$

Note that the constant term of $P(x-x_0)$ is $F(x_0)$, and if F is $C^{[\alpha]}$ in a neighbourhood of x_0, the polynomial P is exactly the Taylor expansion of F at x_0 of order $[\alpha]$. Nonetheless, (1.1) can hold for a large α even if F is not differentiable in a neighbourhood of x_0 (consider for instance the 'chirp' $x^n \sin(x^{-n})$ in a neighbourhood of 0 for a large n).

The *Hölder exponent* $h(x_0)$ is the supremum of all the values of α such that (1.1) holds. This Hölder exponent is a function which is defined point by point and describes the local variations of the irregularity of the function F. We will mention several problems related with the determination of the Hölder exponent in Section 2.

We are interested in analyzing functions whose Hölder exponent may widely change from point to point. This instability makes the task of determining the Hölder exponent $h(x)$ very difficult numerically. Nonetheless the study of such functions has proved important in several domains of physics and signal analysis. An important example is supplied by multifractal functions, which are such that the points with a given Hölder exponent form fractal sets; examples of such functions include plots of random walks, interfaces developing in reaction-limited growth processes or turbulent velocity signals at inertial range (see [3] or [5]). A relevant mathematical tool studied in this context is the *spectrum of singularities* $d(h)$ (the function $d(h)$ associates to each positive h the Hausdorff dimension of the set A_h of the points x where $h(x) = h$). The most important example where one would like to determine the spectrum of singularities of a function is probably the velocity of fully developed turbulence. The reason is that turbulent flows are not spatially homogeneous: the irregularity of the velocity seems to differ widely from point to point. This phenomenon, called *intermittency*, suggests that the determination of the spectrum of singularities of the velocity of the fluid might be a nontrivial function, universal (i.e. independent of boundary conditions in the limit of small viscosity), and thus would yield an important information on the nature of turbulence.

2. Hölder exponent

2.1 The wavelet method

The first problem in the study of a multifractal function is the determination of its pointwise regularity. This is usually done using one of the two following methods; either using directly the definition of the Hölder regularity, or a wavelet criterion. There are several variants for such criteria, see [21], which all relate pointwise regularity with decay conditions of the wavelet coefficients. Let us mention the simplest one.

Let F be a function defined on $I\!R$. One associates with F a two-variables function, its wavelet transform, defined in the 'time-scale' open half plane by

$$C(a,b) = \frac{1}{a}\int F(t)\psi(\frac{t-b}{a})dt \quad a>0, b\in I\!R \tag{2.1}$$

(here ψ is smooth, compactly supported, and has vanishing moments).

A function $F \in C^\alpha(I\!R)$ if (1.1) holds for every x_0, the constant C being independent of x_0. $F \in C^\alpha(I\!R)$ if and only if

$$|C(a,b)| \leq Ca^\alpha. \tag{2.2}$$

Proposition 2.1. *Let $F : I\!R \to I\!R$ be a bounded function. If F is $C^h(x_0)$,*

$$|C(a,b)| \leq Ca^h\left(1+\frac{|b-x_0|}{a}\right)^h. \tag{2.3}$$

Conversely suppose that there exists $h' < h$ such that

$$|C(a,b)| \leq Ca^h\left(1+\frac{|b-x_0|}{a}\right)^{h'}, \tag{2.4}$$

then F is $C^h(x_0)$.

Note that (2.4) implies a minimal uniform regularity of f (using (2.2), we see that f belongs to $C^{h-h'}$ in a neighbourhood of x_0). The other slightly different wavelet criteria of pointwise regularity share all this drawback: they can only be applied if f has some minimal uniform regularity. Hence an important problem is to determine if there exists a simple condition on the wavelet transform which would imply pointwise regularity, even when f has no minimal uniform regularity (one should take into account the counterexamples given in [21] which show in which sense the converse part in Proposition 2.1 is optimal). This problem is important for instance because many functions which have a dense set of discontinuities are known to be multifractal. We will see in Section 2.3 the case of Lévy processes and of some functions of arithmetic nature which are constructed as a superposition of jumps.

2.2 Functions with prescribed Hölder exponents

We consider now the construction of functions with a given Hölder exponent. This problem was first raised by Jacques Lévy Véhel in the context of speech simulation (see [9]); a speech signal has a Hölder exponent with sharp fluctuations (especially within consonants), hence the idea that an efficient way to keep the relevant features of such a signal would be to store its Hölder exponent rather than sampled values of the signal itself. Which functions are Hölder exponents?

Theorem 2.1. *A nonnegative function $h(x)$ is the Hölder exponent of a continuous function if and only if it can be written as a lim inf of a sequence of continuous functions.*

Note that this theorem leaves completely open the problem of determining the Hölder exponents of non-continuous functions. This problem is not a mere refinement, because of the importance of functions with a dense set of singularities, that we will consider later.

There exists different constructive proofs of Theorem 2.1 which are deterministic. One of them relies directly on Proposition 2.1, and is obtained through an explicit positioning of the wavelet coefficients, [17]. Alternative constructions were developed independently by Jacques Lévy Véhel, Yves Meyer and Khalid Daoudi; one of them is based on *Iterated Function Systems*, see [9].

These constructions allow to construct the most general Hölder exponents. This generality must be payed: Functions constructed through these procedures are extremely peculiar and could not be used in any realistic simulation. It would thus be useful to have a natural random construction of such functions. A partial answer to this problem is supplied by the *Multifractional Brownian motion*. Its construction is a straightforward extension of the definition of the Fractional Brownian Motion of order α except that, instead of defining it by the fractional integral of order α of a white noise, we take a fractional integration of order $\alpha(x)$. Thus its Hölder exponent is allowed to change, and has, near the point x, the same features as a fractional brownian motion of order $\alpha(x)$.

Definition 2.1. *Let $\alpha(x)$ be a function defined on $\mathbb{R}^d$ such that $0 < \alpha(x) < 1$. Let $A = \sup \alpha(x)$; we suppose that $\alpha(x) \in C^A(\mathbb{R}^d)$. The Multifractional Brownian motion of order $\alpha(x)$ is defined by*

$$B_\alpha(x) = \int \frac{e^{ix.\xi} - 1}{|\xi|^{\alpha(x)+d/2}} dW(\xi) \tag{2.5}$$

where W is a standard Brownian motion.

This process shares many properties with the fractional Brownian motion; for instance it is locally self-similar (see [6]). It was introduced independently in [24] and [6]. Though the definition does not require Hölder regularity for $\alpha(x)$, it is only under this assumption that we can prove that the Hölder exponent of B_α is actually $\alpha(x)$. Thus, though this construction is more natural for simulations, this process does not completely fulfill our purpose, since the Hölder exponent is required to be smooth. In particular, for speech simulations, one would like to reach Hölder exponents with discontinuities. Thus, it would be important to get rid of this regularity requirement in the construction of the multifractional Brownian motion.

2.3 Functions with a dense set of discontinuities

I am ashamed to say that I was unaware of the existence of some very interesting multifractal functions which have a dense set of discontinuities until the spring of 1994. At that time and almost simultaneously Jean-François Le Gall drew my attention to Lévy processes and Jean-Pierre Kahane to the *integrable Riemann Function.*

A Lévy process X_t ($t \geq 0$) valued in $I\!R^d$ is, by definition, a stochastic process with stationary independent increments: $X_{t+s} - X_t$ is independent of the $(X_v)_{0 \leq v \leq t}$ and has the same law as X_s. If we except the Brownian motion, the compound Poisson processes and their sums, the other Lévy processes have a dense set of discontinuities.

In his famous *Habilitationsshrift* of 1854 on trigonometric series, Bernard Riemann introduced the integral that now bears his name. To show that this new integral did in fact extend the Cauchy integral, Riemann gives the example of a function that is too irregular to be Cauchy integrable but that is nevertheless 'Riemann integrable'

$$R(x) = \sum_{n=1}^{\infty} \frac{(nx)}{n^2}, \tag{2.6}$$

where $(x) = x - m$ if $|x - m| < 1/2$ and $(x) = 0$ if $x = m + 1/2$, $m \in Z$. Riemann remarked that R is continuous except at the rationals $p/2q$ (p and $2q$ having no common factor), with the following right and left limits at these points

$$R(\frac{p}{2q}^{-}) = R(\frac{p}{2q}) + \frac{\pi^2}{16q^2} \quad \text{and} \quad R(\frac{p}{2q}^{+}) = R(\frac{p}{2q}) - \frac{\pi^2}{16q^2}.$$

Thus R is discontinuous on a dense set; but, in contrast with the characteristic function of rationals, the following property holds: For all $\varepsilon > 0$, the set of points where R has a discontinuity of amplitude larger than ε is finite. Thus R is indeed Riemann integrable.

This 'Integrable' Riemann function and Lévy processes may seem completely unrelated. Actually Lévy processes are constructed as a sum of piecewise linear functions which have more and more jumps, just as in Formula (2.6). In order to describe this construction, we need to recall some properties of Lévy processes.

The characteristic function of a Lévy process X_t (taking values in $I\!R^d$) satisfies $I\!E(e^{i\langle\lambda|X_t\rangle}) = e^{-t\psi(\lambda)}$ where

$$\psi(\lambda) = i\langle a|\lambda\rangle + \frac{1}{2}Q(\lambda) + \int_{I\!R^d} \left(1 - e^{i\langle\lambda|x\rangle} + i\langle\lambda|x\rangle 1_{|x|<1}\right) \pi(dx);$$

Q is a positive quadratic form and $\pi(dx)$ is the Lévy measure of X_t, i.e. a positive Radon measure defined on $I\!R^d - \{0\}$ satisfying

$$\int (1 \wedge |x|^2)\pi(dx) < \infty. \tag{2.7}$$

The Lévy measure is usually not integrable in the neighbourhood of the origin; this is in particular the case for stable Lévy processes of index α who satisfy (in polar coordinates) $\pi(dr, d\theta) = r^{-\alpha-1}dr\nu(d\theta)$ where ν is a finite measure on the unit sphere. When $\pi(I\!R^d) = +\infty$, the growth of the Lévy measure near the origin can be estimated using the *upper index*

$$\alpha = \inf\{\alpha : \int_{|x|\leq 1} |x|^\alpha \pi(dx) < \infty\}$$

(when X_t is a stable process, this definition coincides with the definition of the stability index).

Lévy processes are constructed as a superposition of independent compensated Poisson processes X_t^j which have jumps whose size belongs to

$$\Gamma_j = \{x : \ 2^{-j-1} < |x| \leq 2^{-j}\}.$$

Let Y_t^j be the compound Poisson process of Lévy measure

$$\pi_j(dx) = 1_{\Gamma_j}(x)\pi(dx)$$

and let X_t^j be the compound compensated Poisson process

$$X_t^j = Y_t^j - t\int_{\mathbb{R}^d} x\pi_j(dx);$$

then $X_t = \sum_{j=1}^{\infty} X_t^j$; X_t is thus a sum of discontinuous functions with more and more jumps (the expectation of the number of jumps of X_t^j is $\pi(\Gamma_j)$ which usually goes to infinity).

Paul Lévy was actually aware of the similarities of Lévy processes and deterministic functions such as R. Let us quote [23]; the function R *is exactly what Paul Lévy called a compensated jump function: all jumps are negative and their sum is infinite but the continuous parts of* (nx) *provide a shift such that the series converges. Paul Lévy considered the simpler function*

$$\mathcal{L}(x) = \sum_{n=0}^{\infty} \frac{(2^n x)}{2^n}.$$

as an illustration of what occurs frequently in the theory of stochastic processes with independent increments.

The analysis of Riemann's function R and of Lévy processes have many similarities. Consider for instance the very simple following lemma of [18].

Lemma 2.1. *Assume that f is discontinuous on a dense set. For $x_0 \in \mathbb{R}$, assume that $r_n \to x_0$ and that at each point r_n f has right and left limits $f(r_n^+)$ and $f(r_n^-)$ and that*

$$|f(r_n^+) - f(r_n^-)| = s_n > 0.$$

Then

$$h_f(x_0) \leq \liminf \frac{\log s_n}{\log |r_n - x_0|}. \tag{2.8}$$

Let us recall the proof of this lemma. By definition

$$h_f(x_0) = \sup\{\alpha : \ f \ \text{ is } \ C^\alpha(x_0)\}.$$

Take $\alpha < h_f(x_0)$. There exists a polynomial P of degree less than $[\alpha]$ and $C > 0$ such that

$$\forall |x - x_0| \leq 1, \ \ |f(x) - P(x - x_0)| \leq C|x - x_0|^\alpha.$$

Since

$$|(f(r_n^+) - P(r_n^+ - x_0)) - (f(r_n^-) - P(r_n^- - x_0))| = s_n > 0,$$

either $|f(r_n^+) - P(r_n - x_0)| \geq s_n/2$ or $|f(r_n^-) - P(r_n - x_0)| \geq s_n/2$. For sake of argument assume that $|f(r_n^+) - P(r_n - x_0)| \geq s_n/2$. Then, by our assumptions about the limit $f(r_n^+)$ and by the continuity of P, we can, for each n, choose $r'_n > r_n$ such that

$$|f(r'_n) - P(r'_n - x_0)| \geq s_n/4$$

and $1/2 \leq \frac{|r'_n - x_0|}{|r_n - x_0|} \leq 2$. Thus

$$0 < \frac{s_n}{4} \leq |f(r'_n) - P(r'_n - x_0)| \leq C|r'_n - x_0|^\alpha \leq C'|r_n - x_0|^\alpha.$$

We conclude that

$$\liminf \frac{\log s_n}{\log |r_n - x_0|} \geq \alpha.$$

hence Lemma 2.1.

It appears clearly in this proof that the upper bound (2.8) takes into account individual discontinuities but not accumulations of them. It will thus give the right Hölder exponent only if the discontinuities are well 'spread out'. This is the case for Riemann's function, for arithmetic reasons; it is also the case for Lévy processes, for probabilistic reasons. There is probably a common cause for these phenomena, which asserts that, if the discontinuities are ordered by decreasing order of magnitude, Lemma 2.1 gives the exact Hölder exponent everywhere under some assumptions on the discrepancy of the sequence of discontinuities. A precise statement (and proof...) would be important in order to obtain a unifying approach to the study of functions with a dense set of discontinuities.

3. Multifractal functions

When a Hölder exponent jumps from point to point, its determination is either impossible or irrelevant. Standard examples are supplied by *multifractal functions.* In this case, the set of points where the function F has a given Hölder exponent h is a fractal set; we have thus to deal with a whole collection of fractal sets indexed by the parameter h (hence the *'multi'* in multifractal). We want to extract the important information concerning the size and geometry of these sets of singularities. A minimal relevant mathematical notion is the *spectrum of singularities* of F. For each $h \geq 0$, $d(h)$ is the Hausdorff dimension of the set of points x_0 where the Hölder exponent of F is h (we take $dim(\emptyset) = -\infty$). Of course we could be more ambitious and try to obtain more information about the geometry of these sets. An important notion is for instance the *lacunarity* of a fractal set, which was defined by B.Mandelbrot. We could also try to understand better the interplay between the respective geometries of theses sets. For instance, in all examples of multifractal functions, all theses sets are dense in a given Cantor-type set. Is this situation generic in some sense? We are not aware of any results in these directions.

3.1 Multifractal formalism

If one comes back to the definition of the spectrum of singularities, it is impossible to compute numerically the spectrum of a signal since it involves the successive determination of several intricate limits, and a blind application of the formula giving the definition of the Hausdorff dimension would yield enormous, totally unstable calculations. The only method is to find some 'reasonable' assumptions under which the spectrum could be derived using only averaged quantities (which should be numerically stable) extracted from the signal. Therefore U. Frisch and G. Parisi introduced, in the context of fully developed turbulence, a formula which allows to compute the spectrum of singularities of a signal by a Legendre transform of some numerically easily computable function, see [13]. Note that the study of this Legendre transform was already advocated in the seminal paper of Mandelbrot [26]. Alternative formulas were proposed by Arneodo, Bacry and Muzy (see [3]). Let us now describe these so-called *multifractal formalisms for functions.*

The Structure Function method requires first the computation of

$$S_q(l) = \int_{I\!R^m} |F(x+l) - F(x)|^q dx.$$

Supposing that the scaling $S_q(l) \sim |l|^{\zeta(q)}$ holds, the spectrum of singularities is computed using the formula $d(h) = \inf_q (qh - \zeta(q) + m)$.

The Wavelet Transform Integral method: One computes

$$\tilde{Z}(a,q) = \int_{I\!R^m} |C(a,b)|^q db$$

and, if $\tilde{Z}(a,q) \sim a^{\eta(q)}$,

$$d(h) = \inf_q (qh - \eta(q) + m) \tag{3.1}$$

The Wavelet Transform Maxima method requires first the computation of

$$Z(a,q) = \sum_{\ell} \sup_{(b,a')\in\ell} |C(a',b)|^q$$

where ℓ is a line of maxima of the wavelet transform considered on $[a,0]$, and $\sup_{(b,a')\in\ell}$ means that the supremum is taken for (b,a') on this line of maxima (so that $a' \leq a$). If $Z(a,q) \sim a^{\theta(q)}$, then $d(h) = \inf_q (qh - \theta(q))$.

In many cases, the three methods can be shown to yield the same function $d(h)$ [19]. However, numerically, the best method seems to be the last one, probably because the restriction of the computation to the maxima insures that small errors are less taken into account (because, at the maxima, they are relatively less important), see [3]. More generally, methods that involve the wavelet transform are numerically more stable, probably because they involve only averaged quantities and not the direct values of the function.

The multifractal formalism may be surprising at first sight because it relates pointwise behaviors to global estimates. Let us give the heuristic argument from which it is derived. Though this argument cannot be transformed into a correct mathematical proof, it shows at least why these formulas can be expected to hold, and a careful study of its implicit assumptions shows for which type of functions it cannot hold.

Let us calculate the contribution of the Hölder singularities of order h to the integral $\int_{I\!R^m} |C(a,b)|^q db$. Near a singularity of order h, we have, in a small box of size a, $|C(a,b)|^q \sim a^{hq}$ because Proposition 2.1 basically means that if F is $C^h(x_0)$ and not smoother, then the order of magnitude of its wavelet transform in the cone $|b - x_0| \leq a$ is about a^h. If the dimension of these singularities is $d(h)$, there are about $a^{-d(h)}$ such boxes, each of volume a^m, so that the total contribution to the integral is $a^{hq+m-d(h)}$. The real order of magnitude of the integral should be given by the largest contribution, so that

$$\eta(q) = \inf_h (hq + m - d(h)). \tag{3.2}$$

This formula is not the one we are looking for since we know $\eta(q)$ and we look for $d(h)$; but if (3.2) holds and if d is concave (we will see that in general this assumption needs not be verified, however in many cases it is), $d(h)$ is recovered by an inverse Legendre transform formula which yields (3.1). Of course, if d is not concave, one expects (3.1) to yield only the convex hull of the spectrum; this is the 'phase transition' phenomenon.

Though (3.1) does not always hold, the following upper bound is always valid if F has a minimal uniform Hölder regularity (see [21]).

Proposition 3.1. *Let $\varepsilon > 0$. The following upper bound holds for any function $F \in C^\varepsilon(I\!R^m)$:*

$$d(h) \le \inf_p (m - \eta(p) + hp). \tag{3.3}$$

In general, (3.3) is not an equality. We define $\mathcal{C}$ as the class of functions that can be written as the supremum of a countable set of functions of the form $c1_{[a,b]}(x)$ (where we can have $a = b$). Thus, Riemann integrable functions belong to $\mathcal{C}$, but also, for instance, the indicatrix function of the rationals (but not the indicatrix function of the irrationals). Any positive function of $\mathcal{C}$ is a spectrum of singularities, see [19]; the proof of this result is constructive; one defines explicitly the wavelet coefficients of the corresponding function. Nonetheless, we still do not know if any spectrum of singularities necessarily belongs to $\mathcal{C}$.

There exist some mathematical examples which Hölder exponent can be analytically determined, and the validity of the Multifractal Formalism can be tested. These examples include the class of self-similar functions [19], [7], and also a number of specific examples which belong to the history of mathematics, like Riemann's 'non-differentiable' function [14], Polya's 'triangle-filling' curve [20], lacunary wavelet series (see Section 4.3), ... , see also [7], [1], [2], [19], [18], and references therein. A careful analysis of these meaningful examples gives some insight about the conditions a signal must satisfy so that the Multifractal Formalism is valid; in all the examples of this list of apparently unrelated examples there exists a common feature, which is some self-similarity (deterministic or statistic) either of the function or of its wavelet transform. It is therefore reasonable to conjecture that if a function satisfies some self-similarity condition, perhaps in some statistical sense, then the multifractal formalism is likely to hold. But it is impossible to state a reasonably general conjecture (one should be careful to avoid in such a statement the counterexamples exhibited by M. Ben Slimane in [7]). In particular, we are extremely far to determining if it does hold for the velocity of a turbulent fluid or not.

4. Chirps and oscillating singularities

Why does the multifractal formalism sometimes fail? A well-known reason is the 'phase transition' phenomenon that we mentioned earlier; strong local oscillations of the signal can also be a more subtle cause of failure. Such strong oscillations appear when the signal has at many points a chirp-like behavior; chirps are functions that behave like $x^\alpha \sin(1/x^\beta)$ at the origin. This phenomenon had until recently escaped attention because the intuition concerning the multifractal formalism for functions was based on corresponding previous results concerning the multifractal formalism for measures, and of course positive measures, by definition, have no oscillations. Thus the considerations we will develop are relevant for at least two reasons:

- They show that the problems posed by the Multifractal Formalism is of a very different nature for functions and measures.
- They give some insight about possible generalizations of the Multifractal Formalism for functions that would take into account this oscillatory behavior.

In the heuristic calculation that justified it, we interpreted Proposition 2.1 as meaning that if F is $C^h(x_0)$ and not smoother, the order of magnitude of its wavelet transform in the cone $|b - x_0| \leq a$ is about a^h. This holds for cusp-like singularities like $|x - x_0|^h$ but not for 'chirp-like' singularities, which have a very small wavelet transform in the cone.

4.1 Oscillating singularity exponents

First we discuss the mathematical notions of *oscillating singularity* and *chirp*. There is a general agreement on an informal definition: A chirp of type (h, β) at x_0 should oscillate as

$$f(x) = |x - x_0|^h \sin(\frac{1}{|x - x_0|^\beta}) \tag{4.1}$$

in the neighborhood of x_0. In signal analysis, this notion should cover functions whose *instantaneous frequency* increases fast at some time. A remarkable property of (4.1) is that, after each integration, the Hölder exponent of the primitive at x_0 is increased by $1 + \beta$ and not 1 as could be expected. This remark is the starting point of the definition of chirps given by Yves Meyer, see Definition 4.2 below and [21]. We will actually use a slightly different definition which has the additional advantage to be stable with respect to the addition of 'smooth noise', a feature which is mandatory for using this notion in signal processing.

The choice of the definition we use (see [1] and [2]) is based on the following remark. Let B be a brownian motion; consider the function

$$C(x) = x^{1/3} \sin(1/x) + B(x); \tag{4.2}$$

the strongest singularity at 0 is the chirp $x^{1/3} \sin(1/x)$, and one actually observes this oscillatory behavior after magnifying enough the graph near the origin. But after one integration, the random term becomes preponderant, and the oscillatory behavior disappears. These oscillations, which do not exist in the primitive, should be reflected in the chirp exponents.

Let us now show how we can introduce a definition of *oscillating singularities* which agrees with the definition of a chirp for functions such as (4.1), and has the required stability properties with respect to the addition of 'smooth noise'. If

$$F(x) = |x - x_0|^h g(\frac{1}{|x - x_0|^\beta}) + O(|x - x_0|^{h'}) \tag{4.3}$$

where $h' > h$, the first term describes the local behavior of F near x_0. In that case, we would like to say that the type of oscillating singularity at x_0 is (h, β), i.e., the oscillation of F should be the oscillation of the lowest order term of its expansion. Clearly, we need to require that g has one vanishing moment. After one integration, the main term of the primitive of F may be the remaining term, which is $O(|x-x_0|^{h'+1})$. This last remark shows that, in sharp contrast with the definition of chirps given in [21], the oscillation exponent β of an oscillating singularity should not be determined using a large number of integrations, or even one integration.

Let $h^t(x_0)$ denote the Hölder exponent of the fractional primitive of order t at x_0 of the function F defined by (4.3). More precisely, if F is a bounded function, we denote by $h^t(x_0)$ the Hölder exponent at x_0 of the function

$$F_t = (Id - \Delta)^{-t/2}(\phi F) \tag{4.4}$$

where ϕ is a C^∞ compactly supported function satisfying $\phi(x_0) = 1$, and the operator $(Id - \Delta)^{-t/2}$ is the convolution operator which amounts to multiply the Fourier transform of the function with $(1 + |\xi|^2)^{-t/2}$. If one performs a fractional integration of order t small enough, i.e. such that

$$h + (1+\beta)t < h' + t, \tag{4.5}$$

then $h^t(x_0) = h + (1+\beta)t$ (here h, h' and β are the exponents used in (4.3)). The gain of pointwise Hölder regularity at x_0 after a fractional integration of very small order t is $(1+\beta)t$; hence the following definition for exponents of oscillating singularities.

Definition 4.1. *Let $F : \mathbb{R}^d \to \mathbb{R}$ be a bounded function. The oscillating singularity exponents of F at a point x_0 are defined by*

$$(h, \beta) = \left(h(x_0), \left.\frac{\partial}{\partial t} h^t(x_0)\right|_{t=0} - 1\right) \tag{4.6}$$

These exponents belong to $[0, +\infty] \times [0, +\infty]$.

Remarks: This definition makes sense because, for a given x_0, the function $t \to h^t(x_0)$ is differentiable (with a possible derivative of $+\infty$), as is shown in [1]. Note that if $h^t(x_0) = +\infty$ the exponent β is not defined.

4.2 Construction and analysis of functions with oscillating singularities

If we retain the Hölder *and* the oscillation exponents as the relevant parameters of pointwise regularity, the very first question is to determine which functions $(h(x), \beta(x))$: $\mathbb{R} \to \mathbb{R}^2$ are simultaneously the Hölder and oscillation exponents of a function. When we consider only $h(x)$ we saw that the problem is settled; but when both functions are considered, the problem is completely open. This is not interesting only for theoretical reasons. A simple constructive solution could give a more refined solution to the simulation problems that motivated the construction of functions with a prescribed Hölder exponent.

We are naturally led to define the *Spectrum of oscillating singularities* as the dimension $D(h,\beta)$ of the set $E_{h,\beta}$ of points x_0 where the oscillating singularity exponents are (h,β). Which functions $D(h,\beta)$ are the Spectrum of oscillating singularities of a function f?

As regards functions with oscillating singularities, it is natural to expect that the usual spectrum of singularities $d(h)$ can be recovered by

$$d(h) = \sup_\beta D(h, \beta), \tag{4.7}$$

i.e. by projecting the graph of $D(h,\beta)$ on the plane $\beta = 0$. This formula is verified for all known examples of multifractal functions with oscillating singularities (see [1],[2] and [15]) is likely to hold under rather general conditions. It would be important to determine such conditions, since (4.7) is a bridge between the two notions of multifractal functions.

Also, we have no natural candidate for a multifractal formalism that would be adapted to functions with oscillating singularities, though it can be worked out is some particular cases (see [2]). The determination of $D(h,\beta)$ through a multifractal formalism is important for two reasons: it yields more information about the nature of the singularities than the function $d(h)$, and it would be likely to hold in more cases, because the origin of the failure of the multifractal formalism, i.e. oscillating singularities, would precisely be taken into account.

4.3 Lacunary wavelet series

The success of wavelet techniques in many fields of applications is largely due to the following remarkable property: Many signals, images, or mathematical functions can be accurately represented in a wavelet basis using very few nonzero coefficients. It is the case of piecewise smooth signals, of the velocity of turbulent flows [3], of solutions of nonlinear hyperbolic equations, see [11]. The starting point of the denoising algorithm of images called *wavelet shrinkage* is based on the remark that, since an image composed of piecewise smooth parts has few nonzero wavelet coefficients, a noised image, can be denoised by setting to zero all small wavelet coefficients; this amounts to approximate the noisy image by a lacunary wavelet series [10] and [12].

Mathematically, the fact that a function has few non-vanishing wavelet coefficients can be formalized by stating that it belongs to Besov spaces $B_p^{s,p}$ for p close to zero (see [10]).

Though many signals or functions have been shown to be accurately represented by lacunary wavelet series, the properties of these functions have never been investigated. We will consider a simple probabilistic model of such series. Though the model is itself extremely simple, the corresponding random functions have an extremely complicated local structure: they exhibit a whole range of oscillating singularities located on random fractal sets.

Let $\eta < 1$. For each $j \geq 0$ we choose at random and independently $[2^{\eta j}]$ locations $k2^{-j} \in [0,1]$. The corresponding wavelet coefficients $C_{j,k}$ take the value $2^{-\alpha j}$; we set to 0 all other wavelet coefficients. This is the most elementary model of random lacunary wavelet series one can think of.

In order to study theses wavelet series, we now introduce a slightly different notion of *chirp exponents* which differs from the definition of 'oscillating singularity exponent' given in Section 4.1

Definition 4.2. *Let $h \geq 0$ and $\beta > 0$. A function $f \in L^\infty$ is a chirp of type (h,β) at x_0 if the iterated primitives $f^{(-1)}, \dots f^{(-n)}, \dots$ satisfy*

$$f^{(-n)} \in C^{h+n(\beta+1)}(x_0).$$

The following characterization holds [21].

Proposition 4.1. *A function $f \in L^\infty$ is a chirp of type (h,β) at x_0 if and only if there exists a function $r(x) \in C^\infty$ in a neighbourhood of x_0 and $\varepsilon > 0$ such that*

$$f(x) = r(x - x_0) + (x - x_0)^h g_+\left((x - x_0)^{-\beta}\right)$$

if $0 < x < \varepsilon$ *and*

$$f(x) = r(x - x_0) + |x - x_0|^h g_-\left(|x - x_0|^{-\beta}\right)$$

if $-\varepsilon < x < 0$, *the functions* g_+ *and* g_- *being 'infinitely oscillating', i.e. they have bounded primitives of any order.*

The following property allows to define 'chirp exponents'.

Lemma 4.1. *Let* $f \in L^\infty$. *If* f *is a chirp of type* (h, β) *at* x_0 *with* $\beta > 0$ *and* $f \in C^{h'}(x_0)$ *for* $h' > h$, *then* $\forall \beta' < \beta$, f *is a chirp of type* (h', β') *at* x_0.

This is a straightforward consequence of Definition 4.2 and of the characterization given by Proposition 2.1. This lemma means that the interior of the set of couples (h, β) such that f is a chirp of type (h, β) at x_0 is a rectangle: $h \leq h_0$, $\beta < \beta_0$. We can use these two values to define chirp exponents.

Definition 4.3. *The chirp exponents of a function* f *at* x_0 *are*

- $h(x_0)$ *(which is the Hölder exponent at* x_0*),*
- $\beta(x_0) = \sup\{\beta : \exists h$ *such that* f *is a chirp of type* (h, β) *at* $x_0\}$.

Using this definition, we naturally obtain the notion of *chirp spectrum*: The chirp spectrum $d(h, \beta)$ of f is the Hausdorff dimension of the set of points where f has the chirp exponents (h, β).

The reader will have noticed the difference with the notion of *spectrum of oscillating singularities* which was defined in Section 4.2 Let us stress that the notion of oscillating singularities is adapted to the analysis of real signals, since it is not sensitive to perturbations by smooth noise, while the notion of chirp is adapted to mathematically defined functions. Of course the open problems that we mentioned about oscillating singularity exponents and oscillating singularity spectrums can also be considered in this slightly different setting.

Theorem 4.1. *The lacunary wavelet series* F *defined above is almost surely a multifractal function. Its chirp spectrum is supported by the segment*

$$h = \alpha(\beta + 1) \quad \textit{for} \quad h \in [\alpha, \alpha/\eta]$$

and on this segment

$$d(h, \beta) = \eta(\beta + 1).$$

This very specific example suggests that functions which have 'most' of their wavelet coefficients vanishing are likely to be 'generically' multifractal functions, if the positions of the big wavelet coefficients are decorrelated between scales.

4.4 Back to functions with a dense set of discontinuities

The Riemann integrable function R is multifractal. More precisely, let x_0 be an irrational number and p_n/q_n be the sequence of convergents of the continued fraction expansion of x_0. We define τ_n by

$$|x_0 - \frac{p_n}{q_n}| = \frac{1}{q_n^{\tau_n}}.$$

It is an elementary result from continued fractions that $\tau_n \geq 2$. Let

$$\tau(x_0) = \limsup_{n \in A} \tau_n \tag{4.8}$$

where A is the set of integers n such that q_n is even (if this set contains only a finite number of elements, we set $\tau(x_0) = 2$).

Theorem 4.2. *The Hölder exponent of the integrable Riemann function R at x_0 is*

$$h_R(x_0) = \frac{2}{\tau(x_0)},$$

and its spectrum of singularities is given by

$$\begin{aligned} d(\alpha) &= \alpha \quad \text{for} \quad \alpha \in [0,1] \\ &= -\infty \quad \text{elsewhere.} \end{aligned}$$

Note that this Hölder exponent is strikingly similar to the Hölder exponent of another function attributed to Riemann:

$$\mathcal{R}(x) = \sum \frac{1}{n^2} \sin(\pi n^2 x).$$

Indeed, let

$$\tau'(x_0) = \limsup_{n \in B} \tau_n$$

where B is the set of integers n such that p_n and q_n are not both odd, see [14]. The Hölder exponent of $\mathcal{R}$ is

$$h_{\mathcal{R}}(x_0) = \frac{1}{2} + \frac{1}{2\tau'(x_0)},$$

see [18]. $\mathcal{R}$ is also a multifractal function with spectrum of singularities

$$\left.\begin{aligned} d_{\mathcal{R}}(\alpha) &= 4\alpha - 2 \quad \text{if} \quad \alpha \in [\frac{1}{2}, \frac{3}{4}] \\ &= 0 \qquad\quad \text{if} \quad \alpha = \frac{3}{2} \\ &= -\infty \qquad \text{elsewhere.} \end{aligned}\right\} \tag{4.9}$$

It would certainly be interesting to determine if there is a deeper relationship between these two functions introduced by Riemann.

The trigonometric series $\mathcal{R}$ can be generalized in 2 dimensions by

$$\mathcal{R}_\alpha(x) = \sum \frac{1}{(n^2+m^2)^\alpha} \sin(\pi n^2 x + m^2 y).$$

The multifractal properties of this function have been investigated by H.Oppenheim in his thesis. However its spectrum of singularities is not completely known yet. It would also be of interest to investigate the functions

$$\sum \frac{1}{n^p} \sin(\pi n^p x),$$

or more generally trigonometric series related with arithmetic functions.

Lévy's function $\mathcal{L}$ a is another example of a multifractal function. The reader can consult [18] where he will find other similar examples of multifractal functions with a dense set of discontinuities.

In the determination of the spectrum of singularities of Lévy processes we can deal immediately with four trivial cases:

- X_t is deterministic; then $X_t = Ct$ and $d(h) = -\infty \ \forall h$.
- X_t is a compound Poisson process; then X_t is piecewise linear, with a finite number of jumps on any bounded interval, so that $d(0) = 0$ and $d(h) = -\infty$ else.

- X_t is a Brownian motion; then $d(1/2) = 1$ and $d(h) = -\infty$ else.
- X_t is the superposition of a Brownian motion and a compound Poisson process; one easily checks that $d(0) = 0$, $d(1/2) = 1$ and $d(h) = -\infty$ else.

Let

$$\begin{aligned} d_\alpha(h) &= \alpha h \quad \text{if } h \in [0, 1/\alpha] \\ &= -\infty \quad \text{else;} \end{aligned}$$

$$\begin{aligned} \overline{d_\alpha}(h) &= \alpha h \quad \text{if } h \in [0, 1/2] \\ &= 1 \quad \text{if } h = 1/2 \\ &= -\infty \quad \text{else.} \end{aligned}$$

The following theorem yields the spectrum of singularities of Lévy processes, see [16].

Theorem 4.3. *Let X_t be a Lévy process satisfying*

$$\limsup C_j = +\infty$$

and

$$\sum 2^{-j} \sqrt{C_j \log(1 + C_j)} < \infty.$$

- *If X_t has no Brownian component ($Q \equiv 0$), the spectrum of singularities of almost every sample path of X_t is $d_\alpha(h)$.*
- *If X_t has a Brownian component ($Q \not\equiv 0$), the spectrum of singularities of almost every sample path of X_t is $\overline{d_\alpha}(h)$.*

It is striking that the proofs of Theorems 4.2 and 4.3 use similar arguments, based on the very regular repartitions of the discontinuities either of Lévy processes, or of Riemann's function. I strongly suspect that, under some equirepartition assumption of this type, one could prove that functions with a dense set of discontinuities are generically multifractal, with a spectrum of singularities $d(h) = ah$; the invert of the number a measuring how fast the rate of the number of jumps of small size increases (when the size of the jumps goes to 0).

4.5 Partial Differential Equations

Let us now come back to the problem initially raised by U.Frisch and G.Parisi, namely to determine if 'generically' the velocity of a turbulent fluid is a multifractal function. From a mathematical point of view, this problem can be translated as follows: We can consider a partial differential equation which modelizes the evolution of the velocity of a fluid, and solve it taking for initial value a random process. It is extremely difficult to obtain some information of the random process thus obtained at time $t > 0$; to our knowledge, the only information available concerns Burgers equation

$$\frac{\partial u}{\partial t} + \frac{\partial}{\partial x}\left(\frac{u^2}{2}\right) = 0. \tag{4.10}$$

This equation has been considered with initial values such as Brownian motion, white noise or fractional Brownian motions, see [4], [28], [29]). Some information is available about the statistics of the jumps at time t.

The problem of determining the random process we thus obtain at time $t > 0$ is completely solved in the case of Brownian motion as initial data: J. Bertoin in [8] proved that the increments of the solution at time t is a Lévy process with Laplace exponent (for $t = 1$)

$$I\!E(e^{-q(u(x,1)-u(x,0))}) = \exp(-x(\sqrt{2q+1}-1)).$$

In this case a monofractal initial value (a Brownian motion) immediately becomes multifractal. This is yet a too specific and isolated result to draw some general conclusions concerning 'generic' solutions of nonlinear evolution equations. It would certainly be important to extend this result, and, for instance, show the existence of multifractal solutions to general scalar hyperbolic equations

$$\frac{\partial u}{\partial t} + \frac{\partial}{\partial x}(f(u)) = 0 \tag{4.11}$$

with f convex, and fractional Brownian motions as initial data. Working out several dimensional equations would also be important.

For those who do not feel confident with stochastic processes, a deterministic equivalent of the problems we rose is obtained by taking as initial data a Weierstrass function $\sum a^{-n} \sin b^n x$ instead of a fractional Brownian motion.

4.6 Traces of Multifractal functions and negative dimensions

Let us return to one of our initial motivations: the validity of the multifractal formalism applied to turbulence data. In practice the multifractal formalism cannot be applied to the whole four-dimensional velocity $u(x,t)$. In experimental situations, the velocity is measured at one point of the wind tunnel where turbulence is created. So we do not have access to the velocity itself but rather to a one dimensional trace of four dimensional function. This rises many mathematical problems that are far from having received a satisfactory answer. For instance, if the multifractal formalism holds for the a four dimensional function, does it hold for almost every one dimensional trace?

The following phenomenon is observed when the multifractal formalism is applied to one dimensional turbulence data: it yields for some values of h a negative value of $d(h)$, see [3]. This can be explained as follows: A standard ergodicity argument suggests that the multifractal formalism when applied to this one-dimensional trace yields the three dimensional spectrum minus 2. Results of Kahane and Mattila, see [22] and [27], show that 'generically' the intersection of a three dimensional fractal set of dimension d with a line is of dimension $d-2$. When d is less than 2, the multifractal formalism still yields $d-2$, which becomes negative and has been interpreted by Benoit Mandelbrot in [25] as 'rare' singularities that do not appear on all traces.

The multifractal formalism can be restated in a way which is easier to work with for mathematicians: Let $W^{s,p}(I\!R^d)$ be the space of L^p functions whose derivatives of order s belong to L^p (when s is not an integer, one should use fractional derivatives) The function $\eta(p)$ can be defined as

$$\eta(p) = \sup\{s: \quad F \in W^{s/p,p}(I\!R^d)\}.$$

Thus another related question is to determine to which Sobolev spaces $W^{s',p'}(I\!R)$ does the trace of a function in $W^{s,p}(I\!R^d)$ belongs. A partial answer is given by this theorem, see [21].

Theorem 4.4. *Let $f \in W_p^s(I\!R^d)$, $m < d$ and $a \in I\!R^{d-m}$; let $f_a(x) = f(x,a)$ ($x \in I\!R^m$). If s' satisfies $s - (d-m)/p \leq s' \leq s$ and $s' > 0$, then $f_a \in W_p^{s'}(I\!R^m)$ except for a set of values of a of dimension at most $d = d - m - p(s - s')$, and this result is optimal.*

This theorem only gives upper bounds; our understanding of the traces of multifractal functions is still very poor. Suppose for instance that the function we consider has oscillating singularities; can they be detected on traces? Is almost every trace of an oscillating singularity oscillating?

References

1. Arneodo, A., Bacry, E., Jaffard, S., Muzy, J.F.: *Oscillating Singularities on Cantor sets. A grand-canonical multifractal formalism.* To appear in J. of Stat. Phys..
2. Arneodo, A., Bacry, E., Jaffard, S., Muzy, J.F.: *Singularity spectrum of multifractal functions involving oscillating singularities.* To appear in J. of Fourier Anal. and App..
3. Arneodo, A., Bacry, E., Muzy, J.F. (1995): *The thermodynamics of fractals revisited with wavelets.* Physica A, Vol. 213, 232–275.
4. Avellaneda, M.W.E. (1995): *Stastical properties of shocks in Burgers turbulence.* Comm. Mat. Phys., Vol. 172, 13–38.
5. Bacry, E., Arneodo, A., Muzy, J.F. (1993): *Singularity spectrum of fractal signals from wavelet analysis: exact results.* Jour. Stat. Phys., 70, N. 314.
6. Benassi, A., Jaffard, S., Roux, D.: *Elliptic Gaussian Random Processes.* To appear in Revista Mathematica Iberoamericana.
7. Benslimane, M. (1996): *Etude du formalisme multifractal pour les fonctions.* Thèse de l'ENPC.
8. Bertoin, J. (Preprint 1997): *The inviscid Burgers equation with Brownian initial velocity.*
9. Daoudi, K., Lévy Véhel, J., Meyer, Y. (Preprint 1995): *Construction of continuous functions with prescribed local regularity.* To appear in Constr. Approx.
10. DeVore, R., Lucier, B. (1992): *Fast wavelet techniques for near-optimal image processing.* Proc. IEEE Mil. Commun. conf..
11. DeVore, R., Lucier, B. (Preprint 1993): *On the size and smoothness of solutions to nonlinear hyperbolic conservations laws.*
12. Donoho, D.: *Denoising via soft thresholding.* IEEE Transactions on information theory, to appear.
13. Frisch, U., Parisi, G. (1985): *Fully developped turbulence and intermittency.* Proc. Int. Summer school Phys. Enrico Ferrmi, North Holland, 84–88.
14. Jaffard, S. (1996): *The spectrum of singularities of Riemann's function.* Revista Mathematica Iberoamericana, Vol. 12, N. 2, 441–460.
15. Jaffard, S. (Preprint): *On lacunary wavelet series T.*
16. Jaffard, S. (Preprint): *The multifractal nature of Lévy processes.*
17. Jaffard, S. (1995): *Functions with a prescribed Hölder exponent.* Applied and Computational Harmonic Analysis, Vol. 2, 400–401.
18. Jaffard, S. (1997): *Old friends revisited. The multifractal nature of some classical functions.* J. Fourier Anal. and App., V. 3, N. 1, 1–22.
19. Jaffard, S.: *Multifractal formalism for functions. Part I and Part II.* To appear in S.I.A.M. Journal of Mathematical Analysis.

20. Jaffard, S., Mandelbrot, B. (1996): *Local regularity of nonsmooth wavelet expansions and application to the Polya function.* To appear in Adv. in Math., Vol. 120, N. 2, 265–282.
21. Jaffard, S., Meyer, Y. (1996): *Wavelet methods for pointwise regularity and local oscillations of functions.* Memoirs of the A.M.S., Vol. 123, N. 587.
22. Kahane, J.P. (Preprint 1986): *Sur la dimension des intersections.* Aspects of Mathematics and its applications, Elsevier science publishers.
23. Kahane, J.P., Lemarié, P.G. (1996): *Fourier series and Wavelets.* Gordon Breach.
24. Lévy Véhel, J., Peltier, R. (1995): *Multifractional Brownian Motion : definition and preliminary results.* INRIA Technical Report.
25. Mandelbrot, B. (1991): *Random Multifractals: negative dimensions and the resulting limitations of the thermodynamic formalism.* Proceedings of the Royal Society of London A, V. 434, 79–88.
26. Mandelbrot, B. (1962): *Intermittent turbulence in selfsimilar cascades; divergence of high moments and dimension of the carrier.* J. Fluid Mech., 331–358.
27. Mattila. P. (1985): *On the Hausdorff dimension and capacities of intersections.* Mathematika, 32, 213–217.
28. She, Z., Aurell, E., Frisch, U. (1992): *The inviscid Burgers equation with initial data of Brownian type.* Comm. Mat. Phys., 148, 623–641.
29. Sinai, Y. (1992): *Statistics of shocks in solutions of inviscid Burgers equation* Comm. Mat. Phys., 148, 601–621.

Integral Transforms of Multifractal Measures

Closed-form Results and Engineering Applications

M. Giona[1,3] and O. Patierno[2,3]

[1] Dipartimento di Ingegneria Chimica Universitá di Cagliari
piazza d'Armi, 09123 Cagliari, ITALY
[2] Dipartimento di Ingegneria Chimica, Universitá di Roma "La Sapienza"
via Eudossiana 18, 00184 Roma, ITALY
[3] Centro Interuniversitario sui Sistemi Disordinati e sui Frattali nell'Ingegneria Chimica c/o Dip. Ingegneria Chimica Universitá di Roma "La Sapienza"
via Eudossiana 18, 00184 Roma, ITALY

Summary. The properties of integral transforms (Laplace, Fourier, Stieltjes) of multifractal measures associated with affine IFSP are analyzed in detail. Closed form expressions are obtained for a particular class of affine IFSP labeled unimodular. The theory is applied to such engineering problems as scattering from fractals and reactions in continuous mixtures, and to integral (Volterra-type) equations involving singular (multifractal) kernels.

1. Introduction

One of the main achievements of fractal theory is related to the study and characterization of singular non-atomic distributions, usually referred to as multifractal measures [1, 2, 3].

In the multifractal scaling theory, the support $\mathcal{C}$ of a measure μ, can be regarded as composed by the union of subsets $\mathcal{C}_\alpha$, $\mathcal{C} = \cup_\alpha \mathcal{C}_\alpha$, each of which is characterized by a local Hölder exponent equal to α. The multifractal spectrum $f(\alpha)$ is the Hausdorff dimension of the subset $\mathcal{C}_\alpha$, and is related to the spectrum of generalized dimensions $D(q)$ through a Legendre transform [2, 3].

Multifractal distributions arise in many natural phenomena (e.g. turbulence, fluid mixing) frequently associated with multiplicative (cascade) stochastic processes or chaotic dynamical systems. Multifractal concepts have already been successfully applied to engineering problems such as image and signal processing [4, 5].

The simplest (and most studied) family of multifractal measures is the invariant measures of affine Iterated Function Systems with Probability (IFSP) [6, 7]. It should be observed that the theory of IFSP represents an extremely powerful theoretical framework to address the properties of multifractal measures. By applying IFSP theory, the statistical properties (moment analysis) of multifractal measures have been studied and applied to the inverse problem of image reconstruction [8, 9, 10].

This article focuses on mathematical physical applications of singular (multifractal) measures, and specifically on the properties of integral transforms of invariant measures associated with affine IFSP. We present closed-form results and uniformly convergent approximation methods to obtain Laplace, Fourier and Stieltjes transforms. These results are applied to some physical problems such as scattering theory from fractals and chemical reaction kinetics in a continuous mixture.

Finally, the concept of a Volterra integral equation in the presence of singular multifractal kernels is introduced and developed.

As probability measures associated with affine IFSP are analyzed throughout this article, we present a brief review of the basic concepts of IFS theory, which will also prove useful for further developments.

An affine IFSP is a system $\{w_h(x), p_h\}_{h=1}^N$, where $w_h(x) = A_h x + b_h$ are affine contractive maps on a metric space (X, d), (the base space X is a compact subspace of R^n, and d a distance function defined on it), and p_h the corresponding probability weights, $\sum_{h=1}^N p_h = 1$.

In the set of probability measures $\mathcal{M}(X)$ on the σ-algebra of Borel subsets of X, the Markov operator $M : \mathcal{M}(X) \to \mathcal{M}(X)$ associated with the IFSP $\{w_h, p_h\}_{h=1}^N$ is defined by the equation

$$M[\nu] = \sum_{h=1}^{N} p_h \nu \circ w_h^{-1} , \qquad \nu \in \mathcal{M}(X) , \tag{1.1}$$

where $\circ$ indicates the composition operator. The space $(\mathcal{M}, d_H)$, equipped with the Hutchinson metrics d_H, is a complete metric space, and the Markov operator turns out to be a contraction within it [6]. Consequently, there exists a unique invariant measure μ of the IFSP, which is a fixed point of the Markov operator

$$M[\mu] = \mu . \tag{1.2}$$

The set $\mathcal{C}$, constituting the support of the measure μ, satisfies the relation

$$\mathcal{C} = \bigcup_{h=1}^{N} w_h(\mathcal{C}) , \tag{1.3}$$

which is referred to as the Hutchinson equation.

An important consequence of the definition of the Markov operator, eqs. (1.1)–(1.2) is the averaging theorem for continuous functions. Let $f : X \to C$ be a generic continuous function attaining complex values. The average of f with respect to the unique invariant measure of the IFSP satisfies the equation

$$\int_{\mathcal{C}} f(x) d\mu(x) = \sum_{h=1}^{N} p_h \int_{\mathcal{C}} f \circ w_h(x) d\mu(x) , \tag{1.4}$$

which is the basic relation for the analysis developed below.

2. Integral Transforms

In this article, we consider linear integral transforms of invariant measures associated with affine IFSP of the form

$$\mathcal{T}(z) = \int_{\mathcal{C}} K(z, x) d\mu(x) , \tag{2.1}$$

where $K(z, x)$ is a continuous kernel (with respect to its arguments).

In particular, we consider the Fourier transform

$$\mathcal{F}(k) = \int_{\mathcal{C}} \exp(i < k, x >) d\mu(x) , \tag{2.2}$$

where i is the imaginary unit, $k \in R^n$ and $< \cdot, \cdot >$ indicates the inner product. Eq. (2.2) can be regarded as the Fourier-Stieltjes transform associated with μ whose support is $\mathcal{C}$.

In the one-dimensional case, let us assume without loss of generality that $\mathcal{C} \subseteq [0,1]$. Let us order the maps w_h for increasing values of b_h, i.e. $b_h < b_{h+1}$ and let us assume $b_1 = 0$, and $a_h > 0$.

The mixed Stieltjes transform $\mathcal{M}(s,z)$ of μ can be defined as

$$\mathcal{M}(s,z) = \int_{\mathcal{C}} \frac{\exp(-sx)}{1+zx} d\mu(x) , \tag{2.3}$$

where s is a complex variable and z a real variable. For $z = 0$, the mixed Stieltjes transform reduces to the Laplace transform $\mathcal{M}(s,0) = \mathcal{L}(s)$. For $s = 0$, it reduces to the Stieltjes transform $\mathcal{M}(0,z) = \mathcal{S}(z)$. An application of this transform is discussed in section 4.2.

3. General Properties of Integral Transforms

In this section we briefly present the main results obtained for integral transforms of invariant measures of affine IFSP.

By applying eq. (1.4) to the kernel $f(x) = \exp(i < k, x >)$, it follows that

$$\int_{\mathcal{C}} \exp(i < k, x >) d\mu(x) = \sum_{h=1}^{N} p_h \int_{\mathcal{C}} \exp(i < k, A_h x + b_h >) d\mu(x)$$

$$= \sum_{h=1}^{N} p_h \exp(i < k, b_h >) \int_{\mathcal{C}} \exp(i < A_h^t k, x >) d\mu(x) , \tag{3.1}$$

where A_h^t is the transpose of A_h, i.e.

$$\mathcal{F}(k) = \sum_{h=1}^{N} p_h \exp(i < k, b_h >) \mathcal{F}(A_h^t k) . \tag{3.2}$$

Eq. (3.2) can be regarded as the spectral representation of the Markov operator, eqs. (1.1)–(1.2).

In a similar manner, by plugging $f(x) = \exp(-sx)/(1+zx)$ into eq. (1.4), a functional equation is obtained for the mixed Stieltjes transform

$$\mathcal{M}(s,z) = \sum_{h=1}^{N} p_h \frac{\exp(-b_h s)}{1+b_h z} \mathcal{M}\left(a_h s, \frac{a_h z}{1+b_h z}\right) . \tag{3.3}$$

In particular, for the Laplace transform, it follows that

$$\mathcal{L}(s) = \sum_{h=1}^{N} p_h \exp(-b_h s) \mathcal{L}(a_h s) . \tag{3.4}$$

Eqs. (3.2)–(3.4) constitute the starting point for further analysis. The results obtained can be organized into the following subtopics.

3.1 Closed-form Results and Uniformly Convergent Approximations

Closed-form results have been obtained for a particular family of affine IFSP, referred to as *unimodular*. An affine IFSP is said to be unimodular, if $w_h(x) = Ax + b_h$, ($w_h(x) = ax + b_h$ in one dimension), i.e. if $A_h = A$ for all the maps w_h. In particular, in one dimension the family of unimodular IFSP coincides with that of self-similar IFSP.

For unimodular IFSP, the Laplace transform of AES invariant measures can be expressed in closed form as a convergent infinite product [11]:

$$\mathcal{L}(s) = \prod_{n=0}^{\infty} \left[\sum_{h=1}^{N} p_h \exp(-a^n b_h s) \right] \quad , \tag{3.5}$$

where $\mathcal{L}(s)$ is analytic in the half-plane $\text{Re}[s] > 0$. Moreover, the sequence of approximants $\{L_n(s)\}$

$$L_n(s) = \prod_{j=0}^{n} \left[\sum_{h=1}^{N} p_h \exp(-a^j b_h s) \right] \quad , \tag{3.6}$$

converges uniformly towards $\mathcal{L}(s)$.

An analogous result holds for the Fourier transforms. In the case of unimodular IFSP, the Fourier transform can be expressed as

$$\mathcal{F}(k) = \prod_{n=0}^{\infty} \left[\sum_{h=1}^{N} p_h \exp\left(i < (A^t)^n k, b_h >\right) \right] \quad , \tag{3.7}$$

and the sequence $\{F_n(k)\}$

$$F_n(k) = \prod_{j=0}^{n} \left[\sum_{h=1}^{N} p_h \exp(i < (A^t)^j k, b_h >) \right] \quad . \tag{3.8}$$

converges uniformly to $\mathcal{F}(k)$ in any bounded set of k.

In the case of generic affine IFSP, the integral transforms can be obtained from the recursion schemes reported in Table 3.1, which converge towards the corresponding integral transform for $n \to \infty$.

Fig. 3.1 shows the convergence properties of the sequence $F_n(k)$ towards $\mathcal{F}(k)$ obtained by applying the recursion shown in Table 3.1 in the case of the IFPS generating the two-dimensional Sierpinski gasket, $A = \text{diag}(1/2, 1/2)$, $b_1 = (0,0)^t$, $b_2 = (1/2, 0)^t$, $b_3 = (0, 1/2)^t$, with $p_h = 1/3$. This figure shows the maximum distance $\Delta(n) = \sup_k |F_n(k) - \mathcal{F}(k)|$, for k_i $(i = 1, 2)$ in the range (-100,100). An exponential dependence $\Delta(n) \sim \exp(-\lambda n)$ can be observed.

3.2 Scaling Analysis

Let us first consider the Laplace transform, bearing in mind the assumption $b_h < b_{h+1}$, $b_1 = 0$ and $a_h > 0$. It has been shown [11] that the Laplace transform $\mathcal{L}(s)$ of invariant measures of affine IFSP displays asymptotically for large s, $\text{Re}[s] \gg 1$ a power-law behaviour

Integral transform	Recursion scheme
Laplace transform	$L_{n+1}(s) = \sum_{h=1}^{N} p_h \exp(-b_h s) L_n(a_h s)$ $L_0(s) = 1$
Fourier transform	$F_{n+1}(k) = \sum_{h=1}^{N} p_h \exp(i < k, b_h >) F_n(A_h^t k)$ $F_0(k) = 1$
mixed Stieltjes transform	$M_{n+1}(s,z) = \sum_{h=1}^{N} p_h \frac{\exp(-b_h s)}{1+b_h z} M_n\left(a_h s, \frac{a_h z}{1+b_h z}\right)$ $M_0(s,z) = 1$

Table 3.1. Recursion schemes for integral transforms.

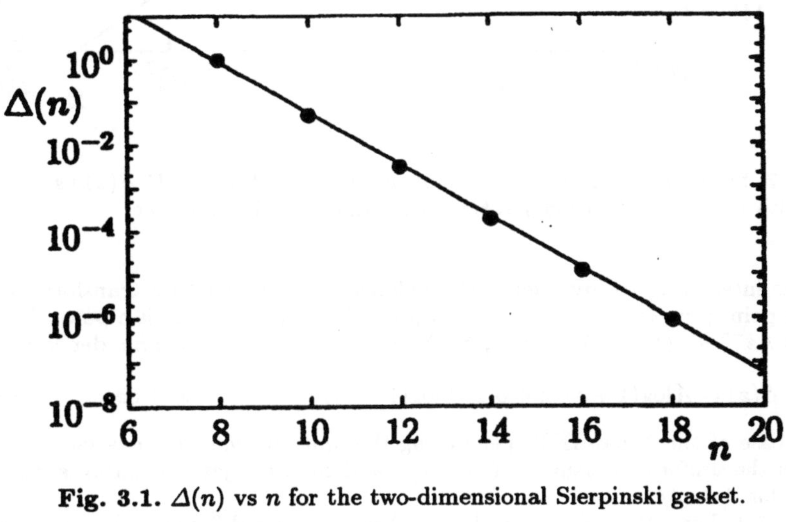

Fig. 3.1. $\Delta(n)$ vs n for the two-dimensional Sierpinski gasket.

$$\mathcal{L}(s) \sim s^{-\beta} \qquad \text{with} \qquad \beta = \frac{\log p_1}{\log a_1} \,. \tag{3.9}$$

An interesting scaling behaviour is related to the Stieltjes transform $\mathcal{S}(z) = \mathcal{M}(0,z)$, which applies in some physical problems (see e.g. section 4.2). The Stieltjes transform of a one-dimensional family of measures parameterized with respect to the probability weights $\{p_h\}$ displays a phase transition in its asymptotic behaviour, i.e. for large z. There is a critical value $p_c = a_1$ of p_1 such that for $p_1 < p_c$: $\mathcal{S}(z) \sim 1/z$, and for $p_1 > p_c$: $\mathcal{S}(z) \sim 1/z^\gamma$, with $\gamma = \log p_1 / \log a_1$, i.e.

$$\mathcal{S}(z) \sim z^{-\gamma} \qquad \text{with} \qquad \gamma = \min\{1, \log p_1/\log a_1\} \,. \tag{3.10}$$

Fig. 3.2 shows the behaviour of $\mathcal{S}(z)$ vs z in the case of the 2-map IFSP, $w_1(x) = x/3$, $w_2(x) = (x+2)/3$, $p_1 = p$, $p_2 = 1-p$, generating the Cantor middle-third set. The critical value is $p_c = 1/3$. The derivative of γ with respect to p is discontinuous at $p = p_c$.

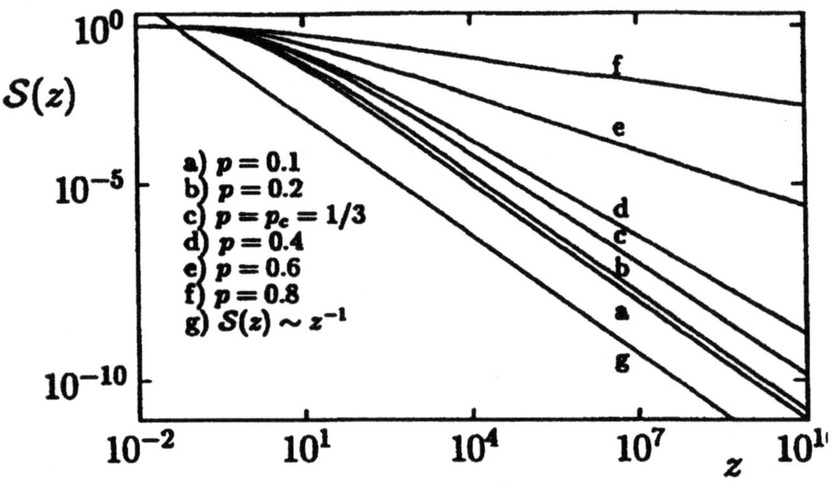

Fig. 3.2. Stieltjes transform of invariant measures of affine IFSP. $\mathcal{S}(z)$ vs z associated with the 2-map IFSP generating the Cantor middle-third set.

It is interesting to investigate the behaviour of the Stieltjes transform at the critical point $p = p_c$. If $p \neq p_c$, the asymptotic behaviour of $\mathcal{S}(z)$ is a power law, i.e. $\mathcal{S}(z) = Az^{-\gamma} + o(z^{-\gamma})$. At $p = p_c$, $\mathcal{S}(z)$ displays a weak logarithmic dependence:

$$\mathcal{S}(z) = A\log(1+z)/z + o(z^{-1}\log(1+z)) \,, \qquad z \to \infty \,. \tag{3.11}$$

In the case of the 2-map IFSP generating the unit interval, the invariant measure at p_c is the uniform measure $d\mu(x) = dx$, and the Stieltjes transform is given by $\mathcal{S}(z) = \log(1+z)/z$.

Fig. 3.3 shows the behaviour of $z^\gamma \mathcal{S}(z)$ vs $1+z$ for different values of p in the case of the 2-map IFSP generating the Cantor middle-third set. Numerical analysis (data fitting) indicates a value of the constant A appearing in eq. (3.11) bounded

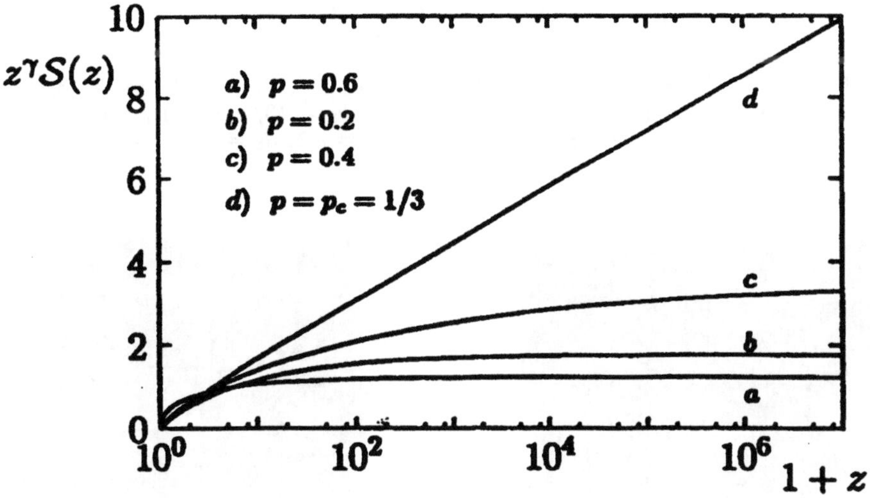

Fig. 3.3. $z^\gamma S(z)$ vs $1+z$ for the 2-map IFSP generating a binomial measure on the Cantor middle-third set.

between 0.60 and 0.64, which seems to support the conjecture (not justified) that in this case $A \simeq d_H$, $d_H = \log 2/\log 3$ being the Hausdorff dimension of the support $\mathcal{C}$.

4. Application to Engineering Problems

In this section we discuss some applications of the theory presented above to physical and engineering problems.

4.1 Scattering from Fractals

Fourier transforms arise in scattering experiments. The study of scattering from fractal structures has been extensively discussed in [12, 13] by applying renormalization methods. Attention should be drawn to the analogy between the recursion scheme to compute the Fourier transform and the renormalization analysis developed in [12, 13].

The application of eqs. (3.7)–(3.8) or of the recursion scheme reported in Table 3.1 provides a useful starting point to investigate scattering in the presence of a multifractal density function and in the important case of scattering from fractal surfaces. To give an example, Fig. 4.1 shows the contour plot of the scattering intensity (structure factor) $I(k) = |F_n(k)|$ ($n = 100$) obtained by applying the recursion scheme of Table 3.1 to two fractal interfaces: a self-similar Koch curve generated by means of 8 contractions

$$w_h(x) = \begin{pmatrix} s & 0 \\ 0 & s \end{pmatrix} + \begin{pmatrix} b_{xh} \\ b_{yh} \end{pmatrix}, \quad h = 1, .., 4 \tag{4.1}$$

$$w_h(x) = \begin{pmatrix} 0 & sc_h \\ sd_h & 0 \end{pmatrix} + \begin{pmatrix} b_{xh} \\ b_{yh} \end{pmatrix}, \quad h = 5, .., 8,$$

with $s = 1/4$, $b_1 = (0,0)^t$, $b_2 = (1/4,1/4)^t$, $b_3 = (1/2,-1/4)^t$, $b_4 = (3/4,0)^t$, $b_5 = (1/4,0)^t$, $b_6 = (1/2,1/4)^t$, $b_7 = (1/2,0)^t$, $b_8 = (3/4,-1/4)^t$, and $c_5 = -1$, $d_5 = 1$, $c_6 = 1$, $d_6 = -1$, $c_7 = 1$, $d_7 = -1$, $c_8 = -1$, $d_8 = 1$, $p_h = 1/8$, and the

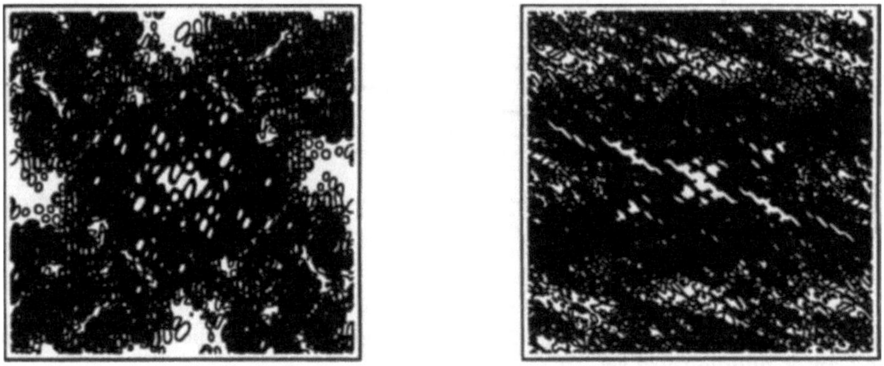

Fig. 4.1. Contour plots of $I(k) = |F_n(k)|$, ($n = 100$, $-200 \leq k_i \leq 200$ $i = 1,2$) for two-dimensional fractal model surfaces, as discussed in section 4.1: self similar Koch curve (left); self-affine fractal interface (right).

self-affine curve generated by means of a 5-map IFSP with

$$w_h(x) = \begin{pmatrix} s_{xh} & 0 \\ 0 & s_{yh} \end{pmatrix} \begin{pmatrix} x \\ y \end{pmatrix} + \begin{pmatrix} b_{xh} \\ b_{yh} \end{pmatrix} \tag{4.2}$$

where $s_{xh} = 1/5$ for $h = 1, .., 5$, $s_{yh} = 1/3$ for $h = 1, .., 4$ and $s_{y5} = -1/3$, $b_1 = (0,0)^t$, $b_2 = (2/5,0)^t$, $b_3 = (3/5,1/3)^t$, $b_4 = (4/5,2/3)^t$, $b_5 = (1/5,1/3)^t$, $p_h = 1/5$.

Preliminary analyses have been performed on the properties of Fourier transforms of self-similar and self-affine interfaces for different orientations of the wave-vector of k. In the case of self-similar model interfaces the scaling behaviour of the intensity $I(k)$ is independent of the orientation, while self-affine structures display a continuum dependence on the direction of k.

In experiments, the scattered intensity $\Sigma(k)$ in the Fourier plane is proportional to the square of $I(k)$,

$$\Sigma(k) = A[I(k)]^2 \phi(k), \tag{4.3}$$

where $\phi(k)$ is the form factor corresponding to the intensity scattered by an elementary unit ($\phi(k)$ is usually of sine-shape i.e. $\phi(k) = [(\sin \alpha k)/k]^2)$. The integral transform theory developed in section 3. agrees with the diffraction experiments performed by Allain and Cloitre [12] and Reed [13], and can be regarded as a generalization of their analyses.

The closed form results for the Fourier transform may prove a particularly useful way to attain a rigorous mathematical analysis of the physical theory of scattering from pore fractal structure developed by Bale and Schmidt [14]. The scaling theory of Bale and Schmidt is still the subject of discussion and investigation and the application of the integral transform theory to some model structures of pore fractals may help towards a better understanding of this issue.

The theory developed in section 3. applies also to wavelet transforms [15] $\Psi(k)$

$$\Psi(k) = \int_C \psi(x;k)d\mu^*(x)\,, \tag{4.4}$$

where k is a set of parameters, and the wavelet function ψ is generally supposed to be of zero mean for the wavelet transform to be invertible. Wavelet transforms can be regarded as a mathematical microscope highlighting the local properties of the object by a suitable choice of the wavelet ψ.

To make an example let us consider the one-dimensional complex Morlet wavelet defined by the kernel $\psi(x;k,\lambda) = \exp(ikx - \lambda x^2)$. By applying eq. (1.4) to this wavelet transform in the case of linear IFSP, it follows that

$$\Psi(k,\lambda) = \sum_{h=1}^{N} p_h \exp(ikb_h - \lambda x^2)\Psi(ka_h + 2i\lambda a_h b_h, \lambda a_h)\,. \tag{4.5}$$

The same numerical methods applied to $\mathcal{F}(k)$ in the case of non unimodular IFSP (Fig. 4.1) applies also to $\Psi(k,\lambda)$.

4.2 Reactions in Continuous Mixtures

The results developed for integral transforms of multifractal measures find an application in all the problems involving distribution functions characterized by an highly singular structure.

An example is given by chemical reaction kinetics in continuous mixtures. Continuous thermodynamics and kinetics are concerned with complex chemical mixtures, for which different chemical species can be labeled with respect to a continuous parameter (e.g. a reaction rate coefficient or boiling point temperature) [16]. Such a description is useful to describe complex fuel mixtures, polymeric solutions, wastewater solutions, etc. For example, Moore and Anthony [17] have experimentally observed by means of chromatographic methods that diesel fuel mixtures are characterized by a very singular composition that can hardly be described by means of smooth continuous distributions.

Let us consider the simple case of a first-order reaction evolving in a Completely mixed Stirred Tank Reactor, CSTR (the concentration inside the reactor is uniform and equal to the outlet concentration) under isothermal conditions. Let $c(x,t)$ be the concentration inside the reactor (volume V, inlet volumetric flowrate Q) parameterized with respect to the dimensionless reaction rate coefficient $x = k/k_{max} \in [0,1]$, where k is the rate coefficient [s^{-1}] and k_{max} is the maximum value attained by k. Let $c_o(x) = c_o\rho(x)$ the inlet concentration, defined on the support C, with $\int_C \rho(x)dx = \int_C d\mu(x) = 1$.

In dimensionless form, the balance equation inside the reactor reads as

$$\frac{\partial u(x,\theta)}{\partial\theta} = \frac{\rho(x) - u(x,\theta)}{T} - xu(x,\theta)\,, \tag{4.6}$$

where $\theta = tk_{max}$ is the dimensionless time, $u(x,\theta) = c(x,\theta/k_{max})/c_o$ and $T = Vk_{max}/Q$ the dimensionless residence time of the reactor. The total reactant concentration $U(\theta) = \int_C u(x,\theta)dx$ can be expressed as

$$U(\theta) = \int_C \frac{d\mu(x)}{1+Tx}(1-\exp(-\theta/T - x\theta))$$
$$= \mathcal{S}(T) - \exp(-\theta/T)\mathcal{M}(\theta,T) . \tag{4.7}$$

At steady state ($\theta \to \infty$), the total reactant concentration is equal to the Stieltjes transform of the inlet distributions. In cases where the inlet distribution can be approximated by means of a singular measure associated with an affine IFSP, the analysis developed in section 3. can therefore be applied in straightforward fashion.

The case of the dynamics response of a CSTR is just a simple, fairly academic application. A similar approach can be extended also to transport schemes involving continuously parameterized complex mixtures (e.g. with respect to the diffusivity or to the absorption constants), as in chromatographic and adsorption experiments.

5. Volterra Integral Equations

Another application of the Laplace transforms of multifractal densities is related to the solution of convolutional Volterra-type linear equations involving multifractal kernels. In the analysis of these equations, a formal difficulty occurs related to the proper definition of a density function $\rho(x)$ associated with the invariant measure of affine IFSP. This problem has some analogy with the formal definition of Dirac's delta functions within the theory of generalized functions.

As in the case of the theory of generalized functions, a definition of density functions associated with multifractal measures can be expressed in terms of their integral properties. Let μ be the invariant measure of an affine one-dimensional IFSP and let $\{m_n\}$ be the associated moment hierarchy. Its density function $\rho(x)$, indicated formally as $d\mu(x) = \rho(x)dx$, can be defined from its integral properties:

- if $f(x)$ is a continuous function $f(x) = \sum_n a_n x^n$ in C, then

$$\int_C f(x)\rho(x)dx = \sum_n a_n m_n ; \tag{5.1}$$

- the Laplace transform $\widehat{\rho}(s)$ of $\rho(x)$ is the Laplace-Stieltjes transform associated with μ, i.e.

$$\widehat{\rho}(x) = \int_C \exp(-sx)\rho(x)dx = \int_C \exp(-sx)d\mu(x) = \mathcal{L}(s) , \tag{5.2}$$

where $\mathcal{L}(s)$ is defined by eq. (3.4).

We are fully aware of the fact that this definition of density needs a formal improvement (which may constitute the basis of an extended theory of generalized functions involving multifractal distributions). The approach developed here can, however, be applied to solve some problems, as in the case of the Heaviside operational calculus, involving convolutional integrals and Laplace transforms of density functions.

Bearing in mind this difficulty, a generalized linear Volterra equation can be introduced as

$$f(x) = (1-\lambda)\rho_1(x) + \lambda \int_0^x f(x-y)\rho_2(y)dy , \tag{5.3}$$

where $\lambda \in [0,1]$, and $\rho_i(x)$ ($i = 1,2$) are generic density functions for which a Laplace transform can be defined, either in the classical way or through eqs. (5.2) and (3.4). Let $\widehat{f}(s)$, $\widehat{\rho}_i(s)$ be the Laplace transforms respectively of $f(x)$, $\rho_i(x)$. In the Laplace domain, the solution of eq. (5.3) is given by

$$\widehat{f}(s) = \frac{(1-\lambda)\widehat{\rho}_1(s)}{1-\lambda\widehat{\rho}_2(s)} , \tag{5.4}$$

and $f(x)$ is a density function, $\widehat{f}(0) = 1$, with $f(x) \geq 0$.

If we take eq. (5.4) as our starting point and perform some algebra, the moment hierarchy associated with $f(x)$ can be calculated in closed-form as a function of the moment hierarchies of $\rho_1(x)$, $\rho_2(x)$.

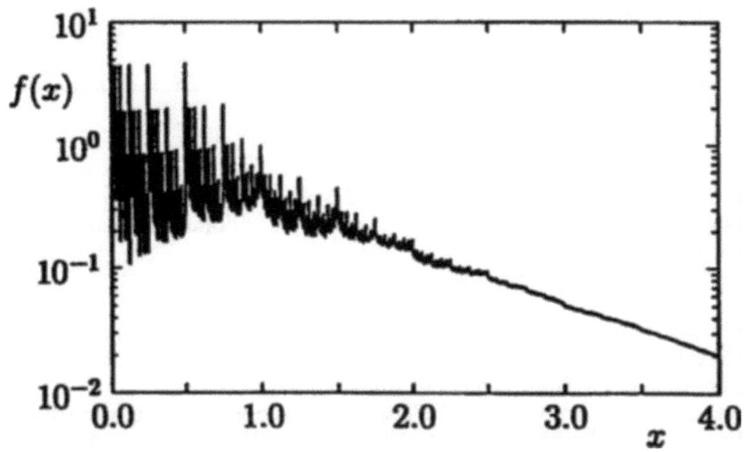

Fig. 5.1. Solution $f(x)$ of the convolutional Volterra equation, eq. (5.3) involving multifractal densities $\rho_1(x)$, $\rho_2(x)$ associated with binomial measures on the unit interval respectively for $p = 0.7$ and $p = 0.3$.

Figs. 5.1–5.2 shows two examples of density functions $f(x)$, solution of eq. (5.3) ($f(x)$ has been evaluated numerically by means of a fixed-point algorithm starting from an approximation for ρ_1, ρ_2), in the presence of multifractal densities ρ_1, ρ_2 generated by the 2-map binomial IFSP on the unit interval ($w_1(x) = x/2$, $p_1 = p$; $w_2(x) = (x+1)/2$, $p_2 = 1-p$). Fig. 5.1 refers to $\rho_1(x)$ and $\rho_2(x)$, corresponding respectively to 0.7 and $p = 0.3$, and Fig. 5.2 to $\rho_1(x) = \rho_2(x)$, corresponding to $p =$ 0.7. As can be observed from this figure, the singularity structure of the resulting measure ν, $d\nu(x) = f(x)dx$ appears to depend on position. The solution of eq. (5.3) in the presence of multifractal ρ_i appears to be the counterpart for measures of dimensional heterogeneous fractal curves discussed by Tricot [18], for which the local Hausdorff dimension is position dependent (as in the case of the Weierstrass-Tricot function). This observation is purely qualitative and should be taken as a stimulus for further study of Volterra equations. The quantitative characterization of the solution of eq. (5.3) is under investigation at present.

The results of this section and their perspectives can be summarized as follows.

- An operational calculus can be defined for multifractal measures (densities) starting from the definition of the integral transforms of invariant measures of affine IFSP.

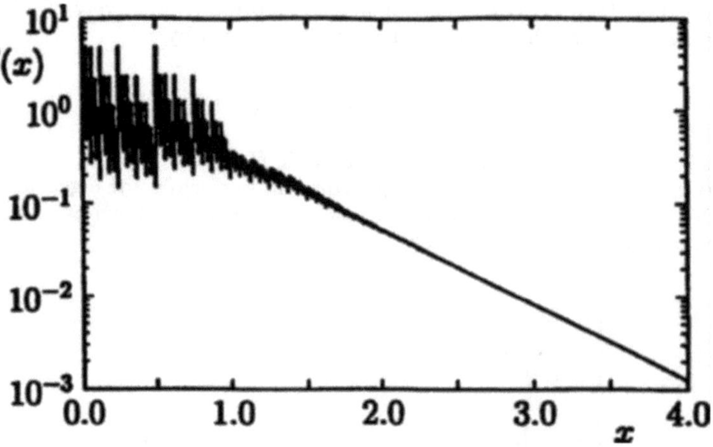

Fig. 5.2. Solution $f(x)$ of the convolutional Volterra equation, eq. (5.3) involving multifractal densities $\rho_1(x) = \rho_2(x)$ associated with binomial measures on the unit interval for $p = 0.7$.

- The solution of linear Volterra convolutional equations provides a new class of multifractal distributions displaying qualitative properties which are, at a first sight, different from those typical of multinomial measures.
- The introduction of convolutional Volterra equations involving multifractal measures may lead to the development of a new class of linear filters for multifractal signal processing.

6. Concluding remarks

In this article we have analyzed the properties of integral transforms of multifractal measures in some detail, also presenting some applications to physical and engineering problems.

Multifractal measures can be regarded as a useful archetype in engineering applications, intermediate between continuous and atomic distributions, in all those cases, frequently encountered in practice, for which the distribution functions associated with a physical observable display highly singular features.

There is one topic in this analysis, namely the definition of integral equations involving multifractal kernels, which, although physically correct, still needs a further level of formal mathematical refinement. The concept of density functions, $\rho(x)$, associated with generic invariant measures of affine IFSP, $d\mu(x) = \rho(x)dx$, is to some extent not fully rigorous. The analysis presented in section 5. can be regarded as analogous to the extension of the Heaviside operation calculus, involving generalized functions, to multifractal distributions.

As a final remark, we wish to point out some other engineering applications of the theory presented, especially in the field of image and signal processing involving either IFSP or multifractal concepts.

The Fourier analysis of multifractal measures can be further developed in order to explore new approaches to the inverse problems of image reconstruction involving affine IFSP as an alternative to moment analysis. Indeed, eq. (3.2) represents a compact and formally tractable expression for the Markov equation, eqs. (1.1)–(1.2), in the spectral domain (it can be labeled the *spectral Markov equation*), and it could be worthwhile to explore its applicability in the optimization of the inverse reconstruction problem.

Laplace transforms and the analysis of linear Volterra equations involving multifractal distributions can be further developed within the framework of multifractal approaches to signal processing.

References

1. Mandelbrot, B.B. (1974): Intermittent turbulence in self-similar cascade: divergence of high moments and dimension of the carrier. *J. Fluid Mech.*, 62, 331.
2. Halsey, T.C., Jensen, M.H., Kadanoff, L.P., Procaccia, I., Shraiman, B.I. (1986): Fractal measures and their singularities: The characterization of strange sets. *Phys. Rev. A*, 33, 1141.
3. Mandelbrot, B.B. (1988): An Introduction to multifractal distribution functions. In: Stanley, H.E., Ostrowsky, N. (eds), *Fluctuations and Pattern Formation*, Kluwer Academic, Dordrecht.
4. Lévy Véhel, J. (1996): Fractal Approaches in Signal Processing, In *Fractal Geometry and Analysis, The Mandelbrot Festschrift, Curacao 1995, C.J.G. Evertsz, H.-O. Peitgen & R.F. Voss Editors, World Scientific.*
5. Lévy Véhel, J. (1996): Introduction to the Multifractal Analysis of Images In *Fractal Image Encoding and Analysis, Yuval Fisher Editor, Springer Verlag.*
6. Barnsley, M. (1988): *Fractals Everywhere.* Academic Press, Boston.
7. Forte, B., Vrscay, E.R. (1994): Solving the inverse Problem for function/image approximation using iterated function systems I. Theoretical basis. *Fractals*, 2, 325.
8. Abenda, S., Turchetti, G. (1989): Inverse problem for fractal sets on the real line via the moment method. *Il Nuovo Cimento*, 104 B, 213.
9. Bessis, D., Demko, S. (1991): Stable recovery of fractal measures by polynomial sampling. *Physica D*, 47, 427.
10. Forte, B., Vrscay, E.R. (1994): Solving the inverse problem for measures using iterated function systems: a new approach. *Appl. Prob.*, in press.
11. Giona, M. (1996): Analytic expression for the structure factor and for the moment-generating function of fractal sets and multifractal measures. *J. Phys. A*, submitted.
12. Allain, C., Cloitre, M. (1986): Optical diffraction on fractals. *Phys. Rev. B*, 33, 3566.
13. Reed, L.D. (1989): Scattering from ordered fractals. MS Thesis, University of North Dakota.
14. Bale, H.D., Schmidt, P.W. (1984): Small-angle X-ray scattering investigation of submicroscopic porosity with fractal properties. *Phys. Rev. Lett.*, 53, 596.
15. Farge, M., Hunt J.C.R., Vassilicos, J.C. (eds) (1993): *Wavelets, Fractals and Fourier Transforms* Claredon Press, Oxford.
16. Aris, R., Gavalas, G.R. (1966): On the theory of reactions in continuous mixtures. *Phil. Trans. Roy. Soc. London*, A-260, 351.
17. Moore, P.K., Anthony, R.G. (1989): The continuous-lumping method for vapor-liquid equilibrium calculations. *AIChE J.*, 35, 1115.
18. Tricot, C. (1993): *Curves and Fractal Dimension.* Springer Verlag, Berlin.

Texture characterization of ERS-1 images by Regional Multifractal analysis

A tool applied to the study of South Cameroon Rain Forest landscape and bio-geo-hydrological processes

P. Sailhac[1] and F. Seyler[2]

[1] IPGP, Laboratoire de Gomagnetisme
4 Place Jussieu, 75252 Paris Cedex 05, FRANCE
Tel: (33) 01 44 27 73 45 Fax: (33) 01 44 27 73 40
sailhac@ipgp.jussieu.fr

[2] ORSTOM, CENA/USP
Avenida Centenario 303, C.P. 96, 13400-970 Piracicaba S.P., BRASIL
Tel: (55) 19 433 27 33
fseyler@aguia.cena.usp.br

Summary. Texture characterization aiming at landscape description was performed by multifractal analysis on radar images of the South Cameroon rain forest. The global characterization has been done with the Legendre spectrum. The local characterization has been done via the singularity maps of pointwise Hölder exponents; then, multifractal spectra were obtained via a normalized estimator based upon the statistics of the singularity maps. The nature of local texture was therefore characterized by singularity spectra. Preliminary interpretations in terms of bio-geo-hydrological processes are given.

1. Introduction

Texture analysis has been of great interest for the last twenty five years. The averaged parameters of an image, its isotropy or anisotropy for instance, and their local variations determination and contouring are some of the mathematical rules developed to determine the intrinsic characters of an image or to model the differences between images. The common purpose is to make possible its recognition and segmentation. In Geophysics, the goal is finally to understand the geological and/or hydrological properties of the observed region. This lies under what is called Geostatistics [1]. On one hand, the main characters are explored by studying the values of the entropy, the first and second moments, and other statistical parameters [2]; the variations of their local values can be analyzed as a way to show spatial changes [3]. On the other hand, since the nature usually shows scaling behaviors [4], fractal analysis has been applied to the measure of texture diversity [5, 6, 7]. Fractal dimensions show the scaling behavior of the image taken as a continuous surface. It is possible to enlarge the scope of fractal analysis by representing the dimensions D_t of the part of the surface at some specified levels $[t, t + dt)$ of its intensity for instance [8]. By this way, the doors of the multifractal approach have been opened to image analysis.

Within the multifractal analysis as initially developed for chaos, turbulence and percolation [9, 10], the t-levels correspond to q-levels where q is the order of a statistical moment of the image intensity. The "generalized dimension" D_q characterizes the scaling behavior of the moment order q (whose scaling exponent is

$\tau_q = (q-1)D_q$). As for the multifractal spectrum $f(\alpha)$ associated with these generalized dimensions D_q (via a Legendre transform $f(\alpha) = \inf_{q\in\Re}\{q\alpha - \tau_q\}$), the t-levels correspond to α-levels where α is the scaling exponent of the image intensity; so $f(\alpha)$ is under some assumptions the fractal dimension of the α-level sets. Multifractal analysis has been generalized from sum measures (as image intensity for instance) to any measures, capacities and functions ; the choice of the measure versus image properties has been discussed and clearly presented in [11].

The applications of the multifractal analysis to contour detection, image segmentation and filtering are encouraging [11, 12, 13, 14]. Nevertheless, one may argue that the multifractal spectrum has been applied only as the global spectrum of an average multifractal distribution because it has been computed in the image as a whole. We propose to analyze the inner multifractal structure of an image with a set of regional multifractal spectra, each of them being computed in a given area. Actually, this follows the same procedure as the one used on 1D studies for road traffic structures [15] and for geomagnetic temporal behaviors [16].

The main field of applications to images tackled by Lévy Véhel et al. [11, 12, 13, 14] has been the change detection and the extraction of objects previously defined by a priori knowledge. Similarly, convolution-based methods such as Fourier [17, 18] or wavelet analysis [19, 20] have been generally restricted to objects having well defined boundaries. In our case, focused on three sub-scenes of ERS-1 radar images from the South of Cameroon rain forest, three main characteristics impose the use and the development of the regional multifractal analysis:

- The local slope and drainage pattern changes progressively and continuously from a region to another.
- The size of an elementary textural unit is variable following the type of slope/drainage pattern to characterize and cannot be easily predicted. Moreover, the size of the neighbourhood usually chosen in texture analysis (3×3 or 5×5 pixels) is not suitable for radar images which are affected by a random noise known as "speckle" (see [21]);
- The elementary unit of the texture or basic primitive may be a characteristic of the region to be extracted and perhaps other order arrangements as well. It is the reason why, although well suited for the description of geological structures or other periodic linear patterns within the image, the frequency power spectra (Fourier or derived transforms as Hough transform, or bi-dimensional wavelets) seem to be not adapted to our problem.

2. Nature of the Analyzed Images

Our work has been focused on three sub-scenes of C band radar images (frequency $5,3Ghz$, wavelength $5.7cm$) of dense homogeneous rain forest from the South of Cameroon. Their nominal resolution is of $25m$ range by $22m$ azimuth which means a pixel size of $30m \times 30m$ after regridding; their size is about 800×700 pixels.

Each figure represents a typical landscape of this forested area with a distinct kind of hydrological network. They are typical of the region limiting the Nyong and the Boumba basins. The first, called *grdres* (for *"grand réseau"*, great network), shows characteristic features of the tributary network of the upper Nyong basin (Abong Mbang town area); they are large streams covered with palms trees separated by elongated low hills. The second, called *marec* (for *"marécages"*, swamps), is an area of low circular hills surrounded by swamps. The third, called *incis* (for *"incisif"*, sharp), is a region of medium altitude hills jagged into a semi-rectangular network. The images, actually very irregular with high irregularities on a very flat background, have been displayed in a base 5 logarithmic map in order to enhance the details (color plate C.4).

3. Fractal Analysis by Pointwise Hölder exponents

The maps of the pointwise Hölder exponents have been computed as the scaling exponent of the sum measure $\mu_{2\epsilon}(X_{i,j})$ defined on a square of length 2ϵ centered on the pixel $X_{i,j}$:

$$\mu_{2\epsilon}(X_{i,j}) = \sum_{i-\epsilon \leq k \leq i+\epsilon} \sum_{j-\epsilon \leq l \leq j+\epsilon} I(X_{k,l}) \tag{3.1}$$

where $I(X_{k,l})$ is the intensity of the pixel at the position $X_{k,l}$.

The singularity exponent is then computed by linear regression as the slope of the log-log plot of $\mu_{2\epsilon}$ for ϵ ranging from 1 to ν:

$$\alpha_{2\nu}(X_{i,j}) = Slope\,\{\log[2\epsilon], \log[\mu_{2\epsilon}(X_{i,j})]\} \tag{3.2}$$

The maps for $\nu = 36$ and 50 have been computed: the local regions are squares of respective surface $2190m \times 2190m$ and $3030m \times 3030m$ (color plates C.5 and C.6). The dependence on ν in the computed maps corresponds to an averaging from the smaller to the higher ν values, associated with a loss of spatial resolution and a gain in the correlation. We have chosen the optimal value $\nu = 36$.

These singularity maps enhance and properly match with the textures perceived by eyes in the base 5 logarithmic maps of the intensity. For *grdres*, the large flat valleys are colored in soft green (values near 2) while the low hills are in blue or yellow-green (values other than 2); some more complex areas, in dark blue or yellow (extreme values) may be associated with transition zones between valleys and hills or may indicate some heterogeneity of the palms trees coverage. For *marec*, the shape of the singularity maps indicates the circular geometry of hills and flat valleys (as in *grdres*) ; it is the closer map to that of its corresponding intensity map. For *incis*, the distribution of the hills is evident, but the visual aspect is different from that of the intensity map. It demonstrates alternating oriented structures of high and low values. *grdres* is intermediate : the upper left part of the image shows structures analogous to those of *incis* and actually, this part of the image represents a landscape pattern which is a transition to the landscape represented in the *incis* map.

Conventional image processing filters which estimate the heterogeneity of the intensities give results similar to the singularities. Power angular spectrum were calculated following a procedure described in [22]. The results are matching for the three textures, i.e. there is no significant structure in *marec*, and the angular directions found for *grdres* and *incis* are matching those demonstrated by the alternative high and low values in the maps. It seems then that the singularity maps give a good visual representation,

- first of the geometrical configuration of the target: while singularity values are medium for the large flat parts of the images, they are low or high for the hills where the slope induces constant changes of orientation in comparison with the view angle (low singularities for illuminated sides which are in front view to the radar, and high singularities for shadowy sides).
- secondly of the heterogeneity of the image: the structured aspect of the singularity map (the range of singularities in a given area) is increasing with the heterogeneity of the image texture.

In other respects, one may remark the great quality of the logarithmic maps of the intensity (color plate C.4). In reality, they are also (degenerated) singularity maps ; in this case the singularity is a normalized value of the local coarse grained Hölder exponent estimated for one pixel (defined with a similar expression to the pointwise Hölder exponent by replacing the linear regression on the log-log plot by a simple ratio $\log[\mu_{2\epsilon}(X_{i,j})]/\log[2\epsilon] =: \alpha_\epsilon(X_{i,j})$).

4. Global Multifractal Analysis by Image Singularity spectrum

First, in order to clarify the meaning of the singularity spectrum, let us briefly recall the intrinsic statistical nature of the multifractal analysis as shown in [23, 24]. One considers the local singularity $\alpha(x)$ at point x as the limit function for $\varepsilon \to 0$ of $\alpha_\epsilon(x)$ computed on some data. To exhibit the statistics of $\alpha(x)$, one may try to display its distribution histogram $N(\alpha)$. Nevertheless when the data follow a multifractal distribution, $\alpha(x)$ is itself a kind of fractal function, so that the histogram $N_\lambda(\alpha)$ actually depends on the x-step λ chosen for sampling the function $\alpha(x)$. And a "stable" quantity which converges towards the singularity spectrum, or multifractal spectrum f(α) for $\lambda \to 0$, is :

$$f_\lambda(\alpha) = -\frac{\log[N_\lambda(\alpha)]}{\log \lambda}. \tag{4.1}$$

In practice, there are different ways to estimate f(α) (see [11, 23] for a review) and the methods are generally not equivalent, except in particular cases such as the multinomial measure [13, 24]. For procedural simplicity, only the *Moment Method* (or *Legendre Method*) has been performed.

The resulting spectra (figure 4.1) show approximately the same more probable behaviors for the three images: they are identical for α between 1.75 and 2.4 , with more probable value $\alpha_0 = 2.02$ and values ranging from about 1.16 to 2.55. The spreading to the left reported by Lévy Véhel for radar images [11] does not appear at first order (at α_0). But one observes an asymmetry to the left which can be interpreted in this case, with regard to the singularity maps, rather as an effect of the local textural diversity than as a speckle effect. Indeed, the first order intensity variation ratio due to the speckle is theoretically of 0.57 in our images (number of looks 3) while it is 1 for a single-look radar image [25]; this may explain why the shifting reported by Lévy Véhel has not been observed here.

Thus, this three images show the same general features : they are all three C band radar images of the dense homogeneous rain forest from the South of Cameroon.

The actual difference between the three regions is seen on the left-side spreading of their multifractal spectra[1] : for α below 1.75 for *incis* and 1.65 for *grdres* and *marec*. These differences are statistically significant with regard to formula (4.1) because the f(α) values corresponding with these α-levels are near 2. From left to right, there is *marec*, then *grdres* and *incis*. This can be linked with the shape of the intensity images (figure 1) and with the "structured aspect" of the singularity map as previously mentioned. Indeed, when using human sight, one gets the same texture classification from the less to the more "straight-lined". Thus, the spreading of the multifractal spectrum to the left is associated with the heterogeneity and anisotropy in the image.

[1] Notice that the left side of the spectrum is usually easier to estimate than its right side.

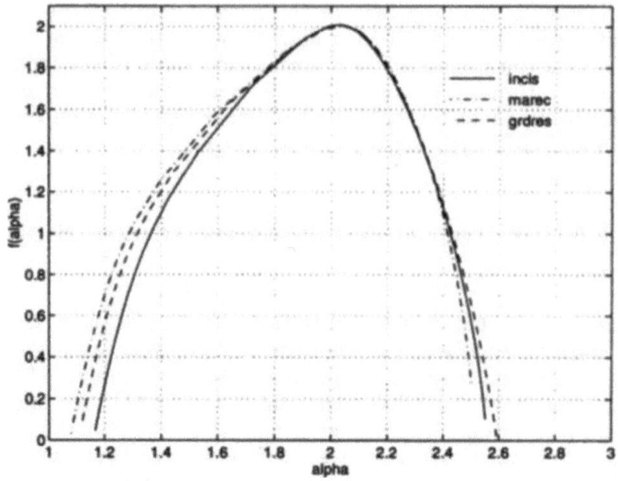

Fig. 4.1. Global singularity spectra f(α) of the three images (*Legendre method*)

5. Regional Multifractal Analysis by Sub-Image Singularity spectrum

The fractal analysis by singularity mapping has given one characteristic parameter $\alpha_{2\nu}(X_{i,j})$ for each region of typical size ν and center $X_{i,j}$. The multifractal spectrum f(α) has given its averaged distribution for the whole image. Our purpose now is to find more than one characteristic parameter for each regions: this is the sub-image singularity spectrum.

First, the images have been cut into sub-images of 100 × 100 pixels and their multifractal spectra have been computed by the Legendre transform as in the previous section. The resulting multifractal spectra have all given only degenerated spectra to the single point $\{\alpha = 2.0, f(\alpha) = 2.0\}$. The same degenerated spectra have been obtained with sub-images of 200 × 200 pixels. We conclude that our radar images contain a kind of inner noise which degenerates the multifractal spectrum, at least when estimated using the Legendre method. It should be linked with the speckle, characteristic of radar images ; indeed, the same procedure on Landsat near infrared images of similar landscape has given classical convex spectra.

A different f(α)-estimator has been used. It is a normalized estimator based upon the convergence of $f_\lambda(\alpha)$ in formula (4.1). Indeed, for λ small enough, $f_\lambda(\alpha)$ is assumed to be stable in the neighborhood of its limit f(α). Thus, as soon as one knows that $f(\alpha_0) = D_0 = 2$ is the support dimension and that α_0 is the most probable singularity value, one gets for λ small enough :

$$2 = -\frac{\log[N_\lambda(\alpha_0)]}{\log \lambda}. \tag{5.1}$$

Combination of equations (4.1) and (5.1) yields a simple normalized f(α)-estimator :

$$f_{normal}(\alpha) = 2\frac{\log[N(\alpha)]}{\log[N_{\max}]} \tag{5.2}$$

where $N(\alpha)$ has been calculated as $N_\lambda(\alpha)$ for λ small, and $N_{\max}$ is its maximum value, equal to $N_\lambda(\alpha_0)$.

The histogram $N(\alpha)$ and the associated normalized f(α)-estimator have been computed on the singularity maps made of 73 × 73 pixels scopes, first on the global images to test the procedure (figures 5.1 and 5.2), then on some sub-images of 100 × 100 pixels (figures 5.3, 5.4, and 5.5).

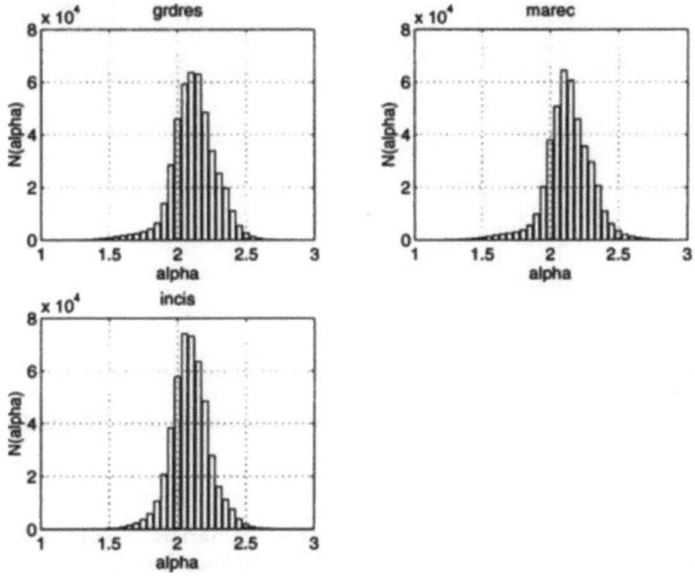

Fig. 5.1. Global histograms N(α) for the three images

For the global images, the left side fits well with the singularity spectra computed by the Legendre transform (figure 4.1); and the anisotropy/heterogeneity classification is still available. Nevertheless, the right side is different, going to $\alpha = 3$; and the more probable value is $\alpha_0 = 2.10 \pm 0.05$: This comes from the different meanings of the estimators and possibly from a size effect.

For the sub-images, the visual texture diversity seen on intensity maps (color plate C.4) or singularity maps (color plates C.5 and C.6). is translated in terms of regional singularity spectra (figures 5.3, 5.4, and 5.5). For instance, it shows the differences between the large flat valleys of *grdres* ($X_{516,616}$) and its low hills ($X_{516,366}$) : the homegeneity of the large flat valleys contrasts with the complexity of the hills. The same rule holds for the spectrum shape of the other images : while Λ shape indicates the homogeneity of the geometry distribution of the singularities, and systematically corresponds with the texture of the flat valleys, possibly inundated ($X_{148,349}$ in *marec* and $X_{580,696}$ in *incis*), $\bigcap$ shape indicates heterogeneity and generally corresponds with the texture of the hills ($X_{498,599}$ in *marec* and $X_{530,246}$ in *incis*).

As the multifractal spectrum depends on other parameters as well, this rule is valid at first order only. For instance, the existence of an inflexion point for α near 1.7 may increase the precision of the characterization ; while there are similitudes between the hill region centered on $X_{398,149}$ and the valley region centered on $X_{398,349}$ in *marec*, there are differences between the valley region of *incis* centered on $X_{530,246}$ and the valleys of *grdres* and *marec*. On the one hand, the case of a few

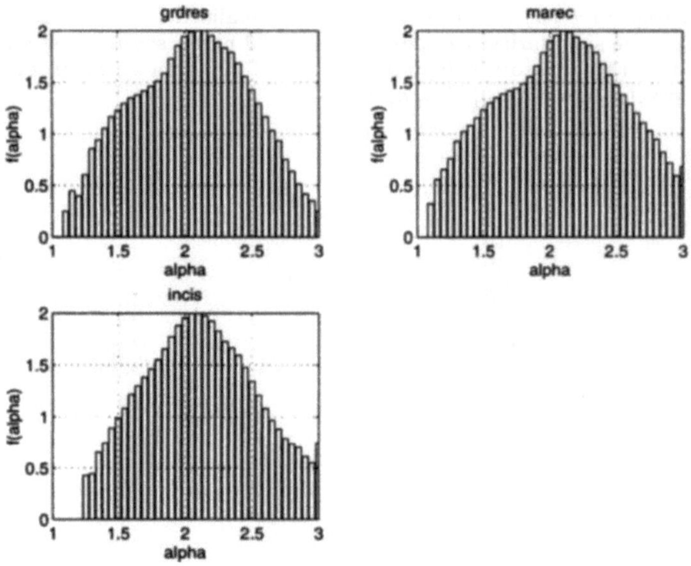

Fig. 5.2. Global singularity spectra f(α) for the three images (Normalized estimator)

isolated hills surrounded by valleys and the case of narrow cutting valleys instead of flat valleys result in multifractal spectra between the Λ and the $\bigcap$. On the other hand, the modulation of typical spectra may be due to other parameters such as vegetation and hydrology of the analyzed region. The results of such a superposition procedure could be similar to the lumping or the sum of two binomial measures analyzed by Lévy Véhel [11].

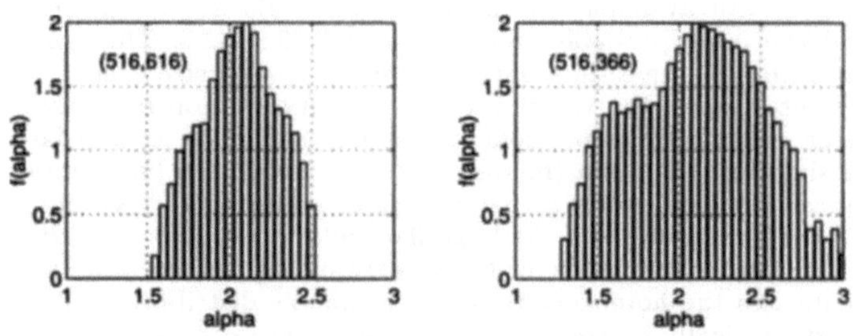

Fig. 5.3. Local singularity spectra f(α) for *grdres* sub-images of 100 × 100 pixels centered on $X_{516,616}$ and $X_{516,366}$ (Normalized estimator)

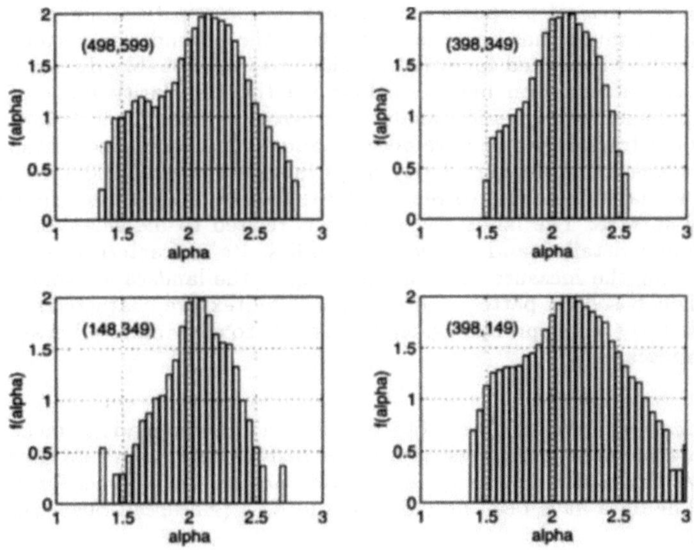

Fig. 5.4. Local singularity spectra f(α) for *marec* sub-images of 100 × 100 pixels centered on $X_{498,599}$, $X_{398,349}$, $X_{148,349}$ and $X_{398,149}$ (Normalized estimator)

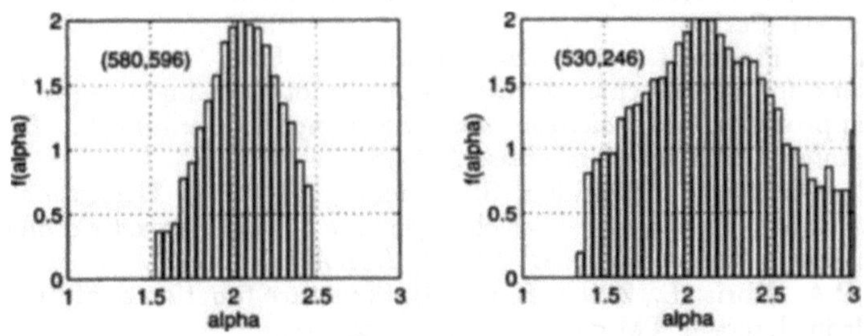

Fig. 5.5. Local singularity spectra f(α) for *incis* sub-images of 100 × 100 pixels centered on $X_{580,696}$ and $X_{530,246}$ (Normalized estimator)

6. Conclusion

A multifractal analysis for texture characterization via radar images has been performed for geophysical purposes. The global characterization, for images of more than 600 × 600 pixels, has been successful with the Legendre spectrum ; nevertheless, the method is limited for smaller radar images, probably due to the speckle. The local characterization has been done via the singularity maps of Pointwise Hölder exponents by regression analysis in boxes of 73 × 73 or 101 × 101 pixels. Regional multifractal spectra have been obtained via a normalized estimator based upon the statistics of the singularity maps in sub-images of 100 × 100 pixels. By this way, we have been able to represent the nature of local texture with help of singularity spectra. The landscape pattern is related to bio-geo-hydrological processes through detailed and costly field studies. Its characterization has a great importance for the measure and the modeling of the landscape evolution, and its segmentation based on patterns recognition. The texture characterization carried out in this study seems promising, since it is able to discriminate between different types of landscape.

Acknowledgement. We are grateful to Jacques Dubois who made us meet each other at a meeting he has organized last year in the *Institut de Physique du Globe de Paris*. As well, we thank anonymous reviewers and Jacques Lévy Véhel for constructive remarks. The data have been provided by the *ESA* (*European Space Agency*).

References

1. Dimitrakopoulos, R. (1993): *Quantitative Geology and Geostatistics for the next century*. Kluwer, Montréal.
2. Sali, E., Wolfson, H. (1991): Texture classification in aerial photographs and satellite data. *Int. J. Remote Sensing,vol.13, n⁰ 18, 3395–3408.*
3. He, D.-C, Wang, L. (1991): Texture features based on texture spectrum. *Pattern Recognition, vol.24, n⁰ 5, 391–399.*
4. Mandelbrot, B.B. (1975): *Les objets fractals: forme, Hasard et Dimension.* Flammarion, Paris.
5. Giberti, L., Picello, L., Dellepiane, S., Serpico, S.P., Vernazza, G. (1988): A framework for region characterization in remote sensing images by fractal-based approach. *Image analysis and processing II, Cantoni V., Di Gesu V. and Levialdi S. Eds, Plenum press, 443–451.*
6. Cuq, F., Durand, P., Hamidou, S., Simonin, A. (1990): Utilisation de filtres géostatistiques pour l'analyse des changements d'échelle à partir d'images satellitaires. *Photo-Interprétation n⁰ 1990-6, ESKA Eds, 33–38.*
7. Brivio, P.A., Doria, L., Zilioli, E. (1993): Structure function analysis of natural scenes from Landsat TM data. *ITC Journal, 6 p.*.
8. De Cola, L. (1993): Multifractals in image process images. *In fractals in Geography, Lam and De Cola Eds, Prentice Hall, 282–304.*
9. Hentschel, H.G.E., Procaccia, I. (1983): The Infinite number of generalized dimensions of fractals and strange attractors. *Physica 8D, 435–444.*
10. Frisch, U., Parisi, G. (1985): Fully developed turbulence and intermittency. *In turbulence and Predictability of Geophysical Flows and Climate Dynamics. Ghi, Benzi and Parisi Eds, North-Holland, p. 84.*

11. Lévy Véhel, J. (1996): Introduction to the multifractal analysis of images. *In Fractal Image Encoding and Analysis, Yuval Fisher Ed., Springer, 52 p.*
12. Lévy Véhel, J., Mignot, P., Berroir, J.P. (1992): Multifractals, texture, and image analysis. *In CVPR.*
13. Lévy Véhel, J., Mignot, P. (1994): Multifractal Segmentation of Images. *In Fractals, vol.2, Issue 3, 371–377.*
14. Canus, C., Lévy Véhel, J. (1996): Change detection in sequences of images by multifractal analysis. IEEE *In Proc. ICASSP-96, 4 p.*
15. Lévy Véhel, J., Vojak, R., Danech-Pajouh M. (1994): Multifractal description of road traffic stucture. *INRIA and INRETS Report, 15 p.*
16. Hongre, L., Sailhac, P., Alexandrescu, M., Dubois, J. (1997): Chaotic and Multifractal Approachs to the Geomagnetic Temporal Fluctuations. *In preparation.*
17. Haralick, R.M. (1979): Statistical and structural approaches to texture. *Proc.* IEEE, *67, 786–804.*
18. D'Astous, F., Jernigan, M.E. (1984) Texture discrimination based on detailed measures of the power spectrum. *Proc.* IEEE, *Comput. Soc. Conf. Pattern Recognition Image Process, 83–86.*
19. Clausi, D.A. (1996): Texture Segmentation of SAR Sea Ice Imagery. PhD thesis. University of Waterloo, Ontario, Canada.
20. Fatemi-Ghomi, N., Palmer, P.L., Petrou, M. (1996): Performance Evaluation of Texture Segmentation Algorithms based on Wavelets. *Proceeding of the worshop on Performance Charateristics of Vision Algorithms, ECCV-96, Cambridge, England, 19 April 1996.*
21. Reeves, R.G., Ansen, A., Landen, D. (1975): *Manual of Remote Sensing.* American Society of Photogrammetry, Eds. Keuffel and Esser Company, Falls Church, Virginia.
22. Robert, R., Seyler, F. (1996): Texture Analysis of Tropical Rain-forest Infrared Satellite Images. *Submited to Photogrammetric Engineering and Remote Sensing, paper n^{0}95-078.*
23. Sailhac, P. (1996): Introduction to Nonlinear Geophysics, Chaos and Fractals: transitions characterizing seismic; and volcanic events need other methods when linear physics is not adapted to represent the processes; specific development on the Gutenberg-Richter law. *Cahiers du Centre Européens de Géodynamique et de Séismologie, vol. 13/14, 41 p.*
24. Evertsz, C.J.G., Mandelbrot, B.B. (1992): Multifractal Measures. *In Chaos and Fractals: New Frontiers in Science, Appendix B. H.-O.Peitgen, H.Jürgen and D.Saupe, Springer, 921–953.*
25. Laur, H. (1989): Analyse d'images radar en télédétection : discriminateurs radiométriques et texturaux. Thèse de l'Université Paul Sabatier, n^0 403, 244p.

Multifractal Analysis of Dipmeter Well Logs for Description of Geological Lithofacies

A. Saucier, O.K. Huseby, and J. Muller

Institutt for energiteknikk,
Box 40
N-2007 Kjeller, NORWAY

Summary. We use multifractal analysis as a geostatistical tool for characterization of microresistivity signals produced by dipmeter well-logging tools. The signal is divided into windows of fixed length. For each window, several texture indices characterizing the irregularity of the microresistivity signal are calculated. Plotted as a function of depths, these texture indices form what we call texture logs. We show that these texture logs can be used to distinguish geological lithofacies differing in their degree of heterogeneity.

1. Introduction

The characterisation of geological formations plays an important role in the description of petroleum reservoirs. Information available for the geological characterisation of a reservoir come from well logs, cores, and seismic signals. The core data is not always available and one must therefore rely on various well logs to characterise lithology and geological zones in the vicinity of a bore hole.

It has been recently realized that high resolution microresistivity dipmeter curves (for modern dipmeter tools [1, 2, 3] about 1 measurement every 2.5 mm) contain valuable information not only on the dip angle of the stratigraphic layers, but also on the texture (variability) of the geological formations. The dipmeter microresistivity logs are so abundant and complex that it can become problematic to extract information out of them. Therefore to help to analyse dipmeter curves, it would be useful to have a way of compressing the detailed information of microresistivities i.e. to extract from consecutive segments of the signal a series of parameters describing its local variability. Such parameters combined with information obtained from the conventional open-hole logs could then be used to improve the description of local geology close to a bore hole.

The first objective of this paper is to achieve a quantitative representation of multi-point correlations in the dipmeter microresistivity signal. For this purpose, we show how multifractal analysis can be adapted to obtain a signal processing tool allowing to extract a compact representation of multi-point statistics.

There have been earlier attempts [4, 5] to characterise globally well log signals using statistical methods focusing on 2-point correlations, i.e. R/S analysis, variograms and power spectra. Our objective is to achieve a local characterisation of a signal that goes beyond 2-point correlations. For signals having suitable scaling properties in a fixed range of scales, multifractal analysis allows to represent multi-point correlations in a simple manner.

For a local analysis of multi-point correlations, we segment the microresistivity signal into consecutive equal-size intervals called windows. Our method of analysis allows to extract from the signal in each window a triplet of *texture parameters* $X = (\sigma, D(1), \alpha_c)$ (to be defined below) representing multi-point correlations, where the texture parameters reflect the distribution of heterogeneities in various geological layers. The second objective of this paper is to examine how the texture parameters correlate with the local lithologies, and to see if the texture indices can help to distinguish different lithofacies. This is demonstrated on an example from the North Sea reservoir.

2. The representation of texture for multifractals

In this paper, we define the *texture* of a signal to be a complete description of its multipoint statistics. A definition of texture therefore includes a probability distribution (1-point statistics), 2-point probability distributions (from which autocorrelation functions and variograms can be derived), 3-point probability distributions etc. In practice, a description of multipoint statistics can be only partial because of the limited statistical reliability of multipoint probability distributions estimated from finite samples. In other words, texture is known only partially for finite samples. In addition to such statistical constraints, the description of texture is also complicated by the difficulty of representing n-point joint probability distributions for large n. Within these constraints, our goal is to represent texture as completely as possible and in a relatively simple manner, i.e. by using only a few scalar parameters called *texture indices.*

In the following we will describe the *sliding window method*, which is a simple way of distinguishing intervals with different textures within the well. For a local analysis of texture, we segment the microresistivity signal into equal-size intervals called *windows.* Consecutive windows are equally spaced and may overlap. A triplet of texture indices is computed for each window. These indices depend on the variations of the microresistivity signal in the vertical direction and they should reflect the distribution of heterogeneities in the reservoir along the well. Plotted as a function of depth, the texture indices form what we call *texture logs.* They provide a compressed representation of microresistivity logs and capture information about local geology with a resolution equal to the window size.

A complete description of multipoint statistics is not a trivial assignment because of the difficulty of representing multipoint joint probability distributions. In the following, we show that multipoint statistics can be represented in a simpler manner if the signal is multifractal. We have chosen to illustrate our approach using the formalism given in details by Evertsz and Mandelbrot [9], though other approaches such as those of Lévy Véhel [12] and Jaffard [11] are equally valid. Consider a 1-dimensional random signal $S(x_i) \geq 0$, where the x_i's are equally-spaced discrete spatial coordinates. The *measure* $p_x(\delta)$ of an interval $[x-\delta/2, x+\delta/2]$ is defined by

$$p_x(\delta) = \sum_{x-\delta/2 \leq x_i \leq x+\delta/2} S(x_i) \ . \tag{2.1}$$

If $S(x_i)$ is interpreted to be a mass localized at point x_i, then $p_x(\delta)$ is simply the total mass contained in the interval $[x-\delta/2, x+\delta/2]$. Consider the moments $< [p_x(\delta)]^q >$, where q and δ are real parameters and $< ... >$ denotes an ensemble average. If these moments are known for all q and δ, then the texture of $S(x_i)$ is essentially known. Indeed, all the multipoint probability distributions of the signal S are required to calculate $< [p_x(\delta)]^q >$ for all δ and all q.

If the signal has multifractal properties, the moments $< [p_x(\delta)]^q >$ take approximately the form

$$\langle [p_x(\delta)]^q \rangle = \langle [p_x(\delta_1)]^q \rangle (\delta/\delta_1)^{d_B + \tau(q)} \tag{2.2}$$

in a finite interval $[\delta_{min}, \delta_{max}]$ called *scaling range*, where δ_{min} and δ_{max} are the *inner* and *outer scales* respectively and $\delta_{min} \leq \delta_1 \leq \delta_{max}$. The exponents $\tau(q)$ are called the ***mass exponents*** and d_B is the box-dimension [6] of the support of the measure ($d_B = 1$ if the measure is space-filling in one dimension). The multifractal properties of a signal are usually described in terms of the ***generalized dimensions*** $D(q)$, which are related to the mass exponents by the simple transformation

$$D(q) = \frac{\tau(q)}{q-1}. \qquad (2.3)$$

If the scaling property (equation 2.2) holds, the moments $< [p_x(\delta)^q] >$ are described completely by the prefactors $< [p_x(\delta_1)]^q >$ and the generalized dimensions $D(q)$, thereby simplifying the representation of statistical information. This simplification is our prime motivation for using a multifractal description of texture.

3. Method of Analysis

3.1 Associating a Measure to the Microresistivity Signal

We analyse a resistivity signal denoted by $r(x_i)$, where x_i are equidistant coordinates measuring depth in a well. We chose to focus on the vertical resistivity variations because they contain most of the non-trivial textural information. This was done by analysing the transformed signal

$$s_\omega(x_i) = |r(x_{i+1}) - r(x_i)|^\omega , \qquad (3.1)$$

where $\omega > 0$. Indeed, the transformation (3.1) emphasises vertical variations and eliminates any constant component (background) in the signal $r(x_i)$.

The transformed signal $s_\omega(x_i)$ is positive, which is a requirement for using multifractal analysis, moreover (3.1) satisfies the property $\lim \omega \rightarrow 0 \; s_\omega(x_i) = 1$, and therefore the parameter ω controls the degree of variability of $s_\omega(x_i)$: decreasing ω causes the signal $s_\omega(x_i)$ to become smoother. In that sense, ω is a smoothing parameter.

When computing a difference or a derivative, it is important to pay attention to noise. Indeed, a white noise added to a signal can introduce large errors in the estimate of the derivative. The noise contaminating the signal at small scales [7] was therefore filtered out (using a standard low pass filter [8]) before the transformed signal $s_\omega(x_i)$ was computed.

The signal $s_\omega(x_i)$ is irregular and exhibits sharp singularities, as can be seen in figure 3.1b for the well 30/9-6 (this signal corresponds to the microresistivity signal displayed in figure 3.1a. $s_\omega(x_i)$ looks therefore suitable *a priori* for a multifractal analysis, which is a singularity analysis. We would like to point out that there may exist other transformations which would also smoothen the signal. This is being currently investigated together with alternative methods which reduce the importance of outliers.

3.2 Estimation of the Generalized Dimensions $D(q)$

One of the simplest way to estimate the generalized dimensions $D(q)$ is the method of moments [6, 9], that uses a generating function. In this section, we recall briefly how this method is implemented and we introduce a modified generating function that is advantageous in some respects.

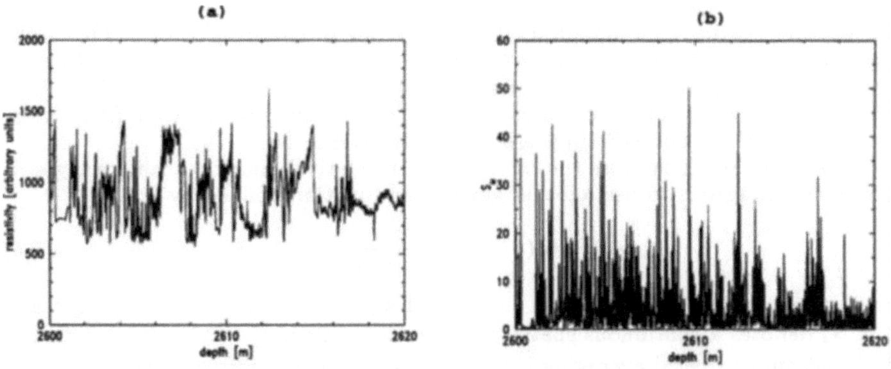

Fig. 3.1. Microresistivity signal $r(x)$ (a) and transformed signal $s_\omega(x) = |r(x) - r(x+\Delta x)|^\omega$ (b) for a 20 m section of well 30/9-6 from the Oseberg oil field in the North Sea. In (a) the irregular nature of $r(x)$ is apparent, and in (b) very localized spikes, looking like needles, are visible.

The window size L is assumed to be a power of two, i.e. $L = 2^N \Delta x$, where N is an integer and Δx is the width of the sampling interval. We define a collection of dyadic length scales $\delta_n = 2^n \Delta x, n = 0, 1, 2, ..., N$. For any δ_n the sample can be split into exactly $N_g(\delta_n) = L/\delta_n = 2^{N-n}$ disjoint and adjacent intervals of size δ_n, forming a grid. Let the index i label all the intervals that have a nonzero measure. Denoting the measure of interval i by $p_i(\delta_n)$, the *generating function* $\chi_q^G(\delta_n)$ is defined for any real q by

$$\chi_q^G(\delta_n) = \Omega^{-q} \sum_{i=1}^{N_B(\delta_n)} [p_i(\delta_n)]^q, \tag{3.2}$$

where $N_B(\delta_n)$ is the number of intervals of size δ_n that have a nonzero measure, and where Ω is the total measure, given by

$$\Omega = \sum_{i=1}^{N_B(\delta_n)} p_i(\delta_n) \ . \tag{3.3}$$

The superscript G in $\chi_q^G(\delta_n)$ indicates that this generating function involves a grid. $\chi_0^G(\delta_n)$ is equal to the number $N_B(\delta_n)$ of intervals of size δ_n needed to cover the support S of the measure. If S is space-filling, then $N_B(\delta_n) = (\delta_n/L)^{-D}$, where $D = 1$ is the dimension of the embedding space. If S is fractal, e.g. a Cantor set, then $N_B(\delta_n) \sim (\delta_n/L)^{-d_B}$ where d_B is the box-dimension [6] of S. If the generating function $\chi_q^G(\delta_n)$ satisfies the power law behaviour

$$\chi_q^G(\delta_n) \sim (\delta_n/L)^{\tau(q)}, \tag{3.4}$$

where $\tau(q)$ is non-linear function of q, then the measure is said to be multifractal. The mass exponent function $\tau(q)$ is concave, i.e. it satisfies $\tau''(q) \leq 0$. It satisfies also $\tau(1) = 0$, as a consequence of the normalization of the measure, i.e. $\chi_1^G(\delta_n) = 1$. Multifractal scaling is usually described in terms of generalized dimensions $D(q)$, defined by equation (2.3). If $\tau(q)$ is an affine function of q, i.e. $\tau(q) = d_B(q-1)$, then the measure is said to be monofractal and d_B is the box-dimension of S. This is the case where $D(q) = d_B$ for all q. In practice, the exponents $\tau(q)$ are obtained for each q by fitting $\chi_q^G(\delta_n)$ to the model (equation (3.4)) with linear regressions ($\ln[\chi_q^G(\delta_n)]$ versus $\ln(\delta_n)$). $D(q)$ is then calculated from $\tau(q)$ according to equation (2.3). More details about these curve-fittings are given in reference [7].

To analyse small samples (i.e. L small) with relatively narrow scaling ranges, it can be advantageous to use another generating function defined by

$$\chi_q^C(\delta) = (\delta/L)^{q-1} \frac{< [p(\delta)]^q >_S}{[< p(\delta) >_S]^q} , \tag{3.5}$$

where $p_x(\delta)$ is the measure of an interval of size δ centered on the point x and $< ... >_s$ denotes a continuous spatial average (the superscript C in $\chi_q^C(\delta)$ indicates that continuous spatial averaging is used). As explained in reference [7], $\chi_q^C(\delta)$ can be regarded as a natural extension of $\chi_q^G(\delta)$. In contrast with $\chi_q^G(\delta)$, $\chi_q^C(\delta)$ does not introduce any restrictions on L and δ and makes a complete use of the data, which is advantageous for small samples. Indeed, all possible intervals of size δ are used by $\chi_q^C(\delta)$ in the estimation of the averages. For large samples, the difference between $\chi_q^G(\delta)$ and $\chi_q^C(\delta)$ is not significant. A small modification to the definition (3.5) must be made if the measure is not space-filling (see reference [7]).

3.3 Representation of Generalized Dimensions

It is convenient to simplify the representation of the generalized dimensions $D(q)$. A partial description of $D(q)$ will be achieved by specifying only two parameters derived from the function $D(q)$: the *information dimension* [9] $D(1)$ and a *curvature parameter* α_c. For reasons explained in reference [7], we define (in analog with the Levy index used by Shertzer and Lovejoy [10]) the curvature index $\alpha_c \geq 0$ by

$$\alpha_c \equiv \frac{2|D'(1)|}{D(0) - D(1)}. \tag{3.6}$$

α_c is a measure of the curvature of the function $D(q)$ in the interval $0 \leq q \leq 1$. As long as $D''(q) \geq 0$ for $0 \leq q \leq 1$, which is a property commonly observed for resistivity signals [7], the inequality $0 \leq |D'(1)| \leq D(0) - D(1)$ holds and consequently the curvature index satisfies $0 \leq \alpha_c \leq 2$. In particular, $\alpha_c = 2$ corresponds to a linear $D(q)$ and $\alpha_c \leq 2$ to a curved $D(q)$. However in general α could be larger than 2.

$D(1)$ is a homogeneity index. A large $D(1)$ corresponds to a homogeneous spatial distribution of the measure, whereas a small $D(1)$ corresponds to a measure that is distributed inhomogeneously in space (clustered). α_c can be regarded as a contrast index: a measure with low α_c tends to be "black and white" (i.e. small or large), while the measure with high α_c exhibits a broader range of "grey tones". Each grey tone has its own dimension, and the tone carrying most of the total measure has dimension $D(1)$. More details about the interpretation of $D(1)$ and α_c in the context of the multifractal formalism are given in reference [7].

4. Characterization of geological structures

4.1 General remarks

As discussed in the previous section, a high $D(1)$ corresponds to a spatially homogeneous measure, while a low $D(1)$ corresponds to a more variable measure. We expect therefore the information dimension $D(1)$ to be correlated with the degree of geological heterogeneity. A more heterogeneous formation should produce a less homogeneous measure and therefore a lower $D(1)$. Conversely, a less heterogeneous formation should produce higher values of $D(1)$. The curvature index α_c quantifies the degree of multifractality of the measure, i.e. the width of the distribution of the total measure Ω around the singularity index [7] $\alpha = D(1)$. This parameter describes a significant part of the resistivity variations in a window, but unfortunately it is less obvious *a priori* to interpret α_c in geological terms. However, this parameter does correlate with some aspects of the local geology and can therefore be useful. For example, we shall see for the well 30/9-6 that α_c picks up clearly boundaries between different geological lithofacies.

4.2 Example 1: Well 30/9-6

The dipmeter microresistivity data were recorded in the Ness formation of well 30/9-6 in the Oseberg field in the North Sea, at depths ranging from 2600 m to 2680 m. At these depths, the large scale environment of the formation has been classified as a lower to upper delta plain. The formation has been classified in terms of three types of lithofacies denoted by N1, N2 and N3. In N1 the sub-environments are dominated by channels, in N2 by crevasses and in N3 by flood plains, swamps and distal crevasse splays. The sub-environment models of the lithofacies indicate that N1 is the least heterogeneous one, while N3 is the most heterogeneous. N2 has a degree of heterogeneity somewhere in between N1 and N2.

The texture logs $D(1)$ and α_c obtained with a 2m window size are shown in figure 4.1.

The texture logs $D(1)$ and α_c are shown together with the geological classification and the natural gamma radiation log. The σ-log (not displayed here) was also obtained by computing the standard deviation of $s_\omega(x)$ for each window.

We see in figure 4.1 that the three N1 lithofacies have the highest levels of $D(1)$ values, which agrees well with the lower degree of heterogeneity in these facies. The two intervals classified as N2 lithofacies are so thin that it is difficult to draw convincing conclusions. The three lithofacies N3 have low values of $D(1)$, in agreement with the higher degree of heterogeneity of these facies.

Even if the values of $D(1)$ are generally lower in N3 than in N1, there are local exceptions within the lithofacies. One example is the large value of $D(1)$ observed within the first N3 lithofacies. Such exceptions are usually due to small layers that differ in their degree of heterogeneity from the rest of the lithofacies. In the case of the first N3 lithofacies, a more detailed core description reveals that the high values of $D(1)$ correspond to a 2.5-3.0 m layer of sandstone. It is emphasised that the variations of α_c often indicate fairly clearly the boundaries between two facies, e.g. the central facies N1.

If we compare the texture logs to the natural gamma log (figure 4.1), we observe a certain degree of anti-correlation. The only major exceptions to this tendency are the first half of the rightmost N3 and the rightmost N1 facies. This anti-correlation can be interpreted as follows. We know that gamma logs usually reflect the content of clay minerals [1, 2, 3]. Since clayey lithologies happen to be the most heterogeneous ones for the well considered here, it follows naturally that low values of $D(1)$ (i.e. inhomogeneity) should be observed when the gamma radiation is high. The anti-correlation between the texture log $D(1)$ and the gamma log is not a universal observation. Indeed, small grained and clayey lithologies, where heterogeneities are evenly distributed in space, would normally produce high gamma radiation and high $D(1)$.

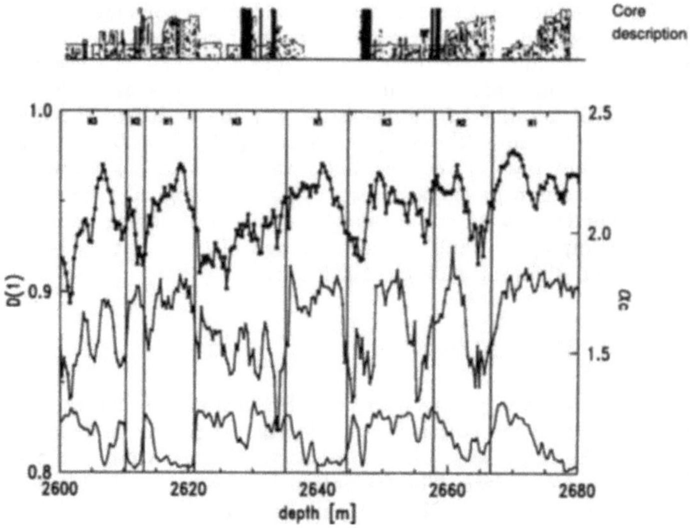

Fig. 4.1. From upper to lower, the curves are texture logs $D(1)$ and α_c and natural gamma radiation log from well 30/9-6 in the Oseberg field in the North Sea. The texture logs are obtained using 2 m windows. The left and right axis give scales for $D(1)$ and α_c respectively, whereas the natural gamma log is given in arbitrary units. The core description on top of the figure is a geological interpretation of the core plug from the well. Lithofacies N1-N3 are indicated with horizontal lines.

To show that the three texture parameters $D(1)$, α_c and σ can be useful to distinguish different facies we have plotted in figure 4.2 the cross plot of $D(1)$ vs. σ. Because of the relatively low number of N2 windows, N2 was not considered as a separate group. The N1 and N2 facies were instead considered as one group because they belong to similar environments. In figure 4.2, we see that the two groups N3 and N1-N2 separate fairly clearly in the parameter space. Indeed, the facies N1-N2 are characterized by high values of $D(1)$ and low values of σ. Considering that many windows overlap between two different facies, and remembering that a facies often contains sub-layers that are not consistent with its general characterisation (a globally heterogeneous facies may well be weakly heterogeneous locally), we conclude that the degree of separation of the two clusters N1-N2 and N3 is satisfactory. Thus the two groups of facies can be distinguished by their texture logs.

4.3 Example 2: Well E

Well E is located in the Troll field in the North Sea. It was analysed in depths ranging from 1870-2030 m and results are displayed in figure 4.3. The gross environment and sub-environment and process interpretation was not available, but the analysed sequence resembles what one would expect for a tide and current dominated continental shelf, as described in Gilreath (1987). Based on the core description, six major sequences was identified, they are denoted 1-6 in figure 4.3.

Sequence 1 contains laminated sand/silt mixtures with some bioturbation. Sequence 2 contains sand/silt mixtures with strong bioturbations, but lacks laminations. Sequence 3 contains cleaner sands, is laminated and it lacks bioturbations.

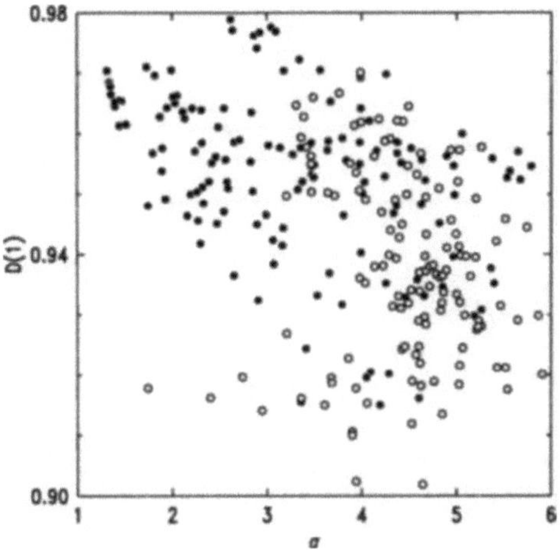

Fig. 4.2. Cross plot of $D(1)$ versus σ from resistivity signals from well 30/9-6, using a 2 m window size. Data points from windows in lithofacies N1-N2 (•) are grouped together, and distinguished from data from windows in lithofacies N3 (∘).

Sequence 4 contains less clean sands, is more bioturbated and laminations are lacking. Sequence 5 resembles sequence 3, and sequence 6 resembles sequence 2 except that the grain size is somewhat smaller.

The general picture indicate that $D(1)$ values are larger in the laminated sequences (3 and 5) than in the laminated / bioturbated (1) and bioturbated (2, 4 and 6) ones. In the figure, we see that the GR-log can be used to distinguish the fine grained from the more coarse lithologies (as expected) and thus can be used to identify sequences 3-6. However, the GR-log does not distinguish very well the sequences 1 and 2. In this case, the additional information provided by the $D(1)$ log might be useful to help distinguish the sequences from each other. To check this, a magnification of the parts corresponding to sequences 1 and 2 are displayed in figure 4.4.

In addition to the core log (top) and the usual texture logs and GR-log plot (bottom), the raw (upper) and transformed (lower) resistivity signals are plotted in the middle of the figure.

First, note the abrupt resistivity change at $\sim$ 1897 m, due to dynamical amplification used by the resistivity tool. This change is reflected in the α_c log whereas the $D(1)$ log is less sensitive to this artificial change in the resistivity signal.

Second, we observe that the GR-log does not discriminate sequence 1 from sequence 2. The sequences are also difficult to see from the raw and transformed signals. In contrast, the $D(1)$ log seems to indicate a small difference between the sequences, which suggest that detection of the two sequences could be assisted using the $D(1)$ log.

5. Conclusions

We have shown that multifractal analysis can be adapted to characterise dipmeter microresistivity well logs. The multifractal description, which is possible when suitable scaling properties are observed, allows a compact representation of multipoint

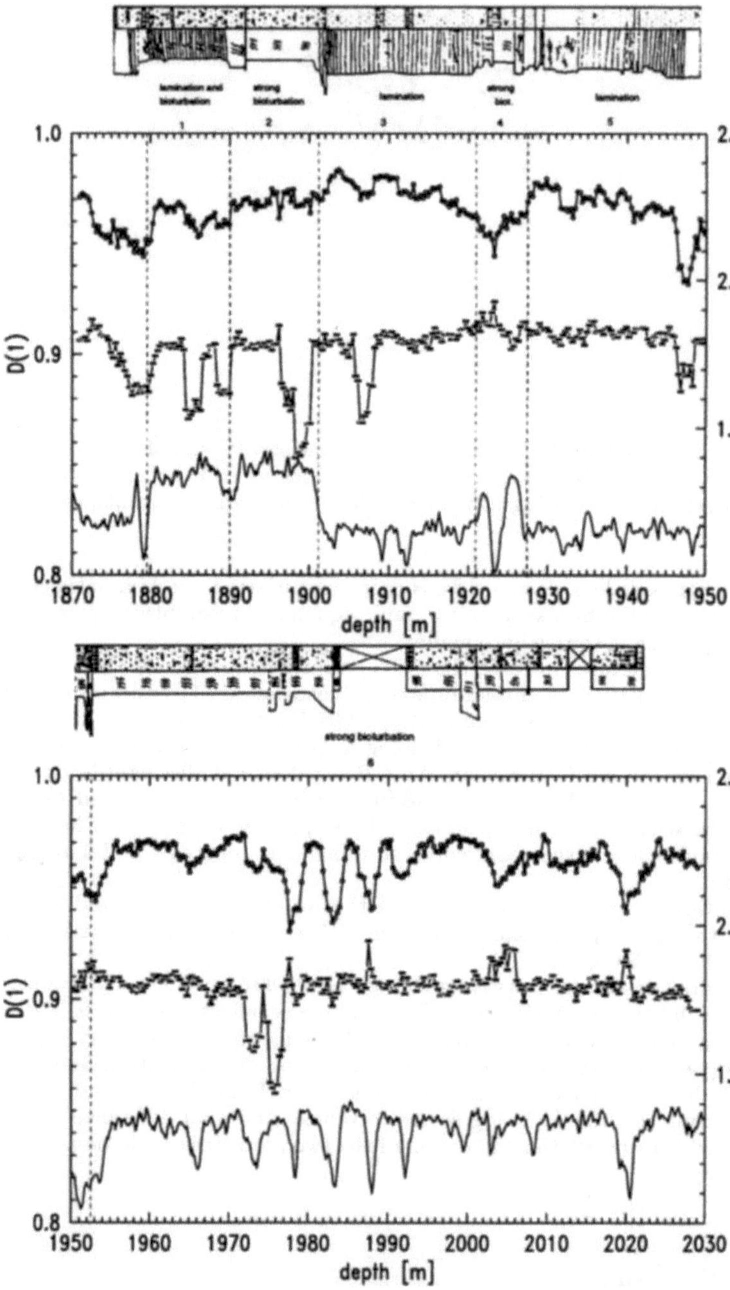

Fig. 4.3. Texture logs $D(1)$, α_c and GR-log from well E in the North Sea. Left and right axis give scales for $D(1)$ and α_c respectively, GR-log in arbitrary units.

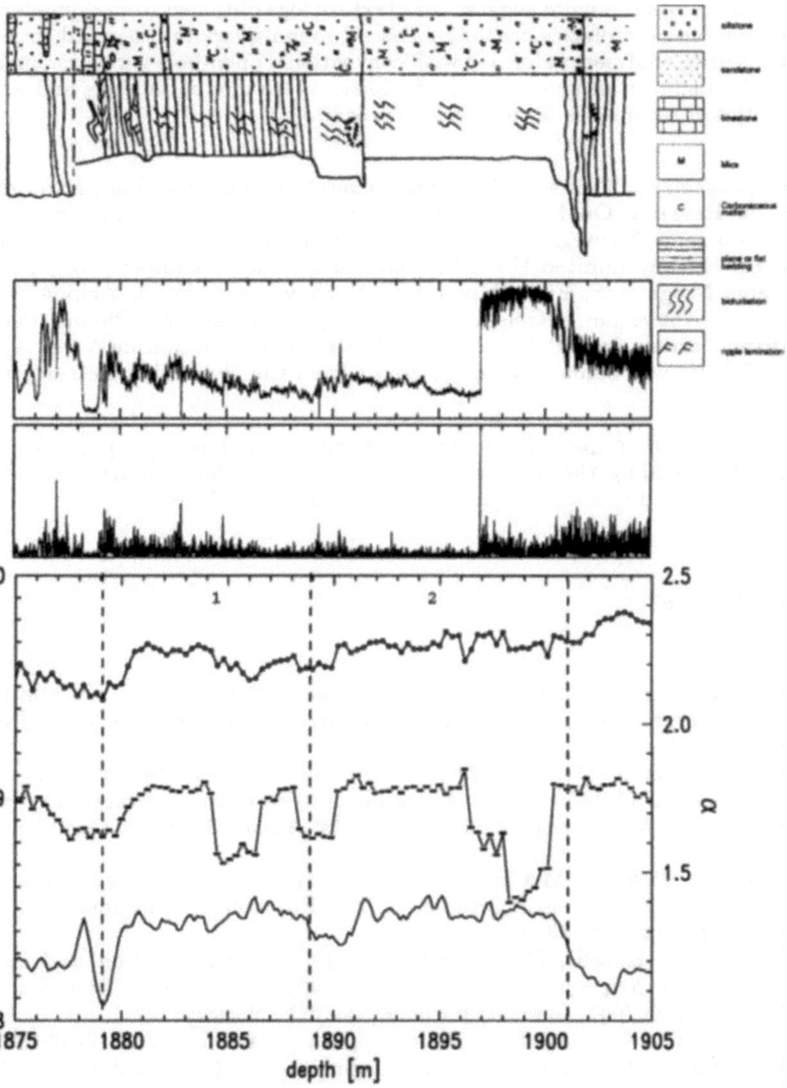

Fig. 4.4. Blow-up of one part of logs from Well E. From top to bottom the figure show core log, raw microresistivity signal, filtered and transformed signal and texture logs $D(1)$ (o) and α_c (▬) together with the GR-log.

statistics. These statistics were represented by a triplet $(\sigma, D(1), \alpha_c)$ of texture parameters. It was found that these parameters reflect the degree of heterogeneity of lithofacies. Our results also suggests that the texture parameters can be used to distinguish different lithofacies within a formation.

These results therefore suggest that texture logs obtained from dipmeter curves could improve electrofacies predictions based on multivariate statistical correlations and clustering analyses performed with conventional open-hole wire logs. Such work, which also includes an investigation of additional texture parameters and advanced segmentation techniques is currently in progress.

Acknowledgement. This work has been financed and performed as part of the Program for Research On Field-Oriented Improved Recovery Technology (PROFIT), which is a joint Norwegian research program between Amoco Norway Oil company, BP Norway Limited U.A., Conoco Norway Inc., Deminex (Norge) A/S, Elf Aquitaine Norge A/S, Fina Exploration Norway u.a.s., Norsk Agip A/S, Norsk Hydro a.s., Norwegian Petroleum Directorate, Petrobras A/S, Phillips Petroleum Company Norway, Saga Petroleum A.S., Statoil and Total Marine Norsk A/S. One of us (O.K.H) was financed through grants from the Norwegian Research Council (NFR)

We thank the PROFIT Board for permission to publish this paper and Norsk Hydro for providing the dipmeter data. We acknowledge stimulating discussions with T.M. Rønningsland, Jacques Lévy Véhel and Jan Sagen.

References

1. Schlumberger (1986): *Formation MicroScanner Service.* Publ. no. M89100.
2. Schlumberger (1983): *Strategraphic High Resolution Dipmeter Tools.* Publ. no. M86300.
3. Schlumberger (1974): *Log Interpretation, vol. II - Applications.* N.Y..
4. Hewett, T.A. (1986): *Fractal distributions of reservoir heterogeneity and their influence on fluid transport.* SPE 15385, 61st Annual Technical Conference, New Orleans.
5. Muller, J., Bokn, I., McCauley, J.L. (1992): *Ann. Geophysicae* **10**, 735.
6. Halsey, T.C., Jensen, M.H., Kadanoff, L.P., Procaccia, I., Shraiman, B.I. (1986). *Physical Review A*, **33**, 1141.
7. Saucier, A., Huseby, O.K., Muller, J. (1995): *Quantification and modelling of reservoir heterogeneities based on multifractal scaling.* Technical report IFE/KR/F-95/047, Institutt for energiteknikk, Norway.
8. Press, H. W., Flannery, B.P., Teukolsky, S.A., Vetterling, W.T. (1986): *Numerical Recipes, the Art of Scientific Computing*, pp. 495, Cambridge University Press, New York.
9. Evertsz, C.J.G., Mandelbrot, B.B. (1992): Multifractal measures, appendix B. In *Chaos and Fractals*, edited by H.O. Peitgen, H. Jurgens and D. Saupe, Springer-Verlag, New York.
10. Schertzer, D., Lovejoy, S. (1987). *J. Geophys. Res. D* **92**, 9693.
11. Jaffard, S.: *Multifractal formalism for functions.* To appear in SIAM Journal of Math. An..
12. Lévy Véhel, J. (1995): *Fractal Approaches in Signal Processing.* Fractals **3**, 755.

IFS THEORY

Correspondence Between Fractal-Wavelet Transforms and Iterated Function Systems With Grey Level Maps

F. Mendivil and E.R. Vrscay

Department of Applied Mathematics
Faculty of Mathematics, University of Waterloo
Waterloo, Ontario, CANADA N2L 3G1
mendivil@augusta.math.uwaterloo.ca
ervrscay@links.uwaterloo.ca
http://links.uwaterloo.ca

Summary. It is well known that the action of a "Fractal Transform" or (Local) Iterated Function System with Grey Level Maps (IFSM) on a function $f(x)$ induces a very simple mapping on its expansion coefficients c_{ij} in the Haar wavelet basis. This is the basis of the "discrete fractal-wavelet transform": subtrees of the wavelet coefficient tree are scaled and copied to lower subtrees. Such transforms, which we shall also refer to as IFSW - IFS on wavelet coefficients - have been introduced into image processing with other (compactly supported) wavelet basis sets in an attempt to remove the blocking artifacts in the standard IFS block encoding algorithms. Although not as straightforward as in the Haar case, we show that there is a relationship between such wavelet transforms and IFSM. In fact, for most such transforms, there is an equivalent IFSM, which provides a further mathematical basis for their use in image processing. We also present results for the case of periodized wavelets, a common implementation in image processing. Finally, we prove some results on the fractal dimension of the graph of an attractor of IFSM or IFSW operators.

1. Introduction

Recently, a number of workers [3, 7, 8, 9, 10] (to name a few) have independently devised "discrete fractal-wavelet transforms" (DFWT) which involve an IFS-type transformation on the wavelet expansion coefficients of signals and images. In the case of Haar wavelet expansions, as was shown in [7], there is a very simple connection between these wavelet transforms and (Local) Iterated Function Systems with Grey Level Maps (IFSM). Due to the non-overlapping nature of the Haar wavelet functions, the action of the IFSM automatically translates into a simple scaling and copying of wavelet coefficient subtrees onto lower subtrees. With suitable restrictions on the scaling coefficients α_i, the transformation is contractive in $l^2(\mathbb{N})$. The wavelet expansion coefficients c_{ij} of a signal or image are then stored to a prescribed level k^* of refinement, along with a set of codes describing the "parent-child" subtree pairs and their corresponding "optimal" parameters α_i. Iteration of the fractal-wavelet transform then generates higher resolution coefficients c_{ij}, $i > k^*$. To repeat: In the case of Haar wavelets, this procedure in wavelet coefficient space is equivalent to the action of an IFSM in "pixel space". It also illustrates the fact that fractals and wavelet bases share the fundamental property of scaling.

Such discrete wavelet transforms were then applied to expansion coefficients for generalized (non-Haar) systems of compactly supported wavelets. In these cases, the supports of contiguous wavelets overlap. A major motivation for the generalization of this method was the attempt to reduce the blockiness exhibited by the usual fractal (= Haar) block-encoding (see [3, 8, 10]).

The purpose of this paper is to show that there is a connection between wavelet transforms for generalized wavelets and IFS-type methods on function spaces. However, the connection is not quite as straightforward as in the Haar case. In Section 2., we consider the case of non-periodized wavelets and show that the discrete wavelet transform is equivalent to a recurrent IFSM. In Section 3., we consider *periodized* wavelets which are used extensively in image processing. The situation is slightly more complicated but a correspondence to IFSM is shown. In Section 4., we briefly present some results on the fractal dimension of attractors for IFSM and IFSW.

We now conclude this section with a brief review of the basic ideas of IFSM. An N-map Iterated Function System with Grey Level Maps (IFSM) is defined by [6]:

1. **The IFS component:** $\mathbf{w} = \{w_1, w_2, \ldots, w_N\}$ where each $w_i : X \to X$ is a contraction and
2. **The grey level component:** $\Phi = \{\phi_1, \phi_2, \ldots, \phi_N\}$ where each $\phi_i : \mathbb{R} \to \mathbb{R}$ is Lipschitz.

Here (X, d) is a complete metric space, the "pixel space", typically $[0,1]$ or $[0,1]^2$ with Euclidean metric. Associated with an IFSM (w, Φ) is an operator T, the *fractal transform*, whose action on a suitable space of functions $\mathcal{F}(X)$ is given by

$$T(f)(x) = \sum_i \phi_i\left(f(w_i^{-1}(x))\right), \quad f \in \mathcal{F}(X). \tag{1.1}$$

For simplicity, both the w_i and the ϕ_i are often assumed to be affine so that $w_i(x) = s_i x + a_i$ and $\phi_i(t) = \alpha_i t + \beta_i$.

Another interesting and special case of an IFSM operator is the following:

$$T(f)(x) = \theta(x) + \sum_i \alpha_i f(w_i^{-1}(x)). \tag{1.2}$$

Here, the function $\theta(x)$ acts as a "condensation" function for the IFSM.

Under suitable conditions on the w_i and the ϕ_i, the operator T is contractive in $\mathcal{F}(X)$ (for example, in $L^p(X)$ it suffices that $\left(\sum_i |s_i \alpha_i^p|\right)^{1/p} < 1$) and so has a unique fixed point $\bar{u} = T\bar{u}$ (see [6]).

Let $\{q_i(x)\}$ be an orthonormal basis of $L^2(X)$. Given an IFSM operator T there exists a corresponding operator M on the sequence space $l^2(\mathbb{N})$ (viewed as sequences of coefficients in expansions of functions with respect to the q_i). The following figure illustrates this procedure:

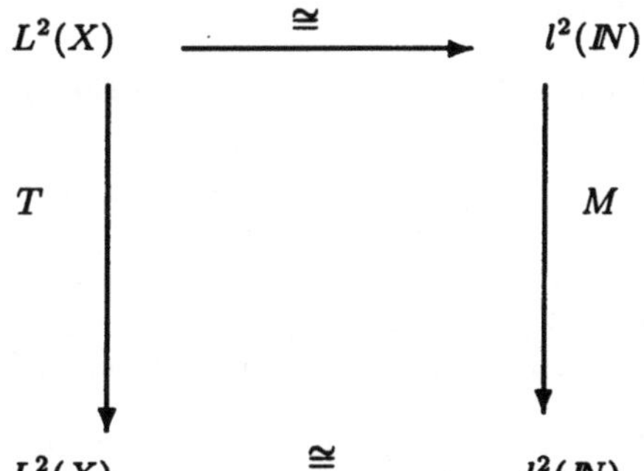

In the event that the IFSM T is affine or linear, then the operator M will also be affine or linear. If T is contractive in $\mathcal{L}^2(X)$, then M is contractive in $l^2(\mathbb{N})$. (See [7] for further discussion.) As mentioned above, when the orthonormal basis is the Haar wavelet basis ψ_{ij}, then the operator M becomes a mapping of wavelet coefficient subtrees to lower subtrees. Such mappings will be introduced in the next section.

2. Fractal-Wavelet Transforms (IFSW) and Their Correspondence with IFSM

For simplicity, we restrict our attention to the one-dimensional case $X = [0,1]$. The results extend in a straightforward manner to the important case $X = [0,1]^2$ (images). Let ϕ be a scaling function with which is associated a multiresolution analysis of $\mathcal{L}^2(R)$, and let ψ be the associated "mother wavelet" function from which are derived the functions

$$\psi_{mn}(x) = 2^{m/2}\psi(2^m x - n), \quad m, n \in Z, \tag{2.1}$$

which form an orthonormal basis of $\mathcal{L}^2(R)$. In particular we shall be concerned with functions $f \in \mathcal{L}^2(R)$ which admit wavelet expansions of the form (in the $\mathcal{L}^2$ sense)

$$f(x) = b_{00}\phi(x) + \sum_{i \geq 0, 0 \leq j \leq 2^i - 1} c_{ij}\psi_{ij}(x), \tag{2.2}$$

where $b_{00} =< f, \phi >$ and $c_{ij} =< f, \psi_{ij} >$. We are interested in wavelets ψ having compact support on $\mathbf{R}$ (implying, in turn, that f has compact support). The expansion coefficients are conveniently displayed in the form of an *infinite* wavelet tree:

<table>
<tr><td colspan="8" align="center">b_{00}</td></tr>
<tr><td colspan="8" align="center">c_{00}</td></tr>
<tr><td colspan="4" align="center">c_{10}</td><td colspan="4" align="center">c_{11}</td></tr>
<tr><td colspan="2" align="center">c_{20}</td><td colspan="2" align="center">c_{21}</td><td colspan="2" align="center">c_{22}</td><td colspan="2" align="center">c_{23}</td></tr>
<tr><td>B_{30}</td><td>B_{31}</td><td>B_{32}</td><td>B_{33}</td><td>B_{34}</td><td>B_{35}</td><td>B_{36}</td><td>B_{37}</td></tr>
</table>

Note that each entry B_{ij} represents a tree of infinite length: we shall also refer to such a tree with apex c_{ij} as the *block* B_{ij}. We now consider some IFS-type operations on this wavelet tree.

Example 1: Using the above notation, consider the following transformation:

$$M : B_{00} \rightarrow \begin{array}{|c|c|} \hline \multicolumn{2}{|c|}{c_{00}} \\ \hline \alpha_0 B_{00} & \alpha_1 B_{00} \\ \hline \end{array}\,, \qquad |\alpha_i| < \frac{1}{\sqrt{2}}. \tag{2.3}$$

The restrictions on the α_i follow from the condition that the wavelet coefficient sequence c_{ij} belong to $l^2(Z)$. This is a simple example of a "discrete fractal-wavelet transform" to which, for simplicity, we shall refer as an *IFS on wavelet coefficients* or IFSW.

The goal is to find an IFSM operator T (as defined by Eq. (1.2)) acting in "physical space", i.e. on functions supported on R, which corresponds to an IFSW operator M acting on the wavelet coefficients. The dilation/translation relations within the wavelet basis provide the key to this goal. Let $w_1(x) = x/2$ and $w_2(x) = x/2 + 1/2$. Then

$$\sqrt{2}B_{00} \circ w_1^{-1} = B_{10} \quad \text{and} \quad \sqrt{2}B_{00} \circ w_2^{-1} = B_{11} \tag{2.4}$$

since

$$\sqrt{2}\psi_{i,j}\left(w_1^{-1}(x)\right) = \psi_{i+1,j}(x) \quad \text{and} \quad \sqrt{2}\psi_{i,j}\left(w_2^{-1}(x)\right) = \psi_{i+1,j+2^i}(x). \tag{2.5}$$

Thus, this simple IFSW operator will correspond to the following two-map IFSM with condensation function $\psi(x)$:

$$T(f)(x) = c_{00}\psi(x) + \sqrt{2}\alpha_0 f\left(w_1^{-1}(x)\right) + \sqrt{2}\alpha_1 f\left(w_2^{-1}(x)\right). \tag{2.6}$$

Note that the IFSM operator depends on the particular wavelet basis chosen.

In the simple case of Haar wavelets, the mother wavelet $\psi(x)$ decomposes into non-overlapping components:

$$\psi(x) = I_{[0,1/2)}(x) - I_{[1/2,1)}(x). \tag{2.7}$$

As such, the IFSM operator T in Eq. (2.6) corresponds to a simple two-map IFSM with IFS maps w_1 and w_2 and grey-level maps $\phi_1(t) = \sqrt{2}\alpha_1 t + 1$ and $\phi_2(t) = \sqrt{2}\alpha_2 t - 1$. If T is contractive, then its fixed point attractor function $\bar{u}$ has [0,1] as support.

However, in the case of other compactly supported wavelets, no such spatial decomposition into separate grey-level maps is possible. As well, the support of the attractor function $\bar{u}$ is necessarily larger than [0,1]. To illustrate, consider the particular IFSW in which $\alpha_1 = 0.4$ and $\alpha_2 = 0.6$. The IFSM operator T in Eq. (2.6) is contractive. Figures (2.1a) and (2.1b) show the IFSM attractor functions for, respectively, the Haar wavelet and "Coifman-6" cases. In both cases, we have chosen $b_{00} = 0$ and $c_{00} = 1$.

Example 2: Consider the fractal wavelet transform with four block maps as follows:

$$W_1 : B_{10} \to B_{20}, \quad W_2 : B_{11} \to B_{21}, \quad W_3 : B_{10} \to B_{22}, \quad W_4 : B_{11} \to B_{23}, \tag{2.8}$$

with associated multipliers α_i, $1 \le i \le 4$. Diagrammatically,

$M \; : \; B_{00} \to$

<table>
<tr><td colspan="4">c_{00}</td></tr>
<tr><td colspan="2">c_{10}</td><td colspan="2">c_{11}</td></tr>
<tr><td>$\alpha_1 B_{10}$</td><td>$\alpha_2 B_{11}$</td><td>$\alpha_3 B_{10}$</td><td>$\alpha_4 B_{11}$</td></tr>
</table>

. (2.9)

(The reader may suspect that this has the aroma of a local IFSM.) Now iterate this process, assuming it converges to a limit $\bar{B}_{00}$ which represents the wavelet expansion of a function $\bar{u}$. Then

$$\bar{u} = c_{00}\psi_{00} \; + \; \bar{v}. \tag{2.10}$$

We need only focus on the function $\bar{v}$ which admits the wavelet expansion

<table>
<tr><td colspan="2">c_{10}</td><td colspan="2">c_{11}</td></tr>
<tr><td>$\alpha_1 B_{10}$</td><td>$\alpha_2 B_{11}$</td><td>$\alpha_3 B_{10}$</td><td>$\alpha_4 B_{11}$</td></tr>
</table>

. (2.11)

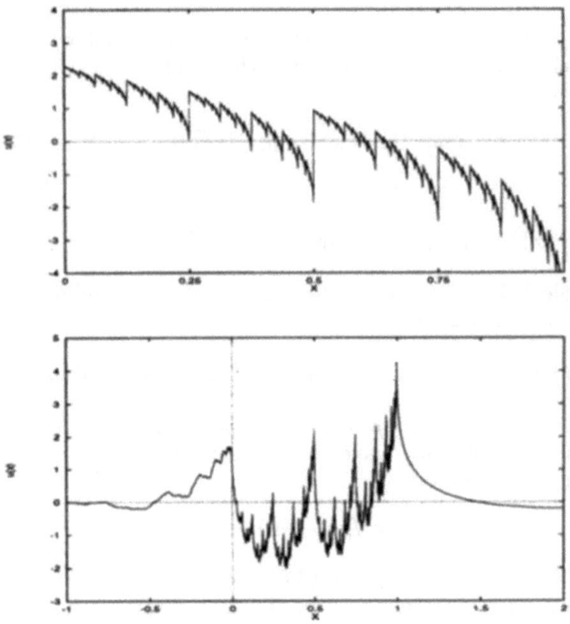

Fig. 2.1. Attractor functions $\bar{u}$ for Example 1: (a) Haar wavelet basis, (b) Coifman-6 wavelet basis.

Since $< \psi_{10}, \psi_{11} >= 0$, etc., we may write

$$\bar{v} = \bar{v}_1 + \bar{v}_2 , \tag{2.12}$$

where the components $\bar{v}_i$ satisfy the relations

$$\begin{aligned} \bar{v}_1(x) &= c_{10}\psi_{10}(x) + \alpha_1\sqrt{2}\bar{v}_1(2x) + \alpha_2\sqrt{2}\bar{v}_2(2x) \\ \bar{v}_2(x) &= c_{11}\psi_{11}(x) + \alpha_3\sqrt{2}\bar{v}_1(2x-1) + \alpha_4\sqrt{2}\bar{v}_2(2x-1). \end{aligned} \tag{2.13}$$

We may consider these equations as defining a kind of *vector IFSM with condensation*. The vector $\bar{v}$ is composed of the orthogonal components $\bar{v}_1$ and $\bar{v}_2$ that satisfy the above fixed point relations in Eq. (2.13). These equations may be written more compactly as:

$$\bar{v}_i(x) = b_i(x) + \sum_{j=1}^{2} \phi_{ij}(\bar{v}_j(w_{ij}^{-1}(x))), \quad i = 1, 2, \tag{2.14}$$

where

$$w_{11}(x) = w_{12}(x) = \frac{1}{2}x, \quad w_{21}(x) = w_{22}(x) = \frac{1}{2}x + \frac{1}{2}, \tag{2.15}$$

$$b_1(x) = c_{10}\psi_{10}(x), \quad b_2(x) = c_{11}\psi_{11}(x) , \tag{2.16}$$

and

$$\phi_{11}(t) = \alpha_1\sqrt{2}t, \quad \phi_{12}(t) = \alpha_2\sqrt{2}t, \quad \phi_{21}(t) = \alpha_3\sqrt{2}t, \quad \phi_{22}(t) = \alpha_4\sqrt{2}t . \tag{2.17}$$

Note that the contractive IFS maps w_{ij} are mappings from the entire base space X into itself. As such, this IFSM is *not* a local IFSM in general. What appeared to be a "local" transform in wavelet coefficient space is a normal IFSM in the base space. (Again, in the special case of the non-overlapping Haar wavelets, the above IFSM may be written as a local IFSM.) The "locality" of the block transform has been passed on to the orthogonal components $\bar{v}_1$ and $\bar{v}_2$ of the function $\bar{v}$. These components may be considered as "non-overlapping" elements of a vector.

This "vector IFSM" is really nothing more than a recurrent IFSM on B_{00} where we split B_{00} as $B_{10} \oplus B_{11}$ and have the IFSM act "between" the components of this splitting.

Example 3: Finally, consider the following modification of the fractal wavelet transform in Example 2:

$$W_1 : B_{10} \to B_{20}, \quad W_2 : B_{11} \to B_{21}, \quad W_3 : B_{11} \to B_{22}, \quad W_4 : B_{10} \to B_{23}, \tag{2.18}$$

with associated multipliers α_i, $1 \le i \le 4$. Graphically,

$M \; : \; B_{00} \;\to$

c_{00}			
c_{10}		c_{11}	
$\alpha_1 B_{10}$	$\alpha_2 B_{11}$	$\alpha_3 B_{11}$	$\alpha_4 B_{10}$

. (2.19)

As above, iterate this process, assuming it converges to a limit $\bar{B}_{00}$ which represents the wavelet expansion of a function $\bar{u}$. Then

$$\bar{u} = c_{00}\psi_{00} \; + \; \bar{v}. \tag{2.20}$$

where $\bar{v}$ admits the wavelet expansion

c_{10}		c_{11}	
$\alpha_1 B_{10}$	$\alpha_2 B_{11}$	$\alpha_3 B_{11}$	$\alpha_4 B_{10}$

. (2.21)

As before, we write

$$\bar{v} \; = \; \bar{v}_1 \; + \; \bar{v}_2 \; . \tag{2.22}$$

and determine the relations satisfied by the components $\bar{v}_i$. It is necessary to determine the affine changes of variables involved in mapping B_{10} to B_{23} and B_{11} to B_{22}. The result is

$$\begin{aligned} \bar{v}_1(x) &= c_{10}\psi_{10}(x) \; + \; \alpha_1\sqrt{2}\bar{v}_1(2x) \; + \; \alpha_2\sqrt{2}\bar{v}_2(2x) \\ \bar{v}_2(x) &= c_{11}\psi_{11}(x) \; + \; \alpha_3\sqrt{2}\bar{v}_2(2x-\frac{1}{2}) \; + \; \alpha_4\sqrt{2}\bar{v}_1(2x-\frac{3}{2}). \end{aligned} \tag{2.23}$$

This is a vector or recurrent IFSM with condensation functions which could be written in the compact form of Eq. (2.14).

3. IFSW for Periodic Wavelets on [0, 1]

In many applications one needs wavelet bases adapted to a compact interval. For this purpose, periodizing the wavelet basis functions is useful. Furthermore, in practical computations, one often deals with finite data sets and periodizing the discrete transform is simple and clean to implement.

For a function $f \in L^2(\mathbb{R})$, define the periodized version of f to be

$$f^*(x) = \sum_i f(x-i) \tag{3.1}$$

where the sum is over all integer i. This process "wraps" the function f around over the interval $[0,1]$ and sums the various contributions.

Let ϕ be a scaling function for an MRA on $\mathbb{R}$ and ψ be the associated "mother" wavelet. Then the periodized wavelets $\psi^*_{ij}(x)$ form an orthonormal basis of $L^2[0,1]$ and we have a nested MRA structure as in the case of wavelets on $\mathbb{R}$ [4]. Unfortunately, the nice translation/scaling relations (represented by Eq. (2.5)) no longer hold for these periodized wavelets. To see this, consider the following simple example. By definition,

$$\begin{aligned}\psi^*_{1j}(x) &= \sum_k \psi_{1j}(x-k) \\ &= \sum_k \psi(2x-2k-j),\end{aligned} \tag{3.2}$$

where the latter equation follows from the scaling property of the wavelets. However, the function

$$\begin{aligned}\psi^*(2x-j) &= \sum_k \psi(2x-j-k) \\ &\neq \psi^*_{1j}(x).\end{aligned} \tag{3.3}$$

As a result, the rather simple scaling analysis of Section 2. is not applicable here. The IFSM associated with an IFSW will have to be different in the periodic case than in the case of functions on $\mathbb{R}$.

Suppose that we are given an IFS on the wavelet coefficient tree for some periodized wavelets on the interval $[0,1]$. This operator induces an operator

$$T : L^2[0,1] \to L^2[0,1].$$

Since we are interested in periodic functions, we think of the unit circle with $[0,1]$ as the "natural" fundamental domain for the periodization.

Let ψ be the mother wavelet for some wavelet decomposition. Suppose that the support of ψ lies in the interval $[0,N]$. Then

$$supp(\psi^j) \equiv supp(\psi(2^n x - j)) \subset 2^{-n}[j, j+N] = [\frac{j}{2^n}, \frac{j+N}{2^n}]$$

with j running from 0 to $2^n - 1$. So, for sufficiently large n we have that their supports have length less than 1 and that all the supports lie in the interval $[0,3/2]$.

We now define the IFS maps $w_1, w_2, \hat{w}_1$ and $\hat{w}_2$ on the circle. These maps will be induced by corresponding maps $W_1(x) = x/2$ and $W_2(x) = x/2 + 1/2$ on $\mathbb{R}$. For the maps w_1 and w_2, we use the coordinate system on the circle induced by the fundamental domain $[0,1]$. Notice that the image of $[0,1]$ under w_1 is $[0,1/2]$ – this map would not be a continuous map on the circle, so we must make a "cut" at the point $x = 0 = 1$. Thus, these maps are, in some sense, only "local" maps. For the maps $\hat{w}_1$ and $\hat{w}_2$, we use the coordinate system on the circle induced by the domain $[1/2,3/2]$. Here the "cut" point is $x = 1/2$ (the point antipodal to the point 0). Thus, the image of the circle under $\hat{w}_1$ is $[1/4,3/4]$ – the "middle" of the circle – and under $\hat{w}_2$ is $[3/4,5/4]$. Again, these are "local" maps.

With these definitions, the usual scaling and translation relations among the wavelets are modified as follows. For $i \geq n$,

$$\sqrt{2}\psi^*_{i,j}(w_1^{-1}(x)) = \psi^*_{i+1,j}(x) \quad \text{and} \quad \sqrt{2}\psi^*_{i,j}(w_2^{-1}(x)) = \psi^*_{i+1,2^i+j}(x) \tag{3.4}$$

for $0 \leq j < 2^{i-1}$ and

$$\sqrt{2}\psi^*_{i,j}(\hat{w}_1{}^{-1}(x)) = \psi^*_{i+1,j}(x) \quad \text{and} \quad \sqrt{2}\psi^*_{i,j}(\hat{w}_2{}^{-1}(x)) = \psi^*_{i+1,2^i+j}(x) \tag{3.5}$$

for $j \geq 2^{i-1}$.

The functions $\psi^*_{i,j}$ are supported on the interval $[0, 1]$ for $j < 2^{i-1}$ and on the interval $[1/2, 3/2]$ for $j \geq 2^{i-1}$. Thus, we use the w_i's for the first ones and the $\hat{w}_i$'s for the second ones. In our block notation, these scaling and translation relations become

$$\sqrt{2}B_{i,j} \circ w_1^{-1} = B_{i+1,j} \quad \text{and} \quad \sqrt{2}B_{i,j} \circ w_2^{-1} = B_{i+1,2^i+j} \tag{3.6}$$

for $0 \leq j < 2^{i-1}$ and

$$\sqrt{2}B_{i,j} \circ \hat{w}_1{}^{-1} = B_{i+1,j} \quad \text{and} \quad \sqrt{2}B_{i,j} \circ \hat{w}_2{}^{-1} = B_{i+1,2^i+j} \tag{3.7}$$

for $j \geq 2^{i-1}$.

These relations are the crucial observation needed to show that the IFS on the wavelet coefficients induces an IFSM operator on $L^2[0, 1]$ (in the usual sense of an IFSM). In order to do this, we break the space $L^2[0, 1]$ into two orthogonal parts. The first part is the space spanned by the first n levels of the wavelet basis:

$$V_1 = span\{\psi^*_{i,j} \mid 0 \leq i < n, 0 \leq j < 2^i\}$$

and the second part is the orthogonal complement of V_1 in $L^2[0, 1]$ so that

$$V_1 \oplus V_2 = L^2[0, 1].$$

Notice that T maps V_1 into both V_1 and V_2. However, $T(V_2) \subset V_2$. Thus, we have a sort of "recurrent" IFSM between V_1 and V_2. We represent the operator T as

$$T = (T_1 + T_2) \oplus T_3 \tag{3.8}$$

with

$$T_1 : V_1 \to V_1 \quad \text{and} \quad T_2 : V_1 \to V_2 \quad \text{and} \quad T_3 : V_2 \to V_2 \tag{3.9}$$

Let v be the fixed point of the operator T, thus $T(v) = v$. Writing $v = v_1 \oplus v_2$, we obtain

$$T_1(v_1) \oplus (T_2(v_1) + T_3(v_2)) = v_1 \oplus v_2 \tag{3.10}$$

and thus

$$T_1(v_1) = v_1 \tag{3.11}$$

$$T_2(v_1) + T_3(v_2) = v_2. \tag{3.12}$$

So the IFS on V_1 is a simple IFS and the IFS on V_2 is an IFS with condensation function $T_2(v_1)$.

The vector v_1 is relatively easy to calculate since V_1 is finite dimensional. One needs only to compute the coefficients of v_1 with respect to the basis $\psi^*_{i,j}$ for appropriate i, j. These coefficients are given by the IFSW operator. From v_1 one obtains the function $T_2(v_1)$ – the condensation function for the IFSM on V_2. The IFSM on V_2 basically corresponds to the case of non-periodized wavelets (i.e. wavelets on $\mathbb{R}$).

Finally, we return to the fractal-wavelet transform of Example 1 applied to coefficient trees for periodized wavelets. In the Haar case, the result is trivially the same as in Figure (2.1a). The periodic attractor function $\bar{u}$ for the Coifman-6 periodized wavelets is shown in Figure (3.1). (Note that the wavelet coefficients c_{ij} for all three graphs are identical.)

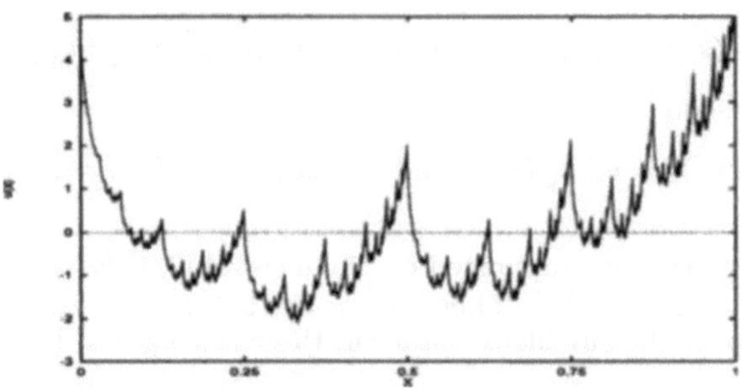

Fig. 3.1. Attractor function $\bar{u}$ for IFSW in Example 1 using Coifman-6 periodized wavelet functions.

4. Fractal Dimension of IFSM and IFSW

In this section we discuss the dimension of the attractor of an IFSM or IFSW (which we think of as an IFSM with condensation). We will primarily be interested in the box counting dimension, although in many cases this is the same as the Hausdorff dimension. Our methods are direct adaptations of the methods in [1, 5], thus we give only the sketches of the proofs.

Recall that our usual IFSM operator has the form

$$T(f)(x) = \sum_i \alpha_i f(w_i^{-1}(x)) + \beta_i \chi_i, \tag{4.1}$$

where χ_i is the characteristic function of $X_i = w_i([0,1])$ and $w_i(x) = s_i x + b_i$.

We can think of the piecewise constant parts $\beta_i \chi_i$ as being a piecewise-defined "condensation function" for the IFSM. Thus, the more general form of the IFSM is given by

$$T(f)(x) = \sum_i \alpha_i f(w_i^{-1}(x)) + \theta(x) \tag{4.2}$$

We obtain the following theorem on the fractal dimension of the graph of the attractor for the IFSM (4.2). Notice that if θ is C^1, then the graph of θ will have dimension 1.

Theorem 4.1. *Suppose that the IFSM operator T as defined in equation (4.2) is non-overlapping and that the dimension of the graph of $\theta(x)$ is 1. Then the dimension of the graph of the fixed point of T is the unique solution D to*

$$\sum_i |\alpha_i||s_i|^{D-1} = 1 \tag{4.3}$$

if $\sum_i |\alpha_i| > 1$ and 1 otherwise.

Proof: As in ([2], p. 226), we give just a heuristic reasoning.

Denote by $N(\epsilon)$ the number of boxes of side length ϵ that intersect the graph of the fixed point function. Then by the self-similarity,

$$N(\epsilon) \approx \sum_i |\frac{\alpha_i}{s_i}| N(\frac{\epsilon}{|s_i|}) + C\epsilon^{-1}$$

where the last term appears because of the "condensation function" θ. Notice that if we substitute $N(\epsilon) = C\epsilon^{-D}$ into this equation, we obtain

$$1 = \sum_i |\alpha_i||s_i|^{D-1}$$

which is the desired relation (the $C\epsilon^{-1}$ term drops out as $\epsilon \to 0$). ∎

The following Corollary is also a special case of results in [1, 5] (the special case where there is no shear in the IFS maps).

Corollary 4.1. *For the affine IFSM operator T as defined in equation (4.1), the dimension of the fixed point of T is the unique solution D to equation (4.3) if $\sum_i |\alpha_i| > 1$ and 1 otherwise.*

Now, in the case of an IFSM corresponding to an IFSW, the mappings are certainly not non-overlapping. Thus, the previous results do not apply. As in the case of a usual IFS, we obtain an upper bound on the dimension.

Theorem 4.2. *Suppose that ψ, the "mother wavelet" for the wavelet MRA, is differentiable. Then for the affine IFSW operator S defined as in equation (2.3) the dimension of the fixed point of S is bounded above by the unique solution D to*

$$\sum_i |\alpha_i||s_i|^{D-1} = 1 \tag{4.4}$$

if $\sum_i |\alpha_i| > 1$ and 1 otherwise.

Proof: Let $N(f, \epsilon)$ be the number of boxes in a "column" that intersect the graph of f. Then

$$N(f + g, \epsilon) \leq N(f, \epsilon) + N(g, \epsilon). \tag{4.5}$$

Using this observation one can obtain the upper bound in the case where the images in the IFSM are overlapping in a similar manner as one obtains the upper bound for the similarity dimension for geometric IFS. ∎

Acknowledgement. This research has been supported by the following grants from the Natural Sciences and Engineering Research Council of Canada (NSERC) which are gratefully acknowledged: an Operating Grant (ERV) as well as an NSERC Collaborative Projects Grant (ERV, with C. Tricot, École Polytechnique, U. de Montréal, B. Forte, U. of Verona, Italy and J. Lévy Véhel, INRIA, Rocquencourt, France).

References

1. Barnsley, M.F. (1986): Fractal Functions and Interpolation. *Constructive Approximation* **2**, 303–329.
2. Barnsley, M.F. (1988): *Fractals Everywhere*. Academic Press, New York.
3. Davis, G. (March 1995): Self-Quantized Wavelet Subtrees: A Wavelet-Based Theory for Fractal Image Compression. In *IEEE Data Compression Conference Proceedings*, edited by J. Storer. IEEE Computer Society Press; A Wavelet-Based Analysis of Fractal Image Compression, (Preprint 1996).
4. Daubechies, I. (1992): *Ten Lectures on Wavelets*. SIAM Press, Philadelphia, PA.
5. Falconer, K.J. (1990): *Fractal Geometry: Mathematical Foundations and Applications*. John Wiley & Sons, Ltd., West Sussex, England.
6. Forte, B., Vrscay, E.R. (July 8-17, 1995): Theory of Generalized Fractal Transforms. To appear in *Fractal Image Encoding and Analysis*, edited by Y. Fisher, Springer-Verlag, New York, 1966. Proceedings of the NATO ASI held in Trondheim, Norway.
7. Forte, B., Vrscay, E.R. (July 8-17, 1995): Inverse Problem Methods for Generalized Fractal Transforms. To appear in *Fractal Image Encoding and Analysis*, edited by Y. Fisher, Springer-Verlag, New York, 1966. Proceedings of the NATO ASI held in Trondheim, Norway.
8. Krupnik, H., Malah, D., Karnin, E. (March 1995): Fractal Representation of Images via the Discrete Wavelet Transform. *IEEE 18th Convention of Electrical Engineering in Israel, Tel-Aviv.*
9. Simon, B. (October 1995): Explicit link between local fractal transform and multiresolution transform. Proceedings of the IEEE Conference on Image Processing.
10. van de Walle, A. (1995): Merging fractal image compression and wavelet transform methods. To appear in *Fractal Image Coding and Analysis*, edited by Y. Fisher, Springer-Verlag, New York, 1996; *Relating Fractal Compression to Transform Methods*, Master of Mathematics Thesis, Department of Applied Mathematics, University of Waterloo.

Modeling fractal shapes using generalisations of IFS techniques

J. Thollot, C.E. Zair, E. Tosan, and D. Vandorpe

LIGIM - Université Claude Bernard
Bât 710 - 43, bd du 11 Novembre 1918
69622 Villeurbanne Cedex, FRANCE
Tel: 04.72.44.80.00 p 40.07 Fax: 04.72.43.13.12
{jthollot,czair}@ligim.univ-lyon1.fr

Summary. One of the major problems in geometric modeling is the control of shape construction. Indeed, one should be able to construct geometrical forms by combining or manipulating simple entities. This problem is even more important when we deal with fractal geometry. In this paper, we propose some methods for increasing the modeling capabilities of fractal shape constructions. We propose two extensions of the IFS model. The first is based on the integration of free form techniques in an IFS formalism, while the second deals with the definition of matrix of attractors to give a constructive approach of fractal shapes.

1. Introduction

The aim of geometric modeling is to model forms by manipulating or combining well-known shapes such as circles or boxes. The resulting model becomes more convenient if it allows us to control and to manipulate the final shape easily. When we deal with fractal geometric modeling, the control on fractal figures is not so easily achieved as with classical smooth ones. The difficulty comes from the fact that fractal shapes are usually generated with iterative or recursive procedures.

In order to have a better control on the shape, we wish to develop a *fractal modeler*. Such a modeler should be constituted by a set of basic shapes, a set of unary operations (shape modifications) and a set of binary operations (shape combinations). We distinguish two kinds of operations: set operations (union, intersection, restrictions) and geometric operations (transformations, deformations, interpolations, extrusions).

Our work focuses on extensions of the IFS model. We have been inspired by existing techniques in classical geometric modeling. One of these techniques is constructive geometry which allows to build complex attractors using set operations (union, intersection, ...). Such a construction is possible using IFS matrices which are able to generate a wide class of fractal shapes, not necessarily self-similar. These matrices can be combined with particular operations to yield complex fractal shapes in a constructive way. Another classical technique is the free form one. We demonstrate, by considering a particular complete metric space and particular IFS, that we can generate IFS attractors that are either smooth or fractal figures. These attractors combine the advantages of free forms (control via a collection of control points) and those of fractals (self-similarity, complex shapes). Thus, fractal shapes can be manipulated with the same flexibility as classical free forms.

2. IFS modeling

Fractals are characterized by their property of self-similarity. Barnsley's idea consists of using this property to encode the different parts of the fractal by contractive operators. These operators use an iterative scheme to transform any initial shape to a union of its reduced parts. The contraction property of the operators is used to prove the existence of a unique invariant set named the attractor. Below, we summarize the basic notions of IFS theory and introduce the notation used in the remainder of the paper.

An IFS-based modeling system is defined by a triple $(E, d, \mathcal{S})$ where:

- (E, d) is a complete metric space. We shall call it the *iterative space.*
- $\mathcal{S}$ is a semigroup acting on points of E such that : $p \mapsto T\ p$, where $T \in \mathcal{S}$ is a contractive operator. $\mathcal{S}$ is called the *iterative semigroup.*

The space $\mathcal{H}(E)$ denoting the set of non-empty compacts of E is a complete metric space, endowed with the Hausdorff distance :

$$d_H(K, K') = \max\{\max_{p\in K}\min_{q\in K'} d(p,q), \max_{q\in K'}\min_{p\in K} d(p,q)\}.$$

An IFS is a finite subset of the semigroup $\mathcal{S}$:

$$\mathcal{T} = \{T_1, ..., T_N\}.$$

The associated Hutchinson operator is:

$$K \in \mathcal{H}(E) \mapsto \mathcal{T}K = T_1\ K \cup ... \cup T_N\ K.$$

This operator is contractive in the metric space $(\mathcal{H}(E), d_H)$ and admits a fixed point $\mathcal{A}(\mathcal{T})$ given by :

$$\mathcal{A}(\mathcal{T}) = \lim_{k\to\infty} \mathcal{T}^k K_0.$$

IFS attractors constitute a particular family of compacts, they are *invariant sets* for the semigroup of IFSs :

$$\mathcal{A} = T_1\ \mathcal{A} \cup ... \cup T_N\ \mathcal{A}, \text{ with } T_i \in \mathcal{S}.$$

IFS techniques present an interesting approach for modeling fractals. In addition to the simplicity of their mathematical formalism, they provide an efficient tool to encode complex shapes with a small set of contractive operators. The structure of the attractor depends strongly on the semigroup of the operators. This dependency suggests a classification of fractal shapes into families. For instance, if the semigroup is taken to be of the following types:

- Affine operators: in this case, the attractors are self-affine.
- Homographies: here, the corresponding attractors are self-homographic (see Lévy Véhel [13]).
- Similitudes: the corresponding attractors are self-similar. A wide variety of fractal curves belongs to this family of attractors. (see Dekking [5], Dubuc [7], Tricot [27], Prusinkiewicz [24] and Hwang [12]).
- Dilatations-shearing : this semigroup permits graphs of functions to be defined as IFS attractors. The *fractal interpolation* of curves introduced by Barnsley [2] and extended to surfaces by Massopust [14] [15] gives examples of such functions.
- Dilatations: this semigroup was used by Dubuc [7] to construct *hyperbolic curves.*

This paper focuses on affine operators.

IFS techniques suffer from important drawbacks when used for geometric modeling purposes. These consist of two essential restrictions. Firstly, the attractors are strictly self-affine. Secondly, there is no convenient way to manipulate these attractors. By analogy with free form techniques, manipulation means the possibility of controlling the global shape of the attractor. Such a deficiency is an important limit of fractal geometric modeling.

We propose two generalisations of IFS techniques to overcome these drawbacks. First, we define the projected IFS attractor

$$K'_\omega = P\mathcal{A}(\mathcal{T}) \text{ with } K'_k = P(\mathcal{T}^k K_0),$$

where P denotes projection.
This generalisation enables the construction of attractors with control points which can be handled as efficiently as classical free form shapes (Bézier, spline,...).

Second, we define the restriction of an IFS attractor

$$K''_\omega \subseteq \mathcal{A}(\mathcal{T}) \text{ with } K''_k \subseteq \mathcal{T}^k K_0.$$

This generalisation extends the family of modeling shapes to non-restricted *self-affine* attractors. A set of operations will be given for a constructive approach of fractal attractors using basic operations, such as union and intersection.

3. IFS generalisations

3.1 Free form shapes

In this section, we present a generalisation of the IFS model which integrates free form techniques. This generalisation distinguishes between the iterative space, in which the attractor is defined, and the modeling space, in which this attractor will be projected using a set of control points. In this way, we can construct smooth or fractal forms characterized by a set of control points.

Let us take as iterative space the I-simplex defined by:

$$\mathcal{B}^I_+ = \{(\lambda_j)_{j\in I} |\, \lambda_j \geq 0\,,\ \sum_{j\in I} \lambda_j = 1\,,\ I = [0, ..., m]\},$$

and Markov matrices as the iterative semigroup:

$$\mathcal{S}(I) = \{T |\, \sum_{j\in I} T_{ij} = 1 \text{ and } T_{ij} \geq 0\}.$$

We build IFS attractors which belong to the I-simplex: $\mathcal{A}(\mathcal{T}) \subseteq \mathcal{B}^I_+$. As we distinguish between the iterative space and the modeling space, we can project this attractor in the modeling space (R^2 or R^3) using a set of control points as follows:

$$P\mathcal{A}(\mathcal{T}) \quad = \quad \{P\lambda |\, \lambda \in \mathcal{A}(\mathcal{T})\} \tag{3.1}$$

where:

$$P\lambda = \sum_{j\in I} \lambda_j p_j \text{ with } P = (p_j)_{j\in I}. \tag{3.2}$$

As the attractor belongs to the I-simplex, we have the convex-hull property of the attractor:

$$P\mathcal{A}(\mathcal{T}) \subseteq \mathrm{conv}(\{p_j |\ j \in I\}),$$

where $\mathrm{conv}(\{p_j |\ j \in I\})$ is the bounding box of the p_i's.

In this way, we generate free form fractal shapes which are controlled with the same flexibility as classical free form shapes [33].

3.2 Matrix of IFS

Another way to generalize the IFS model is to construct subsets of an IFS attractor. This can be done using matrices of IFS. Such matrices are a generalisation of the Hutchinson operator. This approach has been developed by Peitgen, Jűrgen and Saupe in [23]. They consider an IFS matrix of size $n \times n$ to be an operator on the space of vectors of size n with compact components. This matrix has a fixed point vector, which is the attractor of the IFS matrix.

Here we propose a new type of attractor. By considering the space of matrices of size $n \times n$ with compact components as iterative space, we construct a fixed point matrix. This definition of an attractor allows us to simulate a large family of existing models. For more details see [28] [25].

More precisely, we write $M(\mathcal{H}(E))$ for the set of matrices of compacts: for $\tilde{K} \in M(\mathcal{H}(E))$,

$$\tilde{K}_{ij} \in \mathcal{H}(E) \cup \emptyset.$$

In the same way, we write $M(\mathcal{P}(\mathcal{T}))$ for the set of matrices of IFS: for $\tilde{H} \in M(\mathcal{P}(\mathcal{T}))$,

$$\tilde{H}_{ij} = \bigcup_{k \in I_{ij}} T_k \text{ or } \emptyset,$$

with $I_{ij} \subseteq \{1, \ldots, N\}$ and $T_k \in \mathcal{T}$.

Using the union and composition operators, as defined in the IFS model, we define the composition of an IFS matrix with a matrix of compacts:

$$(\tilde{H}\tilde{K})_{ij} = \bigcup_{k=1}^{n} \tilde{H}_{ik}\tilde{K}_{kj}.$$

In the same way, we define the composition of two IFS matrices:

$$(\tilde{H}\tilde{H}')_{ij} = \bigcup_{k=1}^{n} \tilde{H}_{ik}\tilde{H}'_{kj}.$$

It is now possible to define an attractor matrix, in the same way as for IFS. Indeed, given $\tilde{K}$ and $\tilde{K}' \in M(\mathcal{H}(E))$, the distance defined by

$$d_\infty(\tilde{K}, \tilde{K}') = \max_{(i,j) \in R} d_H(\tilde{K}_{ij}, \tilde{K}'_{ij})$$

is a distance on $M(\mathcal{H}(E))$. Moreover, given $\tilde{K} \in M(\mathcal{H}(E))$ and $\tilde{H} \in M(\mathcal{P}(\mathcal{T}))$, one can prove that, under certain hypotheses, the sequence $(\tilde{H}^k\tilde{K})_k$ is a Cauchy sequence on $(M(\mathcal{P}(\mathcal{T})), d_\infty)$.

We thus define the attractor matrix associated with an IFS matrix as the limit of the sequence:

$$\tilde{\mathcal{A}}(\tilde{H}) = \lim_{k\to\infty} \tilde{H}^k \tilde{K}_0.$$

Moreover, instead of constructing an attractor matrix, one can consider the union of certain of its components:

$$\mathcal{A}(\tilde{H}, Q_I, Q_F) = \bigcup_{i\in Q_I, j\in Q_J} \tilde{\mathcal{A}}(\tilde{H})_{ij} \text{ with } Q_I, Q_F \subseteq \{1, \ldots, n\}.$$

Thus, a single matrix leads to many different shapes from which a choice can be made.

This type of union has a relation with regular languages. Writing $\Sigma = \{1, \ldots N\}$ the alphabet associated with an IFS $\mathcal{T} = \{T_1, \ldots, T_N\}$ and $L \subseteq \Sigma^*$ a language over Σ, we have:

$$\bigcup_{i\in Q_I, j\in Q_J} \tilde{\mathcal{A}}(\tilde{H})_{ij} = \mathcal{A}(L) = \lim_{k\to\infty} \{T_{\sigma_1} \ldots T_{\sigma_k}\} K_0.$$

Many fractal models have been developed using formal languages [6] [3] [22] [4] [19] as well as some IFS generalisations based on graphs [20] [1] [10]. Our approach enables us to create the same shapes as these approaches [21] [16] [25].

4. Set operations

The matrix model permits combination of two IFS matrices to obtain a new attractor matrix. We have studied two types of operations, some general operations on matrices, such as union, composition, intersection, tensor product [25]; and some operations associated with formal language theory, such as tensor sum, block matrix [29].

These operations provide us with a constructive approach. Indeed, it is possible to combine two IFS matrices and to obtain a resultant matrix that we combine with another one, and so on. Before illustrating this approach with an example, let us give a brief definition of each operation.

4.1 Matrices operations

The union and intersection operations are defined in the following way,

$$(\tilde{H} \cup \tilde{H}')_{ij} = \tilde{H}_{ij} \cup \tilde{H}'_{ij},$$

$$(\tilde{H} \cap \tilde{H}')_{ij} = \tilde{H}_{ij} \cap \tilde{H}'_{ij}.$$

Composition has already been defined:

$$(\tilde{H}\tilde{H}')_{ij} = \bigcup_k \tilde{H}_{ik}\tilde{H}'_{kj}.$$

Tensor product is a more interesting operation because it increases the size of the matrix:

$$\tilde{H}'' = \tilde{H} \otimes_{\cup} \tilde{H}' = \begin{pmatrix} \tilde{H}_{11} \cup \tilde{H}' & \ldots & \tilde{H}_{1n} \cup \tilde{H}' \\ \vdots & & \vdots \\ \tilde{H}_{n1} \cup \tilde{H}' & \ldots & \tilde{H}_{nn} \cup \tilde{H}' \end{pmatrix},$$

with $\tilde{H}''$ a square matrix of size $n.m$, and

$$\tilde{H}_{ij} \cup \tilde{H}' = \begin{pmatrix} \tilde{H}_{ij} \cup \tilde{H}'_{11} & \dots & \tilde{H}_{ij} \cup \tilde{H}'_{1m} \\ \vdots & & \vdots \\ \tilde{H}_{ij} \cup \tilde{H}'_{m1} & \dots & \tilde{H}_{ij} \cup \tilde{H}'_{mm} \end{pmatrix}.$$

4.2 Language operations

Tensor summation is an operation associated with language theory. This operation corresponds to the shuffle (denoted by ⊔) of two regular languages. For this reason, we denote it "$\diamond_{\sqcup}$". We define it by

$$\tilde{H}'' = \tilde{H} \diamond_{\sqcup} \tilde{H}' = \begin{pmatrix} \tilde{\text{Id}}.\tilde{H}_{11} \cup \tilde{H}' & \tilde{\text{Id}}.\tilde{H}_{12} & \dots & \tilde{\text{Id}}.\tilde{H}_{1n} \\ \tilde{\text{Id}}.\tilde{H}_{21} & \tilde{\text{Id}}.\tilde{H}_{22} \cup \tilde{H}' & \dots & \tilde{\text{Id}}.\tilde{H}_{2n} \\ \vdots & & & \vdots \\ \tilde{\text{Id}}.\tilde{H}_{n1} & \tilde{\text{Id}}.\tilde{H}_{n2} & \dots & \tilde{\text{Id}}.\tilde{H}_{nn} \cup \tilde{H}' \end{pmatrix},$$

with $\tilde{H}''$ a square matrix of size $n.m$ and $\tilde{\text{Id}}$ the identity matrix of size m.

The block matrix operation is a simple combination of two matrices:

$$\tilde{H}'' = \tilde{H} \diamond_{\cup} \tilde{H}' = \begin{pmatrix} \tilde{H} & \emptyset \\ \emptyset & \tilde{H}' \end{pmatrix}.$$

As for the tensor summation, this operation has been inspired by language theory. It corresponds to the union of two regular languages. Using this operation we can define the union of two fractal shapes:

$$\mathcal{A}(\tilde{H} \diamond_{\cup} \tilde{H}', Q_I \cup Q'_I, Q_F \cup Q''F) = \mathcal{A}(\tilde{H}, Q_I, Q_F) \cup \mathcal{A}(\tilde{H}', Q'_I, Q'_F)$$

Moreover, if we replace some $\emptyset$ with the identity in the resulting matrix, we obtain the concatenation of two regular languages. We will denote this operation by "$\diamond$.".

Figure 4.1 shows an example of concatenation on the left and an example of shuffle on the right.

4.3 Constructive approach

Figure 4.2 shows how some complex shapes can be constructed, using the "step by step" approach, with the use of the matrix operations. The construction tree is based on four simple IFS attractors.

The color plate C.2 shows a picture obtained by this approach.

4.4 Classification by inclusion

To facilitate the choice of operations, we have ordered (in the sense of set inclusion) the results of these operations. We consider the compact union of all components of the attractor matrix:

$$\bigcup_{i,j \in \{1,\dots n\}} \tilde{\mathcal{A}}(\tilde{H})_{ij}.$$

Denoting by $\mathcal{T}$ the IFS composed of all the transformations used in both IFS matrices $\tilde{H}_1$ and $\tilde{H}_2$, we can prove the following inclusions [25], where $\downarrow$ denotes inclusion.

Fig. 4.1. Language operations

$$\mathcal{A}(\tilde{H}_1 \cap \tilde{H}_2) \downarrow \mathcal{A}(\tilde{H}_1) \cap \mathcal{A}(\tilde{H}_2) \downarrow$$

$$\mathcal{A}(\tilde{H}_1 \circ \tilde{H}_2) \quad \mathcal{A}(\tilde{H}_1) \cup \mathcal{A}(\tilde{H}_2) = \mathcal{A}(\tilde{H}_1 \Diamond_{\cup} \tilde{H}_2)$$

$$\downarrow \swarrow \qquad \downarrow$$

$$\mathcal{A}(\tilde{H}_1 \cup \tilde{H}_2) \qquad \mathcal{A}(\tilde{H}_1 \Diamond_{\cdot} \tilde{H}_2)$$

$$\downarrow \qquad \downarrow$$

$$\mathcal{A}(\tilde{H}_1 \otimes \tilde{H}_2) \qquad \mathcal{A}(\tilde{H}_1 \Diamond_{\cup} \tilde{H}_2)$$

$$\searrow \qquad \swarrow$$

$$\mathcal{A}(\mathcal{T})$$

Figure 4.3 shows an example of inclusion.

5. Geometrical operations

5.1 Shape operations

In classical geometric modeling, basic operations are defined on shapes to provide sufficient capability of modeling. Such basic operations can for instance be a deformation or a combination of two elementary shapes, to generate a complicated one.

Two types of operations may be distinguished: unary operations (transformation or deformation of a shape):

$$F(A) = \{F(p) |\ p \in A\},$$

and binary operations (combination of shapes):

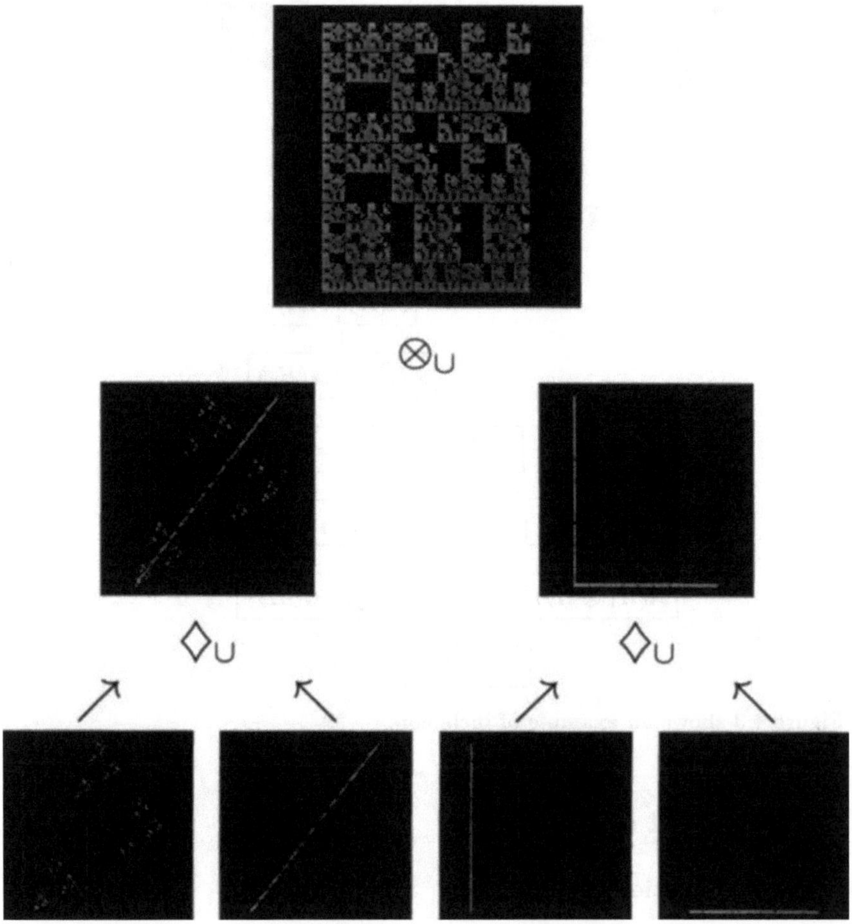

Fig. 4.2. Construction tree

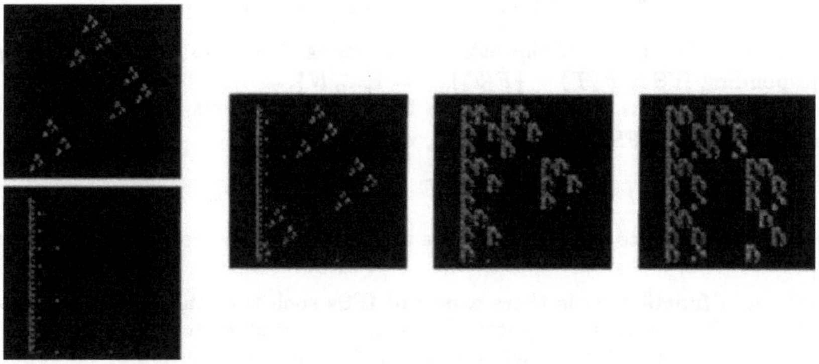

$\mathcal{A}(\tilde{H}_1)$ and $\mathcal{A}(\tilde{H}_2)$ $\subset \mathcal{A}(\tilde{H}_1 \Diamond_{\cup} \tilde{H}_2)$ $\subset \mathcal{A}(\tilde{H}_1 \cup \tilde{H}_2)$ $\subset \mathcal{A}(\tilde{H}_1 \otimes_{\cup} \tilde{H}_2)$

Fig. 4.3. Example of inclusion.

$$A \# A' = \{p \# p' |\ p \in A \text{ and } p' \in A'\}.$$

In classical geometric modeling, curves and surfaces can be obtained by using the two sorts of geometric operations. Parametric curves can be formulated as deformation (or embedding) of the standard interval $[0,1]$ in the modeling space E:

$$A = F([0,1]) = \{F(s)|\ s \in [0,1]\}.$$

Parametric surfaces can be formulated as deformation (or embedding) of the standard set $[0,1]^2$. A classical procedure for patchwork construction is extrusion, i.e. combination of parametric curves :

$$F''(s,t) = F(s) \# F'(t).$$

Such operations cannot be achieved on IFS attractors, since a combination or a deformation of an attractor is not necessarily an IFS attractor. To solve this problem, we introduce the notion of morphism [26]. This allows operations on attractors to be considered as operations on IFS operators [8].

5.2 Operations on attractors

Let us consider three IFS modeling systems $(E,d,\mathcal{S}),(E',d',\mathcal{S}'),(E'',d'',\mathcal{S}'')$. A *unary morphism* is a pair of functions $F : E \to E'$ and $\hat{F} : \mathcal{S} \to \mathcal{S}'$ such that :

$$F(Tp) = \hat{F}(T)F(p).$$

A *binary morphism* is a pair of binary operations $\# : E \times E' \to E''$ and $\hat{\#} : \mathcal{S} \times \mathcal{S}' \to \mathcal{S}''$ such that:

$$(Tp)\#(T'p') = (T \hat{\#} T')(p \# p').$$

Using these morphisms, operations on attractors correspond to operations on the IFS operators. Indeed, if we take an IFS $\mathcal{T} = \{T_1, ..., T_N\}$ and its corresponding attractor $\mathcal{A}$ then:

$$F(\mathcal{A}) = F(\cup_{i=1}^{N} T_i \mathcal{A}) = \cup_{i=1}^{N} \hat{F}(T_i) F(\mathcal{A}). \tag{5.1}$$

Hence, the deformation of the attractor $\mathcal{A}$ using F is also an IFS attractor. Its corresponding IFS is $\hat{F}(\mathcal{T}) = \{\hat{F}(T_i),\ i = 1, ..., N\}$.

These morphisms can be viewed as functional equations, where, for a function F and a couple of IFSs $\{T_i\}$ and $\{T_i'\}$, we have

$$\forall i = 1, .., N,\ \forall p \in E \quad F(T_i\, p) = T_i' F(p). \tag{5.2}$$

It is interesting to notice that there are two ways to solve this functional equation:

1. Given a function F, is there a pair of IFSs such that the functional equation holds ? This leads to polynomial deformation of attractors.
2. Given a pair of IFSs, is there a function F which is a solution of the functional equation ? We demonstrate that, with certain conditions on the IFS, such a function F exists and allows a fractal deformation of IFS attractors.

5.3 Curve deformations

Let us investigate more precisely the functional equation (5.2). Our purpose is to define, using this functional equation, parametric curves which are IFS attractors. Suppose that $E = [0, 1]$, on which we define contractive affine operators

$$\mathcal{S} = \{T = [a, u] \mid \forall t \in [0, 1],\ T\, t = at + u\}.$$

If T_0 and T_1 are two operators such that:

$$T_0\, t = rt; \quad T_1\, t = (1 - r)t + r, \quad 0 < r < 1, \tag{5.3}$$

then, obviously, $[0, 1]$ is the attractor of this IFS since we can write :

$$[0, 1] = T_0\, [0, 1] \cup T_1\, [0, 1].$$

Thus, if a function $\hat{F}$ exists, then $F([0, 1])$ is the attractor of the IFS $\hat{F}(\mathcal{T}) = \{\hat{F}(T_0), \hat{F}(T_1)\}$ which is a self-affine curve defined in the space E'.

This development represents a general framework for modeling curves using IFS [31]. Next we give some convenient hypotheses for generating classical (Bézier, spline) and fractal free form shapes (see Figure 5.1).

5.3.1 Polynomial deformation. Let $E' = \mathcal{B}_+^I$ and $\mathcal{S}' = \mathcal{S}(I)$. Given a function $F(t) = [F_0(t), ..., F_n(t)]^T \in E'$, we are interested in the explicit formulation of the function $\hat{F}$. This establishes a correspondence between attractors in the space E and attractors in E'. If such a function exists, then the curve $F([0, 1])$ is the attractor of the IFS $\hat{F}(\mathcal{T}) = \{\hat{F}(T_0), \hat{F}(T_1)\}$. Its projection $P\,F([0, 1])$ yields a parametric free form curve [32].
For example, if we take $F(t) = B(t)$, the set of Bernstein functions, then we can explicitly define the function $\hat{F}$ as:

$$\begin{array}{rccl} \hat{F}: & \mathcal{S} & \longrightarrow & \mathcal{S}' \\ & T = [u, a] & \longmapsto & \hat{F}(T) = \hat{F}(u, a) = T_1'(u)\, T_0'(a/(1 - u)) \end{array}$$

where

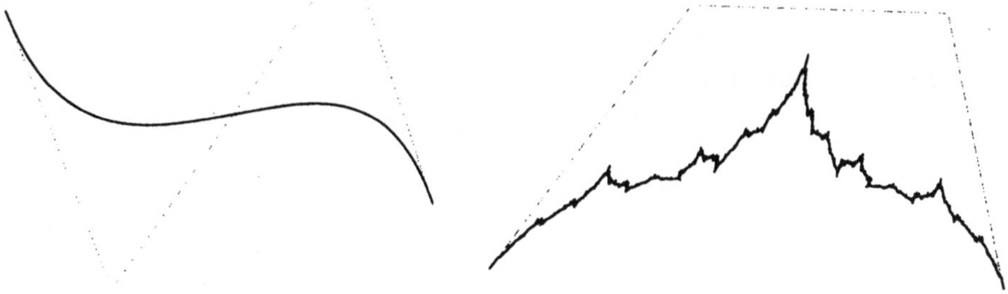

Fig. 5.1. Bézier and fractal curves obtained as IFS attractors. The fractal curve is controlled in the same way as the Bézier curve.

$$T_0' = \begin{pmatrix} 1 & 1-r & (1-r)^2 & (1-r)^3 \\ 0 & r & 2(1-r) & 3r(1-r)^2 \\ 0 & 0 & r^2 & 3(1-r)r^2 \\ 0 & 0 & 0 & r^3 \end{pmatrix},$$

$$T_1' = \begin{pmatrix} (1-r)^3 & 0 & 0 & 0 \\ 3(1-r)^2 r & (1-r)^2 & 0 & 0 \\ 3r^2(1-r) & 2(1-r) & 1-r & 0 \\ r^3 & r^2 & r & 1 \end{pmatrix}.$$

In this way, polynomial deformation of $[0,1]$ yields a classical Bézier curve as the projection of an IFS attractor (see Figure 5.1).

5.3.2 Fractal deformation. It is also interesting to study the inverse procedure to that discussed in (5.2). Suppose that the function F is unknown. Then, for two given IFSs $\mathcal{T} = \{T_0, T_1\}$ (defined as previously) and $\mathcal{T}' = \{T_0', T_1'\} \in \mathcal{S}'$, there is a necessary and sufficient condition for the existence of F. Indeed, if we choose the Markov matrices T_0', T_1' such that $T_0' c_1 = T_1' c_0$ (c_0, c_1 are respectively the fixed points of T_0' and T_1') then a unique function F exists, which is the solution of the functional equation (5.2) [18] [11] [30] [33].

Experiments show that, with general Markov matrices satisfying the condition above, the curves we obtain are fractals, due to the fact that they are nowhere differentiable with a self-similar structure (from the IFS model itself). Thus we see the intimate relationship between the Markov matrices and the smoothness of the attractor (see Figure 5.1).

5.4 Combinations of attractors

Binary geometric operations can be defined on two elementary IFS attractors to create a more complicated one. We can distinguish two types of operations:

- linear operations, such as affine combination or vectorial summation;
- bilinear operations, such as the tensor product.

Both types are discussed below.

5.4.1 Vectorial summations. Under certain conditions, a "special summation" [9] [8] [26], can be associated with vectorial summation of points such that:

$$(Tp) + (T'p') = (T\hat{+}T')(p + p').$$

Summation of attractors is defined by:

$$\mathcal{A}(\mathcal{T}) + \mathcal{A}(\mathcal{T}') = \mathcal{A}(\mathcal{T}\hat{+}\mathcal{T}').$$

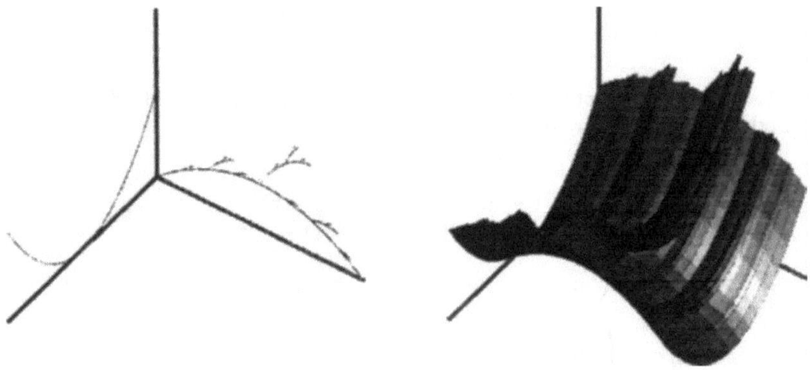

Fig. 5.2. Summation of a tree and a parabola.

Figure 5.2 displays the sum of two attractors, a tree and a parabola. The color plate C.3 displays an other sum of two bidimensional attractors.

5.4.2 Tensor products. A classical method of defining free surface patches is to use tensor products of curves. Given two functions $(F(s), F'(t))$, the free form surface obtained by the tensor product of these two functions is:

$$Q(s,t) = P(F(s) \otimes F'(t)) = \sum_{i=0}^{m} \sum_{j=0}^{m'} F_i(s) F_j'(t) p_{ij} \text{ with } P = (p_{ij}). \tag{5.4}$$

This operation has the following advantage: from two IFS corresponding to two parametric curves, we can find the IFS whose attractor is the surface obtained by the tensor product of these two curves. Indeed, the following property holds [17] [32] [33]:

$$P(\mathcal{A}(\mathcal{T}) \otimes \mathcal{A}(\mathcal{T}')) = P\mathcal{A}(\mathcal{T} \otimes \mathcal{T}'),$$

where $\mathcal{T} \otimes \mathcal{T}'$ is defined as the tensor product of Markov matrices.

There are also two cases of curve deformation: polynomial deformation and fractal deformation. In general, $\hat{F}$ can be defined explicitly in the case of Bézier deformation by:

$$\begin{array}{llll} \hat{F}: & S \simeq \mathcal{S} \otimes \mathcal{S} & \longrightarrow & \mathcal{S}' \otimes \mathcal{S}' \\ & T = \left[\begin{array}{c|cc} u & a & 0 \\ v & 0 & d \end{array} \right] & \longmapsto & \hat{F}(T) = \hat{F}(u,a) \otimes \hat{F}(v,d) \end{array}$$

We can use this function for fractal deformation by replacing the matrices T_0', T_1' by general Markov matrices (see Figure 5.3).

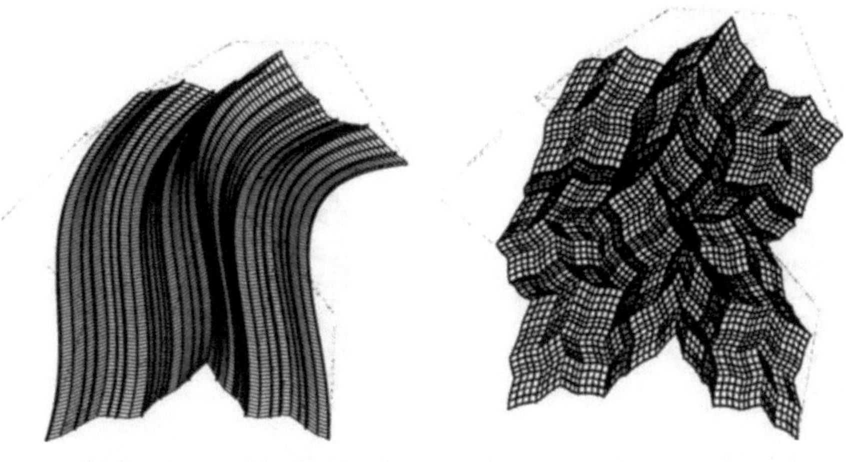

Fig. 5.3. The left figure illustrates a surface obtained by the tensor product of a Bézier curve with a fractal one. The right figure shows the tensor product of two fractal curves.

5.5 Deformations of attractors

The procedure of constructing free form fractals is based essentially on the property which associates another subdivision in E' to each subdivision in the space E through a morphism. In this way, all IFS attractors defined in E can be deformed in E' using the tensor product property. This leads to a straightforward way to construct IFS attractors with control points.

For example, to generate a Sierpinski gasket with control points, we take the Sierpinski IFSs in the spaces E and E':

$$\mathcal{T}_S = \{T_{00}, T_{01}, T_{10}\}, \ \mathcal{T}_S' = \{T_{00}', T_{01}', T_{10}'\},$$

where $T_{ij}' = T_i' \otimes T_j'$.
The deformed attractor is a subset of the surface defined above:

$$\mathcal{A}(\mathcal{T}_S) \subset [0,1]^2 \ \Rightarrow \ \mathcal{A}(\hat{F}(\mathcal{T}_S)) = F(\mathcal{A}(\mathcal{T}_S)) \subset F([0,1]^2).$$

This consists of subdividing the surface into four patches and then omitting one of them (see figure 5.4). However, the generalisation to other IFS attractors can be done by giving the IFS in the space E and then computing its image under the function $\hat{F}$ to yield the IFS in the space E' [33].

Remark:
Previously, we dealt only with polynomial deformation. Such a deformation allows

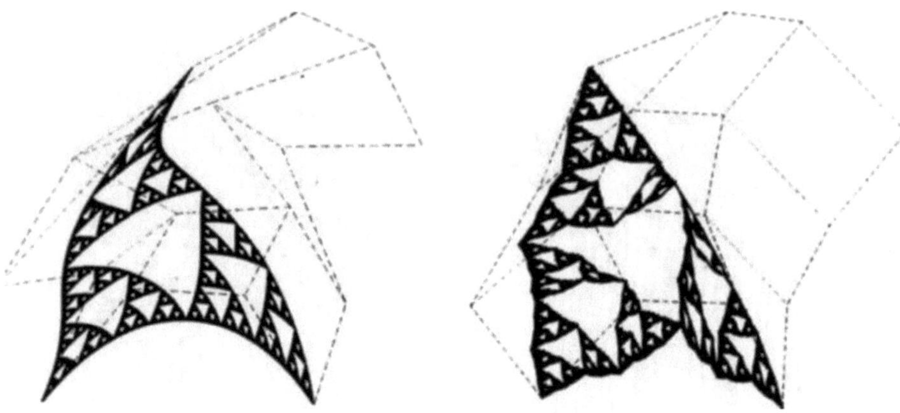

Fig. 5.4. These figures illustrate smooth and fractal deformation of the Sierpinski gasket.

us to construct attractors with control points. However, other types of deformations can also be used. Figure 5.5 illustrates an example of deformation using quadratic functions (elliptic paraboloid and hyperbolic paraboloid) [33].

6. Conclusion

Using generalisations of the classical IFS model, we construct a fractal modeler which manipulates IFS attractors with the same flexibility as classical smooth shapes. However, this modeler manipulates matrix of attractors and free form fractals in a disjoint way. Free form fractals are strictly classical IFS attractors. As an immediate extension of this work, we suggest a further generalisation to construct free form matrices of attractors.

References

1. Bandt, C. (1989): Topological Markov Chaïne and Mixed Self-Similar Sets. *Math. Nachr.*, (142), 107–123.
2. Barnsley, M.F. (1988): *Fractal Everywhere.* Academic press, INC.
3. Berstel, J., Morcrette, M. (1989): Compact Representation of Patterns by Finite Automata. In *Proceedings of Pixim*, 387–395.
4. Culik II, K., Dube, S. (1993): Rationnal and affine expressions for image synthesis. *Discrete Appl. Math.*, 41, 85–120, 1993.

Fig. 5.5. Two embeddings of planar attractor in elliptic paraboloid and in hyperbolic paraboloid.

5. Dekking, F.M. (1980): Variations on Peano. *Nieuw Archief Voor Wiskunde*, (XXVIII), 275–281.
6. Dekking, F.M. (April 1982): Recurrent sets. *Advances in Mathematics*, 44(1), 78–104.
7. Dubuc, S. (1987): Modèles de courbes irrégulières. In *Dimensions Non Entières et Applications*. Masson.
8. Gentil, C. (March 1992): *Les fractales en synthèse d'images: le modèle IFS*. Thèse de doctorat, Université LYON I.
9. Gentil, C., Tosan, E. (December 1992): Description of fractals with IFS. In *COMPUGRAPHICS'92*, 214–221, Lisbon, Portugal. Portuguese ACM Chapter.
10. Hart, J. (May 1992): The object instancing paradigm for linear fractal modeling. In *Proceedings of Graphics Interface'92*.
11. Hata, M. (1991): Topological aspects of self-similar sets and singular functions. In J. Belair and S. Dubuc, editors, *Fractal Geometry and Analysis*, 255–276. Kluwer Academic Publishers.
12. Hwang, S.C., Yang, H.S. (1993): Discrete Approximation of the Koch Curve. *Computer & Graphics*, 17(1), 95–102.
13. Lévy Véhel, J. (1988): Analyse et synthèse d'objets bi-dimensionnels par des méthodes stochastiques. Thèse de doctorat, Université de Paris-sud centre d'Orsay, INRIA.
14. Massopust, P. (April 1990): Fractal surfaces. *Journal of Mathematical Analysis and Applications*, (151), 275–290.
15. Massopust, P. (1994) : *Fractal function and wavelets*. Academic Press.
16. Morcrette, M. (1996): Sur l'équivalence de descriptions de figures itérées. À pararaître dans Theoritical Computer Science.
17. Micchelli, C.A., Prautzsch, H. (1987): Computing surfaces invariant under subdivision. *Computer Aided Geometric Design*, (4), 321–328.
18. Micchelli, C.A., Prautzsch, H. (1989): Union Refinement of Curves. *Linear Algebra and its Applications*, (114-115), 841–870.
19. Merzenich, W., Staiger, L. (1994): Fractals,dimension, and formal languages. *Theoritical Informatics and Application*, 28(3-4), 361–386.

20. Mauldin, R.D., Williams, L.C. (1988): Hausdorff dimension in graph directed constructions. *Trans. Amer. Math. Soc.*, 309, 811–829.
21. Prusinkiewicz, P., Hammel, M.S. (1991): Automata, languages and iterated function systems. In *lecture notes for the SIGGRAPH'91 course : "Fractal modeling in 3D computer graphics and imagery"*.
22. Prusinkiewicz, P., Hammel, M.S. (May 1992): Escape-time Visualization Method for Language-restricted Iterated Function Systems. In *Proceedings of Graphics Interface'92.*
23. Peitgen, H.O., Jurgens, H., Saupe, D. (1992): Encoding images by simple transformations. In *Fractals For The Classroom, New York.*
24. Prusinkiewicz, P., Sandness, G. (Novembre 1988): Koch curves as attractors and repellers. *IEEE Computer Graphics & Applications*, 27–40.
25. Thollot, J. (September 1996): *Extension du modèle IFS pour une géométrie fractale constructive.* Thèse de doctorat, Université Claude Bernard Lyon 1.
26. Tosan, E. (September 1996): *Une Algèbre de Formes fractales utilisant les IFS.* Thèse de doctorat, Université Claude Bernard Lyon 1.
27. Tricot, C. (1993): *Courbes et dimension fractale.* Springer Verlag.
28. Thollot, J., Tosan, E. (December 1993): Construction of fractales using formal languages and matrices of attractors. In Harold P. Santos, editor, *COMPUGRAPHICS'93*, 74–81, Technical University of Lisbon.
29. Thollot, J., Tosan, E. (May 1995): Constructive Fractal Geometry : constructive approach to fractal modeling using language operations. In *GRAPHICS INTERFACE'95, Quebec, Canada*, 196–203.
30. van Overveld, C.W.A.M. (1990): Family of recursively defined curves ,related to the cubic Bézier curve. *Computer-Aided Design*, 22(9), 591–597.
31. Zair, C.E., Tosan, E.: Computer Aided Geometric Design with IFS techniques. In *FRACTALS'97, Denver, USA*, April. to appear.
32. Zair, C.E., Tosan, E. (November 1994): Définition de courbes et surfaces lisses ou fractales à l'aide d'IFS. In *AFIG 94, Toulouse*, 189–200.
33. Zair, C.E., Tosan, E. (August 1996): Fractal modeling using free form techniques. *Computer Graphics Forum*, 15(3), 269–278. EUROGRAPHICS'96 Conference issue.

A Wavelet-based Solution to the Inverse Problem for Fractal Interpolation Functions

K. Berkner *

Department of Mathematics, Rice University
Houston, TX 77005-1892, USA
berkner@cml.rice.edu

Summary. We consider a special class of self-similar functions, the fractal interpolation functions, and provide a constructive solution to the inverse problem: Given a graph of a fractal interpolation function f which is the attractor of an unknown IFS with affine contraction maps $w_1, \ldots, w_N$, find the maps w_i. The solution is based on scaling properties of continuous wavelet transforms. The results of an algorithm are shown in two examples.

Keywords: iterated function systems, inverse problem, wavelets.

1. Introduction

A characteristic property of fractal curves is the self-similarity which is mostly described by a dimension (see e.g. [1]). The dimension is a measure for the regularity of a curve, but does not describe the shape of the curve in detail. Fractal interpolation functions (cf. [2]) are self-affine functions defined on a compact interval. The graph of a fractal interpolation function is the unique attractor of an iterated function systems (IFS) and therefore a fractal curve.

One of the main topics in IFS theory is the inverse problem: Given any compact set A, find an IFS which has A as attractor. The inverse problem has attracted a lot of interest in the context of signal compression. If the IFS can be computed from a given attractor the whole set is encoded by a finite number of maps. Generally, there is no constructive solution to the inverse problem. Even if we assume a knowledge that the compact set A is the attractor of an IFS, it is in general not possible to compute the associated IFS. Therefore the literature on the inverse problem for IFS has focused on the approximation of a given attractor A by an IFS whose attractor is close to the set A. These approximations are based on the collage theorem ([2]). Usually, the number of maps of the approximating IFS differ from the number of maps of the original IFS. Some methods to find approximating IFS are studied in the context of fractal image compression (see e.g. [3], [4]). In [5], the authors develop an algorithm which encodes a graph of a function as an approximate attractor of an IFS. In all papers the algorithms are not able to find the right IFS if the set A is even a "simple" attractor of an IFS.

In [7], the authors present a different approach to the solution of the inverse problem. They use a generalization of IFS to construct a fractal interpolation function to a prescribed regularity. This regularity is given by a function which describes the local Hölder exponent in each point.

In this paper, we present a constructive solution to the inverse problem for fractal interpolation functions without using the collage theorem or a given local

* Supported in part by Texas ATP under grant number TATP 003604-018 and Alexander von Humboldt-Stiftung.

regularity. We use similarity preserving properties of continuous wavelet transforms to find for a given graph of a fractal interpolation function the maps of the corresponding IFS. A priory, we do not have any informations about, e.g., the number of maps or horizontal scaling parameters. We obtain these informations from selected values of wavelet transforms.

2. Fractal interpolation functions

First, we introduce some notation and state important results concerning IFS and fractal interpolation function theory ([2],[1]).

We consider a complete metric space (X, d). The set of nonempty compact subsets of X is typically denoted by $\mathcal{H}(X)$. Equipped with the Hausdorff metric d_H, the space $(\mathcal{H}(X), d_H)$ is a complete metric space.

Let $w_i : X \to X$, $i = 1, \dots, N$, be contraction maps on X, i.e., there exist c_i, $0 < c_i < 1$, such that

$$d(w_i(x), w_i(y)) \leq c_i d(x, y)$$

holds for all $x, y \in X$ and each i. Then it is well known that there exists an unique nonempty set $A \in \mathcal{H}(X)$ with

$$A = \bigcup_{i=1}^{N} w_i(A).$$

A is called the *attractor* of the *Iterated Function System* (IFS) $w = \{w_1, \dots, w_N\}$ ([2],[6]).

Let now $D \in \mathcal{H}(\mathbb{R})$ and consider affine transforms $w_1, \dots, w_N, w_i : [0,1] \times D \to [0,1] \times D$ represented by

$$w_i \begin{pmatrix} x \\ y \end{pmatrix} = \begin{pmatrix} \alpha_i & 0 \\ \gamma_i & \delta_i \end{pmatrix} \begin{pmatrix} x \\ y \end{pmatrix} + \begin{pmatrix} \epsilon_i \\ \zeta_i \end{pmatrix}$$

where $\alpha_i, \gamma_i, \delta_i, \epsilon_i, \zeta_i \in \mathbb{R}$ and such that $0 < \alpha_i < 1$ and $0 < |\delta_i| < 1$ for each i. We give some more conditions on the parameters of the maps w_i. First, let $p_1 = (0, y_1)$ be the fixed point of w_1 and $p_{N+1} = (1, y_{N+1})$ the fixed point of w_N. Choose the parameters such that

$$w_i(p_{N+1}) = w_{i+1}(p_1) \text{ for all } i = 1, \dots, N-1,$$

and that $w_1(p_1), \dots, w_N(p_1), p_{N+1}$ are not all collinear. This condition guarantees that the segments $[w_i(p_1), w_i(p_{N+1})]$ form a nontrivial polygonal curve, i.e., not a straight line. The choice of α_i and δ_i ensures that the maps w_i are contractions with respect to a metric which is equivalent to the Euclidean distance. The attractor of the IFS $= \{w_1, \dots, w_N\}$ is the graph of a continuous function $f : [0,1] \to \mathbb{R}$, which is nowhere differentiable ([1]). The function f is called *fractal interpolation function* with interpolation points $w_1(p_1), \dots, w_N(p_1), p_{N+1}$.

The characterization of a graph of a fractal interpolation function by an IFS leads to a method to construct a fractal interpolation function as uniform limit of a sequence of functions. For a function $g \in C([0,1])$ we consider the operator $H : C([0,1]) \to C([0,1])$ with

$$(Hg)(x) = \gamma_i l_i^{-1}(x) + \delta_i g(l_i^{-1}(x)) + \zeta_i \quad \text{for } x \in [x_{i-1}, x_i] \tag{2.1}$$

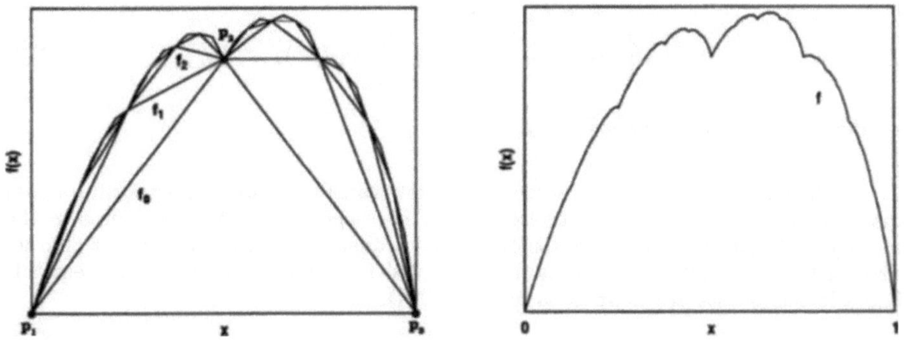

Fig. 2.1. A sequence (f_n) (*left*) converging to a fractal interpolation function f (*right*) with corresponding IFS $\mathbf{w} = \{w_1, w_2\}$, $\alpha_1 = 0.5$, $\delta_1 = 0.3$, $\alpha_2 = 0.5$, $\delta_2 = 0.5$.

with $l_i(x) = w_{x,i}(x, y)$ and $w_i(p_1) = p_i = (x_i, y_i)$. Let f_0 be the piecewise linear function with interpolation points $p_1, \dots, p_{N+1}$. Then the sequence of functions

$$f_{n+1}(x) = (Hf_n)(x) \tag{2.2}$$

is converging to the fixed point f of the operator H which is the fractal interpolation function of the IFS $\mathbf{w} = \{w_1, \dots, w_N\}$. In Figure 2.1 we show an example of a sequence (f_n) which generates a fractal interpolation function f.

3. Zeros of wavelet transforms

Wavelet analysis is a powerful tool to study multi-scale properties of functions and has been applied in many areas such as image compression, edge detection and denoising. In this section we present some definitions and results from wavelet theory (see e.g. [8],[9],[10]).

Definition: A function $\psi \in L^2(\mathbb{R})$ is said to be a *wavelet*, if its Fourier-transform $\hat{\psi}$ satisfies the following condition:

$$\int_0^{\infty} \frac{|\hat{\psi}(\omega)|^2}{|\omega|} d\omega = \int_{-\infty}^{0} \frac{|\hat{\psi}(\omega)|^2}{|\omega|} d\omega = c_\psi < \infty. \tag{3.1}$$

A wavelet ψ is called to have n *vanishing moments*, if

$$\int_{-\infty}^{\infty} \psi(x) x^k dx = 0 \quad \text{for all } k,\ 0 \le k < n \quad \textit{and} \quad \int_{-\infty}^{\infty} \psi(x) x^n dx \neq 0.$$

Definition: Let $f \in L^2(\mathbb{R})$ and ψ be a wavelet. Then the function

$$W_\psi[f] : \begin{cases} \mathbb{R} \times \mathbb{R}_+ & \rightarrow \mathbb{R} \\ (b, a) & \mapsto \frac{1}{\sqrt{a}} \int\limits_{-\infty}^{\infty} \psi(\frac{x-b}{a}) f(x) dx, \end{cases} \tag{3.2}$$

is called the *wavelet transform of f*.

As shown, e.g., in [11],[10], zeros of wavelet transforms are important for signal analysis, especially for edge detection. They correspond to special features in the underlying function. Selected zeros of wavelet transforms are important for characterizing the regularity of a function, which will be demonstrated in Theorem 3.1. The selected points are defined as follows ([10]):

Definition: A point $(b_0, a) \in \mathbb{R} \times \mathbb{R}_+$ is called a *modulus maximum point of* $W_\psi[f]$, if the map $b \mapsto |W_\psi[f](., a)|$ has a local maximum at b_0.

If the wavelet ψ is differentiable and $\psi' \in L^2(\mathbb{R})$ is a wavelet, it is obvious that modulus maximum points of $W_\psi[f]$ are selected zeros of $W_{\psi'}[f]$. That follows from the equation:

$$\frac{\partial W_\psi[f]}{\partial b}(b, a) = -\frac{1}{a} W_{\psi'}[f](b, a). \tag{3.3}$$

Definition: A function $\phi \in L^2(\mathbb{R})$ is called a *smoothing function*, if it satisfies the following two conditions:

1. $\int\limits_{-\infty}^{\infty} \phi(x) \neq 0$
2. $\phi(x) = O(1/(1 + x^2))$.

The derivatives of a smoothing function $\phi \in C^n(\mathbb{R})$ can be used to construct a family of wavelets (see e.g. [12]). For those wavelets the following theorem, which is proved in a more general setting in [10], shows how selected zeros of wavelet transforms are related to singularities of the function f.

Theorem 3.1. *Let $n > 0$ and $\phi \in C^n(\mathbb{R})$ be a smoothing function with compact support. Then $\psi = \phi^{(n)}$ is a wavelet with n vanishing moments. Furthermore, for a function $f \in L^1([c, d])$, where $[c, d]$ is an interval, we get:*

$$\begin{aligned} cl\{x \in (c,d) \;\; | \;\; & f(x) \text{ is in } x \text{ hölder continuous with exponent } s < 1\} \subseteq \\ cl\{x \in (c,d) \;\; | \;\; & \forall \epsilon > 0 \; \exists \text{ modulus maximum point } (b,a) \text{ of } W_\psi[f] \text{ with} \\ & |(b,a) - (x,0)| < \epsilon\}. \end{aligned}$$

Modulus maximum points form continuous curves (*modulus-maxima lines*), which can be parameterized in the scale parameter a. If the wavelet has compact support each point where f is not hölder continuous with exponent $s < 1$ is a starting point of a modulus-maxima line and can be localized by following the line from a coarse scale down to smallest scale. If ψ is differentiable the lines are parts of the zero-set of $W_{\psi'}[f]$. It is shown in [13],[10] that it is possible to measure the local Hölder exponents of f in a neighborhood of the singularity from the scaling behavior of the wavelet transform on modulus-maxima lines. But we do not need this characterization of the regularity of f in the context of this paper and refer the reader to [14],[15] for various applications.

In [16], Mallat and Zhong studied the problem of reconstructing a signal from modulus maximum points of wavelet transforms at dyadic scales. They present an algorithm which reconstructs the original signal by following all modulus maximum points from coarsest to finest scale. In the next section, we show how fractal interpolation functions are reconstructed only from a finite set of zeros of a wavelet transform.

4. Encoding of a fractal interpolation function into the affine maps

From Theorem 3.1 of the previous section, we know that zeros of a wavelet transform contain information about singularities of the function. Now we show how these zeros can be used to find the affine maps $w_1, \ldots, w_N$ which have created the graph of the corresponding fractal interpolation function. The analysis of fractal objects with the help of wavelet transforms has been developed in many areas during the last years. Topics like Hölder exponents, multifractal spectra and image compression are discussed (see e.g. [17],[13],[18]). The inverse problem of Bernoulli-invariant measures with two maps is studied in [19], [20]. In this paper, we assume a knowledge of a given graph which comes from a fractal interpolation function. Let us emphasize that we even do not have any information on the number of maps necessary to generate the graph.

The following theorem explains the self-similarity of zeros of wavelet transforms which can be observed e.g. in Figure 4.1 and 4.3.

Theorem 4.1. *Let f be a fractal interpolation function and $w = \{w_1, \ldots, w_N\}$ an associated IFS with parameters $\alpha_i, \epsilon_i, \gamma_i, \delta_i, \zeta_i$. Furthermore, let ψ be a wavelet with compact support $[-K, K]$ and k vanishing moments, $k \geq 2$. Then for $(b, a) \in \mathbb{R} \times \mathbb{R}_+$ with*

$$[-aK + b, aK + b] \subset [0, 1] \tag{4.1}$$

the following equation holds:

$$W_\psi[f](b, a) = \frac{1}{\delta_i \alpha_i} W_\psi[f](\alpha_i b + \epsilon_i, \alpha_i a) \tag{4.2}$$

for each $i = 1, \ldots, N$.

Proof: With condition (4.1) we get:

$$\begin{aligned}
W_\psi[f](b, a) &= \frac{1}{\sqrt{a}} \int_{-aK+b}^{aK+b} \psi\left(\frac{z-b}{a}\right) f(z)dz \\
&= \frac{1}{\alpha_i \sqrt{a}} \int_{l_i(-aK+b)}^{l_i(aK+b)} \psi\left(\frac{l_i^{-1}(x) - b}{a}\right) f(l_i^{-1}(x))dx \\
&= \frac{1}{\alpha_i \sqrt{a}} \int_{l_i(-aK+b)}^{l_i(aK+b)} \psi\left(\frac{l_i^{-1}(x) - b}{a}\right) \frac{1}{\delta_i} \left(f(x) - \gamma_i l_i^{-1}(x) - \zeta_i\right) dx
\end{aligned}$$

$$= \frac{1}{\alpha_i\sqrt{a}} \int\limits_{l_i(-aK+b)}^{l_i(aK+b)} \psi\left(\frac{\frac{x-\epsilon_i}{\alpha_i}-b}{a}\right)\left(\frac{1}{\delta_i}f(x)-\gamma_i\frac{x-\epsilon_i}{\alpha_i}-\zeta_i\right)dx$$

$$= \frac{1}{\alpha_i\delta_i\sqrt{a}} \int\limits_{l_i(-aK+b)}^{l_i(aK+b)} \psi\left(\frac{\frac{x-\epsilon_i}{\alpha_i}-b}{a}\right)f(x)dx +$$

$$\frac{\gamma_i}{\delta_i\sqrt{a}} \int\limits_{-aK+b}^{aK+b} \psi\left(\frac{x-b}{a}\right)xdx + \frac{\zeta_i}{\delta_i a} \int\limits_{-aK+b}^{aK+b} \psi\left(\frac{x-b}{a}\right)dx$$

Since ψ is a wavelet with at least two vanishing moments we get:

$$\begin{aligned} W_\psi[f](b,a) &= \frac{1}{\delta_i\alpha_i\sqrt{a}} \int\limits_{l_i(-aK+b)}^{l_i(aK+b)} \psi\left(\frac{x-(\alpha_i b-\epsilon_i)}{\alpha_i a}\right)f(x)dx \\ &= \frac{1}{\delta_i\alpha_i}W_\psi[f](\alpha_i b+\epsilon_i,\alpha_i a). \end{aligned}$$

□

In the following our main goal is to find special points of the zero set which can be used to compute the parameters of the IFS using Equation (4.2). We assume that the wavelet ψ is the k-th ($k \geq 2$) derivative of a $(k+1)$-times differentiable smoothing function with compact support $[-K,K]$ and f a fractal interpolation function. We consider the partial derivative $\frac{\partial W_\psi[f]}{\partial b}$. Since ψ' is a wavelet with $k+1$ vanishing moments we get with Theorem 4.1:

Corollary 4.1. *With the conditions on the wavelet ψ from above we get:*

$$\begin{aligned} & W_\psi[f](b,a)=0 \quad \textit{and} \quad \frac{\partial W_\psi[f]}{\partial b}(b,a) = -\frac{1}{a}W_{\psi'}[f](b,a)=0 \\ \Longleftrightarrow\quad & W_\psi[f](\alpha_i b+\epsilon_i,\alpha_i a)=0 \quad \textit{and} \\ & \frac{\partial W_\psi[f]}{\partial b}(\alpha_i b+\epsilon_i,\alpha_i a) = -\frac{1}{a}W_{\psi'}[f](\alpha_i b+\epsilon_i,\alpha_i a)=0. \end{aligned}$$

The set

$$S(\psi,f)=\{(b,a)\in\mathbb{R}\times\mathbb{R}_+ \mid W_\psi[f](b,a)=0 \text{ and } \frac{\partial W_\psi[f]}{\partial b}(b,a)=0\}$$

is discrete in the scale-space.

With Theorem 3.1 and Theorem 4.1 we get the following lemma which guarantees the existence of elements of $S(\psi,f)$:

Lemma 4.1. *The set $S(\psi,f)$ is non-empty.*

Proof: We choose a scale $a_0>0$, $b_0\in(0,1)$ and $\rho>0$ such that the rectangle

$$R=[-a_0K+b_0-\rho,a_0K+b_0+\rho]\times(0,a_0]$$

is contained in the set $\{(b,a)\in\mathbb{R}\times\mathbb{R}_+ \mid b>aK \wedge b<1-aK\}$. Since the function f is hölder continuous with exponent $s<1$ in each point and the wavelet ψ has compact support, it follows from Theorem 3.1 that each point of $(0,1)$ must be an origin of a zero-crossing line. Therefore, the rectangle can be chosen so that there are two zero-crossing lines ν, ν' with origins $\nu_0, \nu'_0 \in (-a_0K + b_0 - \rho, a_0K + b_0 + \rho)$ and

$$\begin{aligned} \nu((0,1)) \cap [-a_0K + b_0 - \rho, a_0K + b_0 + \rho] \times \{a_0\} &\neq \emptyset \quad \text{and} \\ \nu'((0,1)) \cap [-a_0K + b_0 - \rho, a_0K + b_0 + \rho] \times \{a_0\} &\neq \emptyset. \end{aligned}$$

Without loss of generality, we can assume that

$$\begin{aligned} \nu((0,1)) \cap \{-a_0K + b_0 - \rho\} \times (0, a_0] &= \emptyset, \\ \nu((0,1)) \cap \{a_0K + b_0 + \rho\} \times (0, a_0] &= \emptyset, \\ \nu'((0,1)) \cap \{-a_0K + b_0 - \rho\} \times (0, a_0] &= \emptyset \quad \text{and} \\ \nu'((0,1)) \cap \{a_0K + b_0 + \rho\} \times (0, a_0] &= \emptyset. \end{aligned}$$

Now we can choose $(a_0, b_1) \in \nu((0,1)) \cap [-a_0K + b_0 - \rho, a_0K + b_0 + \rho] \times \{a_0\}$ and $(a_0, b_2) \in \nu'((0,1)) \cap [-a_0K + b_0 - \rho, a_0K + b_0 + \rho] \times \{a_0\}$ such that

$$W_\psi[f](b,a) \neq 0 \quad \text{for} \quad (b,a) \in (\min(b_1,b_2), \max(b_1,b_2)). \tag{4.3}$$

Since $W_\psi[f]$ is continuously differentiable, $\frac{\partial W_\psi[f]}{\partial b}$ exists for all points in R. If $\frac{\partial W_\psi[f]}{\partial b}(b,a) \neq 0$ for all $(b,a) \in R$ then it follows from the the implicit function theorem that there must be infinitely many zero-crossings in $(\min(b_1,b_2), \max(b_1,b_2)) \times \{a_0\}$ which is a contradiction to Equation (4.3). Thus, points (b,a) with $\frac{\partial W_\psi[f]}{\partial b}(b,a)$ $= 0$ **must occur.** □

The next lemma follows easily from Theorem 4.1 and Lemma 4.1 and presents the theoretical solution to the inverse problem for fractal interpolation functions.

Lemma 4.2. *Let $w = \{w_1, \ldots, w_N\}$ be an IFS with parameters $\alpha_i, \epsilon_i, \gamma_i, \delta_i, \zeta_i$ and f the corresponding fractal interpolation function. Furthermore, let $(b,a) \in S(\psi, f)$. Then there are points $(b_i, a_i) \in S(\psi, f)$, $i = 1, \ldots, N$, such that*

$$a_i = \alpha_i a \tag{4.4}$$

$$b_i = \alpha_i b + \epsilon_i \tag{4.5}$$

$$\sum_{i=1}^{N} \alpha_i = 1 \quad \textit{and} \tag{4.6}$$

$$\alpha_i + \epsilon_i = \epsilon_{i+1}. \tag{4.7}$$

The partition of the unit interval, given by $\epsilon_1, \ldots, \epsilon_N$, who characterize the attractor of the corresponding IFS, is not unique. The graph of a fractal interpolation function can be the attractor of more than one IFS, as the following example shows.

Example: Consider two IFS $\mathbf{w} = \{w_1, \ldots, w_N\}$ and $\mathbf{v} = \{v_1, \ldots, v_N\}$ with $v_1 = w_1 \circ$
w_1,
$v_2 = w_2 \circ w_1, v_3 = w_1 \circ w_2$ and $v_4 = w_2 \circ w_2$. Then $\mathbf{w}$ and $\mathbf{v}$ have the same attractor.

In Figure 4.1 it is shown at the example of the fractal interpolation function from Figure 2.1 how a point $(b^{(0)}, a^{(0)}) \in S(\psi, f)$ is mapped onto $(b_i^{(1)}, a_i^{(1)}) \in S(\psi, f), i = 1, \ldots, N$ by the first components of the affine maps w_i. Moreover, there

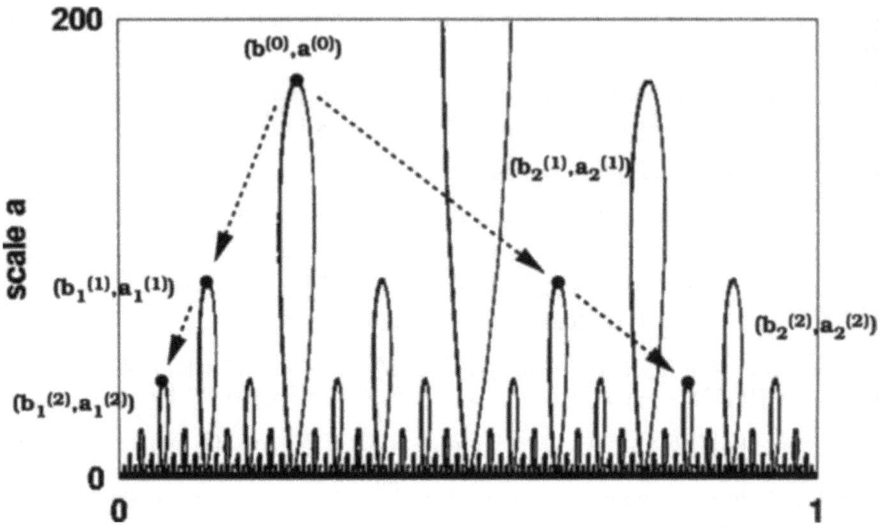

Fig. 4.1. Zeros of the wavelet transform of the fractal interpolation function from Figure 2.1 and the points $(b_i^{(j)}, a_i^{(j)})$ constructed from a given point $(b^{(0)}, a^{(0)}) \in S(\psi, f)$.

are $(b_i^{(j)}, a_i^{(j)}) \in S(\psi, f)$ with $b_i^{(j)} = \alpha_i b_i^{(j-1)} + \epsilon_i$ and $a_i^{(j)} = \alpha_i a_i^{(j-1)}$ for $j = 2, 3, \ldots$ Our algorithm uses these properties to encode the graph of a fractal interpolation function into the smallest possible number of affine maps $w_1, \ldots, w_N$ who have the graph of f as attractor. In the first step we start with a point $(b^{(0)}, a^{(0)}) \in S(\psi, f)$ such that $a^{(0)} = \max\{a \in \mathbb{R}_+ \mid \text{there exists } b \text{ with } (b, a) \in S(\psi, f)\}$ and go down from the scale $a^{(0)}$ to next scale where the points $(b_i^{(1)}, a_i^{(1)}) \in S(\psi, f)$ occur. Then we compute the parameters α_i, ϵ_i from Equations (4.4), (4.5), check the conditions (4.6), (4.7). If necessary we go down to next scale and repeat the steps from above. After a finite number of iterations the right maps must be found. In the second step we compute the parameters γ_i, δ_i and ζ_i from values of f at three x-values which are not all collinear. For more details see [21].

In contrast to the algorithm presented in [16], only very few zero-crossings of wavelet transforms at selected scales are necessary to reconstruct the graph of the fractal interpolation function.

If the smoothing function ϕ is $(k+1)$-times differentiable ($k \geq 2$), in general, the number of zeros of $W_{\phi^{(l)}}[f]$ at fixed scale a increases when l increases from 2 to k. To reduce the amount of computations it is best to choose the wavelet $\psi = \phi^{(2)}$.

The conditions on the smoothing function ϕ restrict the choice of the analyzing wavelet. The numerical computations of wavelet transforms in this paper are done by a wavelet with two vanishing moments. The corresponding smoothing function is a cubic spline (Figure 4.2) (see [9]).

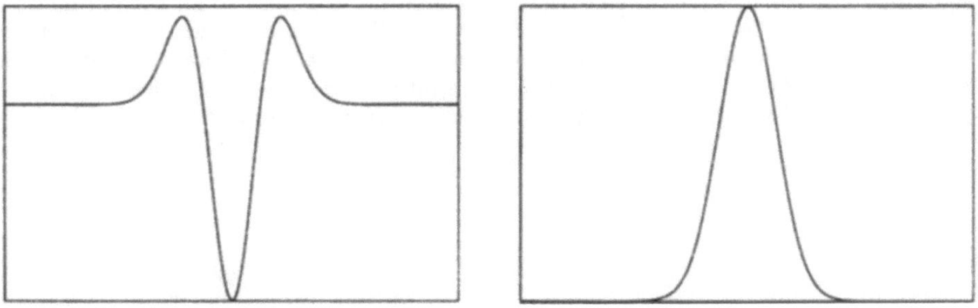

Fig. 4.2. *Left*: The wavelet with two vanishing moments. *Right*: The corresponding smoothing function.

Now we consider the results of our algorithm at an example with scaling factors α_i all equal and one with non-equal scaling factors.

Examples:

1. $\mathbf{w} = \{w_1, w_2, w_3\}$ with

$$w_1\begin{pmatrix} x \\ y \end{pmatrix} = \begin{pmatrix} 1/3 & 0 \\ 54 & 0.5 \end{pmatrix}\begin{pmatrix} x \\ y \end{pmatrix} + \begin{pmatrix} 0 \\ 0 \end{pmatrix}$$
$$w_2\begin{pmatrix} x \\ y \end{pmatrix} = \begin{pmatrix} 1/3 & 0 \\ -5 & -0.5 \end{pmatrix}\begin{pmatrix} x \\ y \end{pmatrix} + \begin{pmatrix} 1/3 \\ 50 \end{pmatrix}$$
$$w_3\begin{pmatrix} x \\ y \end{pmatrix} = \begin{pmatrix} 1/3 & 0 \\ -32.3 & 0.23 \end{pmatrix}\begin{pmatrix} x \\ y \end{pmatrix} + \begin{pmatrix} 2/3 \\ 40 \end{pmatrix}$$

2. $\mathbf{w} = \{w_1, w_2, w_3, w_4\}$ with

$$w_1\begin{pmatrix} x \\ y \end{pmatrix} = \begin{pmatrix} 0.125 & 0 \\ 30 & 0.3 \end{pmatrix}\begin{pmatrix} x \\ y \end{pmatrix} + \begin{pmatrix} 0 \\ 0 \end{pmatrix}$$
$$w_2\begin{pmatrix} x \\ y \end{pmatrix} = \begin{pmatrix} 0.5 & 0 \\ 20 & 0.5 \end{pmatrix}\begin{pmatrix} x \\ y \end{pmatrix} + \begin{pmatrix} 0.125 \\ 30 \end{pmatrix}$$
$$w_3\begin{pmatrix} x \\ y \end{pmatrix} = \begin{pmatrix} 0.25 & 0 \\ -10 & 0.4 \end{pmatrix}\begin{pmatrix} x \\ y \end{pmatrix} + \begin{pmatrix} 0.625 \\ 50 \end{pmatrix}$$
$$w_4\begin{pmatrix} x \\ y \end{pmatrix} = \begin{pmatrix} 0.125 & 0 \\ -40 & 0.2 \end{pmatrix}\begin{pmatrix} x \\ y \end{pmatrix} + \begin{pmatrix} 0.875 \\ 40 \end{pmatrix}$$

Pictures of the attractors of the IFS and the zeros of $W_\psi[f]$ are shown in Figure 4.3. In Table 4.1 the numerical results of our encoding algorithm are shown. Since the program is still not optimized there are systematic numerical errors in the computation. These errors are caused by the discretization of the scale parameter a. Further work is going on to minimize these errors.

As has become clear by the preceeding the algorithm does not work properly if the function f is not self-affine. It is therefore natural to ask for modifications of the algorithm for the general case. A first step would be a closer study of almost self-affine functions, which means that the functions are close to an affine function with respect to a certain metric. For almost self-affine functions one can hope that the set $S(\psi, f)$ can be kept as a set of important 'feature points'. Starting with points of $S(\psi, f)$ should enable one to compute affine maps such that the affine function defined by this maps approximates the given function f very closely.

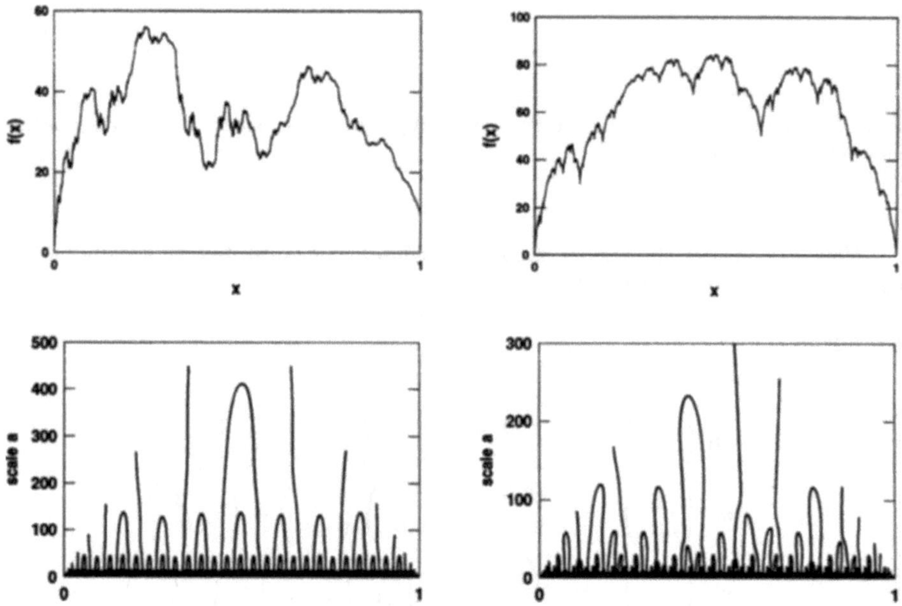

Fig. 4.3. *Top*: Graphs of the fractal interpolation function from examples 1 (*left*) and 2 (*right*). *Bottom*: Corresponding zeros of $W_\psi[f]$.

i	α_i	ϵ_i	γ_i	δ_i	ζ_i
1	0.334143	-0.000358	42.615107	0.538221	0.000000
2	0.334143	0.332925	-4.029680	-0.518349	50.089311
3	0.334143	0.666207	-32.202553	0.232354	39.874109

i	α_i	ϵ_i	γ_i	δ_i	ζ_i
1	0.126882	-0.000692	33.363142	0.281094	0.000000
2	0.501075	0.124484	20.020916	0.476018	31.811389
3	0.251613	0.624179	-33.515252	0.374862	51.465445
4	0.126882	0.874090	-42.215982	0.182836	42.205431

Table 4.1. Parameters computed from $S(\psi, f)$. *Top*: Example 1. *Bottom:* Example 2.

Some first experiments in that regard indicate that an approximation of f based on affine maps generated from $S(\psi, f)$ is indeed possible.

5. Conclusions

We studied zeros of wavelet transforms of fractal interpolation functions. The similarity preserving property of wavelet transforms allowed us to compute the affine maps of the corresponding IFS from a selected finite subset of zeros of wavelet transforms. This finite subset was found by checking special conditions in points of the set $S(\psi, f)$ from maximal scale down to smaller scales.

A resulting algorithm found the right number of maps and gave good approximations of the IFS-maps which determine the graph of the fractal interpolation function uniquely.

Acknowledgement. The author would like to thank Fritz v. Haeseler, Heinz-Otto Peitgen and Raymond O. Wells, Jr. for inspiring ideas and helpful discussions. Furthermore, the author would like to thank Wilhelm Berghorn for providing her with a program for a computation of zeros of wavelet transforms.

Part of this work was done while the author was at CeVis, University of Bremen, Germany. She was at that time supported by Deutsche Forschungsgemeinschaft, Forschergruppe *Dynamische Systeme*.

References

1. Falconer, K. (1994): *Fractal Geometry: Mathematical Foundations and Applications.* John Wiley & Sons, Chichester, UK.
2. Barnsley, M. (1988): *Fractals Everywhere.* Academic Press, Orlando, FL.
3. Fisher, Y. (1994): *Fractal Image Compression, Theory and Application.* Springer-Verlag, New York.
4. Forte, B., Vrscay, E. R. (1994): Solving the Inverse Problem for Function/Image Approximation Using Iterated Function Systems, Part 1,2. *Fractals*, Vol. 2, No. 3, 325–346.
5. Mazel, D.S., Hayes, M.H. (1992): Using Iterated Function Systems to Model Discrete Sequences. *IEEE, Trans. Signal Processing*, Vol. 40, No. 7, 1724–1734.
6. Barnsley, M., Demko, S.G. (1985): *Iterated Function Schemes and the Global Construction of Fractals*, Proc. R. Soc., A 399, 243–275.
7. Daoudi, K., Lévy Véhel, J., Meyer, Y.: Construction of continuous functions from prescribed local regularity. To appear in *Constructive Approximation.*
8. Chui, C.K. (1992): *An Introduction to Wavelets.* Academic Press, Inc..
9. Daubechies, I. (1992): *Ten Lectures on Wavelets.* SIAM, Philadelphia, PA.
10. Mallat, S., Hwang, W.L. (1992): Singularity Detection and Processing with Wavelets. *IEEE Trans. Inform. Theor.*, Vol.38, No.2, 617–643.
11. Witkin, A. (1983): Scale-Space Filtering. in *Proc. Int. Joint Conf. Artif. Intell.* Karlsruhe, West-Germany, 1019-1021.
12. Louis, A.K., Maaß, R., Rieder, A. (1994): *Wavelets.* Teubner Stuttgart.
13. Jaffard, S. (Preprint): *Multifractal Formalism for Functions: Part 1,2.*

14. Berkner, K. (1996): Eine Charakterisierung der Nullstellen-Mengen von Gaußfaltungen und eine daraus resultierende lokale Beschreibung von Kanten in Bildern. Dissertation, Universität Bremen.
15. Evertsz, C.J.G., Berkner, K., Berghorn, W. (1997): A Local Multiscale Characterization of Edges Applying the Wavelet Transform. Appears in: *Proc. Nato A.S.I., Fractal Image Encoding and Analysis, Trondheim, July 1995*, Springer Verlag, New York.
16. Mallat, S., Zhong, S. (1992): Characterization of Signals from Multiscale Edges. *IEEE Trans. Patt. Anal. Mach. Intell.*, Vol.14, No. 7, 710–732.
17. Holschneider, M. (1988): On the wavelet transforms of fractal objects. *J. Stat. Phys.*, 50, 953–993.
18. Davis, G. (1996): A wavelet-based Analysis of Fractal Image Compression. Submitted to *IEEE Trans. Image Processing.*
19. Arnéodo, A., Bacry, E., Muzy, J.F. (1994): Solving the Inverse Fractal Problem from Wavelet Anlaysis. *Europhysics Letters*, Vol. 25, No. 7, 497–484.
20. Muzy, J.F., Bacry, E., Arnéodo, A. (1994): The Multifractal Formalism Revisited with Wavelets. *Intern. J. Bifurc. Chaos*, Vol. 4, No. 2, 245–302.
21. Berkner, K. (1996): A Solution to the Inverse Problem for Fractal Interpolation Functions. Report Nr. 380, Institut für dynamische Systeme, University of Bremen.

Schémas continus de subdivision

S. Ayari et S. Dubuc

Département de mathématiques et de statistique, Université de Montréal,
C.P. 6128, Succ. Centre-ville Montréal (Québec) CANADA H3C 3J7

Résumé. Nous définissons dans un cadre général les schémas de subdivision à base b à l'aide d'une matrice de poids pour engendrer des fonctions d'une variable réelle. Nous déterminons des conditions tantôt nécessaires, tantôt suffisantes pour qu'un schéma de subdivision produise une fonction continue. Une première condition suffisante est inspirée par Dyn, Gregory et Levin et une seconde, par Deslauriers, Dubois et Dubuc. À travers un exemple de schéma de subdivision, nous illustrons ces conditions. Nous indiquons que dans le cas des schémas invariants par translation chacune des ces conditions est également nécessaire pour la continuité du schéma.

Mots-clé : schéma de subdivision, itération, fractale.

1. Schéma de subdivision sur la droite réelle

Nous partons d'une suite $\{y_i\}_{i=-\infty}^{\infty}$ de nombres réels. Nous choisissons la base b du procédé de subdivision et nous disposons d'une matrice de poids $p_{i,j}$. Ceci permet de définir une suite de fonctions $\{f_n\}$ d'une manière récursive.

– f_0 est la fonction définie sur les entiers relatifs à l'aide des données $\{y_i\}$:

$$f_0(i) = y_i. \tag{1.1}$$

– Si n est un entier, f_{n+1} est la fonction définie sur les multiples entiers de $1/b^{n+1}$ par pondération des valeurs $f_n(j/b^n)$ suivant les poids $p_{i,j}$:

$$f_{n+1}(i/b^{n+1}) = \sum_{j=-\infty}^{\infty} p_{i,j} f_n(j/b^n). \tag{1.2}$$

La suite f_n de fonctions obtenue à l'aide de la formule (1.2) représentera le *schéma de subdivision à base b* selon la pondération $p_{i,j}$ et à partir des données y_i. En principe, l'équation (1.2) suppose la convergence d'une série. En pratique, nous supposerons toujours que la matrice des poids admet la propriété suivante :

$$(\forall i) \quad \#\{j : p_{i,j} \neq 0\} < \infty. \tag{1.3}$$

Sous cette hypothèse, la série du membre de droite en (1.2) n'admet en fait qu'un nombre fini de termes, elle est toujours bien définie. Enfin, la fonction f est alors la limite de la suite des fonctions f_n. Pour que cette limite existe, il ne faut pas que les poids $p_{i,j}$ soient arbitraires.

Le modèle que nous donnons généralise les schémas de subdivision qui ont cours depuis une dizaine d'années et qui sont excellemment exposés par Cavareta-Micchelli [1] et par Dyn [6]. Dans le modèle usuel, la base b est nécessairement égale à 2, ce dont nous nous affranchissons. Mais ce qui est encore plus intéressant, c'est que même dans le cas dyadique, le nouveau modèle permet certaines constructions géométriques qui sont absentes dans l'ancien modèle. Une telle illustration est fournie par l'interpolation dyadique d'Hermite proposée par Merrien [8] (professeur à l'INSA de Rennes). Cependant, faute d'espace, nous ne donnons pas ici les détails de cet exemple.

La question que nous voulons traiter est la suivante : comment peut-on déterminer des conditions suffisantes pour que la fonction limite f d'un schéma général de subdivision existe et soit continue ? En conception assistée par ordinateur, c'est la question préalable à l'utilisation d'un schéma spécifique de subdivision.

Nous décrivons maintenant deux classes de schémas de subdivision. La matrice $p_{i,j}$ établit une correspondance entre la suite y_i et la fonction f_1. Considérons la fonction g_1 qui correspond à la suite $z_i = y_{i+1}$. Pour une question d'invariance par translation, il apparaît raisonnable d'exiger la liaison suivante entre f_1 et g_1 : $g_1(i/b) = f_1(i/b+1)$. Que cette relation ait lieu pour toute suite y_i revient à dire que

$$(\forall i)(\forall j) \quad p_{i,j-1} = p_{i+b,j}. \tag{1.4}$$

Définition. Un schéma de subdivision est *invariant par translation* si sa matrice de poids remplit la condition (1.4).

Définition. Un schéma de subdivision à base b est *interpolatoire* si pour toute suite y_i, la première fonction f_1 est telle que $(\forall i) f_1(i) = y_i$; exiger cette condition revient à demander que

$$(\forall i)(\forall j) \quad p_{bi,j} = \delta_{i,j} \tag{1.5}$$

Il s'agit du δ de Kronecker : $(\forall i)\delta_{i,i} = 1$ et $(\forall i)(\forall j \neq i)\delta_{i,j} = 0$.

Nous remarquons que tous les schémas de subdivision que l'on rencontre dans la littérature sont des schémas invariants par translation et sont presque toujours de base 2.

2. Une condition nécessaire de continuité

Pour obtenir une fonction limite continue, nous dégageons une première condition nécessaire.

Proposition 2.1. *Nous supposons que si x_n est un suite de nombres $b-$adiques telles que $b^n x_n$ est toujours entier et $\lim_{n\to\infty} x_n = 0$, alors la suite $f_n(x_n)$ du schéma de subdivision à base b converge vers un nombre $c \neq 0$. Ceci implique que*

$$(\forall i) \quad \sum_{j=-\infty}^{\infty} p_{i,j} = 1.$$

Démonstration. Soit i un entier relatif, choisissons un entier $N > 0$ suffisamment grand pour que $p_{i,j} = 0$ chaque fois que $|j| > N$. On a que

$$f_{n+1}(i/b^{n+1}) = \sum_{j=-N}^{N} p_{i,j} f_n(j/b^n).$$

En faisant tendre n vers l'infini, on obtient que

$$c = \sum_{j=-N}^{N} p_{i,j} c.$$

La simplication du facteur $c \neq 0$ de chaque membre de la dernière équation produit l'identité escomptée. □

Remarque. On peut interpréter la proposition précédente de la manière suivante : le vecteur colonne dont toutes les composantes valent 1 est un vecteur propre pour la matrice des poids ; la valeur propre de ce vecteur propre est égale à 1.

3. Une première condition suffisante de continuité

On considère un schéma de subdivision de base b et dont la matrice des poids est $p_{i,j}$; pour des données initiales y_i, on pose $f_0(i) = y_i$; on définit la suite de fonctions f_n par la récurrence (1.2).

Définition. Le schéma est *continu* si pour toute suite convergente x_n de nombres b−adiques telles que $b^n x_n$ est toujours entier, alors la suite de nombre $f_n(x_n)$ converge. Si $x = \lim_{n\to\infty} x_n$, on pose alors $f(x) = \lim_{n\to\infty} f_n(x_n)$. Il faut que cette convergence se réalise pour chaque choix de données initiales. Dans un schéma continu, on obtient donc une fonction limite f définie sur tout l'axe réel ; dans une telle situation, la fonction f est continue.

Nous recherchons des conditions sur les poids $p_{i,j}$ pour que le schéma de subdivision soit continu. À cette fin, nous introduisons une série de quantités :

$$\Delta_n = \sup_i\{|f_n((i+1)b^{-n}) - f_n(ib^{-n})|\}, \tag{3.1}$$

la matrice

$$q_{i,j} = \sum_{k=-\infty}^{j} p_{i,k} - \sum_{k=-\infty}^{j} p_{i+1,k} \tag{3.2}$$

et le coefficient

$$\kappa = \sup_i\{\sum_{j=-\infty}^{\infty} |q_{i,j}|\} \tag{3.3}$$

Lemme 3.1. *Supposons que*

$$(\forall i) \quad \sum_{j=-\infty}^{\infty} p_{i,j} = 1.$$

Alors les inégalités suivantes sont satisfaites :

$$(\forall n) \quad \Delta_n \leq \kappa^n \Delta_0 \tag{3.4}$$

Démonstration. Démontrons l'inégalité suivante

$$\Delta_{n+1} \leq \kappa \Delta_n \quad \forall n \tag{3.5}$$

Partons d'une valeur déterminée de n. Introduisons 4 vecteurs colonnes X, Y, V, W dont les composantes sont respectivement

$$x_i = f_n(ib^{-n}),$$

$$y_i = f_{n+1}(ib^{-n-1}),$$

$$v_i = x_{i+1} - x_i,$$

$$w_i = y_{i+1} - y_i.$$

Un calcul relativement simple fait voir que la matrice $Q = q_{i,j}$ envoie le vecteur V sur le vecteur W. En effet, si i est donnée, on peut trouver un entier N tel que $p_{i,j} = 0$ et $p_{i+1,j} = 0$ si $|j| > N$. On a que

$$y_i = \sum_{j=-N}^{N} p_{i,j} x_j.$$

Puisque

$$x_j = x_{N+1} - \sum_{k=j}^{N} v_k,$$

on obtient que

$$y_i = x_{N+1} \sum_{j=-N}^{N} p_{i,j} - \sum_{j=-N}^{N} p_{i,j} \sum_{k=j}^{N} v_k.$$

$$y_i = x_{N+1} - \sum_{k=-N}^{N} \sum_{j=-N}^{k} p_{i,j} v_k.$$

$$y_i = x_{N+1} - \sum_{j=-N}^{N} \sum_{k=-N}^{j} p_{i,k} v_j.$$

De la même façon, on a que

$$y_{i+1} = x_{N+1} - \sum_{j=-N}^{N} \sum_{k=-N}^{j} p_{i+1,k} v_j.$$

Si Q est la matrice $\sum_{k=-N}^{j} (p_{i,k} - p_{i+1,k})$, alors $W = QV$.

Or $\Delta_n = ||V||_\infty$, $\Delta_{n+1} = ||W||_\infty$ et $\kappa = \sup\{||QV||_\infty : ||V||_\infty \leq 1\}$. Ceci justifie l'inégalité (3.5). De là, on tire l'inégalité (3.4). □

Théorème 3.1. *Soit un schéma de subdivision de base b, de matrice de poids $p_{i,j}$ et de données de départ y_i. On suppose que les conditions suivantes sont remplies :*

- $(\forall i) \quad \sum_{j=-\infty}^{\infty} p_{i,j} = 1;$
- $(\exists N) \quad |bj - i| > N \Rightarrow p_{i,j} = 0;$
- $(\exists B)(\forall i)(\forall j) \quad |p_{i,j}| \leq B;$
- $\kappa < 1$ *où κ est définie par* (3.2) *et* (3.3).

Alors le schéma est continu.

Démonstration. Nous adaptons la démonstration du lemme 3.1 de Dyn-Gregory-Levin [5]. Le schéma de subdivision produit une suite de fonctions f_n selon la formule (1.2). f_n n'est d'abord définie que sur les multiples entiers de $1/b^n$. Entre deux multiples entiers consécutifs, on peut prolonger f_n par interpolation linéaire ; f_n devient donc définie sur tout l'axe réel. Avant d'aller plus loin, ajoutons une hypothèse supplémentaire : les accroissements $y_{i+1} - y_i$ sont bornés.

Recherchons une majoration de

$$M_n = ||f_{n+1} - f_n||_\infty.$$

$$M_n = \sup_x\{|f_{n+1}(x) - f_n(x)|\} = \sup_i\{|f_{n+1}(i/b^{n+1}) - f_n(i/b^{n+1})|\}.$$

Or

$$f_{n+1}(i/b^{n+1}) - f_n(i/b^{n+1}) = \sum_{j:|bj-i|\leq N} p_{i,j}[f_n(j/b^n) - f_n(i/b^{n+1})] \qquad (3.6)$$

Le point critique de l'argumentation consiste à montrer que M_n est $O(\Delta_n)$ où Δ_n est défini selon (3.1). Si i' est la partie entière de i/b et si $i'' = i'+1$, alors $f_n(j/b^n) - f_n(i/b^{n+1})$ est une combinaison convexe de $f_n(j/b^n) - f_n(i'/b^n)$ et de $f_n(j/b^n) - f_n(i''/b^n)$. Chacune de ces deux quantités est une somme d'accroissements $f_n((k+1)/b^n) - f_n(k/b^n)$ et le nombre de termes dans chacune des sommes ne dépasse pas N/b; d'où $|f_n(j/b^n) - f_n(i/b^{n+1})| \leq (N/b)\Delta_n$. Enfin, le nombre de termes dans le membre de droite de l'équation (3.6) est au plus $2N/b + 1$. D'où $M_n \leq (2N/b+1)B(N/b)\Delta_n$.

D'autre part, le lemme 3.1 indique que $\Delta_n \leq \kappa^n \Delta_0$; l'hypothèse supplémentaire affirme que Δ_0 est un nombre fini ; la série $\sum_{n=0}^{\infty} M_n$ est donc convergente, puisque $\kappa < 1$. Par le critère M de Weierstrass, la suite f_n est uniformément convergente. La limite f de cette suite est nécessairement continue et le schéma est lui-même continu.

Maintenant, mettons de côté l'hypothèse que la suite $y_{i+1} - y_i$ est bornée. Pour y arriver, nous avons besoin du prochain lemme.

Lemme 3.2. *Soit un schéma de subdivision de base b, de matrice de poids $p_{i,j}$ et de données de départ y_i, on admet l'existence d'un nombre N pour lequel*

$$(\forall i)(\forall j) \quad |bj - i| > N \Rightarrow p_{i,j} = 0.$$

Considérons un entier $L > 0$, alors les valeurs prises par chacune des fonction f_n sur $[-L, L]$, ne dépendent aucunement des valeurs y_j lorsque $|j| > L + N/(b-1)$.

Démonstration. Soit $n > 0$, considérons un entier $j > L + N/(b-1)$, demandons-nous pour quel entier i, la valeur y_j peut avoir une influence sur $f_n(i/b^n)$. Or on sait que la quantité $f_n(i/b^n)$ est la somme des termes de la forme

$$p_{i_0,i_1} p_{i_1,i_2} \cdots p_{i_{n-1},i_n} y_{i_n}$$

où $i_0 = i, i_n = j$. Pour que y_j ait une influence sur $f_n(i/b^n)$, il faut que $p_{i_{k-1},i_k} \neq 0, k = 1, 2, \ldots, n$; si bien que $i_{k-1} \geq bi_k - N, k = 1, 2, \ldots, n$ et l'on obtiendra que $i \geq b^n j - N(b^n - 1)/(b-1)$. Si $f_n(i/b^n)$ est influencé par y_j, on a donc que $i/b^n \geq j - N(1 - b^{-n})/(b-1) > L$. C'est ce qu'il fallait vérifier. □

Revenons à la démonstration du théorème 3.1. Soit $L > 0$, on veut démontrer que la suite $f_n(x)$ converge uniformément sur $[L, L]$. Posons $y_j = 0$ si $|j| > L + N/(b-1)$. D'après le lemme 3.2, ce changement n'affecte aucunement les valeurs de f_n sur $[L, L]$. D'autre part, les accroissements des y_i modifiés sont bornés ; on peut se servir de la première partie de la démonstration et en conclure la convergence uniforme de la suite f_n sur $[L, L]$. □

Il est possible d'étendre la portée du théorème 3.1. Pour ce faire, on désigne par P la matrice $p_{i,j}$, par $p_{i,j}^m$ les éléments de la matrice P^m alors que $m > 0$ est un entier. On pose

$$q_{i,j}^m = \sum_{k=-\infty}^{j} p_{i,k}^m - \sum_{k=-\infty}^{j} p_{i+1,k}^m \tag{3.7}$$

afin de déterminer le coefficient

$$\kappa_m = \sup_i \{ \sum_{j=-\infty}^{\infty} |q_{i,j}^m| \} \tag{3.8}$$

Théorème 3.2. *Soit un schéma de subdivision de base b, de matrice de poids $p_{i,j}$ et de données de départ y_i. On suppose que les conditions suivantes sont remplies :*

- $(\forall i) \quad \sum_{j=-\infty}^{\infty} p_{i,j} = 1;$
- $(\exists N) \quad |bj - i| > N \Rightarrow p_{i,j} = 0;$
- $(\exists B)(\forall i)(\forall j) \quad |p_{i,j}| \leq B;$
- $(\exists m > 0) \quad \kappa_m < 1$ *où* κ_m *est définie par* (3.7) *et* (3.8).

Alors le schéma est continu.

Démonstration. Le schéma de subdivision de base b, de matrice de poids $p_{i,j}$ et de données de départ y_i engendre la suite de fonctions f_n. Si m est un entier > 0, le schéma de subdivision de base b^m, de matrice de poids $p_{i,j}^m$ et de données de départ y_i engendre la suite de fonctions $f_0, f_m, f_{2m}, \ldots$. Nous appliquerons le théorème 3.1 à ce second schéma de subdivision. Vérifions que dans ce cas, les hypothèse de ce théorème sont satisfaites.

- En effet, si V est le vecteur dont toutes les composantes valent 1, par hypothèse $PV = V$; d'où $P^m V = V$.
- On sait que si $|bj - i| > N$, alors $p_{i,j} = 0$. Soient i et j deux entiers pour lesquels $p_{i,j}^m \neq 0$. Il existe donc une suite de $m+1$ entiers $i_0, i_1, \ldots, i_m$ tels que $i_0 = i, i_m = j$ et $p_{i_{k-1},i_k} \neq 0, k = 1, 2, \ldots, m$. Ceci montre que $|bi_k - i_{k-1}| \leq N, k = 1, 2, \ldots, m$ et de là que $|b^m j - i| \leq N(b^m - 1)/(b-1)$.
- On sait que $(\forall i)(\forall j)|p_{i,j}| \leq B$. $p_{i,j}^m$ est une somme de termes, chaque terme ne dépasse pas B^m en valeur absolue et le nombre de ces termes différents de 0 ne dépasse pas $(2N+1)^{m-1}$, c'est le nombre de choix d'indices $i_1, i_2, \ldots, i_{m-1}$ tels que $|bi_k - i_{k-1}| \leq N, k = 1, 2, \ldots, m-1$, alors que $i_0 = i$ et $i_m = j$. On a donc montré que $|p_{i,j}^m| \leq B^m (2N+1)^{m-1}$.
- L'hypothèse $\kappa_m < 1$ est précisément la dernière hypothèse du théorème 3.1 à rencontrer.

Par le théorème 3.1, sur tout intervalle $[-L, L]$, la suite de fonctions $f_{nm}, n = 0, 1, 2, \ldots$ convergent uniformément vers une fonction f. Soit r un entier compris entre 0 et $m-1$, soit i un entier, alors

$$f_{nm+r}(i/b^{nm+r}) = \sum_{j=-\infty}^{\infty} p_{i,j}^r f_{nm}(j/b^{nm}).$$

Supposons que le nombre i/b^{nm+r} est voisin de x à ϵ près. Cherchons le lien entre x et j si $p_{i,j}^r \neq 0$. Dans un tel cas, il existe une suite d'indices $i_1, i_2, \ldots, i_{r-1}$ tels que $|bi_k - i_{k-1}| \leq N, k = 1, 2, \ldots, r-1$, alors que $i_0 = i$ et $i_r = j$. D'où $|b^r j - i| \leq N_r = N(b^r - 1)/(b-1)$ et $|j/b^{nm} - i/b^{nm+r}| \leq N_r/b^{nm+r}$. L'écart du nombre j/b^{nm} à x ne dépasse pas $\epsilon + N_r/b^{nm+r}$ et le nombre $f_{nm}(j/b^{nm})$ sera voisin de $f(x)$ d'autant plus que n sera grand. D'autre part, la somme sur j des tous les poids (non nuls) $p_{i,j}^r$ est égale à 1. On a donc montré que $f_{nm+r}(i/b^{nm+r})$ est voisin de $f(x)$. La suite de fonctions f_n convergent uniformément sur $[-L, L]$ vers la fonction f. Le premier schéma de subdivision est continu. □

4. Une seconde condition suffisante de continuité

Deslauriers-Dubois-Dubuc [3] ont déterminé des conditions suffisantes de continuité dans le cas d'interpolation itérative à plusieurs variables. Ce que nous nous proposons de faire dans cette section est de transposer ces conditions dans le contexte des schémas de subdivision à une variable.

Si h est un nombre positif, si $f_n(x)$ est la suite de fonctions produites par le schéma de subdivision de base b et de poids $p_{i,j}$, alors nous posons

$$C(h) = \sup\{\sum_{j=\infty}^{\infty} |p_{i,j} - p_{i',j}| : |i - i'| \leq h\}. \tag{4.1}$$

$$\omega_n(h) = \sup\{|f_n(i/b^n) - f_n(i'/b^n)| : |i - i'| \leq h\}. \tag{4.2}$$

Lemme 4.1. *On suppose que* $|bj - i| > N \Rightarrow p_{i,j} = 0$. *Si* $(\forall i)$ $\sum_{j=-\infty}^{\infty} p_{i,j} = 1$ *et si* $h \geq 2N/(b-1)$, *alors* $\omega_n(h) \leq [C(h)/2]^n \omega_0(h)$.

Démonstration. Démontrons l'inégalité suivante

$$\omega_{n+1} \leq [C(h)/2]\omega_n \quad \forall n. \tag{4.3}$$

Partons d'une valeur déterminée de n. Considérons deux entiers i et i' tels que $|i - i'| \leq h$. Désignons par J la collection d'entiers $\{j : |bj - i| \leq N\}$. Si j est hors de J, alors $p_{i,j} = 0$. D'où

$$f_{n+1}(i/b^{n+1}) = \sum_{j \in J} p_{i,j} f_n(j/b^n).$$

De même, si $J' = \{j : |bj - i'| \leq N\}$, alors

$$f_{n+1}(i'/b^{n+1}) = \sum_{j \in J'} p_{i',j} f_n(j/b^n).$$

Si nous posons $M = \max\{f(j/b^n) : j \in J \cup J'\}, m = \min\{f(j/b^n) : j \in J \cup J'\}$ et $c = (M + m)/2$, alors nous obtenons que

$$f_{n+1}(i/b^{n+1}) - f_{n+1}(i'/b^{n+1}) = \sum_{j \in J \cup J'} (p_{i,j} - p_{i',j})(f_n(j/b^n) - c),$$

puisque l'hypothèse $(\forall i)$ $\sum_{j=-\infty}^{\infty} p_{i,j} = 1$ entraîne que $\sum_{j \in J \cup J'} (p_{i,j} - p_{i',j})c = 0$. D'où

$$|f_{n+1}(i/b^{n+1}) - f_{n+1}(i'/b^{n+1})| \leq \sum_j |p_{i,j} - p_{i',j}|(M - m)/2.$$

Un calcul simple fait valoir que le diamètre de $J \cup J'$ ne dépasse pas $(2N+h)/b$, ce qui est majoré par h. De ce fait, on obtient l'inégalité $M - m \leq \omega_n(h)$ de sorte que

$$|f_{n+1}(i/b^{n+1}) - f_{n+1}(i'/b^{n+1})| \leq [C(h)/2]\omega_n(h)$$

et $\omega_{n+1}(h) \leq [C(h)/2]\omega_n(h)$. La conclusion se vérifie ensuite par récurrence. □

Théorème 4.1. *Soit un schéma de subdivision de base b, de matrice de poids $p_{i,j}$ et de données de départ y_i. On suppose que les conditions suivantes sont remplies :*

- $(\forall i) \quad \sum_{j=-\infty}^{\infty} p_{i,j} = 1;$
- $(\exists N) \quad |bj - i| > N \Rightarrow p_{i,j} = 0;$
- $(\exists B)(\forall i)(\forall j) \quad |p_{i,j}| \leq B;$
- $C(h) < 2$ *où* $h = \max(2N/(b-1), 1)$ *et* $C(h)$ *est définie par* (4.1).

Alors le schéma est continu.

Démonstration. Nous reprenons en bonne partie la démonstration du théorème 3.1. Le schéma de subdivision produit une suite de fonctions f_n selon la formule (1.2). f_n n'est d'abord définie que sur les multiples entiers de $1/b^n$. Entre deux multiples entiers consécutifs, on peut prolonger f_n par interpolation linéaire ; f_n devient donc définie sur tout l'axe réel. Avant d'aller plus loin, ajoutons une hypothèse supplémentaire : la quantité $\omega_0(h)$ est finie.

Recherchons une majoration de

$$M_n = ||f_{n+1} - f_n||_\infty.$$

$$M_n = \sup_i\{|f_{n+1}(i/b^{n+1}) - f_n(i/b^{n+1})|\}.$$

Or

$$f_{n+1}(i/b^{n+1}) - f_n(i/b^{n+1}) = \sum_{j:|bj-i|\leq N} p_{i,j}[f_n(j/b^n) - f_n(i/b^{n+1})] \tag{4.4}$$

Le point critique de l'argumentation consiste à montrer que M_n est $O(\omega_n(h))$ où $h = \max(2N/(b-1), 1)$ et $\omega_n(h)$ est défini selon (4.2). Si i' est la partie entière de i/b et si $i" = i' + 1$, alors $f_n(j/b^n) - f_n(i/b^{n+1})$ est une combinaison convexe de $f_n(j/b^n) - f_n(i'/b^n)$ et de $f_n(j/b^n) - f_n(i"/b^n)$. On vérifie que chacune de ces deux quantités est majorée par $2\omega_n(h)$ étant donné que $|j - i'| \leq 2h$ et $|j - i"| \leq 2h$; d'où $|f_n(j/b^n) - f_n(i/b^{n+1})| \leq 2\omega_n(h)$. D'autre part, le nombre de termes dans le membre de droite de l'équation (4.4) est au plus $2N/b + 1$. D'où $M_n \leq (2N/b+1)2B\omega_n(h)$.

D'autre part, le lemme 4.1 indique que $\omega_n(h) \leq [C(h)/2]^n\omega_0(h)$; l'hypothèse supplémentaire affirme que $\omega_0(h)$ est un nombre fini ; la série $\sum_{n=0}^{\infty} M_n$ est donc convergente, puisque $C(h) < 2$. Par le critère M de Weierstrass, la suite f_n est uniformément convergente. La limite f de cette suite est nécessairement continue, et le schéma est lui-même continu.

Maintenant, mettons de côté l'hypothèse que $\omega_0(h)$ est fini. Soit $L > 0$, on veut démontrer que la suite $f_n(x)$ converge uniformément sur $[L, L]$. Posons $y_j = 0$ si $|j| > L + N/(b-1)$. D'après le lemme 3.2, ce changement n'affecte aucunement les valeurs de f_n sur $[L, L]$. D'autre part, la quantité $\omega_0(h)$ calculée à partir des y_i modifiés est finie ; on peut se servir de la première partie de la démonstration et en conclure la convergence uniforme de la suite f_n sur $[L, L]$. □

Il est possible d'étendre la portée du théorème 4.1. Pour ce faire, on désigne par P la matrice $p_{i,j}$, par $p^m_{i,j}$ les éléments de la matrice P^m alors que $m > 0$ est un entier. On pose

$$C_m(h) = \sup\{\sum_{j=\infty}^{\infty} |p^m_{i,j} - p^m_{i',j}| : |i - i'| \leq h\}. \tag{4.5}$$

$$N_m = \max\{|bj - i| : p^m_{i,j} \neq 0\}. \tag{4.6}$$

Théorème 4.2. *Soit un schéma de subdivision de base* b, *de matrice de poids* $p_{i,j}$ *et de données de départ* y_i. *On suppose que les conditions suivantes sont remplies :*

- $(\forall i) \quad \sum_{j=-\infty}^{\infty} p_{i,j} = 1;$
- $(\exists N) \quad |bj - i| > N \Rightarrow p_{i,j} = 0;$
- $(\exists B)(\forall i)(\forall j) \quad |p_{i,j}| \leq B;$
- $(\exists m > 0) C_m(h) < 2$ *où* $h = \max(2N_m/(b^m - 1), 1)$ *et* $C_m(h)$ *est définie par* (4.5).

Alors le schéma est continu.

Démonstration. Il suffit de reprendre mot à mot la démonstration du théorème 3.2 et de n'y effectuer que quelques changements : remplacer les 3 mentions au théorème 3.2 par des mentions du théorème 4.1 et remplacer l'expression $\kappa_m < 1$ par $C_m(h) < 2$.

5. Exemples de schémas continus

Comme exemple, nous choisissons un schéma d'interpolation dyadique à quatre poids. Les valeurs des poids sont comme suit :

$$p_{2i,j} = \begin{cases} 1 & si \quad j = i, \\ 0 & \text{sinon} \end{cases} \quad \text{et} \quad p_{2i+1,j} = \begin{cases} a & si \quad j = i-1, \\ b & si \quad j = i, \\ c & si \quad j = i+1, \\ d & si \quad j = i+2, \\ 0 & \text{sinon} \end{cases}$$

Nous supposons que la condition nécessaire de continuité est satisfaite : $a+b+c+d = 1$. Si on évalue le paramètre κ donné par les formules (3.2) et (3.3), on obtient que

$$\kappa = |a| + |d| + \max(|a+b|, |c+d|).$$

Si on choisit comme valeurs particulières de (a, b, c, d) respectivement :

$$(-1/16, 9/16, 9/16, -1/16), (-2/10, 6/10, 7/10, -1/10), (4/10, 3/10, 2/10, 1/10),$$

on obtient les trois courbes de la figure 5.1 et dont les paramètres respectifs κ sont : $5/8, 9/10$ et $6/5$. La continuité des courbes engendrées par les deux premiers exemples est justifiée par le théorème 3.1. Pour justifier la continuité de la troisième courbe, il faut calculer le paramètre correspondant de la seconde génération κ_2 qui lui vaut $0,84$ et faire appel au théorème 3.2.

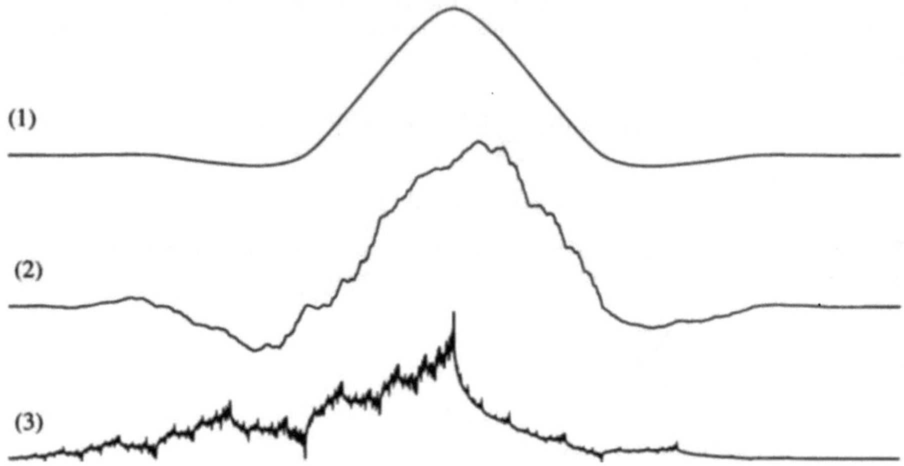

Fig. 5.1. Trois exemples d'interpolations itératives dyadiques.

Maintenant, nous allons tenter d'illustrer par le même exemple la deuxième condition suffisante de continuité. Ici $h = 6$, les paramètres respectifs $C(h)$ sont : 5/2, 16/5 et 2. À la génération suivante, la quantité h vaut encore 6 et les paramètres $C_2(h)$ sont respectivement : 2,40 , 3,06 et 2. Pour justifier la continuité, il faut se rendre au moins à la troisième génération où les valeurs de $C_3(h)$ sont 1,75 , 2,22 et 1,754. La continuité pour le premier et le troisième exemple est alors justifiée par le théorème 4.2. Pour la justification de la continuité du deuxième exemple, il faut calculer le paramètre $C_4(h)$ de la quatrième génération qui lui vaut $1,37$.

Fig. 5.2. Courbe de solidarité. Une fractale dont 3 parties sont semblables au tout, un exemple de subdivision triadique.

Comme autres exemples de schémas continus, nous faisons valoir qu'il est possible d'obtenir une foule de courbes fractales si l'on autorise l'usage de poids complexes. C'est ainsi que la figure 5.2 reproduit les valeurs prises dans le plan complexe par une fonction engendrée par un schéma de subdivision à base 3, dont les poids sont des nombres complexes. Cette courbe a déjà été décrite par le second auteur [4] dans un autre contexte de géométrie fractale. Mais elle représente aussi un cas particulier du schéma de subdivision. Il suffit de prendre une suite y_n où $y_0 = 0$ et $y_1 = 1$, de choisir 3 comme base b et de se servir de la matrice de poids dont les poids non nuls sont $p_{3j,j} = 1$, $p_{3j+1,j} = (2-i)/5, p_{3j+1,j+1} = (3+i)/5$, $p_{3j+2,j} = (3-i)/5, p_{3j+2,j+1} = (2+i)/5$, pour $j = 0, \pm 1, \pm 2, \ldots$ La courbe obtenue est la trace dans le plan complexe de la fonction $t \mapsto f(t)$ lorsque t varie entre 0 et 1.

Autre exemple : la courbe du dragon imaginée par Harter-Heighway [7] s'obtient avec une base $b = 2$ et une matrice de poids dont les poids non nuls sont $p_{2j,j} = 1$, $p_{4j+1,2j} = (1-i)/2, p_{4j+1,2j+1} = (1+i)/2$, $p_{4j+3,2j+1} = (1+i)/2, p_{4j+3,2j+2} = (1-i)/2$, pour $j = 0, \pm 1, \pm 2, \ldots$

6. Conditions nécessaires et suffisantes de continuité

Lorsqu'un schéma de subdivision est invariant par translation, les conditions suffisantes de continuité données par les théorèmes 3.2 et 4.2 deviennent également des conditions nécessaires.

Théorème 6.1. *Soit un schéma de subdivision de base b, dont la matrice de poids* $p_{i,j}$ *satisfait la condition* (1.3). *On suppose que le schéma est invariant par translation et continu. Alors les conditions suivantes sont remplies :*

- $(\forall i) \quad \sum_{j=-\infty}^{\infty} p_{i,j} = 1;$
- $(\exists N) \quad |bj - i| > N \Rightarrow p_{i,j} = 0;$
- $(\exists B)(\forall i)(\forall j) \quad |p_{i,j}| \leq B;$
- $(\exists m > 0) \quad \kappa_m < 1$ *où* κ_m *est définie par* (3.7) et (3.8);
- $(\exists m > 0) \quad C_m(h) < 2$ *où* $h = \max(2N_m/(b^m - 1), 1)$, N_m *est donnée par*(4.6)

La démonstration de ce théorème n'est pas immédiate, mais peut être obtenue en suivant les traces de la démonstration du théorème 8.7 de [3].

7. Conclusion

Les schémas de subdivision forment un outil puissant pour créer toute sorte de fonctions ; c'est ainsi que l'on peut obtenir aussi bien des courbes irrégulières comme les fractales que des fonctions régulières comme les splines. Nous avons élargi la classe de ces schémas. Pour cette classe élargie, nous avons trouvé deux conditions qui permettent de vérifier la continuité d'un schéma de subdivision à une variable. La première condition exige très souvent moins de calculs, cependant la seconde condition se prête mieux à une généralisation aux schémas de subdivision à plusieurs variables. Nous terminons en posant deux questions. Peut-on trouver des conditions nécessaires et suffisantes pour que la fonction limite produite par un schéma de subdivision soit une fonction dérivable ? Quel est le degré de régularité des fonctions produites par un schéma donné ? La réponse à ces questions pourrait fort bien dépendre des travaux de Cohen et de ses collaborateurs sur les ondelettes, nous pensons particulièrement au travail de Cohen, Daubechies et Amos [2]. Effectivement, il a un lien serré entre les opérateurs de subdivision et la théorie des ondelettes, mais ceci est une autre histoire.

Remerciement. Le second auteur reçoit une subvention de recherche du CRSNG (Gouvernement fédéral du Canada).

Références

1. Cavaretta, A.S., Micchelli, C.A. (1989) : The design of curves and surfaces by subdivision algorithms. In Lyche T., Schumaker L.L. (eds) (1989) *Mathematical Aspects of Computer Aided Geometric Design* (Oslo 1988) Academic Press, Boston, 115–153.
2. Cohen, A., Daubechies, I., Amos, R. (1996) : How smooth is the smoothest function in a given refinable space? *Appl. Comput. Harmon. Anal.* **3**, 87–89.
3. Deslauriers, G., Dubois, J., Dubuc, S. (1991) : Multidimensional iterative interpolation. *Canad. J. Math.* **43**, 297–312.
4. Dubuc, S. (1982) : Une foire de courbes sans tangentes. In *Actes du 6ème congrès du Groupement des mathématiciens d'expression latine.* (Luxembourg 1981) Gauthier-Villars, Paris, 99–123.
5. Dyn, N., Gregory, J.A., Levin, D. (1991) : Analysis of uniform binary subdivision schemes for curve design. *Constr. Approx.* **7**, 127–147.
6. Dyn, N. (1992) : Subdivision schemes in computer-aided geometric design. In Light W. (ed) (1992)*Advances in Numerical Analysis Volume II.* Wavelets, subdivision algorithms and radial basis functions. Clarendon Press, Oxford, 36–104.
7. Gardner, M. (1967) : Mathematical games. *Scientific American* **216** mars, 124–125, avril, 119–120.
8. Merrien, J.L. (1992) : A family of Hermite interpolants by bissection algorithms. *Numerical Algorithms* **2**, 187–200.

AMS *Subject Classifications*: 65D10, 65Q05, 28A80.

RANDOM PROCESSES

Fractal percolation with neighbour interaction

R.M. Burton[1], T. Coffey[1], F.M. Dekking[2], and K. Hyman[2]

[1] Department of Mathematics, University of Oregon,
Corvallis, Oregon 97331, USA
burton@math.orst.edu

[2] Technische Universiteit Delft, TWI, SSOR,
Mekelweg 4, 2628 CD Delft, THE NETHERLANDS
f.m.dekking@twi.tudelft.nl

1. Introduction

Fractal percolation was introduced by Benot Mandelbrot in [4] and further analysed in [1]. It can be considered as the most natural way to randomize the classical middle third Cantor set in a two-dimensional setting. Recently Peres [5] has discovered a close connection between fractal percolation and the path of Brownian motion.

The present work is a continuation of [2] were fractal percolation was considered as a natural way to randomize the quadtree algorithm for binary images. It was argued there that although the randomization is natural, the corresponding images are not, due to the fact that the set is totally disconnected if percolation does not occur - which seems to happen for all images which are less than 90% black (it is an open problem to obtain good bounds on the level of blackness were the transition from total disconnectedness to large connected components occurs). A simple form of neighbour interaction that privileges the formation of connected components was defined in [2]. That this scheme is successful is evident from the simulations, the main result of the present paper is that nevertheless there will be infinitely many components in the limit (see Section 3.). We proceed now to a more formal description of fractal percolation - without or with neighbour interaction. Let $I_n(j!)$ be the nth order binary intervals

$$I_n(j) = [\frac{j}{2^n}, \frac{j+1}{2^n}] \qquad j = 0, 1, \ldots, 2^n - 1.$$

We also consider the nth order binary squares

$$I_n(j,k) = I_n(j) \times I_n(k) \qquad j,k = 0, \ldots, 2^n - 1.$$

These squares will be coloured white or black, which we code by assigning zeroes respectively ones to a $2^n \times 2^n$ matrix M_n. The black squares determine a set K_n in the unit square by

$$K_n = \bigcup_{M_n(i,j)=1} I_n(i,j).$$

In fractal percolation the K_n's are obtained recursively by a random mechanism which works as follows. Choose a parameter p with $0 \le p \le 1$. Put $K_0 = [0,1]^2$. Given K_n, K_{n+1} is obtained by determining independently with probability p which of the four subsquares $I_{n+1}(2i,2j), I_{n+1}(2i,2j+1), I_{n+1}(2i+1,2j), I_{n+1}(2i+1,2j+1)$ of a black square $I_n(i,j)$ will be black. (White squares give rise to four white squares). In the matrix representation an example is given by

$$M_0 = 1 \to M_1 = \begin{pmatrix} 1 & 1 \\ 0 & 1 \end{pmatrix} \to M_2 = \begin{pmatrix} 1 & 0 & 1 & 0 \\ 0 & 1 & 1 & 0 \\ 0 & 0 & 0 & 1 \\ 0 & 0 & 1 & 1 \end{pmatrix} \to \dots$$

where the transitions occur with probability $p^3(1-p)$ respectively $p^7(1-p)^5$. The compact sets K_n decrease to a compact set K_∞. The set K_∞ is fractal: conditioned on being non-empty, it has Hausdorff dimension $\log(4p)/\log 2$ almost surely [1].

We call two squares $I_n(i,j)$ and $I_n(\hat{\imath},\hat{\jmath})$ *neighbours* if

$$|i-\hat{\imath}| + |j-\hat{\jmath}| = 1.$$

In fractal percolation with neighbour interaction the colour of the four children $I_{n+1}(2i,2j), \dots, I_{n+1}(2i+1,2j+1)$ of a black square $I_n(i,j)$ is determined by the colours of the neighbours of $I_n(i,j)$. One of the simplest way to do this (and 'favor connectedness') is by the following directional majority rule. Let again p be a parameter, $0 \le p \le 1$.

Let $b = M_n(i-1,j) + M_n(i,j-1)$ be the total blackness of the left and lower neighbours of $I_n(i,j)$. Then the left-lower child $I_{n+1}(2i,2j)$ is coloured black if $b=2$, white if $b=0$ and black with probability p, white with probability $1-p$ in case $b=1$. The colours of the other three children follow by rotational symmetry.

In the matrix representation this looks as follows.

$$\begin{pmatrix} * & * & * \\ * & 1 & 1 \\ * & 1 & * \end{pmatrix} \to \begin{pmatrix} * & * \\ * & 1 \end{pmatrix}, \quad \begin{pmatrix} * & * & * \\ * & 1 & 0 \\ * & 0 & * \end{pmatrix} \to \begin{pmatrix} * & * \\ * & 0 \end{pmatrix}$$

$$\begin{pmatrix} * & * & * \\ * & 1 & 1 \\ * & 0 & * \end{pmatrix} \xrightarrow{p} \begin{pmatrix} * & * \\ * & 1 \end{pmatrix}, \quad \begin{pmatrix} * & * & * \\ * & 1 & 1 \\ * & 0 & * \end{pmatrix} \xrightarrow{1-p} \begin{pmatrix} * & * \\ * & 0 \end{pmatrix}.$$

Here the center of the 3×3 matrix corresponds to the square $I_n(i,j)$ and the 2×2 matrix to the four children $I_{n+1}(2i,2j), \dots, I_{n+1}(2i+1,2j+1)$. The $*'s$ indicate either 0 or 1.

To initialize the process we need to specify the colours of the neighbouring squares of $[0,1]^2 = K_0$. To have more flexibility we rather start at $K_1 = I_1(0,0) \cup I_1(0,1) \cup I_1(1,0) \cup I_1(1,1)$. The following two matrices represent the colourings of the 12 neighbouring squares of the 4 black squares in K_1:

$$J_1 = \begin{pmatrix} 0 & 0 & 0 & 0 \\ 0 & 1 & 1 & 0 \\ 0 & 1 & 1 & 0 \\ 0 & 0 & 0 & 0 \end{pmatrix} \qquad J_2 = \begin{pmatrix} 0 & 1 & 1 & 0 \\ 0 & 1 & 1 & 0 \\ 0 & 1 & 1 & 0 \\ 0 & 1 & 1 & 0 \end{pmatrix}$$

For example, the upper right 0 in J_1 , indicates that $[1, \frac{3}{2}] \times [1, \frac{3}{2}]$ is coloured white. The same evolution rules are applied to these 12 marginal squares. (In case of J_1 this trivially gives a white margin for all n, in case of J_2 the one-sidedness of the rules unambiguously defines the margin. See Figure 1.1 for some simulations of K_8.

The process has both purely random and deterministic characteristics. The most prominent deterministic characteristic is given by the following '4-block lemma'. Let a *4-block* be four squares whose union is a square. See [2] for the (simple) proofs of the following lemmas.

Lemma 1.1. *If K_n contains a 4-block then K_{n+1} contains a 4-block.*

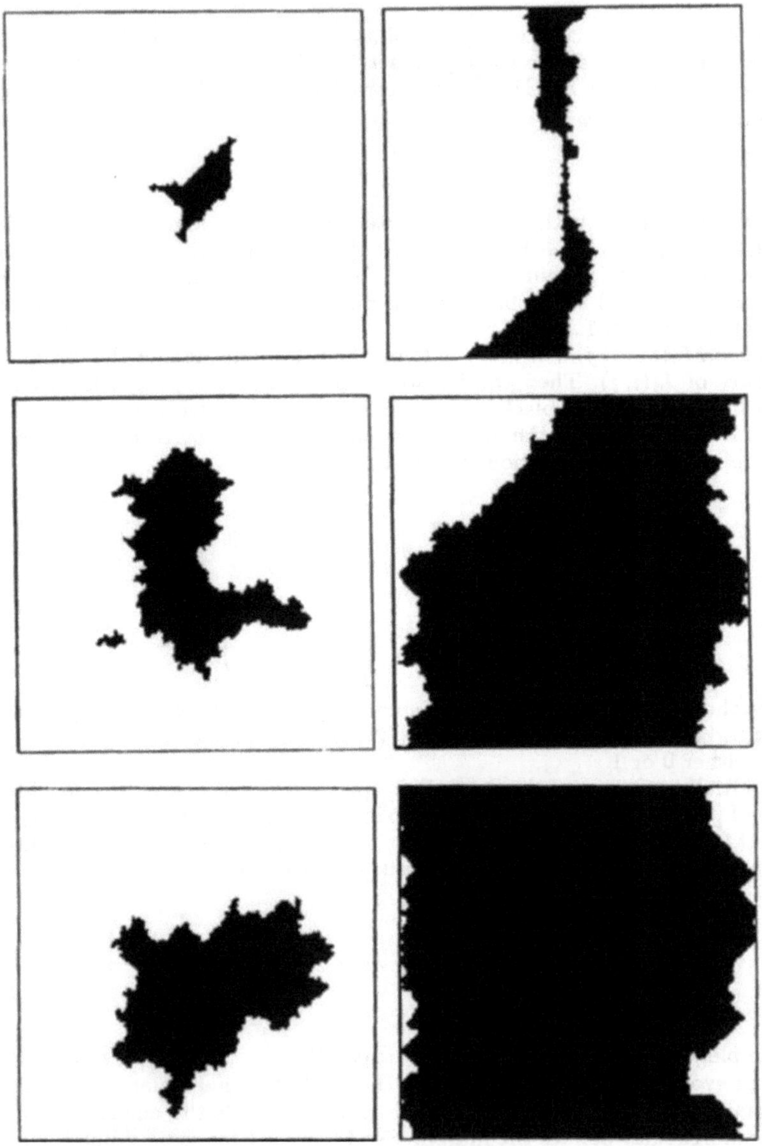

Fig. 1.1. Six realisations of K_8. Left: initial condition $J_1, p = 0.2, 0.5, 0.6$. Right: initial condition $J_2, p = 0.25, 0.7$ and 0.9.

Lemma 1.2. *Let C be any configuration in K_n containing a 4-block, then $C \cap K_\infty$ will have non-empty interior almost surely.*

Lemma 1.3. *If J_2 is the initial condition, then the line segment $\{\frac{1}{2}\} \times [0,1]$ is contained in K_∞.*

It is immediate from Lemma 1.2 that $K_\infty = \cap_{n \geq 1} K_n$ is not a fractal, however, we conjecture that it has a fractal boundary. Some of the properties of K_∞ will be discussed in Section 4. All the proofs there are based on basic properties of branching processes.

The simple interaction rules we have given may of course be generalised. New issues and interesting sets arise when the (K_n) are no longer monotone, which happens if white squares can give rise to black squares. See Figure 4.1 for an example. For related work, see [3] where random quadtrees model binary images of lungslices with a maximum likelihood estimation procedure.

See [6] for a one-dimensional version of neighbour interaction.

2. Connectedness

Theorem 2.1. *Let $0 < p < 1$. Almost surely K_∞ has infinitely many components with non-empty interior.*

Proof: We define an *isolated 4-block* as the 4×4 configuration

$$\begin{pmatrix} 0 & 0 & 0 & 0 \\ 0 & 1 & 1 & 0 \\ 0 & 1 & 1 & 0 \\ 0 & 0 & 0 & 0 \end{pmatrix}$$

Let a *cluster* be a maximal set of black neighbouring squares. A *full cluster* is a cluster that contains a 4-block. By Lemma 1.2 a full cluster generates a component in K_∞ with non-empty interior. What we will show is that the number of full clusters in K_{4n} is bounded from below by a supercritical branching process which has zero extinction probability. Clearly this implies the statement of the theorem (and in fact somewhat more: the full clusters yielding the components increase at an exponential rate almost surely). More specifically, we claim that a full cluster in K_n will produce a full cluster and an isolated 4-block in K_{n+3} or K_{n+4} with positive probability. Since the evolutions of clusters (despite the neighbour interaction) occur independently, this yields a branching process with mean larger than one. Let C be a full cluster in K_n. Consider the highest black square in the right most column intersected by C. We call this square e. Consider e as the center of a 3×3 square. Then, by definition of e, this configuration has the form

$$\begin{pmatrix} * & 0 & * \\ * & 1 & 0 \\ * & * & * \end{pmatrix},$$

where * denotes either 0 or 1.(Here we assume in case of initial condition J_2 that n is so large that $[1 - \frac{1}{2^n}, 1] \times 1 - \frac{1}{2^n}, 1]$ is not in K_n). There are two possibilities for the square below the center square e, yielding

$$S = \begin{pmatrix} * & 0 & * \\ * & 1 & 0 \\ * & 0 & * \end{pmatrix} \quad \text{and} \quad \overline{S} = \begin{pmatrix} * & 0 & * \\ * & 1 & 0 \\ * & 1 & * \end{pmatrix}.$$

Note, that again by the definition of e, we can be more specific:

$$S = \begin{pmatrix} * & 0 & * \\ 1 & 1 & 0 \\ * & 0 & * \end{pmatrix} \quad \text{and} \quad \overline{S} = \begin{pmatrix} * & 0 & * \\ * & 1 & 0 \\ * & 1 & 0 \end{pmatrix}.$$

We first show that $\overline{S}$ in K_n will produce an S-configuration in K_{n+1} with probability at least $p^2(1-p)$. Indeed the last two columns of $\overline{S}$ give rise to the configuration

$$C_0 = \begin{pmatrix} 0 & 0 & * & * \\ 0 & 0 & * & * \\ * & 0 & 0 & 0 \\ 1 & 1 & 0 & 0 \\ * & 0 & 0 & 0 \\ * & * & 0 & 0 \end{pmatrix}$$

in K_{n+1} with probability at least $p^2(1-p)$, and we recognize S, or rather an instance of S with a right column of zeroes, in C_0. The proof of the claim will be finished if we show that S (in K_n) produces an isolated 4-block in K_{n+3} with positive probability. To start with, S will produce with probability at least $p^3(1-p)^2$ configuration C_1 in K_{n+1}, where

$$C_1 = \begin{pmatrix} * & * & 0 & 0 & * & * \\ * & 0 & 0 & 0 & * & * \\ * & 1 & 1 & 0 & 0 & 0 \\ * & 1 & 0 & 0 & 0 & 0 \\ * & * & 0 & 0 & * & * \\ * & * & 0 & 0 & * & * \end{pmatrix}$$

In C_1 we find the configuration $\begin{pmatrix} 0 & 0 \\ 1 & 1 \\ 1 & 0 \end{pmatrix}$, which produces with probability at least $p^2(1-p)^3$ configuration C_2 in K_{n+2}, where

$$C_2 = \begin{pmatrix} 0 & 0 & 0 & 0 \\ 0 & 0 & 0 & 0 \\ * & 0 & 0 & * \\ * & 1 & 1 & * \\ * & 0 & 0 & 0 \\ * & * & 0 & 0 \end{pmatrix}$$

Finally the central 3×2 configuration in C_2 produces with probability at least $p^4(1-p)^4$ the configuration

$$C_3 = \begin{pmatrix} 0 & 0 & 0 & 0 \\ 0 & 0 & 0 & 0 \\ 0 & 1 & 1 & 0 \\ 0 & 1 & 1 & 0 \\ 0 & 0 & 0 & 0 \\ 0 & 0 & 0 & 0 \end{pmatrix}$$

in K_{n+3}, and C_3 contains an isolated 4-block. Clearly, this 4-block is not generated (in the way of Lemma 1.1) from the 4-block contained in the full cluster C in K_n we started with. Hence we will have *two* full clusters in K_{n+4} with probability at least $p^{11}(1-p)^{10}$. □

3. Bottleneck property

As remarked in the introduction each K_∞ will contain the line segment $\{\frac{1}{2}\} \times [0,1]$ if the initial condition equals J_2, which implies that vertical percolation is always possible. The following result shows that there is a $p_b \approx 0.29$ such that for $p < p_b$ percolation is 'infinitely slow' with positive probability, while for $p \geq p_b$ percolation is 'easy' almost surely. We define the horizontal line segments L_a by $L_a = [0,1] \times \{a\}$ for $0 \leq a \leq 1$.

Proposition 3.1. *Let J_2 be the initial condition. For $p < 1 - \frac{1}{2}\sqrt{2}$ there is a positive probability that $K_\infty \cap L_a = \{(\frac{1}{2}, a)\}$ for infinitely many $a \in [0,1]$. For $p \geq 1 - \frac{1}{2}\sqrt{2}$ the intersection $K_\infty \cap L_a$ will contain for all $a \in [0,1]$ a non-empty interval a.s.* .

Proof: All the squares $[\frac{1}{2} - \frac{1}{2^n}, \frac{1}{2}] \times I_n(j)$ and $[\frac{1}{2}, \frac{1}{2} + \frac{1}{2^n}] \times I_n(j), j = 0, \ldots, 2^n - 1$ in the two columns next to $x = \frac{1}{2}$ are black. Let Z_n be the number of pairs of squares

$$([\frac{1}{2} - \frac{2}{2^n}, \frac{1}{2} - \frac{1}{2^n}] \times I_n(j), [\frac{1}{2} + \frac{1}{2^n}, \frac{1}{2} + \frac{2}{2^n}] \times I_n(j))$$

$j = 0, \ldots, 2^n - 1$ in the next two columns which are *both* white. For example, if K_2 is represented by the matrix

$$\begin{pmatrix} * & * & 1 & 1 & 1 & 1 & * & * \\ * & * & 0 & 1 & 1 & 1 & * & * \\ * & * & 0 & 1 & 1 & 0 & * & * \\ * & * & 1 & 1 & 1 & 0 & * & * \\ * & * & 0 & 1 & 1 & 0 & * & * \\ * & * & 0 & 1 & 1 & 0 & * & * \\ * & * & 1 & 1 & 1 & 1 & * & * \\ * & * & 0 & 1 & 1 & 1 & * & * \end{pmatrix}$$

then $Z_2 = 3$. Since the middle two columns consist solely of black squares, a moment of thought learns that (Z_n) is a branching process whose offspring distribution is binomially distributed with parameters 2 and $(1-p)^2$ (here we also use the observation that the evolutions left and right of $x = \frac{1}{2}$ occur independently). Since the mean of Z_1 equals $2(1-p)^2$, this branching process is supercritical if and only if $p < 1 - \frac{1}{2}\sqrt{2}$. Each element of Z_n corresponds to a row in K_n which has only two black squares (the two central squares). Moreover, such rows are nested as n increases. Thus there are infinitely many lines L_a intersecting K_∞ merely at $\{(\frac{1}{2}, a)\}$ on $Z_n \to \infty$, which event has positive probability if $p < 1 - \frac{1}{2}\sqrt{2}$.

On the other hand, if (Z_n) dies out at time N, then there will be 4-blocks all along $x = \frac{1}{2}$ (either at the leftside on the rightside) in K_{N+1}. As in the proof of Lemma 1.2 this will imply that $K_\infty \cap L_a$ will contain a non-empty interval for all $a \in [0,1]$ almost surely.

4. Horizontal percolation

Let the initial condition be J_2, then vertical percolation is certain, c.f. the previous subsection. However, almost surely horizontal percolation, i.e., the presence of a connected subset in K_∞ which intersects both $x = 0$ and $x = 1$, will not happen for p small enough. This will in fact not occur if $p \leq \frac{1}{2}$. To see this, consider the number of black squares of K_n in the first column. These can generate 1 or 2 black squares in K_{n+1}, each with probability at most p. So this number is bounded from above by a branching process with offspring distribution with mean $2p$, which dies out almost surely if $p \leq \frac{1}{2}$.

Proposition 4.1. *Let the initial condition be J_2. For all $p > .902$ the random set K_∞ percolates horizontally with positive probability.*

Proof: We will actually show that K_∞ contains a.s. infinitely many line segments of the type $[0, \frac{1}{2}] \times \{a\}$ for such p. By symmetry, the same will be true (independently) on the left side of $\{\frac{1}{2}\} \times [0,1]$, and percolation occurs via a path consisting of three line segments (recall that $\{\frac{1}{2}\} \times [0,1]$ is always a subset of K_∞).

An *m-band* in K_n is a configuration of the type

$$\begin{pmatrix} 1 & 1 & . & . & . & . & . & 1 & 1 \\ 1 & 1 & . & . & . & . & . & 1 & 1 \\ . & . & & & & & & . & . \\ . & . & & & & & & . & . \\ 1 & 1 & . & . & . & . & . & 1 & 1 \\ 1 & 1 & . & . & . & . & . & 1 & 1 \end{pmatrix}$$

of size $m \times 2^{n-1}$, occurring between $x = 0$ and $x = \frac{1}{2}$ in K_n. Now note that a 3-band in K_n generates a 3-band in K_{n+1} or 4-band in K_{n+1} with probability $2p^3(1-p)$, respectively p^4 (there is randomness in the first column only). The 4-band in K_{n+1} generates two 3-bands in K_{n+2} with probability at least p^6 or (this takes some counting) one 3-band in K_{n+2} with probability at least $3p^3(1-p) + p^3 - p^6$. It follows that the number of 3-bands in K_{2n} is bounded from below by a branching process whose offspring distribution has mean

$$\begin{aligned} EZ_1 &= p^4(2p^6 + 4p^3 - 3p^4 - p^6) + [2p^3(1-p)]^2 \\ &= p^{10} + p^8 - 4p^7 + 4p^6. \end{aligned}$$

Since $EZ_1 > 1$ for $p = .902$, the number of 3-bands in K_{2n} will tend to infinity with positive probability for all $p \geq .902$. This implies the claim at the beginning of the proof. □

References

1. Chayes, J.T., Chayes, L., Durrett, R. (1988): Connectivity properties of Mandelbrot's percolation process. *Probab. Theory Rel. Fields 77*, 307–324.
2. Dekking, F.M. (1995): Quadtrees, Mandelbrot Percolation and the Modelling of Random Sets. *Fractals 3(3)*, 777–784.
3. Dekking, F.M., Kraaikamp, C., Schouten, J.G. (1996): Binary images and inhomogeneous tree-indexed Markov chains. *Reports of the Faculty of Technical Mathematics and Informatics no. 96-144*, TU Delft.
4. Mandelbrot, B. (1974): Intermittent Turbulence in Self-similar Cascades: Divergence of High Moments and Dimension of the Carrier. *J. Fluid Mech. 62*, 331–358.

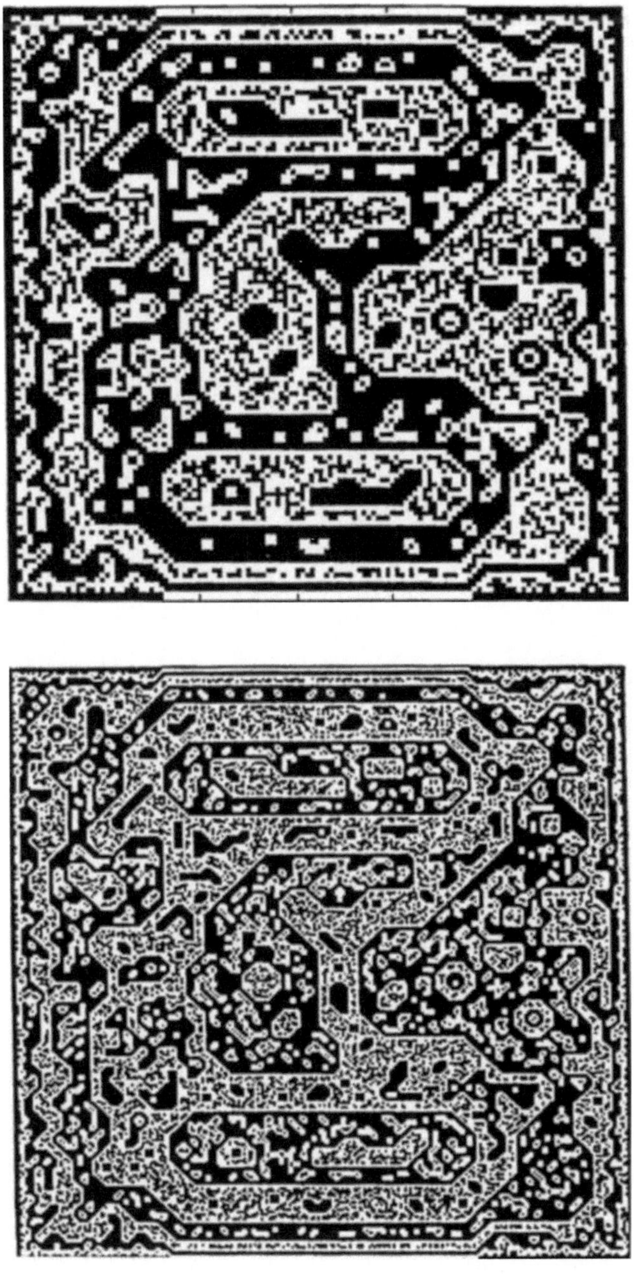

Fig. 4.1. A realization of K_7 and K_8 for an example of a non-monotone neighbour interaction rule.

5. Peres, Y. (1996): Intersection-Equivalence of Brownian Paths and Certain Branching Processes. *Comm. Math. Phys. 177*, 417–434.
6. Siebesma, A.P., Tremblay, R.R., Erzan, A., Pietronero, L. (1989): Multifractal cascades with interactions. *Physica A 156*, no. 2, 613–627.

Identification of Elliptic Gaussian Random Processes

A. Benassi[1], S. Cohen[2], J. Istas[3], and S. Jaffard[4]

[1] Université Blaise Pascal, U.R.A. 1501 C.N.R.S,
Les Cézeaux, 63170 Aubières, FRANCE
[2] Département de Mathématiques, Université de Versailles St Quentin en Y.
45, Avenue des Etats Unis, 78035 Versailles, France,
and C.E.R.M.I.C.S., E.N.P.C., Noisy le Grand, FRANCE
[3] Laboratoire de Biométrie, I.N.R.A., Domaine de Vilvert,
78350 Jouy-en-Josas, FRANCE
[4] Département de Mathématiques, Université Paris 12,
61 Avenue du Général de Gaulle, 94010 Créteil Cedex, FRANCE,
and C.M.L.A., E.N.S. Cachan, FRANCE

Abstract We consider the identification of the relevant parameters of the symbol of the pseudo-differential operator defining an Elliptic Gaussian Random Process. This identification is done by using the observation of the discretization of one sample path of the process. Constructive estimators based on generalization of quadratic variations are given. Almost sure convergence of these estimators is proved.

Keywords: Gaussian processes, pseudo-differential operators, identification.

1. Introduction

Elliptic Gaussian Random Processes (in short EGRP) are generalizations of the Fractional Brownian Motions introduced by [7]. Their purpose was to propose a model that would fit the statistical characteristics of several random signals such as the level of water flows, economic series and some noises. Since then, Fractional Brownian Motions have also been used as a model for several natural phenomena, including for instance natural image textures ([10]). Fractional Brownian Motion of order α $(0 < \alpha < 1)$ are examples of Gaussian process whose spectral density takes the form $\lambda^{1+2\alpha}$, but one of the limitations of such stationary increments processes is that they cannot model local modulations of the parameters: stationarity implies uniformity. We model spatial changes of the parameters by introducing a dependency in t on $\sigma(t,\xi)$ the symbol of the pseudo-differential operator associated with the EGRP. This is the point of view of [1], where the local properties of EGRP are studied.

The aim of this paper is to identify the relevant parameters of the symbol of the EGRP from the discrete observation of one sample path. Let us briefly introduce the EGRP. Let X_t be a centered Gaussian process. Two isomorphic Hilbert spaces are associated with X: the space $\mathcal{H}$ defined as the closure of the random variables $Z = \sum_i \alpha_i X(t_i)$ for the scalar product $(Z/T) = \mathbb{E}(ZT)$ and the Reproducing Kernel Hilbert Space H composed of the functions which can be written $f_Z(t) = \mathbb{E}(X_t Z)$, with $Z \in \mathcal{H}$. The scalar product in H is $(f_Z, f_Y)_H = \mathbb{E}(ZY)$. For EGRP H is continuously embedded in L^2 and we can define a self-adjoint operator A, by $(f,g)_H =< A(f)/g >_{L^2}$.

Following [1], we call EGRP special instances of centered Gaussian processes which Reproducing Kernel Hilbert Space are topologically equivalent to a Sobolev space H^s. This can be interpreted as an ellipticity assumption on the operator A.

We specify slightly more the setting by requiring A to be a pseudo-differential operator associated with a symbol $\sigma(t,\xi)$

$$A(f)(t) = \int_{\mathbf{R}} e^{it\xi}\sigma(t,\xi)\hat{f}(\xi)d\xi ,$$

where $\hat{f}$ is the Fourier transform of f. The properties of X are driven by the associated symbol σ. For instance, X has the Markov field property iff A is differential; X has stationary increments iff the symbol $\sigma(t,\xi)$ does not depend on t and Fractional Brownian Motion of order α are obtained for $\sigma(t,\xi) = |\xi|^{1+2\alpha}$.

All the informations on X are contained in the operator A and therefore in the symbol $\sigma(t,\xi)$. As pointed out by [1], the local properties of an EGRP are contained in the principal part of the symbol. Therefore we assume in this paper the following expansion, for $|\xi|$ large enough

$$\sigma(t,\xi) = a(t)|\xi|^{1+2\alpha} + b(t)|\xi|^{1+2\beta}. \tag{1.1}$$

The asymptotical expansion (1.1) implies in particular that the corresponding EGRP behaves locally like an α-fractional Brownian motion: EGRP are locally self-similar. We focus in this paper on the non-differentiable case (i.e. $0 < \alpha < 1$) but our results can be easily extended to the differentiable case.

The process X_t is observed at sampling points $k\Delta, k = 0,\dots,N$ with $N\Delta = 1$. The main result of this paper is to give strongly consistent estimators of parameters α and β and of the function $a(t)$ as $N \to \infty$. The estimation of $b(t)$ could be done in the same way, but this involves then a lot of technical complications. Therefore we only briefly mention how to estimate function $b(t)$.

The paper is organized as follows.

First we recall the relevant properties of EGRP. We introduce the discrete quadratic functionals of the sample path of process X that are needed for the estimation. We state the main result. For the proof we first recall the identification results available for the *Filtered White Noises* (in short FWN), a class of Gaussian processes defined through stochastic integrals. We prove an equivalence result between FWN and EGRP. This equivalence result imply identification results for EGRP.

2. Preliminaries

2.1 A short review on EGRP

In this section we sketch an account of the relevant properties of EGRP. Let us suppose $0 < \alpha < 1$. Let $\sigma(t,\xi)$ be the symbol of a pseudo-differential operator

$$A(f)(t) = \int_{\mathbf{R}} e^{it\xi}\sigma(t,\xi)\widehat{f}(\xi)d\xi ,$$

satisfying

A 2.1. (α)
– $\exists R > 0$, $\forall t \in \mathbb{R}$, $\forall |\xi| > R$, for $i = 0$ to 2

$$\left| \frac{\partial^i \sigma(t,\xi)}{\partial \xi^i} \right| \leq C_i (1 + |\xi|)^{2\alpha+1-i}$$

– $\exists \epsilon > \epsilon'$ such that

$$|\sigma(s,\xi) - \sigma(t,\xi)| \leq (1 + |\xi|)^{2\alpha+1+\epsilon'} |s-t|^{\epsilon}$$

–

$$C(1 + |\xi|)^{2\alpha+1} \leq \sigma(t,\xi) \leq \frac{1}{C}(1 + |\xi|)^{2\alpha+1}. \tag{2.1}$$

Assumption **A 2.1**(α) corresponds to assumption $HA(2\alpha + 1)$ in [1]. Under assumption **A 2.1**(α), [1] show that an Hilbert space H_A is associated with the pseudo-differential operator $A(f)(t) = \int_{\mathbb{R}} e^{it\xi} \sigma(t,\xi) \widehat{f}(\xi) d\xi$ where $f \in L^2(\mathbb{R})$, by defining $< f,g >_{H_A} = < Af, g >_{L^2}$. Since H_A is a Reproducing Kernel Hilbert Space, all the properties of H_A can be translated in properties of a Gaussian Process X called an *Elliptic Gaussian Random Process.*

An orthonormal wavelet basis Φ_λ of H_A indexed by the dyadic intervals is exhibited, with the help of $A^{-1/2}$, which is defined as follows: (ψ_λ) is the orthonormal basis of L^2 defined by Lemarié and Meyer (e.g. [8]), $\psi_\lambda(x) = 2^{j/2}\psi(2^j x - k)$ where λ stands for the couple $(j,k) \in \mathbb{Z}^2$: $\Phi_\lambda = A^{-1/2}(\psi_\lambda)$. Using the canonical isomorphism between $\mathcal{H}$ the Gaussian space of X and H_A we get

$$X = \sum_\lambda \xi_\lambda \Phi_\lambda \text{ a.s. and in } L^2 ,$$

where the ξ_λ are independent normalized centered Gaussian variables.

The aim of this paper is to identify the relevant parameters of the symbol defining an EGRP. Therefore, we assume the following asymptotic expansion as $|\xi| \to \infty$

A 2.2.

$$\sigma(t,\xi) = a(t)|\xi|^{1+2\alpha} + b(t)|\xi|^{1+2\beta} + p(\xi) ,$$

where $0 < \beta < \alpha < 1$; a, b are C^1 strictly positive functions on $[0,1]$, and p is a C^∞ function such that p is identically 1 on $|\xi| \leq 1$ and is vanishing if $|\xi| > 2$.

In assumption **A** 2.2, p is supposed to satisfy assumption **A 2.1**(α), but we shall understand that only the asymptotic behavior of σ when $|\xi| \to \infty$ is important for the identification.

2.2 Generalizations of quadratic variations

We want to identify the relevant parameters of σ using a discrete regularly spaced sampling of X_t on $[0,1]$. This identification is performed with the help of some quadratic functionals QF_N and $\widetilde{QF_N}$ of the discretized sample path. These quadratic functionals are in some sense generalizations of classical quadratic variations. Let Δ be the discretization step and let N be the number of observations on the time interval $[0,1]$, so that $N\Delta = 1$. For the identification of parameters we fix a real matrix $D = (d_{k,k'})$, $k, k' = 0, \dots, K$ such that

A 2.3.

$$d_{k,k'} = d_k \, d_{k'} \quad \sum_{k=0}^{K} d_k = 0 \quad \sum_{k=0}^{K} k d_k = 0\,.$$

We consider the related quadratic variations with step $\frac{1}{N}$

$$\begin{aligned} QF_N(\phi) &= \sum_{p=0}^{N-K} \sum_{k,k'=0}^{K} d_{k,k'} \phi\left(\frac{k+p}{N}\right) X_{\frac{k+p}{N}} X_{\frac{k'+p}{N}}\,, \\ \widetilde{QF_N} &= QF_{N/2} - \frac{QF_{N^2/2}}{QF_{N^2}} QF_N \end{aligned} \tag{2.2}$$

where ϕ is a non-random function that will be useful for estimating function a. We denote by QF_N the functional $QF_N(\phi)$ when ϕ is identically 1. The matrix $D = (d_{k,k'})$ acts like a quadratic difference operator on the sample path of (X_t). For instance the discrete classical quadratic variation of X

$$V_N = \sum_{p=0}^{N-1} (X_{\frac{p+1}{N}} - X_{\frac{p}{N}})^2 \tag{2.3}$$

is a special example of QF_N with

$$K = 1 \quad d_0 = -1, \;\; d_1 = 1.$$

V_N does not satisfy **A 2.3** and we use more general variation.

For instance,

$$K = 2 \quad d_0 = 1, \;\; d_1 = -2, \;\; d_2 = 1 \tag{2.4}$$

is a discrete second derivative satisfying **A 2.3**.

3. Main result

We can now state the main theorem.

Theorem 3.1. *Assume that X is an EGRP with symbol σ satisfying* **A 2.1–A 2.3**. *Furthermore assume* $\sup(0, \frac{3\alpha-1}{2}) < \beta < \alpha < 1$. *Then the estimators*

$$\widehat{\alpha}_N = \frac{1 + log_2 \frac{QF_{N/2}}{QF_N}}{2}\,, \tag{3.1}$$

$$\widehat{I}_N(\phi) = \frac{N^{2\widehat{\alpha}_N - 1} QF_N(\phi)}{F_{2\widehat{\alpha}_N}} \tag{3.2}$$

with $\phi \in C^1[0,1]$, $supp(\phi) \subset]0,1[$, *and*

$$\widehat{\beta}_N = \frac{1}{2} + \frac{1}{2} log_2 \frac{QF_{N^2/2}}{QF_{N^2}} - log_2 \frac{\widetilde{QF_N/2}}{\widetilde{QF_N}} \tag{3.3}$$

converge respectively to α, $I(\phi) = \int_0^1 a^2(t)\phi(t)dt$ *and* β *almost surely as* $N \to \infty$.

One can similarly construct an estimator of $b(t)$ through $J(\phi) = \int_0^1 a(t)b(t)\phi(t)dt$.
This estimator then involves a pointwise estimator of $a(t)$ (constructed in [5]) that we however will not construct in this paper.

4. Proofs

The main idea of the proof is to give a link between EGRP and processes defined through stochastic integrals called *Filtered White Noise* (in short FWN). Identification results on FWN are available (cf. [3]). A short review of these results is given in the following section. Identification results for EGRP are then deduced through the link between EGRP and FWN.

4.1 Identification results for FWN

We introduce the so called *Filtered White Noise* defined as follows

$$Y_t = \int_{\mathbb{R}} g(t,\xi)(e^{it\xi} - 1)dW(\xi) . \tag{4.1}$$

We suppose that, as $|\xi| \to +\infty$, function g has the following asymptotical expansion

$$g(t,\xi) = \frac{a(t)}{|\xi|^{\frac{1}{2}+\alpha}} + \frac{b(t)}{|\xi|^{\frac{1}{2}+\beta}} + R(t,\xi) \quad (\alpha < \beta) , \tag{4.2}$$

where the remainder term $R(t,\xi)$ should be negligible when compared with the first terms of the expansion. The properties of the process Y defined by (4.1) will be strongly related with the properties of the pseudo-differential operator

$$A(f)(t) = \int_{\mathbb{R}} e^{it\xi} g(t,\xi)\widehat{f}(\xi)d\xi . \tag{4.3}$$

First we present the assumptions needed on the expansion of $g(t,\xi)$ given by formula (4.2) to perform the identification of (α, a, β).

A 4.1. $1 > \beta > \alpha > 0$ and $R(t,\xi) \in C^{1,2}([0,1] \times \mathbb{R})$ is a function such that,

$$\left| \frac{\partial^{m+n}}{\partial t^m \partial \xi^n} R(t,\xi) \right| \leq \frac{C}{|\xi|^{\frac{1}{2}+\eta+n}} ,$$

for $m = 0, 1$ and $n = 0, 1, 2$ with $\eta > \beta$. C is a generic constant and may change from one occurrence to another.

A 4.2. $a, b \in C^1([0,1])$ and $\forall t \in [0,1], \quad a(t)b(t) \neq 0$.

The following identification Theorem for FWN can be found in [3].

Theorem 4.1. *Assume A 2.3, A 4.1 and A 4.2. Then the estimators*

$$\widehat{\alpha}_N = \frac{1 + log_2 \frac{QF_{N/2}}{QF_N}}{2} , \tag{4.4}$$

$$\widehat{I}_N(\phi) = \frac{N^{2\widehat{\alpha}_N - 1} QF_N(\phi)}{F_{\widehat{2\alpha}_N}}, \tag{4.5}$$

$$\widehat{\beta}_N = \frac{1}{2} - \frac{1}{2} log_2 \frac{QF_{N^2/2}}{QF_{N^2}} + log_2 \frac{\widetilde{QF_N/2}}{\widetilde{QF_N}} \tag{4.6}$$

with $\phi \in C^1[0,1]$ and $supp(\phi) \subset]0,1[$, converge respectively to α, $I(\phi) = \int_0^1 a^2(t)\phi(t)dt$ and β almost surely as $N \to \infty$.

4.2 Links between EGRP and FWN

Define the pseudo-differential operator L by

$$L(f)(t) = \int_{\mathbb{R}} \frac{e^{it\xi}}{\sqrt{\sigma(t,\xi)}} \widehat{f}(\xi) d\xi \, .$$

Define process Y by $Y_t = \sum_{\lambda \in \mathbb{Z}^2} \xi_\lambda L(\psi_\lambda)(t)$ where ψ_λ is the Lemarié-Meyer's basis defined in Section 2.1. We point out that Y is indeed a FWN defined by

$$Y_t = \int_{\mathbb{R}} \frac{e^{it\xi}}{\sqrt{\sigma(t,\xi)}} dW(\xi) \, ,$$

where $dW(\xi)$ is the Brownian measure such that

$$\forall f \in L^2(\mathbb{R}) \quad \int_{\mathbb{R}} f(\xi) dW(\xi) = \sum_\lambda \xi_\lambda < f, \psi_\lambda >_{L^2} .$$

Our first claim is that EGRP X and FWN Y are the same up to a more regular process.

Proposition 4.1. *Under assumption A 2.1(α) $X - Y$ belongs to C^γ almost surely for each $\gamma < \alpha + 1/2$.*

Proof of Proposition 4.1

In order to prove that $X - Y$ is a smooth process we first check that

$$L(\psi_\lambda) - A^{-1/2}(\psi_\lambda) = 2^{-(\alpha+1)j} \phi_\lambda \, , \tag{4.7}$$

where ϕ_λ are "vaguelettes" ([9]): this means that the ϕ_λ have the same smoothness and localization as an ordinary orthonormal wavelet basis. This is similar to Theorem 1.2 of [1] except that our conclusion is more precise. In [1], the key Lemma is Proposition 2.4 which asserts a kind of symbolic calculus for the algebras of matrices $\mathcal{M}^\gamma$. This proposition can be sharpened by extending this symbolic calculus to the algebras of matrices $\mathcal{M}^{\gamma,\gamma'}$ defined in [4].

The proof of this result basically follows the proof of Proposition 2.4. in [1] and the proof of the Proposition 4.1 is now standard. Let us just sketch it. (4.7) implies that

$$X_t - Y_t = \sum_\lambda \xi_\lambda 2^{-(\alpha+1)j} \phi_\lambda$$

where ϕ_λ are "vaguelettes". We estimate separately the regularity of each frequency. Let

$$D_j(t) = \sum_k \xi_\lambda 2^{-(\alpha+1)j} \phi_\lambda ,$$

then for $n = 0,\ 1,\ 2$

$$\begin{aligned} |D_j^{(n)}(t)| &\le \sum_k |\xi_\lambda| \frac{C2^{-(\alpha-n+1/2)j}}{(1+|2^j t - k|)^2} \\ &\le Cj2^{-(\alpha-n+1/2)j} . \end{aligned} \tag{4.8}$$

Let j_0 be such that

$$2^{-j_0} \le |t - t_0| \le 2.2^{-j_0} .$$

If $0 < \alpha < 1/2$, using (4.8) with $n = 1$ and the mean value theorem

$$\begin{aligned} \sum_{j=-\infty}^{j_0} |D_j(t) - D_j(t_0)| &\le \sum_{j=-\infty}^{j_0} Cj|t-t_0|2^{(1/2-\alpha)j} \\ &\le Cj_0|t-t_0|2^{(1/2-\alpha)j_0} \\ &\le C|t-t_0|^{(\alpha+1/2)} Log(|t-t_0|) , \end{aligned}$$

and using (4.8) with $n = 0$

$$\begin{aligned} \sum_{j=j_0}^{\infty} |D_j(t) - D_j(t_0)| &\le \sum_{j=j_0}^{\infty} Cj2^{-(\alpha+1/2)j} \\ &\le Cj_0 2^{-(\alpha+1/2)j_0} \\ &\le C|t-t_0|^{(\alpha+1/2)} Log(|t-t_0|) , \end{aligned}$$

so that $X_t - Y_t$ is a.s. C^γ for each $\gamma < \alpha + 1/2$. If $1/2 \le \alpha < 1$, we estimate $\sum_{j=-\infty}^{j_0} |D_j'(t) - D_j'(t_0)|$ by the same technique, using (4.8) with $n = 1$ and $n = 2$.

In order to use the identification results for FWN when asymptotical expansion (4.2) is available, we define σ_∞ such that

$$\frac{1}{\sqrt{\sigma_\infty(t,\xi)}} = \frac{1}{a^{1/2}(t)} \frac{1}{|\xi|^{1/2+\alpha}} - 1/2 \frac{b(t)}{a^{3/2}(t)} \frac{1}{|\xi|^{1/2+3\alpha-2\beta}} ,$$

is the asymptotic development of order two of $\dfrac{1}{\sqrt{\sigma(t,\xi)}}$ as $|\xi| \to \infty$. If $\alpha - \beta < \frac{1-\alpha}{2}$, $\dfrac{e^{it\xi}-1}{\sqrt{\sigma_\infty(t,\xi)}} \in L^2(\mathbb{R})$ and $X_t^\infty = \displaystyle\int_{\mathbb{R}} \frac{e^{it\xi}-1}{\sqrt{\sigma_\infty(t,\xi)}} dW(\xi)$ is a FWN whose parameters $(\alpha, a^{-1/2}, 3\alpha - 2\beta, -1/2\dfrac{b(t)}{a^{3/2}(t)})$ can be identified by the estimators of Theorem 4.1. Since these estimators do not distinguish processes that differs only by a more regular process, the identification for X will be a consequence of the identification of $X^\infty$and of the following proposition.

Proposition 4.2. *$X - X^\infty$ is almost C^γ with $\alpha + 1/2 > \gamma > 3\alpha - 2\beta$.*

Proof

Because of Proposition 4.1 we can replace X_t by

$$Y_t = \int_{\mathbf{R}} \frac{e^{it\xi} - 1}{\sqrt{\sigma(t,\xi)}} dW(\xi)$$

its FWN counterpart. Then we define $\tilde{\sigma}_\infty(t,\xi)$ by

$$\frac{1}{\sqrt{\tilde{\sigma}_\infty(t,\xi)}} = \frac{1}{a^{1/2}(t)} \frac{1}{|\xi|^{1/2+\alpha} + p(\xi)} - 1/2 \frac{b(t)}{a^{3/2}(t)} \frac{1}{|\xi|^{1/2+3\alpha-2\beta} + p(\xi)}$$

which is similar to σ_∞. But $\frac{1}{\sqrt{\tilde{\sigma}_\infty(t,\xi)}}$ always belongs to L^2. We let

$$Y_t^\infty = \int_{\mathbf{R}} \frac{e^{it\xi} - 1}{\sqrt{\tilde{\sigma}_\infty(t,\xi)}} dW(\xi).$$

$$Y_t - Y_t^\infty = \sum_\lambda \xi_\lambda [\int_{\mathbf{R}} e^{it\xi} (\frac{1}{\sqrt{\sigma(t,\xi)}} - \frac{1}{\sqrt{\tilde{\sigma}_\infty(t,\xi)}}) \bar{\hat{\psi}}_\lambda(\xi) d\xi].$$

In order to prove that $Y - Y^\infty$ is C^γ with $\gamma > 3\alpha - 2\beta$ we first prove that

$$\tilde{\psi}_\lambda(t) = 2^{j(1/2+\gamma)} \int_{\mathbf{R}} e^{it\xi} \omega(t,\xi) \hat{\psi}_\lambda(\xi) d\xi$$

are "vaguelettes" (cf [9]) where $\omega(t,\xi) = \frac{1}{\sqrt{\sigma(t,\xi)}} - \frac{1}{\sqrt{\tilde{\sigma}_\infty(t,\xi)}}$. This is a direct consequence of the following estimates on $\omega(t,\xi)$:

$$\left| \frac{\partial^{l+m}}{\partial t^l \partial \xi^m} \omega(t,\xi) \right| \leq \frac{C}{|\xi|^{1/2+\gamma+m-l}} \quad \text{if } |l| \leq 1 \text{ and } \quad m \leq 1 .$$

We can thus write

$$Y_t - Y_t^\infty = \sum_\lambda \xi_\lambda 2^{1/2+\gamma} \tilde{\psi} .$$

We conclude that $Y - Y^\infty$ is C^γ with $\gamma > 3\alpha - 2\beta$ as in the proof of Proposition 4.1.

We are left to consider

$$\begin{aligned} X_t^\infty - Y_t^\infty &= \frac{1}{a^{1/2}(t)} \int [\frac{e^{it\xi}}{|\xi|^{1/2+\alpha} + p(\xi)} - \frac{e^{it\xi} - 1}{|\xi|^{1/2+\alpha}}] W(d\xi) \\ &- 1/2 \frac{b(t)}{a^{3/2}(t)}) \int_{\mathbf{R}} [\frac{e^{it\xi}}{|\xi|^{1/2+3\alpha-2\beta} + p(\xi)} - \frac{e^{it\xi} - 1}{|\xi|^{1/2+3\alpha-2\beta}}] W(d\xi) . \end{aligned} \tag{4.9}$$

We conclude by claiming that $X^\infty - Y^\infty$ have the same regularity as the two FWN in the right member of (4.9) which can be shown to be C^γ with the "vaguelette" tools.

As a corollary of this Proposition, we deduce the

Proof of Theorem 3.1

Clearly one has

$$N^{2\gamma'-1} QF_N(Z) \to 0 \text{ a.s.}$$

for each processes Z belonging to C^γ with $\gamma > \gamma'$. Using Cauchy-Schwartz inequality this result can be extended to weighted quadratic functionals $QF_N(\phi)$. Then it is clear that the estimators $\widehat{\alpha}_N$, $\widehat{I}_N(\phi)$ and $\widehat{\beta}_N$ are converging to the same limit applied on a sample path of X or of X^∞ because of Proposition 4.2.

References

1. Benassi, A., Jaffard, S., Roux, D. (1994a): Gaussian processes and Pseudodifferential Elliptic operators. To appear in *Revista Mathematica Iberoamericana.*
2. Benassi, A., Cohen, S., Jaffard, S. (1994b): Identification de processus gaussiens elliptiques. *Cr. Acad. Sc. Paris, Série I.* **319**, 877–880.
3. Benassi, A., Cohen, S., Istas, J., Jaffard, S. (Preprint 1996): Identification of Filtered White Noises.
4. Ezzine, A. (1996): Méthodes d'ondelettes pour la résolution numérique d'équations intégrales (Thèse de l'Ecole Nationale des Ponts et Chaussées). To appear.
5. Istas, J. (1996): Estimating the singularity function of a Gaussian process with applications. *Scand. J. Statist. Vol.23*, **5**, 1996, 581–596.
6. Istas, J., Lang, G. (1997): Quadratic variations and estimation of the local Hölder index of a Gaussian process. *Ann. Inst. H. Poincaré.* To appear.
7. Mandelbrot, B., Van Ness, J. (1968): Fractional Brownian Motions, Fractional Noises and Applications. *SIAM Review.* **10**, 422–437.
8. Meyer, Y. (1990): *Ondelettes et Opérateurs.* volume 1. Hermann, Paris.
9. Meyer, Y. (1990): *Ondelettes et Opérateurs.* volume 2. Hermann, Paris.
10. Pentland, A. (1984): *Fractal-based description of natural scenes.* IEEE Trans., PAMI, **6**, 661–674.

Modeling and Estimation of Fractional Brownian Motion using Multiresolution Stochastic Processes

M.M. Daniel and A.S. Willsky

Laboratory for Information and Decision Systems at the Massachusetts Institute of Technology, Room 35-425, Cambridge, MA, 02139, USA
cafepre@mit.edu

Abstract In this paper, we present a comprehensive framework for synthesizing, processing, and analyzing fractional Brownian motion (fBm). fBm is modeled using the class of multiresolution processes introduced in [2]. The structure of these processes, which are indexed by the nodes on a tree, is naturally suited to the modeling of fBm and other $1/f$ processes. These processes can reproduce and improve upon the random midpoint displacement and the wavelet-based approximations of fBm. Furthermore, these multiresolution processes are also accompanied by a general set of algorithms for estimation and likelihood calculation. While wavelet-based models for fBm assume dense measurements of the finest-scale process and a constant variance for the measurement noise, the multiscale algorithms make no such assumptions. [1]

1. Introduction

A wide variety of physical phenomena are described by random processes whose power spectral densities behave as $1/f^{\alpha}$. For example, $1/f$ processes can be used to describe average temperature distributions [7, 10], annual flow rates in rivers [7], the noise in vacuum tubes and electrical components [10], biological time series like heartbeats [17], economic time series like the Dow Jones Industrial Average [17], and traffic in communications networks [16]. Processes with $1/f$-like power spectra are also used to generate images that model real world objects like clouds and mountain ranges [1, 14]. These processes possess two common characteristics, statistical self-similarity and long-range dependence. A popular model that possesses these two characteristics is the class of fractional Brownian motions [12], which are a generalization of Brownian motion. A fractional Brownian motion (fBm) is a Gaussian process $x(t)$ with zero mean and covariance [12]

$$E\big[x(t)x(s)\big] = \frac{\sigma^2}{2}\big[|t|^{2H} + |s|^{2H} - |t-s|^{2H}\big], \qquad 0 < H < 1. \tag{1.1}$$

This class of processes is completely characterized by $E\big[x(1)^2\big] = \sigma^2$ and the Hurst exponent H.

The problem with fBm, due in part to the non-stationarity, is that it does not lend itself naturally to traditional methods of signal synthesis, signal estimation,

[1] This research was supported by the Office of Naval Research under grant N00014-91-J-1004 and the Air Force Office of Sponsored Research both under grant F49620-95-1-0083 and through Boston University under subcontract GC123919NGN.

and parameter estimation. Two common methods for approximately synthesizing fBm are random midpoint displacement and wavelet synthesis. The midpoint displacement algorithm provides very accurate syntheses for large values of H, but it is not a particularly useful tool for analyzing and estimating fBm. Another method for approximately synthesizing fBm is based on the observation that the wavelet transform approximately whitens fBm [6, 15], as long as the wavelet function is sufficiently regular relative to H. In addition, Wornell [17] also used the wavelet-based framework for accomplishing signal processing tasks like the estimation of fBm from dense and regularly spaced samples of the finest-scale process. However, common processing tasks like the interpolation of fBm from sparse or irregularly spaced samples of $x(t)$ is difficult to accomplish in this framework, especially if the noise process is non-stationary. In this paper, we provide a comprehensive framework for the synthesis, analysis, and estimation of fBm using the class of multiresolution stochastic processes first introduced in [2]. The application of the multiresolution models of [2] is motivated by three factors. First, statistically self-similar processes like fBm seem naturally suited to multiresolution modeling, e.g., with a wavelet basis. Second, the class of models in [2] has been successfully applied to both the modeling of Brownian motion [11] and the estimation of the Hurst exponent from sample paths of fBm [5]. Third, these multiresolution models are accompanied by an efficient estimator and a likelihood calculator that do not require uniform sampling intervals or uniform variance for the measurement noise, and thus a variety signal processing tasks can be performed.

2. Fractional Brownian Motion

Fractional Brownian motion (fBm) was first proposed in [12] as a class of random processes with strong interdependence between distant samples. These processes are statistically self-similar (SSS) in the sense that

$$x(at) \stackrel{\mathcal{P}}{=} a^H x(t), \qquad a > 0, \tag{2.1}$$

where $\stackrel{\mathcal{P}}{=}$ denotes equality in (finite) distributions. A process $x(t)$ is a fBm with Hurst exponent $0 < H < 1$ *if and only* if $x(t)$ is Gaussian and has zero mean, $x(t)$ has stationary increments, meaning that $E\left[\left(x(t+s)-x(t)\right)^2\right] = g(s)$, and $x(t)$ has self-similar increments, i.e., $g(as) = a^{2H} g(s)$. The function $g(s)$ is called the *structure function*. The self-similarity of the increments implies that the structure function $g(s)$ must have the form $g(s) = \sigma^2 |s|^{2H}$. The covariance function in Eq. (1.1) follows from this structure function and the stationary increments property.

The stationary and self-similar increments properties are important attributes of fBm. However, fBm does not have independent increments, except for the special case when $H = 1/2$. For $H = 1/2$, $x(t)$ is Brownian motion and a Markov process. For $H \neq 1/2$, the significant correlation between increments separated by large distances is what makes the modeling and analysis of fBm such a difficult problem.

A discrete-time version of fBm can be derived from samples of $x(t)$. If $x[n] \stackrel{\Delta}{=} x(n\Delta t)$, then $x[n]$ is a zero-mean Gaussian process with covariance function

$$E\left[x[n]x[m]\right] = \frac{\sigma^2(\Delta t)^{2H}}{2}\left(|n|^{2H} + |m|^{2H} - |n-m|^{2H}\right), \tag{2.2}$$

and the variance of the increments process is $E\left[(x[n+m]-x[n])^2\right] = \sigma^2(\Delta t)^{2H}|m|^{2H}$. Discrete-time fBm can be described in terms of a first-order autoregression

$$x[n] = x[n-1] + w[n], \qquad x[0] = 0, \tag{2.3}$$

where the first-order increments process $w[n]$ is white only for $H = 1/2$.

2.1 Random Midpoint Displacement for Synthesizing Brownian Motion

The random midpoint displacement algorithm synthesizes Brownian motion by generating samples at successively finer sampling intervals [1]. Midpoint displacement follows from the Markovianity of Brownian motion, which holds that, *conditioned* on any two samples $x(t_1)$ and $x(t_2)$ for $t_1 < t_2$, the signal $x(t)$ on the interval $t \in (t_1, t_2)$ is independent of the signal outside this interval. The implication is that if samples of Brownian motion have been generated for times $t_0, t_1, \ldots, t_N$, then the midpoints of the N intervals partitioned by these samples can be generated independently.

To be more precise, assume that we want to synthesize a path of Brownian motion on the unit interval $t \in [0,1]$. The endpoint $x(0)$ is always zero and the endpoint $x(1)$ can be generated from a sample of the distribution[2] $\mathcal{N}(0, \sigma^2)$. The midpoint $x(1/2)$ can be decomposed as[3]

$$x(1/2) = E\big[x(1/2) \,\big|\, x(1), x(0)\big] + \tilde{x}(1/2). \tag{2.4}$$

Because Brownian motion is Gaussian, $E\big[x(1/2) \,\big|\, x(1), x(0)\big]$ is the linear least-squares error (LLSE) estimate of $x(1/2)$ from $x(1)$ and $x(0)$. Also, $\tilde{x}(1/2)$ is the estimation error, which must be independent of $x(1)$ and $x(0)$. Using standard LLSE estimation equations, the elements of Eq. (2.4) follow as

$$E\big[x(1/2) \,\big|\, x(1), x(0)\big] = \frac{1}{2}\, x(1), \qquad \mathrm{var}[\tilde{x}(1/2)] = \sigma^2/4. \tag{2.5}$$

The midpoint $x(1/2)$ can be expressed as an interpolation—$x(1)/2$—and a displacement $\tilde{x}(1/2)$ that can be generated independently of $x(1)$ and $x(0)$.

The next step of the displacement algorithm is to generate the midpoints $x(1/4)$ and $x(3/4)$. These midpoints can be decomposed as

$$x(1/4) = E\big[x(1/4) \,\big|\, x(1/2), x(0)\big] + \tilde{x}(1/4), \tag{2.6}$$

$$x(3/4) = E\big[x(3/4) \,\big|\, x(1), x(1/2)\big] + \tilde{x}(3/4). \tag{2.7}$$

Using both the independent increments and the independence of Bayes' least-squares estimation errors from conditioning information, the displacements $\tilde{x}(1/4)$ and $\tilde{x}(3/4)$ are independent of each other and the samples $x(0)$, $x(1/2)$, and $x(1)$. Again invoking the LLSE estimation formulas, Eqs. (2.6) and (2.7) become

$$\begin{aligned} x(1/4) &= \tfrac{1}{2}\, x(1/2) + \tilde{x}(1/4), & \mathrm{var}[\tilde{x}(1/4)] &= \sigma^2/8, \\ x(3/4) &= \tfrac{1}{2}\big(x(1) + x(1/2)\big) + \tilde{x}(3/4), & \mathrm{var}[\tilde{x}(3/4)] &= \sigma^2/8. \end{aligned}$$

To compute the midpoint displacements at finer intervals, the LLSE estimator can be repeatedly applied. However, we can also make use of the statistical self-similarity of Brownian motion. Self-similarity implies that

---

[2] The notation $x \in \mathcal{N}(m_x, P_x)$ indicates that x is a normal random vector with mean m_x and covariance P_x.

[3] The notation $E\big[x \,|\, y\big]$ denotes the expected value of x conditioned upon y.

$$\begin{aligned} \operatorname{var}\left[x(2^{-(m+1)})\right] &= \left(\frac{1}{2}\right)^2 \operatorname{var}\left[x(2^{-m})\right], \\ E\left[x(2^{-(m+1)})x(2^{-m})\right] &= \left(\frac{1}{2}\right)^2 E\left[x(2^{-m})x(2^{-(m-1)})\right]. \end{aligned}$$

Applying these relationships recursively to Eq. (2.1) yields

$$x\left(2^{-(m+1)}\right) = \frac{1}{2}x(2^{-m}) + \widetilde{x}\left(2^{-(m+1)}\right), \tag{2.8}$$

$$x\left(3/2^{m+1}\right) = \frac{1}{2}\left(x(2^{-(m-1)}) + x(2^{-m})\right) + \widetilde{x}\left(3/2^{m+1}\right), \tag{2.9}$$

$$\operatorname{var}\left[\widetilde{x}\left(2^{-(m+1)}\right)\right] = \operatorname{var}\left[\widetilde{x}\left(3/2^{m+1}\right)\right] = \sigma^2\left(\frac{1}{2}\right)^{m+2} \tag{2.10}$$

Furthermore, the independent increments property of Brownian motion means that the interpolation and displacement at any given scale are shift-invariant, i.e., the interpolation at the midpoint is always the average of the endpoints and the variance of the displacement is constant for a given scale m and decreases geometrically with increasing scale m. The shift-invariance for Brownian motion greatly simplifies the midpoint displacement algorithm. However, as shown in the next section, the shift-invariance does not apply to fBm for $H \neq 1/2$, since fBm for $H \neq 1/2$ does not have independent increments.

2.2 Random Midpoint Displacement for fBm

As shown in [1], random midpoint displacement can be used to generate sample paths of a distribution that is approximately equal to the distribution of fBm. To illustrate this algorithm, and to highlight the approximations made, assume again that we want to synthesize fBm on the interval $t \in [0, 1]$. As for Brownian motion, the first step is to set $x(0) = 0$ and to generate $x(1)$ from a sample of the distribution $\mathcal{N}(0, \sigma^2)$. If the midpoint is again decomposed as in Eq. (2.4), the LLSE estimator equations provide

$$E\left[x(1/2)\,|\,x(1), x(0)\right] = \frac{1}{2}x(1) \quad \text{and} \quad \operatorname{var}\left[\widetilde{x}(1/2)\right] = \sigma^2\left[\left(\frac{1}{2}\right)^{2H} - \left(\frac{1}{2}\right)^2\right],$$

where $\widetilde{x}(1/2)$ is independent of $x(1)$ and $x(0)$. Using the self-similarity of fBm, we can then derive

$$E\left[x(2^{-(m+1)})\,|\,x(2^{-m}), x(0)\right] = \frac{1}{2}x(2^{-m}), \tag{2.11}$$

$$\operatorname{var}\left[\widetilde{x}(2^{-(m+1)})\right] = \sigma^2\left(\frac{1}{2}\right)^{2mH}\left[\left(\frac{1}{2}\right)^{2H} - \left(\frac{1}{2}\right)^2\right]. \tag{2.12}$$

The midpoint displacement algorithm for fBm given in [1] is based upon Equations (2.11)-(2.12), and is directly analogous to the midpoint displacement algorithm for Brownian motion. Assume that the samples $x(i\Delta t_{m-1})$, where $\Delta t_{m-1} = (1/2)^m$, are known for $i = 0, \ldots, 2^m$. The samples at the midpoints of the intervals $[k\Delta t_{m-1}, (k+1)\Delta t_{m-1}]$ are at

$$t_k = (2k+1)\,\Delta t_m\,, \qquad k = 0, \ldots, 2^m - 1,$$

where $\Delta t_m = \Delta t_{m-1}/2$. For the algorithm in [1], each sample $x(t_k)$ is synthesized as a linear interpolation from the nearest samples $x(t_k - \Delta t_m)$ and $x(t_k + \Delta t_m)$, plus a displacement whose variance is given by Eq. (2.12). This interpolation and displacement are

$$x(t_k) = \underbrace{\frac{1}{2}\left(x(t_k+\Delta t_m)+x(t_k-\Delta t_m)\right)}_{\text{interpolation}} + \widetilde{x}(t_k), \tag{2.13}$$

$$\mathrm{var}[\widetilde{x}(t_k)] = \sigma^2\left(\frac{1}{2}\right)^{2mH}\left[\left(\frac{1}{2}\right)^{2H}-\left(\frac{1}{2}\right)^{2}\right]. \tag{2.14}$$

All of the displacements $\widetilde{x}(t_k)$ are assumed to be independent. The result is that $x(t)$ is generated for $t = i\Delta t_m$, $i = 0, 1, \ldots, 2^{m+1}$. Finer sampling intervals can be achieved by repeating this process.

This algorithm, while simple and efficient, is an exact representation of fBm only for $H = 1/2$. For $H \neq 1/2$, two approximations are made. First, the interpolation and displacement given by Eq. (2.13)-(2.14) are statistically correct only for $H = 1/2$ or $k = 0$. (Note that Eq. (2.13) reduces to Eq. (2.11) for $k = 0$, i.e., when one of the endpoints is $x(0)$.) Unless $k = 0$ or $H = 1/2$,

$$E\left[x(t_k)\,\middle|\,x(t_k+\Delta t_m), x(t_k-\Delta t_m)\right] \neq \frac{1}{2}\left(x(t_k+\Delta t_m)+x(t_k-\Delta t_m)\right),$$

meaning that the conditional expectation is not equal to a simple average of the endpoint samples. Furthermore, the variance of the prediction error $\widetilde{x}(t_k)$ varies as a function of k, and is equal to the variance in Eqs. (2.14) only for $k = 0$ or $H = 1/2$. The second approximation made by the midpoint displacement algorithm of [1] is that the displacements $\widetilde{x}(t_k)$ are not independent. For instance, conditioned on $x(0)$, $x(1/2)$, and $x(1)$, the displacements $\widetilde{x}(1/4)$ and $\widetilde{x}(3/4)$ are correlated when $H \neq 1/2$. In Section 3., we propose multiscale models that improve upon these two approximations by using statistically accurate descriptions of the interpolation and displacement.

2.3 Wavelet Decompositions

Because fBm is both non-stationary and self-similar, the wavelet transform seems to provide a natural decomposition of fBm. The (orthonormal) wavelet synthesis of fBm is given by

$$x(t) = \sqrt{2^J}\sum_{k=-\infty}^{\infty} a_J[k]\,\phi(2^J t - k) + \sum_{m=J}^{\infty}\sqrt{2^m}\sum_{k=-\infty}^{\infty} d_m[k]\,\psi(2^m t - k),$$

where $\psi(t)$ is the wavelet function, $\phi(t)$ is the scaling function, $d_m[k]$ are the detail coefficients, and $a_J[k]$ are the approximation coefficients at scale J. The index k is the spatial offset and m is the scale index. Note that we have made the convention that m increases for finer resolutions.

As shown in [17], fBm is made stationary when filtered by any ideal bandpass filter. The wavelet function $\psi(t)$ is the impulse response of a bandpass filter, albeit not ideal. Because $1/f$ processes are made stationary when filtered by ideal bandpass filters [17], it should not be surprising that the detail coefficients at any given scale m are stationary [6]. The variance of the stationary process at scale m is [6]

$$\mathrm{var}\left[d_m[k]\right] = \frac{\sigma^2}{2}\,V(H,\psi)\left(\frac{1}{2}\right)^{(2H+1)m}, \tag{2.15}$$

where $V(H, \psi)$ is a constant that depends on only the Hurst exponent and the wavelet function. The variance of the wavelet detail coefficient decreases geometrically with scale at a rate proportional to H, as do the displacements in random midpoint displacement.

If the wavelet coefficients are to be used to approximate fBm, the detail coefficients must be sufficiently close to independent. Mutual independence leads to efficient synthesis algorithms, since the detail coefficients can be generated independently. As shown in [6, 15, 17], the detail coefficients at any given scale are independent only for $H = 1/2$. For $H \neq 1/2$, the correlation between detail coefficients at any given scale decays asymptotically as [6]

$$E\big[d_m[k]\, d_m[l]\big] \sim \mathcal{O}\big(|k-l|^{2(H-R)}\big), \tag{2.16}$$

where R is the number of vanishing moments of the wavelet function, i.e., the regularity of the wavelet function. The correlation between detail coefficients at different scales behaves similarly, i.e., they are approximately uncorrelated if H and R are chosen appropriately. Note that even for $H = 1/2$ the detail coefficients at different scales are not all uncorrelated.

The wavelet decomposition can be used to approximately synthesize fBm if the wavelet detail coefficients are assumed to all be independent [6]. Also, because the wavelet transform is an approximate Karhunen-Loève transform of fBm, it can be used to efficiently estimate fBm from dense samples of $y(t) = x(t) + v(t)$, where the $v(t)$ is white noise with constant variance [17]. While this approach is quite attractive, especially since the wavelet decomposition can be applied to $1/f$ processes other than fBm, there are some drawbacks. First, unless the wavelet regularity R is large enough, the fBm synthesized assuming independent detail coefficients will have a number of artifacts due to correlation among the detail coefficients. Also, one must deal with the boundary effects which arise from a wavelet with large R and a finite-length data interval. Finally, there are applications for which the measurements are not dense, the measurements are not all at the finest scale, or the variance of the measurement noise is not constant. An alternative model for fBm, based on the Haar wavelet but using the multiscale stochastic models of [2], is proposed in the following section and overcomes many of the aforementioned difficulties.

3. Multiscale Models and fBm

In this section, we show how the random midpoint displacement algorithm and the wavelet decomposition naturally lead to multiscale stochastic processes in the form of [2]. These processes not only provide more accurate models for fBm, but they allow for efficient processing and parameter estimation.

3.1 Multiscale Modeling and Estimation

The multiscale random processes introduced in [2] are indexed by the nodes of trees organized into scales. The coarsest scale is indexed by the root node, while the finest scale is indexed by the set of leaf nodes. For example, the multiscale process defined on the binary tree illustrated in Figure 3.1a consists of a set of random vectors $z(s)$ for each node s on the tree. The *scale* of node s, which we denote by $m(s)$, is the distance between node s and the root node of the tree. Define $\overline{\gamma}$ to be the upward (in scale) shift operator, so that $s\bar{\gamma}$ is the parent of any node s, as illustrated in Figure 3.1b. The multiscale processes satisfy the following autoregression *in scale*

$$z(s) = A_s\, z(s\bar{\gamma}) + w(s)\,, \tag{3.1}$$

$$w(s) \sim \mathcal{N}(0, Q_s)\,. \tag{3.2}$$

The autoregression is initialized at the root node $s = 0$ by

$$z(0) \sim \mathcal{N}(0, P_0)\,. \tag{3.3}$$

Since $z(0)$ and $w(s)$ are zero-mean, every process value $z(s)$ will be a zero-mean[4] random vector.

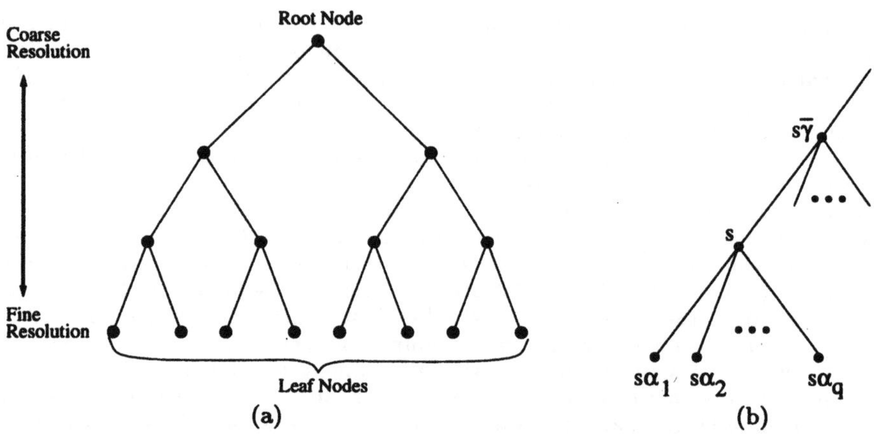

Fig. 3.1. (a) A binary tree used to index a random process at multiple resolutions. (b) The local labeling of the $q+1$ nodes connected to node s.

The process noise $w(s)$ is assumed to be a white-noise process uncorrelated across scale and space and also uncorrelated with the root node variable, i.e., $E\left[w(s)\, z(0)^T\right] = 0$. The whiteness of the process noise implies that a multiscale tree model is characterized completely by P_0 — the root node covariance — and the autoregression parameters A_s and Q_s for all nodes $s \neq 0$. More importantly, the whiteness of the process noise implies a Markov property similar to the Markov property for 1D autoregressive processes driven by white noise [2]. This Markov property leads to an efficient algorithm for the estimation of the tree process from measurements, each of which is a noise-corrupted observation of $z(\cdot)$ at some node of the tree, i.e.,

$$y(s) = C_s z(s) + v(s)\,, \qquad v(s) \sim \mathcal{N}(0, R_s) \tag{3.4}$$

where $v(\cdot)$ is white and uncorrelated with $z(\cdot)$ at all nodes on the tree. The result of the estimator is that the LLSE estimate $\hat{z}(s)$ of the variable at every node in the tree is computed in $\mathcal{O}(Nd^3)$ computations for a tree which has N nodes at the finest scale and constant variable dimension d. The estimator also produces the estimation error covariance $E\left[\left(z(s) - \hat{z}(s)\right)\left(z(s) - \hat{z}(s)\right)^T\right]$ at every node.

[4] The zero-mean assumption is made for simplicity and is easily relaxed by adding a deterministic term to Eq. (3.1) or Eq. (3.3).

Of interest for this paper are internal models. An internal multiscale model is one for which each variable of the process is a linear function of the finest-scale process, which includes all variables at the leaf nodes of the tree [8]. If f is a vector containing the finest-scale process of the tree, then each variable of an internal realization can be expressed as

$$z(s) = V_s f \,. \tag{3.5}$$

The matrices V_s are called *internal matrices*. For the remainder of this paper, the vector f will contain samples of fBm, i.e., $f_n = x[n]$ from Eq. (2.3). Internal realizations are of interest because the tree parameters P_0, A_s, and Q_s can be derived directly from the internal matrices and the covariance of f, P_f. Specifically, substituting Eq. (3.5) into $P_0 = E\big[z(0)\, z(0)^T\big]$ yields

$$P_0 = V_0 P_f {V_0}^T \,. \tag{3.6}$$

The parameters A_s and Q_s can be computed by noting that Eq. (3.1) is just the optimal prediction of $z(s)$ based upon $z(s\overline{\gamma})$, plus the associated prediction error, i.e.,

$$z(s) = E\big[z(s) \,|\, z(s\overline{\gamma})\big] + w(s) \,. \tag{3.7}$$

Using standard equations from LLSE estimation, the model parameters follow as

$$A_s = V_s P_f {V_{s\overline{\gamma}}}^T \, (V_{s\overline{\gamma}} P_f {V_{s\overline{\gamma}}}^T)^{-1} \,, \tag{3.8}$$

$$Q_s = V_s P_f {V_s}^T - V_s P_f {V_{s\overline{\gamma}}}^T \, (V_{s\overline{\gamma}} P_f {V_{s\overline{\gamma}}}^T)^{-1} V_{s\overline{\gamma}} P_f {V_s}^T \,. \tag{3.9}$$

3.2 Multiscale Models for fBm

3.2.1 Improved Midpoint Displacement. The random midpoint displacement algorithm of Section 2.2 can be directly translated into a multiscale autoregression satisfying Eq. (3.1) and (3.2). Assume that we are interested in modeling fBm on the unit interval, and that $\Delta t_M = (1/2)^{M+1}$ is the finest sampling interval of interest. The $(2^{M+1} + 1)$ samples $x(n\Delta t_M)$ can be synthesized from $M + 1$ iterations of the midpoint displacement algorithm, where the first iteration generates $x(0)$, $x(1/2)$, and $x(1)$. The $2^{m+1} + 1$ samples of $x(t)$ generated by completion of the m-th iteration ($m = 0, \ldots, M$) can be mapped to the 2^m nodes at the m-th scale of a binary tree. This mapping is illustrated in Figure 3.2 for $\Delta t_M = 1/16$, i.e., $M = 3$. Each variable $z(s)$ at the finest scale (M) contains three consecutive samples of $x(t)$. The states at the first two scales of the binary tree are given by $z(0) = [x(0),\ x(1/2),\ x(1)]^T$, $z(0\alpha_1) = [x(0),\ x(1/4),\ x(1/2)]^T$, and $z(0\alpha_2) = [x(1/2),\ x(3/4),\ x(1)]^T$. Mimicking midpoint displacement, the autoregression for the first two scales is

$$z(s\alpha_1) = \begin{bmatrix} 1 & 0 & 0 \\ 1/2 & 1/2 & 0 \\ 0 & 1 & 0 \end{bmatrix} z(s) + w(0\alpha_1) \,, \tag{3.10}$$

$$z(s\alpha_2) = \begin{bmatrix} 0 & 1 & 0 \\ 0 & 1/2 & 1/2 \\ 0 & 0 & 1 \end{bmatrix} z(s) + w(0\alpha_2) \,, \tag{3.11}$$

$$Q_{0\alpha_1} = Q_{0\alpha_2} = \mathrm{diag}\Big(0,\ \sigma^2 \left(\frac{1}{2}\right)^{2H(m(s)+1)} \Big[\left(\frac{1}{2}\right)^{2H} - \left(\frac{1}{2}\right)^2\Big],\ 0\Big) \,. \tag{3.12}$$

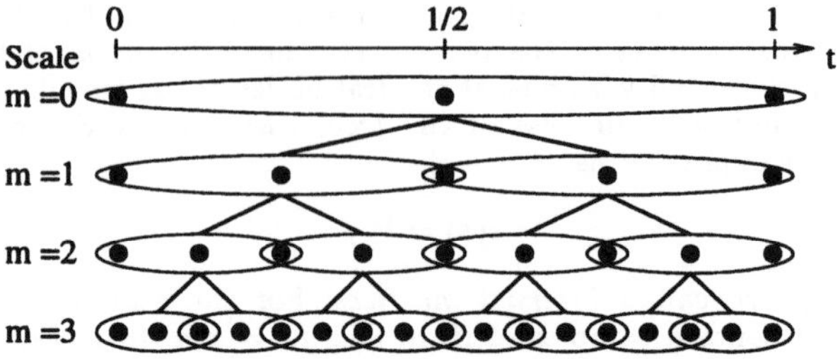

Fig. 3.2. A multiscale tree representation of sampled fBm for $\Delta t_M = 1/16$. Each ellipse illustrates the samples represented by a tree variable.

The process noise variances are obtained from Eq. (2.14). For $H = 1/2$, the tree model will have the correct finest-scale covariance [11]. We refer to the multiscale model defined by Eqs. (3.10)-(3.12) as the **simple displacement model**.

For $H \neq 1/2$, the distribution of the process at the finest scale of the tree will differ from that of sampled fBm. One reason for the discrepancy, as mentioned in Section 2.2, is that the displacements are not truly independent for $H \neq 1/2$, whereas the multiscale process noise is independent by assumption. The other reason is that the interpolation and the displacement variance should vary across a given scale. To derive an interpolation and displacement that produce the correct variance for each sample $x(t_k)$, the following LLSE prediction equations can be used

$$E\big[x(t_k)\,|\,X(t_k)\big] = P_{x(t_k)X(t_k)}P_{X(t_k)}^{-1}X(t_k)\,, \tag{3.13}$$

$$X(t_k) \triangleq \big[x(t_k+\Delta t_m), x(t_k-\Delta t_m)\big]^T, \tag{3.14}$$

$$\mathrm{var}\big[\widetilde{x}(t_k)\big] = P_{x(t_k)} - P_{x(t_k)X(t_k)}P_{X(t_k)}^{-1}P_{X(t_k)x(t_k)}\,, \tag{3.15}$$

$$\widetilde{x}(t_k) \triangleq x(t_k) - E\big[x(t_k)\,|\,X(t_k)\big]\,. \tag{3.16}$$

The covariance matrices follow from sampling Eq. (1.1).

The tree model representing fBm can be improved without changing the interpretation of the state variables from those of the simple displacement model. Namely, we assume that each $z(s)$ at scale $m(s)$ represents three samples of $x(t)$ spaced by $\Delta t_m = (1/2)^{m+1}$. The model parameters based on the optimal prediction of the samples in $z(s)$ from those in $z(s\overline{\gamma})$ are given by Eqs. (3.6), (3.8) and (3.9). Call the resulting model the **enhanced displacement model**. For this model, the process noise representing the midpoint displacements is no longer constant across any scale (unless $H = 1/2$), and the interpolation is no longer the simple average of the endpoints. However, the enhanced model will still be an approximate representation of fBm, since the displacements still have correlations not captured by the tree model.

The two approximations of fBm, the simple displacement model defined by Eqs. (3.10)-(3.12) and the enhanced displacement model based on optimal prediction, can be compared to illustrate the effect of the enhancement. Because the finest-scale processes of these two models are meant to represent fBm, the covariance at the finest scale of the two models can be compared with the exact covariance of fBm. Consider fBm with $H = 0.3$ and $\sigma^2 = 1$ on the unit interval $t \in [0, 1]$. Figure 3.3a shows the absolute value of the difference between the finest-scale covariance of the simple displacement model and the true covariance in Eq. (2.2). The absolute value of the difference between the finest-scale covariance of the enhanced displacement

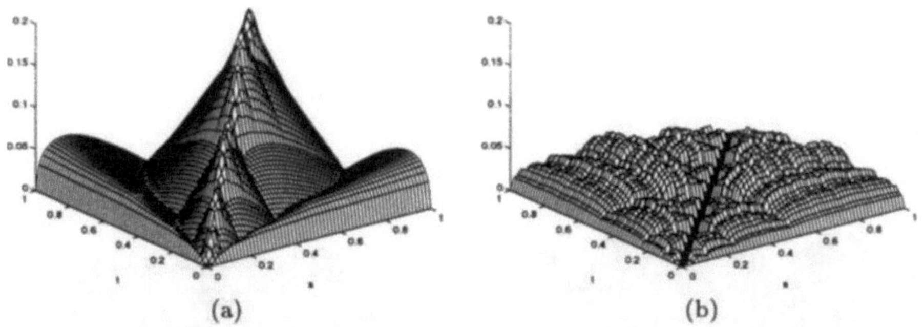

Fig. 3.3. A comparison of the simple and enhanced displacement models $H = 0.3$ and $\sigma^2 = 1$. (a) The absolute difference between Eq. (2.2) and the finest-scale covariance of the simple displacement model and (b) the absolute difference for the enhanced displacement model.

model and the true covariance in Eq. (2.2) is shown in Figure (3.3b). Note that the enhanced displacement model is a more accurate representation of fBm, especially in terms of representing the variance. The variances are represented exactly by the enhanced displacement model because the variances at the finest scale of any multiscale model computed from Eqs. (3.6), (3.8) and (3.9) will be exact, i.e., equal to the diagonal of P_f. The finest-scale covariances of the simple and enhanced displacement models are compared for $H = 0.7$ in Figures 3.4a and Figure 3.4b. Again, the enhanced displacement model is an improvement over the simple displacement model, especially in terms of representing the variance of fBm. However, the errors in general are much smaller for $H = 0.7$ than for $H = 0.3$. The reason is that, the larger the value of H, the smaller the variances of the displacements at fine-scales. Therefore, even if the displacement variances are modeled incorrectly, they will have less influence on the covariance of the finest-scale process. The bottom line is that both of the displacement models produce better approximations of fBm for $H > 1/2$ than for $H < 1/2$, but the enhanced model is more accurate for all values of H.

Approximating fBm within the multiscale framework allows one to accomplish a number of signal processing tasks. Consider the estimation of $x(t)$ from noisy measurements. A sample path of fBm (for $H = 0.3$ and $\sigma^2 = 1$) is illustrated in Figure 3.5a. Consider estimating $x(t)$ for $t \in [0,1]$ from sparse and noisy measurements near the endpoints of the interval. The noisy measurements $y(t_i) = x(t_i) + v_i$, where $\text{var}[v_i] = 0.05$, are illustrated by o's in Figure 3.5a. The LLSE estimate of $x(t)$ based on the exact fBm prior model is illustrated by the solid line in Figure 3.5b. The estimates (of the finest-scale processes) produced by the two multiscale models are given by the dashed (simple displacement model) and dotted (enhanced displacement model) lines. As would be expected from the covariances in Figure 3.3, the enhanced displacement model more closely approximates the estimates based on the fBm prior distribution. However, the finest-scale estimates produced by both models are well within a standard deviation of the estimation error for the exact LLSE estimate.

3.2.2 Improved Wavelet-Based Models. As noted in Section 2.3, the wavelet transform provides an approximate Karhunen-Loève transform of fBm. In analogy with the previous section, the (approximately) independent detail coefficients can be mapped to the process noise of a multiscale tree. A multiscale model

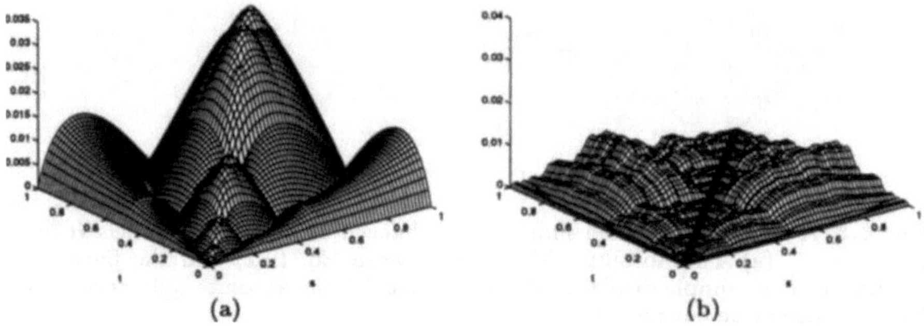

Fig. 3.4. A comparison of the simple and enhanced displacement models $H = 0.7$ and $\sigma^2 = 1$. (a) The absolute difference between Eq. (2.2) and the finest-scale covariance of the simple displacement model and (b) the absolute difference for the enhanced displacement model.

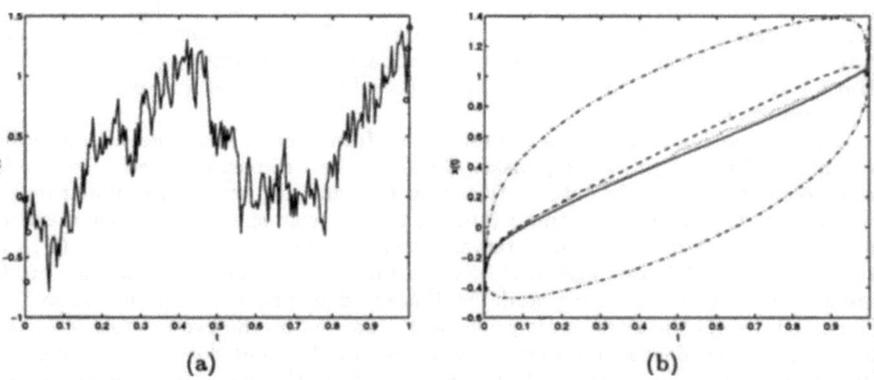

Fig. 3.5. (a) A sample path of fBm for $H = 0.3$ and $\sigma^2 = 1$. (b) The exact LLSE estimate (solid line), the estimates produced by the two multiscale models (dashed and dotted lines), and the one standard deviation error bars for the exact LLSE estimate (dash-dot line).

proposed in [5], which was successfully employed in the estimation of the Hurst exponent, is a binary tree with variables derived from the discrete-time Haar basis. For this multiscale model, each variable $z(s)$ represents a detail and an approximation coefficient at scale $m(s)$. The autoregression for this multiscale model is

$$z(s) = \frac{1}{\sqrt{2}} \begin{bmatrix} 1 & \mathrm{mod}(s) \\ 0 & 0 \end{bmatrix} z(s\bar{\gamma}) + \underbrace{\begin{bmatrix} 0 \\ 1 \end{bmatrix} w_d(s)}_{w(s)}, \tag{3.17}$$

$$\mathrm{mod}(s) \triangleq \begin{cases} 1, & s \text{ is a left descendent}, \\ -1, & s \text{ is a right descendent}, \end{cases} \tag{3.18}$$

when s is not a leaf node. Because $z(s)$ represents an approximation coefficient and a detail coefficient at a particular scale, i.e., $z(s) = \left[a_m[k], \, d_m[k]\right]^T$ for some value of k, $w_d(s)$ must represent $d_m[k]$. The transition to the finest scale is simply

$$z(s) = \frac{1}{\sqrt{2}} \left[1 \;\; \mathrm{mod}\,(s)\right] z(s\bar{\gamma}), \tag{3.19}$$

so that $z(s)$ at the finest scale represents an approximation coefficient at scale M, which for the Haar basis is equal to $2^{M/2}$ times the average value of $x(t)$ over some interval of width $1/2^M$.

The only parameter not specified in Eq. (3.17) is the variance of $w_d(s)$, which is also the variance of $d_m[k]$. In [6, 17], the detail coefficients for the continuous-time wavelet transform are shown to obey Eq. (2.15). However, as shown in [4] for the Haar basis, this strict geometric progression in the variance of the detail coefficients is not appropriate for a discrete-time wavelet transform, i.e., when the finest-scale process is discrete-time fBm. The reason is that sampling fBm entails considerable aliasing, especially for $H < 1/2$. The variances of the detail coefficients can instead be derived directly from the Haar wavelet basis and the covariance in Eq. (2.2). Call the corresponding multiscale model that uses these variances the **simple wavelet model**. These multiscale models have been shown to be very accurate for estimating the value of H from measurements of the finest-scale process. Unfortunately, these models are not entirely appropriate for synthesizing or estimating fBm. The covariance of the finest-scale process for $H = 0.3$ is illustrated in Figure 3.6a. This covariance is quite different from that in Eq. (2.2). First of all, the finest-scale process has constant variance, which is due to the bandpass filtering implied by the wavelet transform. Also, any sample path generated by this model will have distracting artifacts which are due to the "blockiness" of the covariance function; this blockiness is manifested in highly non-stationary increments. Therefore, while this model is appropriate for estimating the Hurst exponent of fBm, it is not appropriate for synthesizing or processing fBm.

A more accurate multiscale representation of fBm is given by noting that, if the states satisfy $z(s) = \left[a_m[k], \, d_m[k]\right]^T$, then the multiscale model is internal. The parameters of the resulting internal model can be derived from Eqs. (3.6), (3.8) and (3.9), where P_f is given by Eq. (2.2). Call the resulting multiscale model the **enhanced wavelet model**. The covariance of the finest-scale process for $H = 0.3$ is illustrated in Figure 3.6b, and is a significant improvement over the simple wavelet model. The variance of the enhanced wavelet model exactly matches $|t|^{0.6}$. The enhanced wavelet model will also provide a close approximation to fBm for any $H > 1/2$, even though Haar wavelet basis is too irregular to represent such smooth processes using wavelet synthesis. The primary difference between the enhanced

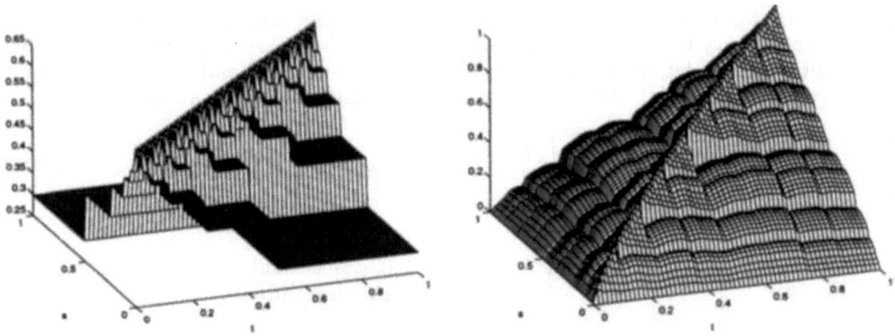

(a) (b)

Fig. 3.6. Assuming $H = 0.3$ and $\sigma^2 = 1$: (a) the covariance at the finest scale of the simple wavelet model; (b) the covariance at the finest scale of the enhanced wavelet model.

wavelet model and the simple wavelet model is that the enhanced model is able to represent local correlations among detail coefficients, whereas the simple wavelet model assumes that the detail coefficients are completely uncorrelated. Namely, the correlation among the detail coefficient represented by $z(s)$ and the detail coefficients represented by the children of $z(s)$ are exactly captured by the enhanced model.

4. Conclusion

This paper demonstrates how to represent the random midpoint displacement and wavelet synthesis approximations of fractional Brownian motion within the multiscale framework proposed by [2]. There are multiple advantages provided by the multiscale framework. First, the multiscale framework allows for more accurate representations of fBm than does midpoint displacement or wavelet synthesis. Second, multiscale processes can be efficiently estimated from an arbitrary set of measurements, i.e., the measurements need not be dense measurements of the finest-scale process with constant measurement noise variance as required by the wavelet synthesis approach [17].

Given the multiscale models proposed in this paper, natural questions are how to extend this framework to provide more accurate models of fBm and how to use multiscale models to represent other $1/f$ and self-similar processes. One approach is to represent the wavelet synthesis for a more regular wavelet basis within the multiscale framework. More regular wavelets lead to an increase in the state dimensions of the corresponding multiscale models. Another approach, described in [3, Chapter 7], is to develop multiscale models for fBm directly from the stationary increments and statistical self-similarity properties of fBm. These multiscale models of fBm can be efficiently realized and can be as accurate as desired.

Another method worth noting for representing fBm is the recently introduced synthesis model provided by [13] that is based on a bi-orthogonal synthesis of fBm. This approach is worth comparing to the multiscale framework in the context of synthesis, signal estimation, and parameter estimation. Such a comparison is beyond the scope of this work, but will be provided in a future publication.

References

1. Barnsley, M.F., Devaney, R.L., Mandelbrot, B.B., Peitgen, H.-O., Saupe, D., Voss, R.F. (1988): *The Science of Fractal Images*. Springer-Verlag.
2. Chou, K.C., Willsky, A.S., Benveniste, A. (1994): Multiscale recursive estimation, data fusion, and regularization. *IEEE Trans. Auto. Cont.*, 39(3), 464–478.
3. Daniel, M.M. (December 1996): *Multi-resolution Statistical Modeling with Application to Modeling Groundwater Flow*, PhD thesis, M.I.T..
4. Fieguth, P.W. (June 1995): *Application of Multiscale Estimation to Large Scale Multidimensional Imaging and Remote Sensing Problems*. PhD thesis, M.I.T..
5. Fieguth, P.W., Willsky, A.S. (May 1996): Fractal estimation using models on multiscale trees. *IEEE Trans. Sig. Proc.*, 1297–1300.
6. Flandrin, P. (March 1992): Wavelet analysis and synthesis of fractional Brownian motion. *IEEE Trans. Info. Theory*, 38(2), 910–917.
7. Hosking, J.M.R. (December 1984): Modeling persistence in hydrological time series using fractional differencing. *Water Resources Res.*, 20(12), 1898–1908.
8. Irving, W.W. (August 1995): *Multiscale stochastic realization and model identification with applications to large-scale estimation problems*. PhD thesis, MIT.
9. Kaplan, L.M., Kuo, C.-C. J. (December 1993): Fractal estimation from noisy data via discrete fractional Gaussian noise (DFGN) and the Haar basis. *IEEE Trans. on Signal Processing*, 41(12), 3554.
10. Keshner, M.S. (March 1982): $1/f$ noise. *Proc. of the IEEE*, 70(3), 212–218.
11. Luettgen, M.R., Karl, W.C., Willsky, A.S., Tenney, R.R. (December 1993): Multiscale representations of Markov random fields. *IEEE Trans. Sig. Proc.*, 41(12), 3377.
12. Mandelbrot, B., van Ness, J. (1968): Fractional Brownian motions, fractional noises, and applications. *SIAM Review*, 10, 422–437.
13. Abry, P., Sellan, F. (1996): The wavelet-based synthesis for fractional brownian motion proposed by F. Sellan and Y. Meyer: remarks and fast implementation. *Applied Computational and Harmonic Analysis*, 3, 377–383.
14. Stoksik, M.A., Lane, R.G., Nguyen, T.D. (May 1995): Practical synthesis of accurate fractal images. *Graphical Models and Image Processing*, 57(3), 206–219.
15. Tewfik, A.H., Kim, M. (March 1992): Correlation structure of the discrete wavelet coefficients of fractional Brownian motion. *IEEE Trans. Info. Theory*, 38(2), 904–909.
16. Willinger, W., Taqqu, M., Leland, W.E., Wilson, D.V. (1995): Self-similarity in high-speed packet traffic: analysis and modeling of ethernet traffic measurements. *Statistical Science*, 10, 67–85.
17. Wornell, G.W. (1993): Wavelet-based representations for the $1/f$ family of fractal processes. *Proc. of the IEEE*, 81(10), 1428–1450.

2D Self-Similar Processes with Stationary Fractional Increments

B. Pesquet-Popescu and P. Larzabal

Laboratoire d'Electricité, Signaux et Robotique
URA CNRS 1375/Ecole Normale Supérieure de Cachan and GDR ISIS
61, Av. du Président Wilson, 94235 Cachan Cedex, FRANCE
beatrice.popescu@lesir.ens-cachan.fr

Abstract This work is concerned with non-stationary 2D fields having stationary increments. The classical fractional Brownian motion model is limited in some applications, since it is isotropic and depends on a single parameter (the Hurst coefficient H). We introduce herein new models for possibly anisotropic textures, based on fractional orders of differencing in two orthogonal directions of space. Some of the properties of fields with fractional stationary increments are also discussed. We then focus on self-similar processes with stationary increments of order one and stress the role of the structure function in the study of these processes.

1. Introduction

Since its introduction by B. Mandelbrot and J.W. Van Ness [1], the fractional Brownian motion (fBm) has become a very popular model in signal and image processing ([2], [3], [4], [5],...). However, 2D fBm models as defined in [3] are not always sufficient in fully characterizing real world surfaces. One of their limitations is the isotropy, which is rather idealistic in the case of real textures. Another limitation is the low degree of flexibility of these models which depend on a single parameter (the Hurst coefficient H). The aim of this paper is to study the possibility of introducing more general models for anisotropic textures.

The extensions studied in this paper are based on the concept of fractional stationary increments. In particular, it is proved that the 2D fBm and some related processes have stationary increments of a fractional order greater than their Hurst parameter. General properties of fields with fractional stationary increments are also presented. For fields with stationary increments of order 1, we extend to the 2D case the results given in [6] and we show that the structure function plays a central role in the study of this class of processes. We also demonstrate that the choice of the structure function is constrained to certain conditions, which are given in this paper.

In the 1D case, the fBm is the only Gaussian random process which is self-similar and with stationary increments. We investigate in the 2D case the existence of other self-similar processes with stationary increments. As pointed out in [7], the definition of self-similarity in $\mathbb{R}^n$ is analogous to that in $\mathbb{R}$, but the extension of the notion of stationary increments is more delicate. In [7], the stationarity of the increments is defined as an invariance of the statistical properties of the increments with respect to any rigid body motion. In this context, the Lévy fractional Brownian fields are the only self-similar Gaussian fields with stationary increments. In this paper, we adopt a different point of view which enables us to find other self-similar Gaussian fields with stationary increments. We also provide the generic form of the structure functions of these fields.

Although this paper is dedicated to more theoretical issues, we should note that the processes which are considered herein have proved to be useful in modeling the sonar images of the sea bed [8].

2. Definitions

Most of this work is concerned with 2D processes with stationary increments. As mentioned above, the manner of defining increments in $\mathbb{R}^2$ is important, as it does lead to fields with different properties. By taking different orders of differencing in two orthogonal directions of space, we are placing emphasis on the anisotropy of the fields under study. An additional degree of flexibility is then introduced by considering fractional orders for the increments.

Definition 2.1. *The increments of fractional order $(D, D') \in \mathbb{R}_+^2$ of a 2D continuous (resp. discrete) random field $F(x, y)$ are defined, when they exist, as: $\forall(x, y, \Delta_x, \Delta_y) \in \mathbb{R}^4$ (resp. $\forall(x, y, \Delta_x, \Delta_y) \in \mathbb{Z}^4$),*

$$\Delta^{(D,D')}F(x, y; \Delta_x, \Delta_y) = \left(1 - q_{\Delta_x}^{-1}\right)^D \left(1 - q_{\Delta_y}^{'-1}\right)^{D'} F(x, y) \tag{2.1}$$

where the symbols $q_{\Delta_x}^{-1}$ and $q_{\Delta_y}^{'-1}$ denote the shift operators:

$$q_{\Delta_x}^{-1} F(x, y) = F(x - \Delta_x, y), \quad q_{\Delta_y}^{'-1} F(x, y) = F(x, y - \Delta_y).$$

The fractional difference operator $\left(1 - q_\Delta^{-1}\right)^D$, $\Delta \in \mathbb{R}$ was defined in [9] by the following expansion:

$$\left(1 - q_\Delta^{-1}\right)^D = \begin{cases} \sum_{k=0}^{\infty} (-1)^k \begin{pmatrix} D \\ k \end{pmatrix} q_\Delta^{-k} & \text{if} \quad D \neq 0 \\ 1 & \text{if} \quad D = 0 \end{cases} \tag{2.2}$$

with

$$\begin{pmatrix} D \\ k \end{pmatrix} = \frac{D(D-1)\dots(D-k+1)}{k!}.$$

In the case of integer increments, (2.2) reduces to the classical binomial formula for $0 \leq k \leq D$ and $0 \leq l \leq D'$. In (2.1), $\Delta^{(D,D')}F(x, y; \Delta_x, \Delta_y)$ may also be interpreted as the result of filtering $F(x, y)$ by a linear filter with transfer function $H(p, s) = \left(1 - e^{-p\Delta_x}\right)^D \left(1 - e^{-s\Delta_y}\right)^{D'}$, $\forall(p, s) \in \mathbb{C}^2$. [1]

With little loss of generality, we will assume throughout this paper that all the processes considered are zero-mean and zero-valued at the origin. More specifically, we will focus on random fields with fractional stationary increments:

Definition 2.2. *A 2D random field has* (wide-sense) stationary increments of order *(D, D') if $\forall(\Delta_x, \Delta_y, \Delta'_x, \Delta'_y)$, $\Delta^{(D,D')}F(x, y; \Delta_x, \Delta_y)$ and $\Delta^{(D,D')}F(x, y; \Delta'_x, \Delta'_y)$*

[1] When $D \in \mathbb{R}_+ \setminus \mathbb{N}$, the function of a complex variable $u \mapsto (1-u)^D$ is defined by $(1-u)^D = e^{D \log(1-u)}$, where a principal determination of the logarithm function is chosen (from $\mathbb{C} \setminus \mathbb{R}_-$ to $\{z \in \mathbb{C} \mid -\pi < \operatorname{Im} z < \pi\}$).

are second-order processes [2] *which are mutually (wide-sense) stationary, i.e. there exists a function* Φ *such that:*

$$\forall(x,y,x',y'), \quad \mathrm{E}\left\{\Delta^{(D,D')}F(x,y;\Delta_x,\Delta_y)\Delta^{(D,D')}F(x',y';\Delta'_x,\Delta'_y)\right\} = \Phi\left(x-x',y-y',\Delta_x,\Delta_y,\Delta'_x,\Delta'_y\right).$$

where $\mathrm{E}\{\cdot\}$ *denotes the expectation operator.*

It can be readily proved that a discrete field $F(n,m)$ has stationary increments of order (D,D') iff $\Delta^{(D,D')}F(n,m;1,1)$ is stationary. In this paper, we will be mainly interested in continuous-space fields, although many of the results (in particular, Proposition 3.4) are also valid for discrete-space fields.

3. Properties of Non-stationary Fields with Stationary Increments

We now state some useful properties of 2D processes with fractional stationary increments:

Proposition 3.1. *If* $F(x,y)$ *is a process with stationary increments of order* (D,D')*, it also has stationary increments of order* (p,q)*, where* $p \geq D$ *and* $q \geq D'$.

Proof: To prove this result, we remark that the increments of order (p,q) can be expressed as the outputs of a linear filter whose inputs are the stationary increments of order (D,D') (see [11]).

It follows that if $F(x,y)$ is a process with stationary increments of order $(D,0)$ or $(0,D')$, it also has stationary increments of order (D,D'). The reciprocal statement is generally false.

We will now consider processes which increments of order $(D,0)$ and $(0,D')$ are stationary and mutually stationary. For conciseness, such random fields will be referred to as processes with stationary increments of order $(D,0)/(0,D')$. By proceeding similarly as for the proof of Proposition 3.1, it can be shown that:

Proposition 3.2. *If* $F(x,y)$ *is a process with stationary increments of order* $(D,0)/(0,D')$*, it also has stationary increments of order* $(p,0)/(0,q)$*, where* $p \geq D$ *and* $q \geq D'$.

We have, in particular, when $D' = D$:

Proposition 3.3. *A process* $F(x,y)$ *is with stationary increments of order* $(D,0)/(0,D)$ *if and only if it has stationary increments of order* D *in all the space directions and provided these increments are mutually stationary.*

Proof: See Appendix A.

As a consequence of this result, the property of stationarity of the increments of order $(D,0)/(0,D)$ is preserved by a 2D rotation.

[2] This implies that the expansion defining the increments should converge in the mean square sense [10]. The increments considered in this paper will be implicitly assumed harmonizable processes whenever necessary.

Remark 3.1. Using the above definitions, one can readily observe that an isotropic 2D fBm [3] $B_H(x, y)$ has stationary increments of order $(1,0)$. This can be checked by noting that the cross-correlation between $\Delta^{(1,0)} B_H(x, y; \Delta_x, 0)$ and $\Delta^{(1,0)} B_H(x', y'; \Delta'_x, 0)$ only depends on $x - x'$, $y - y'$, Δ_x and Δ'_x. For similar reasons, it has stationary increments of order $(1,0)/(0,1)$. Then, it can be deduced from Proposition 3.3 that an isotropic fBm has stationary increments of order 1 in all the space directions.

In the remainder of this paper, we will focus on processes with stationary increments of order $(1,0)/(0,1)$. This class of processes is of primary interest as it includes the 2D fBm as well as some of its extensions which will be studied in this paper. In particular, we have just noted that any process with stationary increments of order $(D,0)/(0,D')$ with $D \leq 1$ and $D' \leq 1$ belongs to this class. In the 1D case, it was shown [6] that the processes with stationary increments (of order 1) are fully characterized by their structure function. Next, we generalize these results to the 2D case, by introducing a two-dimensional structure function giving the variance of the increments in any arbitrary direction (Δ_x, Δ_y). A similar definition was used in [12] to characterize the roughness of machined surfaces which are non-stationary and highly anisotropic.

Here, we study some fundamental properties of the two-dimensional structure function related to processes with stationary increments of order $(1,0)/(0,1)$. We first show that for such fields, the structure function represents the variance of the increments and can be used to express the correlation function of the field and its increments.

Proposition 3.4. *The following conditions are equivalent :*

1. *A 2D field $F(x, y)$ has stationary increments of order $(1,0)/(0,1)$.*
2. *There exists a function Ψ such that: $\forall(x, y, x', y', \Delta_x, \Delta_y, \Delta'_x, \Delta'_y)$,*

$$\mathrm{E}\left\{[F(x,y) - F(x-\Delta_x, y-\Delta_y)]\left[F(x',y') - F(x'-\Delta'_x, y'-\Delta'_y)\right]\right\}$$
$$= \Psi\left(x-x', y-y', \Delta_x, \Delta_y, \Delta'_x, \Delta'_y\right). \quad (3.1)$$

3. *The variance of the increments does not depend on the spatial origin, i.e. there exists a function φ such that:*

$$\forall(\Delta_x, \Delta_y), \quad \mathrm{var}\{F(x,y) - F(x-\Delta_x, y-\Delta_y)\} = \varphi(\Delta_x, \Delta_y). \quad (3.2)$$

4. *The correlation of the field can be expressed as:*

$$\mathrm{E}\left\{F(x,y)F(x',y')\right\} = \frac{1}{2}\left[\varphi(x,y) + \varphi(x',y') - \varphi(x-x', y-y')\right]. \quad (3.3)$$

Proof: See Appendix B.

Here, $\varphi(x, y)$ is called the structure function. It follows in a straightforward manner that the structure function has the following properties:

1. $\varphi(0,0) = 0$.
2. $\forall(x, y) \in \mathbb{R}^2$, $\varphi(-x, -y) = \varphi(x, y)$.
3. $\forall(x, y) \in \mathbb{R}^2$, $\varphi(x, y) \in \mathbb{R}_+$.

Moreover, it is clear that for such fields, the structure function must satisfy additional constraints in order to lead to admissible correlation functions; namely, it cannot increase more rapidly than the square of a 2D norm:

Proposition 3.5. *Let $F(x,y)$ be a 2D field with stationary increments of order $(1,0)/(0,1)$ and let φ be its structure function. Then, $\forall \chi \in \mathbb{R}_+^*$, there exists $A \in \mathbb{R}_+^*$ such that $\forall(\Delta_x, \Delta_y) \in \mathbb{R}^2$, $|\Delta_x| + |\Delta_y| > \chi$, we have $\varphi(\Delta_x, \Delta_y) \leq A\mathcal{N}(\Delta_x, \Delta_y)^2$, where $\mathcal{N}(\cdot,\cdot)$ is any norm of $\mathbb{R}^2$.*

Proof: See Appendix C.

It is clear that by restricting our study to the Gaussian case, we can determine more precisely the form of the structure function of any 2D self-similar process with stationary increments:

Proposition 3.6. *A 2D Gaussian zero-mean random field $F(x,y)$ with stationary increments of order $(1,0)/(0,1)$ is self-similar if and only if there exists a bounded π-periodic function $f : \mathbb{R} \to \mathbb{R}$ such that $\varphi(x,y) = \rho^{2H} f(\theta)$, where $0 < H < 1$, $\rho = \sqrt{x^2+y^2}$, $\theta = arg(x + \imath y)$.*

Proof: A Gaussian random field is determined by its first and second-order statistics; thus, a zero-mean Gaussian field is completely characterized by its structure function.
"$\Leftarrow$": If $\varphi(x,y) = \rho^{2H} f(\theta)$, according to (4), $F(ax,ay)$ and $a^H F(x,y)$ have the same probability distribution. Consequently, $F(x,y)$ is self-similar.
"$\Rightarrow$": If the process with stationary increments of order $(1,0)$ $/(0,1)$ is self-similar, then by using (3.2), there exists $H \in \mathbb{R}$ such that $\forall a > 0, \forall(x,y)$, $\varphi(ax,ay) = a^{2H}\varphi(x,y)$. We then have
$\forall \rho > 0$, $\forall \theta \in (-\pi, \pi]$, $\varphi(\rho\cos\theta, \rho\sin\theta) = \rho^{2H}\varphi(\cos\theta, \sin\theta)$.
Furthermore, according to Proposition 3.5, H must belong to $(0,1)$ and $\varphi(\cos\theta, \sin\theta)$ must be a bounded function of θ. Then, φ takes on the expected form by setting $f(\theta) = \varphi(\cos\theta, \sin\theta)$.

As claimed in the introduction, we have obtained the general form of the structure function of a self-similar Gaussian field with stationary increments of order $(1,0)/(0,1)$. The possible anisotropy of these fields is reflected by the function $f(\theta)$: great spatial variations in some privileged directions will be reflected by a non-homogeneous behaviour with respect to θ. Some particular forms will be discussed in Example 4.3.

In the next section, we will provide some examples of 2D fields with stationary fractional increments.

4. Examples

Example 4.1. As already mentioned, the 2D fBm is probably the most famous 2D self-similar field with stationary increments of order (1,0)/(0,1). More precisely, we can prove that:

Proposition 4.1. *A 2D fBm with parameter $0 < H < 1$ has stationary fractional increments of order $(D,0)/(0,D)$ for all $D \in \mathbb{R}$, $D > H$.*

Using Proposition 3.2, it is obvious that if the process has stationary fractional increments of order $(D,0)/(0,D)$ for $D \in (H,1)$, it also has stationary increments of order $(1,0)/(0,1)$, thereby re-demonstrating the well-known property of the fBm. But this proposition highlights a finer connection between the fractal character (given by the Hurst exponent) and the property of stationarity of the increments.

Some more general "$1/f$" fields with stationary increments of order (1,0)/(0,1) exhibit the same behaviour:

Proposition 4.2. *Consider a field with stationary increments of order (1,0)/(0,1) and let the Fourier transform of its structure function (in the sense of Schwartz distributions) be denoted by* $\hat{\varphi}(\cdot,\cdot)$. *If there exists a bounded function* $B(\cdot,\cdot)$ *such that* $\hat{\varphi}(\omega_x,\omega_y) = B(\omega_x,\omega_y)(\omega_x^2+\omega_y^2)^{-H-1}$, *the random field has stationary increments of order* $(D,0)/(0,D)$ *with* $D > H$.

The proofs of the two above propositions are provided in Appendix D..

Example 4.2. We now consider other examples of 2D fields which are easy to generate since they are derived from 1D processes with stationary increments.

Proposition 4.3. *If* $(A_i)_{i\in\{1,\dots,n\}}$ *are* $n \in \mathbb{N}^*$ *random 1D processes with stationary increments of order D and their increments of order D are mutually uncorrelated, then the 2D random field:*

$$F(x,y) = \sum_{i=1}^{n} A_i\left(\alpha_i x + \beta_i y\right),$$

where $(\alpha_1,\dots,\alpha_n,\beta_1,\dots,\beta_n) \in \mathbb{R}^{2n}$ *has stationary increments of order* $(D,0)/(0,D)$.

Proof: The increments of order $(D,0)$ of $F(x,y)$ can be expressed as:

$$\begin{aligned}\Delta^{(D,0)}F(x,y;\Delta_x,0) &= \sum_{k=0}^{\infty}(-1)^k\binom{D}{k}\sum_{i=1}^{n}A_i\left[\alpha_i(x-k\Delta_x)+\beta_i y\right]\\ &= \sum_{i=1}^{n}\Delta^D A_i\left(\alpha_i x+\beta_i y\,;\,\alpha_i\Delta_x\right)\end{aligned}$$

Therefore, the correlation of these increments is:

$$\begin{aligned}&\mathrm{E}\left\{\Delta^{(D,0)}F(x,y;\Delta_x,0)\Delta^{(D,0)}F(x',y';\Delta'_x,0)\right\}\\ =\;& \sum_{i=1}^{n}\mathrm{E}\left\{\Delta^D A_i\left(\alpha_i x+\beta_i y\,;\,\alpha_i\Delta_x\right)\Delta^D A_i\left(\alpha_i x'+\beta_i y'\,;\,\alpha_i\Delta'_x\right)\right\}\end{aligned}$$

As the increments of order D of each function A_i are stationary processes and mutually uncorrelated, there exists a function ψ such that:

$$\mathrm{E}\left\{\Delta^{(D,0)}F(x,y;\Delta_x,0)\Delta^{(D,0)}F(x',y';\Delta'_x,0)\right\} = \psi(x-x',y-y',\Delta_x,\Delta'_x)$$

and thus the increments of order $(D,0)$ of $F(x,y)$ are stationary.

Similarly, one can prove that the process has stationary increments of order $(D,0)/(0,D)$.

Remark 4.1. If $D \le 1$, Proposition 3.2 guarantees that the field $F(x,y)$ also has stationary increments of order $(1,0)/(0,1)$ and, consequently, the structure function of $F(x,y)$ exists, as well as the structure functions of the processes $(A_i)_{i\in\{1,\dots,n\}}$. We will denote them by $\varphi_F(x,y)$ and $\varphi_i(x)$ respectively. Then, using Statement (3) in Proposition 3.4, one can easily verify that:

$$\begin{aligned}\varphi_F(\Delta_x,\Delta_y) &= \mathrm{var}\left\{F(x,y)-F(x-\Delta_x,y-\Delta_y)\right\}\\ &= \mathrm{var}\sum_{i=1}^{n}\left\{A_i\left(\alpha_i x+\beta_i y\right)-A_i\left[\alpha_i(x-\Delta_x)+\beta_i(y-\Delta_y)\right]\right\}\end{aligned}$$

By proceeding in the same fashion as the proof of Proposition 3.1, it can be shown that the processes $(A_i)_{i\in\{1,\dots,n\}}$ have stationary increments of order 1 which are mutually uncorrelated and therefore we obtain:

$$\begin{aligned} &\operatorname{var}\sum_{i=1}^{n}\{A_i(\alpha_i x+\beta_i y)-A_i[\alpha_i(x-\Delta_x)+\beta_i(y-\Delta_y)]\} \\ =\; &\sum_{i=1}^{n}\operatorname{var}\{A_i(\alpha_i x+\beta_i y)-A_i(\alpha_i x+\beta_i y-\alpha_i\Delta_x-\beta_i\Delta_y)\} \end{aligned}$$

This yields a simple formula for computing the structure function of $F(x,y)$ from the structure functions of the one-dimensional processes A_i:

$$\varphi_F(\Delta_x,\Delta_y)=\sum_{i=1}^{n}\varphi_i(\alpha_i\Delta_x+\beta_i\Delta_y).$$

Remark 4.2. When all the processes $(A_i)_{i\in\{1,\dots,n\}}$ are 1D fBms, they have (according to Proposition 4.1) stationary increments of orders $(D_i)_{i\in\{1,\dots,n\}}$, $\forall D_i > H_i$ and, consequently, $F(x,y)$ have stationary increments of order $(D,0)/(0,D)$, $\forall D > H$, $H=\max_{i\in\{1,\dots,n\}} H_i$.

Remark 4.3. If the processes $(A_i)_{i\in\{1,\dots,n\}}$ are 1D fBms with the same Hurst parameter H, the 2D process $F(x,y)$ is also self-similar.

We synthesized such a field, when $(A_i)_{i\in\{1,\dots,n\}}$ are 1D fBms and $n=2$, $\alpha_1=\beta_2=1$, $\alpha_2=\beta_1=0$. The fBms were generated through an accurate wavelet synthesis method [13], [14]. A typical realization is given in Fig. 4.1.

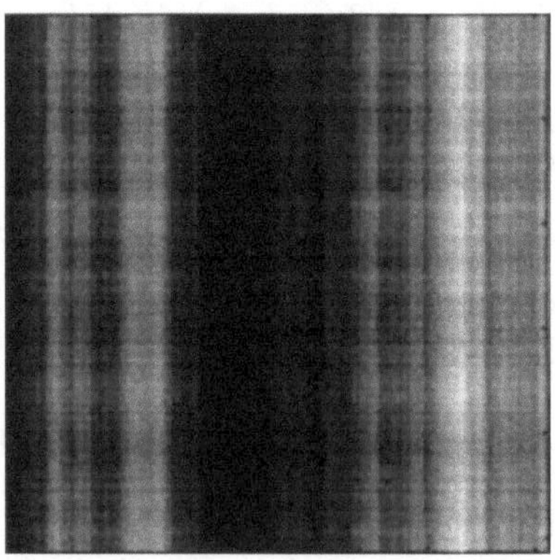

Fig. 4.1. A realization of the anisotropic field in Example 4.2 ($H_1=0.3$, $H_2=0.6$).

Example 4.3. An anisotropic field with stationary increments of order (1,0)/(0,1) can be generated by making the structure function dependent on the orientation.

For example, an anisotropic self-similar process with stationary increments is obtained with the following structure function:

$$\varphi(x,y) = \rho^{2H} f(\theta), \quad \rho = (x^2+y^2)^{1/2}, \; \theta = \arg(x+\imath y)$$
$$\forall\theta \in [0,\pi), \; f(\theta) = f(\theta+\pi)$$
$$\forall\theta \in [0,\pi), \; f(\theta) = 0 \quad \text{if } \theta \notin \mathcal{S}$$

with $0 < H < 1$ and $\mathcal{S} \subset [0,\pi)$.

Another interesting example is given by:

$$\varphi(x,y) = \rho^{2H(\theta)}$$
$$H(\theta) = H_1 \cos^2(\theta-\theta_0) + H_2 \sin^2(\theta-\theta_0)$$

with $0 \leq \theta_0 < \pi$, $0 < H_1 < 1$ and $0 < H_2 < 1$. An application to the analysis of biomedical images of a similar stochastic model can be found in [15].
These fields can be generated by using the incremental Fourier synthesis [16] and by appropriately modifying the correlation function of the increments according to the chosen structure function. A realization of the latter field is presented in Fig. 4.2. The anisotropy of this field can be observed if it is compared with an isotropic 2D fBm (see Fig. 4.3).

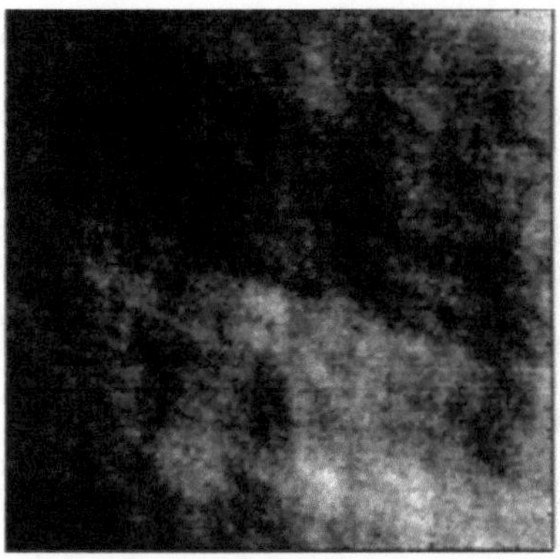

Fig. 4.2. A realization of the random field with structure function $\varphi(x,y) = \rho^{2H(\theta)}$, $H_1 = 0.01$, $H_2 = 0.37$, $\theta_0 = \pi/6$.

Fig. 4.3. A realization of a 2D fBm with $H = 0.19$.

5. Conclusion

In this paper, we have defined non-stationary anisotropic fields with fractional stationary increments and proceeded to study some of their properties. A particularly interesting result is the stationarity of the fractional increments of the 2D fBm for orders higher than its Hurst parameter.

For stationary increments of order $(1,0)/(0,1)$, we have defined the structure function which entirely characterizes the process. A bounding condition has been determined for this function and, in the case of Gaussian self-similar processes, its general form has been provided.

The given results can be useful in many applications. As shown in this paper, they provide a theoretical basis for the synthesis of these fields. They also allow us to analyze many anisotropic fractal phenomena from a statistical viewpoint. For example, one can use these models to derive optimal interpolation methods or to build decision rules. We are currently investigating some of these problems.

References

1. Mandelbrot, B.B., Van Ness, J.W. (1968): Fractional Brownian Motions, Fractional Noises and Applications. *SIAM Rev.* 10, 422–437.
2. Flandrin, P. (1992): Wavelet Analysis and Synthesis of Fractional Brownian Motion. *IEEE Trans. on Information Theory* 38, 910–917.
3. Reed, I.S., Lee, P.C., Truong, T.K. (1995): Spectral Representation of Fractional Brownian Motion in n Dimensions and its Properties. *IEEE Trans. on Information Theory* 41, 1439–1451.
4. Barnsley, M.F., Devaney, R.L., Mandelbrot, B.B. et al. (1988): *The Science of Fractal Images.* Springer Verlag, New York.
5. Mandelbrot, B.B. (1983): *The Fractal Geometry of Nature.* W.H. Freeman and Company, New York.
6. Yaglom, A.M. (1986): *Correlation Theory of Stationary and Related Random Functions* Springer-Verlag, New York.
7. Samorodnitsky, G., Taqqu, M.S. (1994): *Stable Non-Gaussian Random Processes: Stochastic Models with Infinite Variance.* Chapman & Hall, New York.
8. Popescu, B., Costa, P., Larzabal, P., Clergeot, H. (1996): Extraction of Parameters from a Wavelet Packet Analysis for Terrain Matching. *IEEE Symp. Time-Frequency, Time-Scale Analysis, Paris, June 18-21, 1996,* 493–496.
9. Hosking, J. R. M. (1981): Fractional differencing. *Biometrika,* 68:165–176
10. Doob, J. L. (1967): *Stochastic Processes,* Wiley, New York
11. Pesquet-Popescu, B., Larzabal, P. (1996): Wavelet Packet Analysis of 2D Processes with Stationary Fractional Increments. Technical Report, LESIR, ENS Cachan, Jan 1997.
12. Sayles, R.S., Thomas, T.R. (1977): The Spatial Representation of Surface Roughness by Means of the Structure Function: a Practical Alternative to Correlation. *Wear.* 42, 263–276.
13. Sellan, F. (1995): Synthèse de mouvements browniens fractionnaires l'aide de la transformation par ondelettes. *C.R.A.S., Série I* 321, 351–358.
14. Abry, P., Sellan, F. (1996): The Wavelet-Based Synthesis for the Fractional Brownian Motion Proposed by F. Sellan and Y. Meyer: Remarks and Fast Implementation. *Applied Computational Harmonic Analysis* 3, 377–383.
15. Loussot, T., Harba, R., Jacquet, G., Benhamou, G.L., Lespessailles, E., Julien, A. (1996): An oriented Fractal Analysis for the Characterization of Texture, Application to Bone Radiographs. *Proc. EUSIPCO, rieste, 10-13 sept. 1996,* 371–374.
16. Kaplan, L.M., Kuo, C.C.J. (1996): An Improved Method for 2-D Self-Similar Image Synthesis. *IEEE Trans. on Image Processing* 5, 754–761.
17. Guelfand, I.M., Chilov, G.E. (1962): *Les distributions.* Dunod, Paris.
18. Blanc-Lapierre, A., Picinbono, B. (1981): *Fonctions Aléatoires.* Masson, Paris.

Appendices

A. Proof of Property 3.3

Let the field $F(x,y)$ have stationary increments of order D in all the space directions and let these increments be mutually stationary. Note that the increments of order D in the direction given by $(\cos\theta, \sin\theta)$, $\theta \in [0,\pi)$ can be defined as:

$$\forall a \in \mathbb{R},\ \Delta_\theta^D F(x,y;a) = \left(1 - q_{a\cos\theta}^{-1} q'^{-1}_{a\sin\theta}\right)^D F(x,y).$$

Since the increments $\Delta_\theta^D F(x,y;a)$ are mutually stationary $\forall\theta \in [0,\pi)$, $\forall a \in \mathbb{R}$, then this property holds in particular for $\theta = 0$ and for $\theta = \pi/2$. Besides, we have:

$$\begin{aligned} \Delta_0^D F(x,y;a) &= \Delta^{(D,0)} F(x,y;a,0) \\ \Delta_{\pi/2}^D F(x,y;a) &= \Delta^{(0,D)} F(x,y;0,a), \end{aligned}$$

which proves that $\forall a \in \mathbb{R}$ the increments $\Delta^{(D,0)}F(x,y;a,0)$ and $\Delta^{(0,D)}F(x,y;0,a)$ are stationary and mutually stationary.
The reciprocal statement is proved by remarking that the increments $\Delta_\theta^D F(x,y;a)$ can be interpreted in terms of a linear filtering of $F(x,y)$ with transfer function $G(p,s) = \left(1 - e^{-p\,a\cos\theta} e^{-s\,a\sin\theta}\right)^D$. Therefore, we can write:

$$\begin{aligned} \Delta_\theta^D F(x,y;a) &= G_x\left(q_x,q_y\right) \Delta^{(D,0)} F(x,y;a\cos\theta,0) \\ &\quad + G_y\left(q_x,q_y\right) \Delta^{(0,D)} F(x,y;0,a\sin\theta), \end{aligned}$$

where $G_x\left(q_x,q_y\right)$ and $G_y\left(q_x,q_y\right)$ correspond respectively to the filters with frequency responses:

$$\begin{aligned} \widehat{G}_x\left(e^{j\omega_x},e^{j\omega_y}\right) &= \frac{G\left(j\omega_x,j\omega_y\right)}{\widehat{D}\left(\omega_x,\omega_y\right)} \frac{|1-e^{-j\,\omega_x\,a\cos\theta}|^D}{\left(1-e^{-j\,\omega_x\,a\cos\theta}\right)^D} \\ \widehat{G}_y\left(e^{-j\omega_x},e^{j\omega_y}\right) &= \frac{G\left(j\omega_x,j\omega_y\right)}{\widehat{D}\left(\omega_x,\omega_y\right)} \frac{|1-e^{-j\,\omega_y\,a\sin\theta}|^D}{\left(1-e^{-j\,\omega_y\,a\sin\theta}\right)^D} \end{aligned}$$

with $\widehat{D}\left(\omega_x,\omega_y\right) = |1-e^{-j\,\omega_y\,a\cos\theta}|^D + |1-e^{-j\,\omega_y\,a\sin\theta}|^D$. These frequency responses are such that:

$$|\widehat{G}_x\left(e^{j\omega_x},e^{j\omega_y}\right)| = |\widehat{G}_y\left(e^{j\omega_x},e^{j\omega_y}\right)| = \frac{|G\left(j\omega_x,j\omega_y\right)|}{\widehat{D}\left(\omega_x,\omega_y\right)}.$$

Since $|G\left(j\omega_x,j\omega_y\right)| \le 2^D \left[|\sin\left(\frac{a}{2}\,\omega_x\,\cos\theta\right)| + |\sin\left(\frac{a}{2}\,\omega_y\,\cos\theta\right)|\right]^D$, we have:

$$\frac{|G\left(j\omega_x,j\omega_y\right)|}{\widehat{D}\left(\omega_x,\omega_y\right)} \le \frac{\left[|\sin\left(\frac{a}{2}\,\omega_x\,\cos\theta\right)| + |\sin\left(\frac{a}{2}\,\omega_y\,\cos\theta\right)|\right]^D}{|\sin\left(\frac{a}{2}\,\omega_x\,\cos\theta\right)|^D + |\sin\left(\frac{a}{2}\,\omega_y\,\cos\theta\right)|^D} \le \max\left(1, 2^{D-1}\right).$$

Hence, we conclude that $\Delta_\theta^D F(x,y;a)$ is the sum of two stationary, mutually stationary (and harmonizable) fields ($\Delta^{(D,0)}F(x,y;a\cos\theta,0)$ and $\Delta^{(0,D)}F(x,y;0,a\sin\theta)$), each processed by a filter with a bounded frequency response. This enables us to obtain the desired properties for $\Delta_\theta^D F(x,y;a)$ (see [18]) .

B. Proof of Proposition 3.4

(1)⇒(2): This property is shown to be obvious by noting that:

$$F(x,y) - F(x-\Delta_x, y-\Delta_y) = \Delta^{(1,0)}F(x,y;\Delta_x,0) + \Delta^{(0,1)}F(x-\Delta_x,y;0,\Delta_y).$$

(2)⇒(3): The desired result follows by taking $x' = x, y' = y, \Delta'_x = \Delta_x, \Delta'_y = \Delta_y$ and $\Psi(0,0,\Delta_x,\Delta_x,\Delta_y,\Delta_y) = \varphi(\Delta_x,\Delta_y)$.
(3)⇒(4): Since we have assumed in Section 2. that $F(0,0) = 0$, we can write:

$$\begin{aligned}
& \mathrm{E}\left\{F(x,y)F(x',y')\right\} \\
= & \frac{1}{2}\left[\mathrm{E}\left\{F(x,y)^2\right\} + \mathrm{E}\left\{F(x',y')^2\right\} - \mathrm{E}\left\{[F(x,y)-F(x',y')]^2\right\}\right] \\
= & \frac{1}{2}\left[\varphi(x,y) + \varphi(x',y') - \varphi(x-x',y-y')\right].
\end{aligned}$$

(4)⇒(1): This result is obtained by writing:

$$\begin{aligned}
& \mathrm{E}\left\{[F(x,y)-F(x-\Delta_x,y-\Delta_y)]\left[F(x',y')-F(x'-\Delta'_x,y'-\Delta'_y)\right]\right\} \\
= & \frac{1}{2}\left[\varphi(x-x'-\Delta_x,y-y'-\Delta_y) + \varphi(x-x'+\Delta'_x,y-y'+\Delta'_y)\right. \\
& \left.-\varphi(x-x',y-y') - \varphi(x-x'+\Delta'_x-\Delta_x,y-y'+\Delta'_y-\Delta_y)\right],
\end{aligned}$$

which depends only on $x-x', y-y', \Delta_x, \Delta'_x, \Delta_y, \Delta'_y$.

C. Proof of Proposition 3.5

We remark that:

$$\begin{aligned}
\varphi(\Delta_x+\Delta'_x,\Delta_y+\Delta'_y) &= \mathrm{E}\left\{\left[F(x,y)-F(x-\Delta_x-\Delta'_x,y-\Delta_y-\Delta'_y)\right]^2\right\} \\
&= \mathrm{E}\left\{[F(x,y)-F(x-\Delta_x,y-\Delta_y)+F(x-\Delta_x,y-\Delta_y)\right. \\
&\quad \left.-F(x-\Delta_x-\Delta'_x,y-\Delta_y-\Delta'_y)]^2\right\}.
\end{aligned}$$

According to the Cauchy-Schwarz inequality, we have:

$$\varphi(\Delta_x+\Delta'_x,\Delta_y+\Delta'_y) \le \varphi(\Delta_x,\Delta_y) + \varphi(\Delta'_x,\Delta'_y) + 2\sqrt{\varphi(\Delta_x,\Delta_y)\varphi(\Delta'_x,\Delta'_y)}.$$

It follows that $\varphi(\Delta_x+\Delta'_x,\Delta_y+\Delta'_y)^{1/2} \le \varphi(\Delta_x,\Delta_y)^{1/2} + \varphi(\Delta'_x,\Delta'_y)^{1/2}$. This implies that $\forall(n,m) \in \mathbb{Z}^2$,

$$\begin{aligned}
\varphi(n,m)^{1/2} &\le \varphi(n,0)^{1/2} + \varphi(0,m)^{1/2} = \varphi(|n|,0)^{1/2} + \varphi(0,|m|)^{1/2} \\
&\le |n|\varphi(1,0)^{1/2} + |m|\varphi(0,1)^{1/2}.
\end{aligned}$$

Besides, $\forall(\Delta_x,\Delta_y) \in \mathbb{R}^2$ there exists $(n,m) \in \mathbb{Z}^2$ and $(x_0,y_0) \in [0,1)^2$ such that:

$$\Delta_x = \begin{cases} n+x_0 & \text{if } \Delta_x \ge 0 \\ n-x_0 & \text{if } \Delta_x < 0 \end{cases} \quad \text{and} \quad \Delta_y = \begin{cases} m+y_0 & \text{if } \Delta_y \ge 0 \\ m-y_0 & \text{if } \Delta_y < 0 \end{cases}.$$

Therefore,

$$\begin{aligned}
\varphi(\Delta_x, \Delta_y)^{1/2} &= \varphi(n \pm x_0, m \pm y_0)^{1/2} \\
&\leq \varphi(n,m)^{1/2} + \varphi(\pm x_0, \pm y_0)^{1/2} \\
&\leq |n|\varphi(1,0)^{1/2} + |m|\varphi(0,1)^{1/2} + \varphi(x_0,0)^{1/2} + \varphi(0,y_0)^{1/2} \\
&\leq (|n| + |m| + 2) \sup_{(x,y)\in[0,1]^2} \varphi(x,y) \\
&\leq (|\Delta_x| + |\Delta_y| + 2) \sup_{(x,y)\in[0,1]^2} \varphi(x,y)
\end{aligned}$$

For the norm $\mathcal{N}(x,y) = |x| + |y|$, the desired inequality is obtained by taking, for example, $A = \sqrt{\left(1 + \frac{2}{\chi}\right) \sup_{(x,y)\in[0,1]^2} \varphi(x,y)}$.

The final result follows by recalling that all norms in a vector space of finite dimension are equivalent.

D. Fractional Increments of the 2D fBm

As mentioned in [3], for $0 < H < 1$, the correlation between $B_H(\mathbf{x})$ and $B_H(\mathbf{y})$ of any two position-vectors $(\mathbf{x},\mathbf{y}) \in (\mathbb{R}^2)^2$ for a two-dimensional fBm is:

$$R_{B_H}(\mathbf{x},\mathbf{y}) = \mathrm{E}\left\{B_H(\mathbf{x})B_H(\mathbf{y})\right\} = \frac{1}{2}V_H(\|\mathbf{y}\|^{2H} + \|\mathbf{x}\|^{2H} - \|\mathbf{y}-\mathbf{x}\|^{2H}). \quad \text{(D.1)}$$

Using the Fourier transform of the tempered distribution $\|\mathbf{x}\|^D$ ([17]):[3]

$$\|\mathbf{x}\|^D \longrightarrow \frac{2^{D+2}\pi\Gamma(\frac{D+2}{2})}{\Gamma(-D/2)}\|\boldsymbol{\omega}\|^{-(D+2)}, \ D \notin 2\mathbb{Z}$$

one can obtain the Fourier transform of the correlation function of the fBm (for $D = 2H$):

$$S_{B_H}(\boldsymbol{\omega},\boldsymbol{\nu}) = \frac{1}{2}V_H C\left[\frac{1}{\|\boldsymbol{\omega}\|^{2H+2}}\delta(\boldsymbol{\nu}) + \frac{1}{\|\boldsymbol{\nu}\|^{2H+2}}\delta(\boldsymbol{\omega}) - \frac{1}{\|\boldsymbol{\omega}\|^{2H+2}}\delta(\boldsymbol{\omega}-\boldsymbol{\nu})\right] \quad \text{(D.2)}$$

with

$$C = \frac{2^{2H+2}\pi\Gamma(H+1)}{\Gamma(-H)}.$$

We denote $\boldsymbol{\omega} = (\omega_x, \omega_y)$ and $\boldsymbol{\nu} = (\nu_x, \nu_y)$. One can observe that the 2D operator $\Delta^{(D,0)}$ that allows calculating fractional increments of order $(D,0)$ corresponds to a filter with frequency response $G_{\Delta_x}(\omega_x) = \left(1 - e^{-j\omega_x\Delta_x}\right)^D$. Thus, using (D.2) and the formula of interferences, the Fourier transform of the correlation function of the increments of order $(D,0)$ of the fBm, $\mathrm{E}\left\{(\Delta^{(D,0)})B_H(\mathbf{x};\Delta_x,0)\Delta^{(D,0)}B_H(\mathbf{y};\Delta'_x,0)^*\right\}$ is:

---

[3] In the context of distributions, this Fourier transform is a pseudo-function.

$$G_{\Delta_x}(\omega_x)G_{\Delta'_x}(\nu_x)^* S_{B_H}(\boldsymbol{\omega},\boldsymbol{\nu}) = \frac{V_H}{2}C\left[\frac{G_{\Delta_x}(\omega_x)G_{\Delta'_x}(0)^*}{\|\boldsymbol{\omega}\|^{2H+2}}\delta(\boldsymbol{\nu}) + \frac{G_{\Delta_x}(0)G_{\Delta'_x}(\nu_x)^*}{\|\boldsymbol{\nu}\|^{2H+2}}\delta(\boldsymbol{\omega}) - \frac{G_{\Delta_x}(\omega_x)G_{\Delta'_x}(\nu_x)^*}{\|\boldsymbol{\omega}\|^{2H+2}}\delta(\boldsymbol{\omega}-\boldsymbol{\nu})\right].$$

Since $G_{\Delta_x}(0) = G_{\Delta'_x}(0) = 0$, we conclude that the cross-spectral densities of the processes corresponding to the increments of order $(D,0)$ of the fBm depend on Δ_x and Δ'_x, and can be expressed as:

$$\begin{aligned} S_{\Delta^{(D,0)}B_H}(\boldsymbol{\omega},\boldsymbol{\nu}) &= -\frac{V_H}{2}C\frac{G_{\Delta_x}(\omega_x)G_{\Delta'_x}(\nu_x)^*}{\|\boldsymbol{\omega}\|^{2H+2}}\delta(\boldsymbol{\omega}-\boldsymbol{\nu}) \\ &= -\frac{V_H}{2}C\frac{(1-e^{-j\omega_x\Delta_x})^D(1-e^{j\nu_x\Delta'_x})^D}{\|\boldsymbol{\omega}\|^{2H+2}}\delta(\boldsymbol{\omega}-\boldsymbol{\nu}) \\ &= -\frac{V_H}{2}C\frac{(1-e^{-j\omega_x\Delta_x})^D(1-e^{j\omega_x\Delta'_x})^D}{\|\boldsymbol{\omega}\|^{2H+2}}\delta(\boldsymbol{\omega}-\boldsymbol{\nu}). \end{aligned}$$

Now, by passing into polar coordinates:

$$\omega_x = \omega_r \cos\omega_\theta, \quad \omega_y = \omega_r \sin\omega_\theta, \quad \omega_r > 0$$

we obtain:

$$-\frac{V_H}{2}C\frac{(1-e^{-j\omega_x\Delta_x})^D(1-e^{j\omega_x\Delta'_x})^D}{\|\boldsymbol{\omega}\|^{2H+2}} = -\frac{V_H}{2}C\frac{(1-e^{-j\omega_r\cos\omega_\theta\Delta_x})^D(1-e^{j\omega_r\cos\omega_\theta\Delta'_x})^D}{\omega_r^{2H+2}}.$$

We remark that for $D > H$:

$$-\frac{V_H}{2}C\frac{(1-e^{-j\omega_r\cos\omega_\theta\Delta_x})^D(1-e^{j\omega_r\cos\omega_\theta\Delta'_x})^D}{\omega_r^{2H+2}} \sim_{\omega_r\to 0} \mathcal{O}\left(\omega_r^{2D-2H-2}\right)$$

is summable, which is the required condition for the correlation function to be harmonizable [18]. As the spectral density is proportional to $\delta(\boldsymbol{\omega}-\boldsymbol{\nu})$, it is shown in [18] that this provides a sufficient condition for having a stationary process. This proves the stationarity of the increments of order $(D,0)$. The same arguments can be invoked to prove the stationarity of increments of order $(D,0)/(0,D)$, $D > H$.

The above result holds for more general processes with stationary increments. Indeed, if a random process $F(\mathbf{x})$ has stationary increments of order (1,0)/(0,1), its correlation function can be expressed using its structure function as:

$$R_f(\boldsymbol{x},\boldsymbol{y}) = \frac{1}{2}\left[\varphi(\mathbf{x}) + \varphi(\mathbf{y}) - \varphi(\boldsymbol{x}-\boldsymbol{y})\right], \quad (\boldsymbol{x},\boldsymbol{y}) \in (\mathbb{R}^2)^2.$$

As previously noted, the spectral density of this process is:

$$S_F(\boldsymbol{\omega},\boldsymbol{\nu}) = \frac{1}{2}\left[\widehat{\varphi}(\boldsymbol{\omega})\delta(\boldsymbol{\nu}) + \widehat{\varphi}(\boldsymbol{\nu})\delta(\boldsymbol{\omega}) - \widehat{\varphi}(\boldsymbol{\omega})\delta(\boldsymbol{\omega}-\boldsymbol{\nu})\right],$$

where $\widehat{\varphi}$ is the Fourier transform (possibly, in a distributional sense) of φ. In this case, the Fourier transform of the correlation function of the increments of order $(D,0)$ $(\mathrm{E}\left\{(\Delta^{(D,0)})F(\mathbf{x};\Delta_x,0)\Delta^{(D,0)}F(\mathbf{y};\Delta'_x,0)^*\right\})$ is:

$$G_{\Delta_x}(\omega_x)G_{\Delta'_x}(\nu_x)^* S_F(\boldsymbol{\omega},\boldsymbol{\nu}) = \frac{1}{2}\left[G_{\Delta_x}(\omega_x)G_{\Delta'_x}(0)^*\widehat{\varphi}(\boldsymbol{\omega})\delta(\boldsymbol{\nu})\right.$$
$$\left.+G_{\Delta_x}(0)G_{\Delta'_x}(\nu_x)^*\widehat{\varphi}(\boldsymbol{\nu})\delta(\boldsymbol{\omega}) - G_{\Delta_x}(\omega_x)G_{\Delta'_x}(\nu_x)^*\widehat{\varphi}(\boldsymbol{\omega})\delta(\boldsymbol{\omega}-\boldsymbol{\nu})\right].$$

When there exists a bounded function $B(\cdot)$ such that $\widehat{\varphi}(\boldsymbol{\omega}) = B(\boldsymbol{\omega})\|\boldsymbol{\omega}\|^{-2H-2}$, we can conclude that the random process has stationary increments of order $(D,0)/(0,D)$, $D > H$.

Obviously, similar properties hold for n–dimensional signals, with $n \geq 1$.

NETWORK TRAFFIC ANALYSIS

Four approaches to the fractional Brownian storage

I. Norros

VTT Information Technology
P.O. Box 1202, 02044 VTT, FINLAND
ilkka.norros@vtt.fi

Abstract Different approaches to study the storage with fractional Brownian input process are presented: scaling relations, theory of large deviations, a Beneš formula, and use of a Girsanov type formula.

Keywords: fractional Brownian motion, queue, storage, self-similarity, large deviations.

1. Introduction

Extensive data traffic measurements conducted at Bellcore since 1989 and a lot of subsequent work by many authors on many kinds of traffic data have provided convincing evidence that modern teletraffic is often approximately self-similar in the second order sense [11]. If such traffic is aggregated from a sufficiently large number of individual streams to justify a Gaussian approximation, its variation can be modeled by fractional Gaussian noise. In continuous time, the cumulative traffic process A_t, that is, the amount of traffic arriving between 0 and t, can be modelled with "fractional Brownian traffic" [13, 14]

$$A_t = mt + \sqrt{am}Z_t, \quad t \in (-\infty, \infty), \tag{1.1}$$

where Z is a normalized fractional Brownian motion (fBm). The process Z is defined as the centered Gaussian process with stationary increments, continuous paths and variance function $v(t) = EZ_t = |t|^{2H}$. It is a self-similar process:

$$(Z_{\alpha t},\ t \in \Re) \stackrel{\mathcal{D}}{=} (\alpha^H Z_t,\ t \in \Re) \tag{1.2}$$

for every $\alpha > 0$. The model has three parameters: the mean rate m, the coefficient a (index of dispersion at time unit), and the self-similarity parameter H of Z. The traffic is long range dependent when $H \in (\frac{1}{2}, 1)$. For $H \in (0, \frac{1}{2})$, increments on disjoint intervals are negatively correlated. Teletraffic research has almost entirely focused on the long range dependent case, but most mathematical results considered here hold for $H < \frac{1}{2}$ as well.

The discovery of "self-similar traffic" has provided a considerable challenge and stimulus for queueing theory, since most of its traditional methods are based on structures like independent increments, Markov chains, renewal times etc. which are not available when the traffic is self-similar. In fact, the exact queue length distribution (in explicit form) seems to be unknown for any queueing system with a long-range dependent input process. In this paper, we review different approaches to study the queueing process or, rather, storage process arising when traffic defined by (1.1) is fed to an unlimited buffer with constant leak rate $c > 0$. The stationary storage process with fractional Brownian traffic as input is defined as

$$V_t^{(m,a,H,c)} = \sup_{s\le t}\left(A_t - A_s - c(t-s)\right), \quad t \in (-\infty, \infty). \tag{1.3}$$

We also denote

$$V_t^{(b)} = \sup_{s\le t}\left(Z_t - Z_s - b(t-s)\right), \quad t \in (-\infty, \infty). \tag{1.4}$$

Figure 1.1 shows a piece of a simulated realization of the process $V^{(1)}$ with $H = 0.8$. The effect of long range dependence on the storage occupancy is strong — with $H = 1/2$, the mean storage occupancy would be as small as $1/2$.

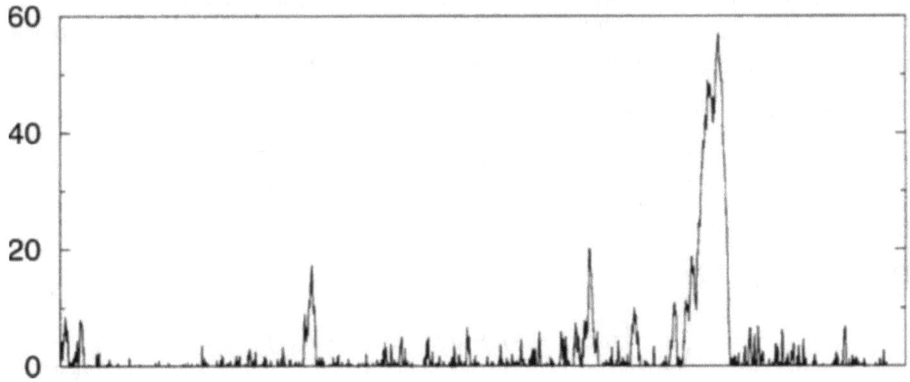

Fig. 1.1. A piece of length 5000 from a realization of $V^{(1)}$ with $H = 0.8$.

Using the self-similarity of Z, it is easy to see that it is sufficient to study the process $V^{(1)}$. This is made explicit in Section 2. $V^{(1)}$ is a stationary process because Z has stationary increments. Since Z is time reversible, $V_0^{(1)}$ is distributed as $\sup_{t\ge 0}(Z_t - t)$ — we shall often use this time-reversed definition for convenience.

The a.s. finiteness of $V^{(1)}$ follows from the ergodicity of fractional Gaussian noise (e.g., [4], Theorem 14.2.1). The formula (1.3) is similar to "Reich's formula" ([2, 17]) for the amount of work (or virtual waiting time) in a queueing system. For $H = \frac{1}{2}$, the process is well known as "reflected Brownian motion", and it has been used in queueing theory as a diffusion approximation for a queueing process (e.g., [9]).

Even the marginal distribution of $V^{(1)}$ is unknown. As already mentioned, many standard methods of storage theory are unapplicable, since there is nothing Markovian with fBm. However, several general techniques are able to yield approximations and bounds. We discuss four such approaches: the self-similarity property alone allows for deriving interesting scaling laws (Section 2.), the theory of large deviations gives logarithmic asymptotics of the queue length distribution (Section 3.), the Beneš approach requires local time analysis but gives something as well (Section 4.), and the use of a Girsanov formula for fBm yields yet another type of approximations (Section 5.). Very recently, L. Massoulie and A. Simonian obtained accurate (better than logarithmic) asymptotics using a fifth method, the theory of extremal values of Gaussian processes[1].

2. Scaling laws

Some properties of $V^{(m,a,H,c)}$ can be deduced from the self-similarity assumption alone. These results do not depend on the fact that Z is Gaussian and hold thus also for other input processes with self-similar variation.

Let us first look at $V^{(m,a,H,c)}$ at different time scales. The storage process is not self-similar, but more flat at larger time scales:

Theorem 2.1. *For any $\alpha > 0$,*

$$\left(V_{\alpha t}^{(m,a,H,c)}\right)_{t\in\Re} \stackrel{\mathcal{D}}{=} \left(\alpha^H V_t^{(m,a,H,c')}\right)_{t\in\Re},$$

where $c' = m + \alpha^{1-H}(c-m)$.

Proof: By (1.2),

$$\begin{aligned}(V_{\alpha t})_{t\in\Re} &= \left(\sup_{s\le t}\left(m\alpha(t-s) + \sqrt{ma}(Z_{\alpha t} - Z_{\alpha s}) - c\alpha(t-s)\right)\right)_{t\in\Re} \\ &\stackrel{\mathcal{D}}{=} \left(\alpha^H \sup_{s\le t}\left(A_t - A_s - (m+\alpha^{1-H}(c-m))(t-s)\right)\right)_{t\in\Re}\end{aligned}$$

□

A similar argument reduces the study of $V^{(b)}$ to that of $V^{(1)}$:

Theorem 2.2. *For any $b > 0$,*

$$\left(V_t^{(b)}\right)_{t\in\Re} \stackrel{\mathcal{D}}{=} \left(b^{-H/(1-H)} V^{(1)}_{b^{1/(1-H)}t}\right)_{t\in\Re}.$$

In particular,

$$P\left(V_t^{(b)} > x\right) = P\left(V_t^{(1)} > b^{H/(1-H)}x\right).$$

---

[1] They showed that $P(V_0^{(1)} > x) \approx Kx^{-\gamma}\exp(-x^{2-2H}/(2\kappa(H)^2))$, where $\gamma = (1-H)(2H-1)/H$. A. Simonian, personal communication.

Scaling laws with practical significance can be obtained by considering the set of system parameters satisfying the storage threshold exceedance criterion

$$\epsilon = P\left(V^{(c)} > x\right). \tag{2.1}$$

Theorem 2.3. *Denote*

$$f(y) = P\left(\sup_{t\geq 0}(Z(t) - yt) > 1\right).$$

Then (2.1) is equivalent to the "buffer dimensioning formula"

$$x = f^{-1}(\epsilon)a^{1/(2(1-H))}c^{(2H-1)/(2(1-H))}\frac{\rho^{1/(2(1-H))}}{(1-\rho)^{H/(1-H)}}, \tag{2.2}$$

where $\rho = m/c$, and to the "bandwidth allocation rule"

$$c = m + f^{-1}(\epsilon)a^{1/(2H)}x^{-(1-H)/H}m^{1/(2H)}. \tag{2.3}$$

Proof: Consider the function

$$q(x,\beta) = P\left(\sup_{t\geq 0}(Z(t) - \beta t) > x\right).$$

By the self-similarity of $Z(t)$ we have

$$q(\alpha x,\beta) = P\left(\sup_{t\geq 0}[Z(\frac{t}{\alpha^{1/H}}) - \frac{\beta}{\alpha}t] > x\right) = q(x, \alpha^{\frac{1-H}{H}}\beta),$$

so that

$$q(x,\beta) = q(1, x^{(1-H)/H}\beta) = f(x^{(1-H)/H}\beta).$$

Since $f(y)$ is obviously strictly decreasing for $y \geq 0$, $f(0) = 1$ and $f(\infty) = 0$, we may write

$$\begin{aligned}\epsilon &= P\left(V^{(c)} > x\right) = P\left(\sup_{t\geq 0}\left[Z(t) - \frac{c-m}{\sqrt{am}}t\right] > \frac{x}{\sqrt{am}}\right)\\ &= f\left(\left(\frac{x}{\sqrt{am}}\right)^{(1-H)/H}\cdot\frac{c-m}{\sqrt{am}}\right)\end{aligned} \tag{2.4}$$

and obtain the desired equations by solving for x and c. □

In the Brownian case $H = \frac{1}{2}$, (2.2) reduces to

$$x = \frac{\rho}{1-\rho}\cdot const. \tag{2.5}$$

In this traditional case we may roughly say that reducing the relative free capacity $1-\rho$ by half costs doubling the storage size. Note that the service rate c has disappeared from the equation!

With $H \neq 1/2$ the situation is different. It is seen from (2.2) that when H is high (the Bellcore measurements mostly give for H values in the region $(0.8, 0.9)$), a substantial increase in utilization, say again halving the free capacity, requires a tremendous amount more storage space. Now, however, the absolute service rate c also effects the relative utilization ρ. The important practical consequence of equation (2.3) is that transmission links with higher capacity can be used with higher utilization without increasing the buffers. The reason for this is the improved multiplexing efficiency.

3. Theory of large deviations

The following theorem gives a lower bound for the tail of the stationary distribution of $V_t^{(1)}$ (i.e., the storage occupancy distribution). Note that a similar bound is obtained in the same way if Z is replaced by some other self-similar process.

Theorem 3.1. *For any $x \geq 0$, we have*

$$P\left(V_0^{(1)} > x\right) \geq \overline{\Phi}\left(\frac{x^{1-H}}{\kappa(H)}\right), \tag{3.1}$$

where

$$\kappa(H) = H^H(1-H)^{1-H}$$

and $\overline{\Phi}(y) = P(Z(1) > y)$ is the residual distribution function of the standard Gaussian distribution.

Proof: The maximum at the right hand side of the obvious inequality

$$P\left(V_0^{(1)} > x\right) \geq \sup_{t \geq 0} P\left(Z_t - t > x\right) = \sup_{t \geq 0} \overline{\Phi}\left(\frac{t+x}{t^H}\right) \tag{3.2}$$

is obtained at

$$t = \frac{Hx}{1-H},$$

yielding the assertion of the theorem. □

Duffield and O'Connell [5] showed that this trivial bound from [13] is in fact accurate in the logarithmic sense. The proof is based on the technique of the theory of large deviations. Define the "free energy function"

$$\mu_t(\beta) = \log E e^{\beta Z_t} = \frac{\beta^2 t^{2H}}{2}$$

and the "entropy function"

$$\mu_t^*(x) = \sup_{\beta}(\beta x - \mu_t(\beta)) = \frac{x^2}{2t^{2H}}.$$

For every t we have

$$\frac{\mu_t(t^{2-2H}\frac{\beta}{t})}{t^{2-2H}} = \frac{\beta^2}{2} = \mu_1(\beta) \doteq \mu(\beta).$$

Denote also $\mu^*(x) = \mu_1^*(x) = x^2/2$.

Theorem 3.2.

$$\lim_{x\to\infty} \frac{1}{x^{2-2H}} \log P\left(V_0^{(1)} > x\right) = -\frac{1}{2\kappa(H)^2}.$$

Proof: **Lower bound:** This part follows immediately from Theorem 3.1, but the following derivation from [5] is more interesting. For any $y > 0$, we have

$$\begin{aligned}
&\liminf_{x\to\infty} \frac{1}{x^{2-2H}} \log P\left(\sup_{t\geq 0}(Z_t - t) > x\right) \\
\geq\ &\liminf_{x\to\infty} \frac{1}{x^{2-2H}} \log P\left(Z_{x/y} - \frac{x}{y} > x\right) \\
=\ &\frac{1}{y^{2-2H}} \liminf_{x\to\infty} \frac{1}{(x/y)^{2-2H}} \log P\left(\frac{Z_{x/y}}{x/y} > 1 + y\right) \\
=\ &-\frac{\mu^*(1+y)}{y^{2-2H}} \\
=\ &-\frac{(1+y)^2}{2y^{2-2H}},
\end{aligned}$$

since

$$\lim_{t\to\infty} \frac{1}{t^{2-2H}} \log P\left(\frac{Z_t}{t} > z\right) = -\mu^*(z),$$

which can be either checked directly (using $\overline{\Phi}(y) \sim \exp(-y^2/2)$) or deduced from the Gärtner-Ellis theorem. It remains to compute that the maximal value of the expression $-(1+y)^2/(2y^{2-2H})$ is $-1/(2\kappa(H)^2)$, obtained at $y = (1-H)/H$.

Upper bound: We prove the upper bound result only for the discrete time version of the process. The reference [5] contains a proof for the intuitively rather obvious fact that the logarithmic limit is the same as in the continuous time case. Denoting $\theta = 1/(2\kappa(H)^2)$, we have

$$\begin{aligned}
&P\left(\sup_{n\geq 1}(Z_n - n) > x\right) \\
\leq\ &\sum_{n=1}^{\infty} P(Z_n - n > x) \\
\leq\ &\sum_{n=1}^{\infty} e^{-\mu_n^*(x+n)}
\end{aligned}$$

$$= e^{-\theta x^{2-2H}} \sum_{n=1}^{\infty} \exp(-\mu_n^*(x+n) + \theta x^{2-2H})$$

$$= e^{-\theta x^{2-2H}} \sum_{n=1}^{\infty} \exp\left(-x^{2-2H}\left(\frac{(1+(n/x))^2}{2(n/x)^{2H}} - \theta\right)\right),$$

where the second step is an application of Chernoff's upper bound

$$P(Z_n > z) \le e^{-\mu_n^*(z)}.$$

To show that the last sum does not make the logarithmic limit larger, note first that its behaviour is similar to that of the expression

$$x \int_0^{\infty} \exp\left(-x^{2-2H}\left(\frac{(1+u)^2}{2u^{2H}} - \theta\right)\right) du.$$

Now, θ is also the minimum of $(1+u)^2/(2u^{2H})$, and the integral converges. Thus, the expression is less than x when x is large. □

The above theorem suggests to approximate the stationary distribution of $V^{(1)}$ by a Weibull distribution:

$$P\left(V_t^{(1)} > x\right) \approx \exp\left(\frac{-x^{2-2H}}{2\kappa(H)^2}\right). \tag{3.3}$$

For the Brownian case $H = 1/2$, this reduces to the exponential distribution with parameter 2, which is the well-known exact result. For $H > 1/2$, however, simulations indicate that the Weibull distribution seems to be stochastically larger than the true distribution. [2]

4. The Beneš method

The so called Beneš' formula (see [2], or [3]) gives a general expression for the probability $P(V_t > x)$ through a partition of the time half-axis $(-\infty, t)$ according to the first (earliest) time point $s < t$ such that the cumulative arrival process A_t satisfies $A_t - A_s - c(t-s) = x$. This first time s is usually characterized by the additional condition that the system be empty at s, i.e. $V_s = 0$ (see [18], pp. 104–108). In the case of a queue with constant service time 1, $A_t - A_s$ is simply the number of customers arriving during the interval $[s,t)$, and Beneš' formula can be written particularly simply as the sum

$$P(V_0 > x) = \sum_{n>x} P(A_{x-n} = n \text{ and } V_{x-n} = 0). \tag{4.1}$$

Simonian and Virtamo [19] generalized this result to the case where A_t is an absolutely continuous process, i.e., the integral of a finite arrival rate process

[2] The new result mentioned in footnote 1 proves this for large x.

A_t. The case that A_t is a fractional Brownian motion was considered in [15] and will be reviewed in this section.

Throughout this section, an expression of the type $P(X \in a + bdx)$ refers to integration with respect to the distribution of $(X - a)/b$.

In our setting, the idea of the Beneš approach is to write the inverse distribution function of $V_0^{(1)}$ as the integral

$$P\left(V_0^{(1)} > x\right) = \int_0^\infty P\left(T^x \in dt\right), \tag{4.2}$$

where

$$T^x = \sup\{t : \ Z_t - t = x\}$$

is the last exit time of the process $Z_t - t$ from the level x (undefined with positive probability for $x > 0$). In the Brownian case $H = \frac{1}{2}$, it is well known that

$$P\left(T^x \in dt\right) = \frac{1}{\sqrt{2\pi t}} e^{-(x+t)^2/(2t)} dt = P\left(Z_t \in x + t + dt\right), \tag{4.3}$$

hence $P(T^x < \infty) = e^{-2x}$ as it should be. In the fractional case, the exact distribution of T^x is (of course!) unknown. However, the "Beneš formula" given in the following theorem suggests useful approximations. The result resembles and is in fact partly inspired by a formula by Durbin [7]. Our technical approach is, however, different from his. Let us denote

$$Z_t^* = \sup\{Z_s : \ s \in [0,t]\}. \tag{4.4}$$

Theorem 4.1.

$$P\left(V_0^{(1)} > x\right) = 2E\left(Z_1^*\right) \int_0^\infty L_H(x,t) P\left(Z_t \in x + t + dt\right), \tag{4.5}$$

where

$$L_H(x,t) = \lim_{u \downarrow t} \frac{1}{(u-t)^{1-H}} P[Z_v < x + v, \forall v > u \,|\, \exists s \in [t,u] : \ Z_s = x + s]$$

is a finite positive function.

Proof: Since T^x obviously has an absolutely continuous distribution, we can write

$$\begin{aligned} & P\left(T^x \in dt\right)/dt \\ = \ & \lim_{u \downarrow t} \frac{1}{u-t} P\left(T^x \in [t,u]\right) \\ = \ & \lim_{u \downarrow t} \left(\frac{1}{(u-t)^H} P\left(\exists s \in [t,u] : \ Z_s = x + s\right) \right. \qquad (4.6) \\ & \left. \cdot \frac{1}{(u-t)^{1-H}} P[Z_v < x + v, \forall v > u \,|\, \exists s \in [t,u] : \ Z_s = x + s] \right). \end{aligned}$$

Now it can be shown that the first factor of the right hand side of (4.6) has the finite limit $2EZ_1^*P(Z_t \in x+t+dt)/dt$ (Theorem 4.2 below). It follows that the limit defining $L_H(x,t)$ also exists and is finite. □

This formula contains two unknown objects, the expectation EZ_1^* and the function $L_H(x,t)$. As regards the first one, it follows from the Sudakov-Fernique inequality ([1], Theorem 2.9) that for $H \in [\frac{1}{2}, 1]$ we have

$$2EZ_1^* \in [\sqrt{2/\pi}, \sqrt{8/\pi}],$$

the lower and upper bounds being achieved at $H = 1$ and $H = 1/2$, respectively. It is no problem to approximate EZ_1^* accurately for practical purposes.

On the other hand, we don't know any reasonable estimate for $L_H(x,t)$. One might think that this makes the whole approach useless. However, it turns out that, as it is often the case with the Beneš approach, the unknown factor does not have a dominating role in the last exit density. Let us replace the whole unknown factor by a constant such that the value of (4.5) is 1 when $x = 0$ (since we have $P(V_0^{(1)} > 0) = 1$). In this way, we obtain the approximation

$$\begin{aligned} P\left(V_0^{(1)} > x\right) &\approx 2(1-H)\int_0^\infty P\left(Z_t \in x+t+dt\right) \\ &= \sqrt{\frac{2}{\pi}}\int_0^\infty \exp\left(-\frac{1}{2}\left[\frac{x}{u^{H/(1-H)}}+u\right]^2\right)du. \end{aligned} \tag{4.7}$$

The expression (4.7) is exact in the case $H = \frac{1}{2}$. Indeed, the homogeneous Markovian structure of Brownian motion implies that $L_{1/2}(x,t)$ is a constant:

Corollary 4.1. *Let W_t denote standard Brownian motion. For any $t > 0$, we have*

$$\lim_{u \downarrow t} \frac{1}{\sqrt{u-t}} P[W_v < x+v, \forall v > u \,|\, \exists s \in [t,u]:\ W_s = x+s] = \sqrt{\frac{\pi}{8}}.$$

The approximation (4.7) is heuristic in the sense that we are able to study its accuracy only numerically. It turns out that (4.7) is smaller than the Weibull approximation (3.3) but the difference is small unless H is rather large, like 0.9. (4.7) seems to be an upper bound for the true value, and thus a better approximation than (3.3).

We conclude this section by stating for fractional Brownian motion a "localization theorem", on which the proof of Theorem 4.1 was based. This result relates the asymptotics of the probabilities of a "localization is space" $\{\exists s \in [t, t+\Delta t]:\ Z_s = x\}$ and a "localization in time" $\{Z_t \in [x, x+\Delta x]\}$. In heuristic infinitesimal stenography, the relation can be written as

$$P\left(\exists s \in [t,t+dt]:\ Z_s = x\right) = \frac{2EZ_1^*}{(dt)^H} P\left(Z_t \in x+dt\right). \tag{4.8}$$

Theorem 4.2. *For any $t > 0$ and $\gamma, x \in \Re$,*

$$\begin{aligned} & \lim_{h\downarrow 0} h^{-H} P\left(\exists s \in [0,h]: \ Z_{t+s} + \gamma(t+s) = x\right) \qquad (4.9) \\ = \ & 2E\left(Z_1^*\right) P\left(Z_t \in x + t + dt\right)/dt. \end{aligned}$$

Here is an outline of the proof [3]

1. We can assume $t = 1$. Moreover, it is sufficient to consider the case $\gamma = 0$. The basic reason for this is that the drift of the scaled process $h^{-H}(Z_{1+ht} - Z_1 + \gamma ht)$, $t \in [0,1]$, goes to zero with h.

2. One can prove the following result, which intuitive content is that the behaviour of Z in small scale is asymptotically independent of its behaviour in large scale. Let $x(h)$ be any bounded real-valued function. Then

$$\mathcal{D} - \lim_{h\downarrow 0} \mathcal{L}\left\{(h^{-H}(Z_{1+ht} - Z_1) : \ t \in [0,1]) \mid Z_1 = x(h)\right\} = \mathcal{L}\left\{Z_t : \ t \in [0,1]\right\}$$

in the sense of weak convergence of probability measures on the space of continuous functions $\mathcal{C}[0,1]$ (equipped with the supremum norm), where $\mathcal{L}\{\cdot\}$ and $\mathcal{L}\{\cdot \mid \cdot\}$ denote the unconditional and conditional law, respectively, of a process having realizations in $\mathcal{C}[0,1]$.

3. Since the mapping $f \mapsto \sup_{t\in[0,1]} f(t)$ is continuous on $\mathcal{C}[0,1]$ and since Z_1^* has a continuous distribution, it follows from the above result that

$$\lim_{h\downarrow 0} P\left[h^{-H} \sup_{t\in[0,1]} (Z_{1+ht} - Z_1) \geq y \,\middle|\, Z_1 = x(h)\right] = P\left(Z_1^* \geq y\right) \qquad (4.10)$$

for any $y \geq 0$ and any bounded function $x(h)$.

4. The assertion follows now by a direct calculation, making a change of integration variable $z = x - h^H y$ and applying the convergence result (4.10) with $x(h) = x - h^H y$ under the integral sign (a dominated convergence argument can be given using estimates of the extremal value theory of Gaussian processes). Denoting $\phi(x) = e^{-x^2/2}/\sqrt{2\pi}$, we have

$$\begin{aligned} & h^{-H} P\left(\exists s \in [0,h]: \ Z_{1+s} = x\right) \\ = \ & h^{-H} \int_{-\infty}^{\infty} \phi(z) P[\exists s \in [0,h]: \ Z_{1+s} = x \mid Z_1 = z]\, dz \\ = \ & \int_{-\infty}^{\infty} \phi(x - h^H y) P[\exists s \in [0,h]: \ Z_{1+s} = x \mid Z_1 = x - h^H y]\, dy \\ = \ & \int_{-\infty}^{\infty} \phi(x - h^H y) P\left[\exists t \in [0,1]: \ \frac{Z_{1+ht} - Z_1}{h^H} = y \,\middle|\, Z_1 = x - h^H y\right] dy \\ = \ & \int_{-\infty}^{0} \phi(x - h^H y) P\left[h^{-H} \inf_{t\in[0,1]} (Z_{1+ht} - Z_1) \leq y \,\middle|\, Z_1 = x - h^H y\right] dy \end{aligned}$$

[3] A rather complete presentation is given in the COST242 temporary document TD(95)004v2 by the same authors as in [15]. It is quite possible that the result can also be found in existing literature on local behaviour of stochastic processes.

$$+\int_0^\infty \phi(x-h^H y) P\left[h^{-H}\sup_{t\in[0,1]}(Z_{1+ht}-Z_1)\geq y \,\middle|\, Z_1 = x-h^H y\right] dy$$

$$\stackrel{h\downarrow 0}{\longrightarrow} \quad 2\phi(x)\int_0^\infty P(Z_1^* \geq y)\, dy$$

$$= \quad 2E(Z_1^*)\,\phi(x).$$

5. A Girsanov formula approach

In this section, we consider the fBm Z only on the positive half-axis, assume that it is defined on some probability space $(\Omega, \mathcal{F}, P)$, and denote by $(\mathcal{F}_t)_{t\geq 0}$ the right continuous, completed version of the history (filtration) generated by Z.

The classical Girsanov theorem for Brownian motion states, in its simplest form, the following. Let W be the standard Brownian motion. Then, for any fixed number a, the process $L_t = e^{aW_t - \frac{1}{2}a^2 t}$ is a martingale with expectation 1 such that, for any $T > 0$, with respect to the measure $dP_{a,T} \doteq L_T dP$, $(W_t)_{t\in[0,T]}$ is a standard Brownian motion with drift a.

A corresponding formula for fractional Brownian motion was given in [12]. The result remained little known[4] and was recently rediscovered in [6], where stochastic analysis of fBm is developed using Malliavin calculus, and in [16].

Let us define a process $(M_t)_{t\geq 0}$ as

$$M_t \doteq c_1 \int_0^t s^{\frac{1}{2}-H}(t-s)^{\frac{1}{2}-H} dZ_s, \tag{5.1}$$

where

$$c_1 = \left[H(2H-1)B(\frac{3}{2}-H, H-\frac{1}{2})\right]^{-1}$$

and $B(\mu,\nu)$ denotes the beta function

$$B(\mu,\nu) = \int_0^1 x^{\mu-1}(1-x)^{\nu-1}dx = \frac{\Gamma(\mu)\Gamma(\nu)}{\Gamma(\mu+\nu)}.$$

The integral in (5.1) can be understood in the L^2 sense, but it can also be defined pathwise almost everywhere (for integration with respect to Gaussian processes, see [10, 8]).

Theorem 5.1. *The centered Gaussian process $(M_t)_{t\geq 0}$ has independent increments and is thus a martingale. It generates, up to null sets, the same history as $(Z_t)_{t\geq 0}$. Its variance function is*

$$EM_t^2 = \langle M, M\rangle_t = c_2^2 t^{2-2H},$$

[4] I am grateful to Prof. G. Molchan for contacting me and giving this reference.

where

$$c_2 = \left(H(2H-1)(2-2H)B(H-\frac{1}{2}, 2-2H)\right)^{-\frac{1}{2}}.$$

We also have the following.

Proposition 5.1. *For $0 \le s \le t$, we have $EM_sZ_t = s$. In particular, the correlation between M_t and Z_t is*

$$Corr\,(M_t, Z_t) = \frac{1}{c_2}. \tag{5.2}$$

The above result is particularly interesting since the constant $c_2 = c_2(H)$ is very close to one through the interval $H \in [1/2, 1)$:

H	0.5	0.6	0.7	0.8	0.9
$1/c_2(H)$	1.0	0.9975	0.9932	0.9907	0.9923

It follows that for $H > 1/2$ the process

$$X_t \doteq \frac{t^{2H-1}}{c_2} M_t$$

can be used as an approximation to Z, with the relative mean square error

$$\frac{Var\,(Z_t - X_t)}{Var\,(Z_t)} = 2\left(1 - \frac{1}{c_2}\right).$$

Note that X is a diffusion process — many quantities related to X can be computed exactly.

Denote by $\mathcal{E}(M)$ the stochastic exponent of M,

$$\mathcal{E}_t(M) \doteq \exp\left(M_t - \frac{1}{2}\langle M, M\rangle_t\right).$$

By a standard result of martingale theory, $\mathcal{E}_t(M)$ is a martingale with expectation 1. Let us restrict to a finite time interval $[0, T]$. For $a \in \Re$, let P_a be the probability on $(\Omega, \mathcal{F}_T, P)$ defined by the relation

$$\left(\frac{dP_a}{dP}\right)_{\mathcal{F}_T} \doteq \mathcal{E}_T(aM).$$

(It is easy to see that if the measure P_a can be extended to the σ-algebra $\mathcal{F}_\infty$, this extension is singular with respect to P_0 if $a \neq 0$.) Now, $\mathcal{E}_T(aM)$ turns out to be the likelihood ratio between the two hypotheses "drift is a" and "drift is 0":

Theorem 5.2. *With respect to the measure P_a, the process Z is a fractional Brownian motion with drift a, i.e., the distribution of Z with respect to P_a is the same as the distribution of $Z_t + at$ with respect to $P = P_0$.*

It is a standard technique to use the Girsanov formula to express the distributions of functionals of processes with drift in terms of the corresponding zero-drift processes. For any $x > 0$, denote

$$T_x = \inf\{t \geq 0 : \ Z_t = x\}.$$

Being a hitting time, T_x is a stopping time with respect to $(\mathcal{F}_t)_{t\geq 0}$. Moreover, it is well known that $T_x < \infty$ with probability one. Now, the Girsanov formula gives the following exact relationship.

Proposition 5.2. *We have*

$$P\left(V_0^{(1)} > x\right) = E\exp\left(-M_{T_x} - \frac{c_2^2}{2}T_x^{2-2H}\right). \tag{5.3}$$

Proof: Since T_x is an almost surely finite stopping time and the Radon-Nikodym density process $L_t = \exp(M_t - \langle M, M\rangle_t/2)$ is a martingale, we have

$$\begin{aligned}
P\left(V_0^{(1)} \geq x\right) &= \lim_{n\to\infty} P\left(\sup_{t\in[0,n]} (Z_t - t) \geq x\right) \\
&= \lim_{n\to\infty} EL_n 1_{\{Z_n^* \geq x\}} \\
&= \lim_{n\to\infty} EL_n 1_{\{T_x \leq n\}} \\
&= \lim_{n\to\infty} E\left(E\left[L_n \,|\, \mathcal{F}_{T_x\wedge n}\right] 1_{\{T_x\leq n\}}\right) \\
&= \lim_{n\to\infty} EL_{T_x\wedge n} 1_{\{T_x \leq n\}} \\
&= \lim_{n\to\infty} EL_{T_x} 1_{\{T_x \leq n\}} \\
&= EL_{T_x} \\
&= E\exp\left(-M_{T_x} - \frac{c_2^2}{2}T_x^{2-2H}\right).
\end{aligned}$$

□

The expression (5.3) contains random variables with unknown distributions. As regards T_x, we can proceed with the following standard trick based on self-similarity.

Proposition 5.3. *The random variable T_x has the same distribution as $(x/Z_1^*)^{1/H}$, where Z^* is defined by (4.4).*

Proof: We have, by the self-similarity of Z,

$$P(T_x \leq t) = P\left(\sup_{s\in[0,t]} Z_s \geq x\right)$$

$$
\begin{aligned}
&= P\left(\sup_{u\in[0,1]} Z_{tu} \geq x\right) \\
&= P\left(\sup_{u\in[0,1]} Z_u \geq \frac{x}{t^H}\right) \\
&= P\left(\left(\frac{x}{Z_1^*}\right)^{1/H} \leq t\right). \qquad \square
\end{aligned}
$$

The random variable M_{T_x} looks more obscure at the first sight. However, by the formula of the conditional expectation in a bivariate Gaussian distribution and Proposition 5.1, we have the exact relations

$$
\begin{aligned}
E\left[M_t \,|\, Z_t\right] &= \frac{Cov\,(M_t, Z_t)}{Var\,(Z_t)} Z_t = t^{1-2H} Z_t, \\
Var\left[M_t \mid Z_t\right] &= \left(1 - \frac{Cov\,(M_t, Z_t)^2}{Var\,(M_t)\, Var\,(Z_t)}\right) Var\,(M_t) \\
&= (c_2^2 - 1)t^{2-2H}.
\end{aligned}
$$

It follows that

$$
E\left[e^{-M_t} \mid Z_t\right] = \exp\left[-t^{1-2H} Z_t + \frac{c_2^2 - 1}{2} t^{2-2H}\right]. \tag{5.4}
$$

Now, restricting to the case $H \geq 1/2$, we suggest the approximation

$$
E\left[e^{-M_{T_x}} \,|\, T_x\right] \approx e^{-xT_x^{1-2H}}, \tag{5.5}
$$

where the motivation is that since the conditional variance of M_t given Z_t is small (because c_2 is close to 1 for $H \geq 1/2$), the condition $T_x = t$ should suggest about the same value for M_t as does the condition $Z_t = x$. We could have kept the variance term in the exponent, but since it turns out to be numerically negligible, we prefer this simpler form. Replacing once more c_2 by 1 and using Proposition 5.3, we obtain

$$
\begin{aligned}
P\left(V_0^{(1)} > x\right) &\approx E\exp\left(-xT_x^{1-2H} - \frac{c_2^2}{2} T_x^{2-2H}\right) \\
&\approx E\exp\left(-xT_x^{1-2H} - \frac{1}{2} T_x^{2-2H}\right) \\
&= Ef(x, Z_1^*),
\end{aligned} \tag{5.6}
$$

where $f(x,z)$ is the function

$$
f(x,z) \doteq \exp\left[-\left(\frac{x}{z}\right)^{(1-H)/H} z - \frac{1}{2}\left(\frac{x}{z}\right)^{(2-2H)/H}\right].
$$

It remains to use a suitable approximation for the distribution of Z_1^*. It is well known that for $H = 1/2$, Z_1^* has the same distribution as $|Z_1|$, and for $H = 1$, we have obviously $Z_1^* = Z_1^+$. The approximation $Z_1^* \overset{\mathcal{D}}{\approx} |Z_1|$ gives

$$\begin{aligned} P\left(V_0^{(1)} > x\right) &\approx Ef(|Z_1|) \\ &= \sqrt{\frac{2}{\pi}} \int_0^\infty \exp\left[-\frac{1}{2}\left(z + \left(\frac{x}{z}\right)^{(1-H)/H}\right)^2\right] dz. \end{aligned}$$

This approximation in fact coincides with (4.7) which was obtained from the Beneš formula. A more accurate simple approximation for the distribution of Z_1^* is the linear interpolation

$$Z_1^* \overset{\mathcal{D}}{\approx} Z_1^+ + 2(1-H)Z_1^-. \tag{5.7}$$

6. Summary

The different approximations are shown in Figure 6.1 in the case $H = 0.8$ together with an empirical distribution of the corresponding discrete time storage system obtained with 100 simulated realizations of fractional Gaussian noise of length 10000 each[5]. The simulated values become unreliable about at queue length 40. One could conjecture that the Weibull distribution gives an upper bound (this has not been proved) and that (5.6) is the most accurate of the approximations presented here[6]. However, none of them is very close to the simulated values for small x.

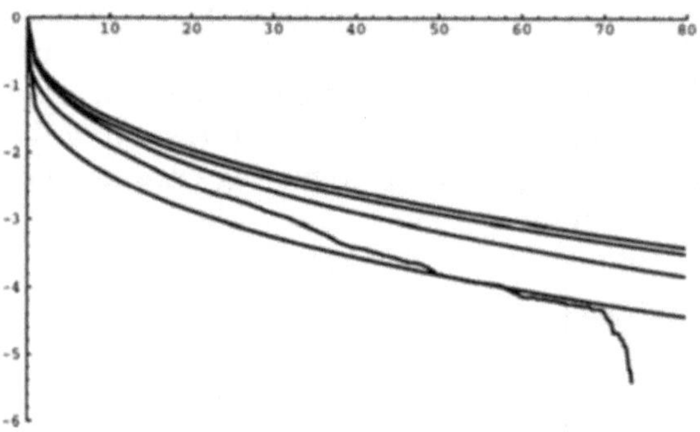

Fig. 6.1. Approximations and a simulation based estimate of $\log_{10} P(V_0^{(1)} > x)$ for $H = 0.8$. Smooth curves from top to bottom: (3.3), (4.7), (5.6)&(5.7), and (3.2). The less smooth curve comes from the empirical distribution function.

[5] I thank Jacques Lévy Véhel for the fBm simulation program which is a part of FRACLAB, a toolbox for signal processing with fractals developed in the FRACTALES team at INRIA.

[6] The new accurate estimate given in footnote 1 (with $K = 1$) would also be rather close to (5.6)&(5.7) in this figure, but a continuation of the plot would show that (5.6)&(5.7) is an underestimate for large x.

References

1. Adler, R.J. (1990): *An Introduction to Continuity, Extrema, and Related Topics for General Gaussian Processes.* Volume 12 of *Lecture Notes-Monograph Series*, Institute of Mathematical Statistics.
2. Beneš, V.E. (1963): *General Stochastic Processes in the Theory of Queues.* Addison Wesley.
3. Borovkov, A.A. (1976): *Stochastic Processes in Queueing Theory.* Springer Verlag.
4. Cornfeld, I.P., Sinai, Y.G., Fomin, S.V. (1982): *Ergodic Theory.* Springer Verlag, Berlin.
5. Duffield, N. G.,O'Connell, N. (1996): Large deviations and overflow probabilities for the general single-server queue, with applications. *Mathematical Proceedings of the Cambridge Philosophical Society.*
6. Decreusefond, L., Üstünel, A.S. (1996): Stochastic analysis of the fractional Brownian motion. Submitted for publication.
7. Durbin, J. (1985) The first-passage density of a continuous Gaussian process to a general boundary. *Journal of Applied Probability*, 22, 99–122, Corrections in J.A.P. 25(1988), p. 840.
8. Feyel, D., de La Pradelle, A. (Preprint 1996): Fractional integrals and Brownian processes.
9. Harrison, J.M. (1985): *Brownian Motion and Stochastic Flow Systems.* Wiley, New York.
10. Huang, S.T., Cambanis, S. (1978): Stochastic and multiple Wiener integrals for Gaussian processes. *Annals of Probability*, 6, 585–614.
11. Leland, W.E., Taqqu, M.S., Willinger, W., Wilson, D.V. (February 1994): On the self-similar nature of Ethernet traffic (extended version). *IEEE/ACM Transactions on Networking*, 2(1), 1–15.
12. Molchan, G.M., Golosov, J.I. (1969): Gaussian stationary processes with asymptotic power spectrum. *Soviet Math. Dokl.*, 10(1), 134–137.
13. Norros, I. (1994): A storage model with self-similar input. *Queueing Systems*, 16, 387–396.
14. Norros, I. (August 1995): On the use of fractional Brownian motion in the theory of connectionless networks. *IEEE Journal on Selected Areas in Communications*, 13(6).
15. Norros, I., Simonian, A., Veitch, D., Virtamo, J. (1995): A Beneš formula for a buffer with fractional Brownian input. In *9th ITC Specialists Seminar '95: Teletraffic Modelling and Measurement.*
16. Norros, I., Valkeila, E., Virtamo, J. (August 1996): A Girsanov type formula for the fractional Brownian motion. Draft version. Presented at the First Nordic-Russian Symposium on Stochastics.
17. Reich, E. (1958): On the integrodifferential equation of Takács I. *Annals of Mathematical Statistics*, 29, 563–570.
18. Roberts, J.W. (editor) (1992): *Performance Evaluation and Design of Multiservice Networks. COST 224 Final Report.* CEC.
19. Simonian, A., Virtamo, J. (October 1991): Transient and stationary distributions for fluid queues and input processes with a density. *SIAM J. Appl. Math.*, 51(6).

Multi-dimensional Fractional Brownian Motion and some Applications to Queueing Theory

S. Belly and L. Decreusefond

École Nationale Supérieure des Télécommunications
46, rue Barrault
75634 Paris, FRANCE
decreuse@res.enst.fr

Abstract Superimposition of different traffic sources are modeled by a sum of fractional Brownian motions (with Hurst parameter varying in the whole range of $(0,1)$) and a supplementary part which can be random. In this setting, it is shown that the overflow probability is non–sensitive to this supplementary part. For the analysis of the fractional Brownian motions, new mathematical tools are provided.

1. Introduction

Designing efficient networks requires a powerful way to determine the optimal size of the different resources like buffers and bandwidth. We address here the problem of determining the convenient size of a buffer, at the input of a switch, which guarantees a fixed loss probability. The model used is that of the single server queue. Since in ATM like networks, data is put into cells of constant length, the service time is taken to be constant. Hence, the problem is that of the statistical characterization of the input stream. Unfortunately, multiplicity of sources (video, data, voice) induce multiplicity of input traffic types. Moreover, the various equipments (switches, bridges and so on) information cells pass through, alter traffic characteristics so that it is almost impossible to define a general model of traffic valid for any classes of sources at any place of a network. Nevertheless, several recent studies showed that the *traffic* seems to possess the property of long-range dependence (see for instance [3], [9]) and self-similarity. This is to say that the evolution of the traffic after time t does not only depend on the situation at time t (as in the Markovian assumption) but also of a long part of the past. It turns out that all the models we dealt with before were Markovian ones, essentially because Markovian models have the good property to be easily analyzed by mathematical tools, but, on the other hand, they do not capture the long-range dependence aspect the real traffic seems to have. Thus, in order to evaluate the impact of long-range dependence on networks performance, new mathematically tractable models are needed. One of the most simple models which can be investigated is the fractional Brownian motion (fBm) (see [10]): A fractional Brownian motion of index (or Hurst parameter) $H \in (0,1)$, $\{B_t^H, t \in R_+\}$, is a centered Gaussian process with autocorrelation function:

$$R_H(s,t) = E_H[B_s^H B_t^H] \triangleq \frac{V_H}{2}(s^{2H} + t^{2H} - |t-s|^{2H}),$$

where,

$$V_H \triangleq \frac{\Gamma(2-2H)\cos(\pi H)}{\pi H(1-2H)}.$$

When $H = \frac{1}{2}$, $B^{\frac{1}{2}}$ is the usual Brownian motion. Using the Kolmogorov criterion, it can be shown that, for any $H \in (0,1)$, the fBm has a continuous version, and with standard techniques it can also be shown that sample paths are nowhere differentiable. B^H has stationary increments and is self-similar: $\{B_{\alpha t}^H, t \in R_+\} \stackrel{(d)}{=} \{\alpha^H B_t^H, t \in R_+\}$. The increments are independent only when $H = \frac{1}{2}$, for $H > \frac{1}{2}$, the increments are positively correlated, and for $H < \frac{1}{2}$ they are negatively correlated. There already exist a few papers on theoretical studies of network performances in the presence of fractional Brownian motion (see [12], [13]). The main assumption of these works is that the input process can be modeled by the sum of a deterministic part and a random process which is an fBm of Hurst parameter greater than 1/2, i.e., the input process has the form:

$$mt + \sqrt{am}B_t^H. \tag{1.1}$$

The aim of this paper is to generalize this analysis when the process mt is replaced by a random process depending eventually on itself and B_t^H is replaced by a superimposition of several fBm with different Hurst parameters. The superimposition allows us to take into account the differences between the different inputs; there is no reason for H to be the same among all the streams.
Actually, the framework is the following: consider a single–server queue with constant service rate (taken to be c unit of work per unit of time) fed by a process $\{X_t,\ t \geq 0\}$, the workload process is denoted by $\{W_t,\ t \geq 0\}$ and W is the workload at stationary regime.

It is known from [13] that the buffer size required, to keep a fixed loss probability, increases dramatically when H is close to 1. Namely, this is due to the fact that, for x large,

$$P(W > x) \simeq \exp(-\frac{(c-m)^{2H}}{2\kappa(H)^2 am}x^{2-2H}), \tag{1.2}$$

where $\kappa(H) = H^H(1-H)^{1-H}$. The problems addressed in this work are the following: 1) is it possible to control the input process so that the tail distribution of W has a nicer behavior than given in Equation (1.2) what happens when we superimpose some fBm's of small (i.e. less than 1/2) Hurst parameter with some fBm's with large (i.e. greater than 1/2) Hurst index ? Furthermore, the tools we will set to answer these questions enable us to derive some quantitative properties of the workload process. Namely, we can give some conditions under which the law of W_t, for all t in R_+, is absolutely continuous with respect to the Lebesgue measure.

The paper is organized as follows : in section 2., we present some basic facts about fBm, in particular we describe a new way to simulate some sample–paths of a fractional Brownian motion. Section 3. is devoted to the theoretical study of questions 1 and 2 for large buffers and in section 4., the absolute continuity of the law of the workload process at each time t is proved.

2. Properties of the fractional Brownian motion

When $H \neq \frac{1}{2}$, the fBm is neither a Markov process, nor a semi-martingale, but only a Gaussian process. Hence, the usual Itô stochastic calculus can not be used. Nevertheless, a stochastic calculus for the fractional Brownian motion has been constructed in [5] using the Malliavin calculus.

2.1 Stochastic Analysis

Let $W = C_0(R_+, R)$ be the real separable Banach space of continuous functions, null at time 0, equipped with the sup norm. For any $H \in (0,1)$, P is the unique probability measure on W such that the canonical process $\{B_t^H, t \in R_+\}$ is an fBm. Let $\mathcal{H}$ be the unique Hilbert space continuously and densely embedded in W such that, for any $\eta \in W^*$,

$$\int_W \exp(i < \eta, \omega >_{W^*,W}) P(d\omega) = \exp(-\frac{1}{2}\|\tilde{\eta}\|_{\mathcal{H}}^2),$$

where $\tilde{\eta}$ stands for the image of η by the injection of W^* into $\mathcal{H}$. The space $\mathcal{H}$ is called the Cameron-Martin space associated with (W, P). Before describing the space $\mathcal{H}$, we give some definitions and notations. Let

$$K_H(t,r) = \frac{(t-r)^{H-\frac{1}{2}}}{\Gamma(H+\frac{1}{2})} F(\frac{1}{2}-H, H-\frac{1}{2}, H+\frac{1}{2}, 1-\frac{t}{r}) 1_{[0,t]}(r),$$

where,

$$F(a,b,c,z) \triangleq \sum_{k=0}^{+\infty} \frac{(a)_k (b)_k}{(c)_k k!} z^k$$

is the Gauss hypergeometric function, $(a)_0 = 1$ and $(a)_k \triangleq \Gamma(a+k)/\Gamma(a) = a(a+1)...(a+k-1)$. Then,

$$R_H(s,t) = \int_0^{s \wedge t} K_H(s,r) K_H(t,r) dr.$$

For more details on Gauss hypergeometric functions see [8] and [11].

Theorem 2.1 (see [5]). $\mathcal{H}$ consists of functions of the form

$$h(t) = K_H \dot{h}(t) \triangleq \int_0^t K_H(t,s)\dot{h}(s)ds, \quad \dot{h} \in L^2(R_+).$$

The scalar product on $\mathcal{H}$ is defined by:

$$(h,g)_{\mathcal{H}} = (\dot{h}, \dot{g})_{L^2(R_+)}.$$

As a consequence:

Corollary 2.1 (see [5]). For all t belonging to R_+, one can construct an fBm by considering the following stochastic integral:

$$B_t^H = \int_0^t K_H(t,s)dB_s^{\frac{1}{2}},$$

where $B^{\frac{1}{2}}$ is a standard Brownian motion.

Remark: Hence, an approximate sample–path on the unit interval of an fBm of Hurst parameter H can be obtained by considering

$$t \to n\sum_{i=1}^{n} \int_{(i-1)/n}^{i/n} K_H(t,s)\ ds.(B^{\frac{1}{2}}_{(i+1)/n} - B^{\frac{1}{2}}_{i/n}),$$

for n sufficiently large. Note that the computation of the integrals are time consuming (see Lavaud-Decreusefond [4] for details), but has to be done only once even for generating a bunch of independent sample paths.
An important property of K_H is:

Proposition 2.1. For any t and r belonging to R_+,

$$K_H(t,r) \leq C_H r^{-|H-\frac{1}{2}|}(t-r)^{-(\frac{1}{2}-H)_+}1_{[0,t]}(r),$$

where C_H is uniform in r and t on any compact set.

2.1.1 Derivative and divergence operators on the Wiener space. Since B^H is not a semi-martingale, there is no immediate notion of stochastic integral with respect to B^H in the sense of Itô. On the other hand, B^H is a Gaussian process. Hence, at least a Wiener integral with respect to B^H can be defined, i.e., a stochastic integral with respect to B^H where the integrands are only allowed to be deterministic. To be able to integrate random processes with respect to B^H, we use here the approach, already initiated in [5], which uses the stochastic calculus of variations to define the Skorohod integral with respect to B^H. For this, we first define a derivative operator called the H-Gross-Sobolev derivative by:

Definition 2.1. For an R-valued smooth cylindrical functional $F \in \mathcal{S}$ of the form, $F = f(B^H_{t_1}, ..., B^H_{t_n})$, where f belongs to the Schwartz space on R^n, the H-Gross-Sobolev derivative of F, denoted by $\nabla_H F$ belongs at ω fixed to $\mathcal{H}$ and is defined by :

$$\nabla_H F(\omega) = \sum_{i=0}^{n} \frac{\partial f}{\partial x_i}(B^H_{t_1}, ..., B^H_{t_n}) R_H(t_i, .).$$

Then, the divergence operator is the adjoint of ∇_H for the L^2 duality pairing, i.e., for any H-valued process u for which there exists C_u such that,

$$|E[(u, \nabla_H F)_{\mathcal{H}}]| \leq C_u \|F\|_{L^2}, \forall F \in \mathcal{S},$$

there exists a L^2 random variable, denoted by $\delta_H u$, such that,

$$E[(u, \nabla_H F)_{\mathcal{H}}] = E[F \delta_H u].$$

Both ∇_H and δ_H are shown to be closable operators and thus can be extended to wider sets of functionals, respectively denoted by $\mathbb{D}_{2,1}$ and $dom(\delta_H)$. By an analogy with the case of the standard Brownian motion, it is convenient to say that δ_H is a stochastic integral with respect to B^H. To mimic the usual notations we set,

$$\int_0^t \dot{u}(s) \delta_H B^H_s \triangleq \delta_H \pi_t(K_H \dot{u}).$$

Proposition 2.2.

i) $\{\int_0^t K_H(t,s) \delta_H B^H_s, t \in R_+\}$ is the fBm of Hurst parameter H.
ii) If u is an adapted process, then $E[|\int_0^1 u_s \delta_H B^H_s|^2] = \int_0^1 E[|u_s|^2] ds$.
iii) For u an adapted process in $L^2(W; \mathcal{H})$, $\{\int_0^t u_s \delta_H B^H_s, t \in R_+\}$ is a martingale and $< M, M >_t = \int_0^t u_s^2 ds$.

Hence, applying the Burkhölder-Davis-Gundy inequality to the martingale $u \mapsto \int_0^u K_H(t,s) u_s \delta_h B^H_s$ and letting $t = u$,

Corollary 2.2. For all $p > 1$, there exists $C_p > 0$ such that,

$$E[|\int_0^t K_H(t,s) u_s \delta_H B^H_s|^p] \leq C_p E[(\int_0^t K_H^2(t,s) u_s^2 ds)^{p/2}].$$

Remark: As the property (i) shows, the behavior of this "integral" is closer to the Wiener integral for Gaussian process than to Itô integral which is defined by Riemann sums. In fact, we do not have $\int_0^t u_s \delta_H B^H_s$ as the limit of

$$\sum_{t_i \in \pi} u_{t_i}(B^H_{t_{i+1}} - B^H_{t_i}), \text{ when} |\pi| \to 0,$$

even for u very regular (see [5] for a complete discussion).

Nevertheless, we still have a Girsanov like theorem:

Theorem 2.2 (Girsanov Theorem). Consider the process $\{X_t, t \geq 0\}$ defined by :

$$X_t = \sum_{i=1}^{n} B_t^{H_i} + \int_0^t K_{H_i}(t,s)u_i(s)ds, \tag{2.1}$$

where $(\{B_t^{H_i}, t \geq 0\})_{i=1}^n$ are n independent fBms and $(u_i)_{i=1}^n$ are n adapted random processes belonging to $L^2(W; L^2([0,1], R))$, such that

$$E\left[\exp\left(\sum_{i=1}^{n} \int_0^t u_i(s)\delta_{H_i} B_s^{H_i} - \frac{1}{2}\sum_{i=1}^{n}\int_0^t u_i^2(s)ds\right)\right] = 1. \tag{2.2}$$

Then the probability Q defined by

$$\left.\frac{dQ}{dP}\right|_{\mathcal{F}_t} = \exp\left(\sum_{i=1}^{n} \int_0^t u_i(s)\delta_{H_i} B_s^{H_i} - \frac{1}{2}\sum_{i=1}^{n}\int_0^t u_i^2(s)ds\right),$$

is such that under Q, $\{X_t, t \geq 0\}$ has the same law as $\{\sum_{i=1}^n B_t^{H_i}, t \geq 0\}$ has under P.

Application: As a byproduct of this result, we are now in position to estimate m in the model of Norros (see Equation (1.1)) through the observation of a sample–path of the input process over a finite time interval: for the sake of simplicity, let X be given by $X_t = mt + B_t^H$ and let P_X denote its law under P. Let φ defined by $(K_H\varphi)(t) = t$. By the Girsanov theorem

$$\left.\frac{dP_X}{dP}\right|_{\mathcal{F}_t}(\omega) = \exp(\int_0^t \varphi(s)\delta_H B_s^H - \frac{1}{2}\int_0^t \varphi^2(s)ds).$$

Hence the P maximum likelihood ratio estimate $\hat{m}_t$ of m is

$$\hat{m}_t = \frac{1}{\int_0^t \varphi^2(s)ds}\int_0^t \varphi(s)\delta_H B_s^H(X(\omega)).$$

A simple calculation shows that

$$\varphi(s) \triangleq \frac{\Gamma(3/2-H)}{\Gamma(2-2H)} s^{1/2-H},$$

Hence,

$$\begin{aligned}\hat{m}_t &= \frac{2^{2-2H}\Gamma(2-H)}{\sqrt{\pi}t^{2-2H}}\int_0^t s^{\frac{1}{2}-H}\delta_H B_s^H(X(\omega)) \\ &= \lim_{|\Pi|\to 0}\frac{1}{3/2-H}\sum_{t_i\in\pi}(t_{i+1}^{3/2-H} - t_i^{3/2-H})(X_{t_{i+1}} - X_{t_i}),\end{aligned}$$

where π is a subdivision of $[0,t]$ which mesh is $|\pi|$.

3. Overflow probability for large buffers

As mentioned in the introduction, we aim at studying the behavior of the queue when the input is controlled by a feedback loop. More precisely, it will be shown that if the control of the input process X acts by adding a so–called drift term to X then the behavior of the queue is globally the same as in the model of Norros, i.e., we still have an equation like (1.2).

3.1 The input stream

Theorem 3.1. Consider n independent fractional Brownian motions, $(\{B_t^{H_i}, t \in R_+\})_{i=1}^n$. We assume that for all

$$i = 1, ..., n, H_i > \frac{1}{2} \tag{3.1}$$

or for all

$$i = 1, ..., n, H_i \le \frac{1}{2}. \tag{3.2}$$

Let for all $i = 1, ..., n$, b_i and σ_i be K-Lipschitz continuous with respect to their second variable, uniformly with respect to their first variable: for all t in R_+, for all x, y in R,

$$|b_i(t,x) - b_i(t,y)| \le K|x-y| \text{ and } |\sigma_i(t,x) - \sigma_i(t,y)| \le K|x-y|.$$

We assume also that for all $i = 1, ..., n$, there exist x_i and y_i in R such that $b(.,x_i)$ and $\sigma(.,y_i)$ belong to $L^q(R_+)$ for some q such that $q^* < 1/|2H-1|$, where $H = \sup_{1\le i\le n} H_i$ and q^* is defined by $\frac{1}{q} + \frac{1}{q^*} = 1$. Then the differential equation

$$X_t = \sum_{i=1}^n \int_0^t K_{H_i}(t,s) b_i(s, X_s) ds + \int_0^t K_{H_i}(t,s) \sigma_i(s, X_s) \delta_{H_i} B_s^{H_i} \tag{3.3}$$

has a unique solution, in the sense that there exists $\{X_t, t \ge 0\}$ satisfying (3.3) and such that $\int_0^\infty E[|X_t|^p] dt < \infty$, for some $p < 1/|1-2H|$.

The proof is based on the Picard iteration (X^k) defined by:

$$\begin{cases} X_t^0 &= x \\ X_t^k &= x + \sum_{i=1}^n \int_0^t K_{H_i}(t,s) b_i(s, X_s^{k-1}) ds + \int_0^t K_{H_i}(t,s) \sigma_i(s, X_s^{k-1}) \delta_{H_i} B_s^{H_i}. \end{cases}$$

By standard calculations, which use Corollary 2.2 and the Hölder inequality, we prove that for all T in R_+ and for some $p < 1/|1-2H|$,

$$\sup_{t\in[0,T]} E[|X_t^{k+1} - X_t^k|^p] \le C^k u_k^{p,H},$$

where $\sum_{k\ge 0} C^k u_k^{p,H}$ converges for all C in $\mathbb{C}$. Hence, by taking subseries, there is a process X such that:

$$X^k \longrightarrow X \quad dP \otimes dt - \text{a.s..}$$

Then, proceeding along similar lines, it can also be proved that the process $\{X_t, t \geq 0\}$ satisfies the differential equation (3.3) and the existence follows. Suppose that there are two processes X and Y satisfying the differential equation (3.3). Using Corollary 2.2, it can be shown that

$$X = Y \quad dP \otimes dt - \text{a.s..}$$

Remark: If we consider the case, for all $i = 1, ..., n, t$ in R_+, x in $R, \sigma_i(t, x) = 1$. The differential equation becomes:

$$X_t = \sum_{i=1}^{n} \int_0^t K_{H_i}(t, s) b_i(s, X_s) ds + B_t^{H_i}. \tag{3.4}$$

And we need no restriction on the values of H_i (see (Hyp3.1) or (Hyp3.2)) to prove the existence and the uniqueness of the solution.
Furthermore, suppose that for all $i = 1, ..., n$ there exists g_i in $L^2(R_+)$ such that for all t, x in R_+,

$$|b_i(t, x)| \leq g_i(t).$$

Then X has a continuous version. In the sequel, only the differential equation (3.4) will be considered. Hence, we have no restriction on the values of the H_i's.

3.2 The workload of the queue

We turn now to the study of the workload, i.e., the virtual waiting time (or the time needed to serve all the requests waiting in the queue at time t), in a deterministic single server queue, where the input process is the process $\{X_t, t \geq 0\}$, the solution of the differential equation (3.4). It is first unclear that there exists a stationnary regime and that $W = \sup_{t \geq 0}(X_t - ct)$. For this to hold, the following properties are required (see Borovkhov [2]):

i) $X_t - ct$ converges in probability to $-\infty$ as t grows to infinity,
ii) $\sum_{i=1}^{n} B_{-t}^{H_i} + ct$ converges dP-a.s. to infinity as t grows to infinity,
iii) $\lim_{V \to \infty} \lim_{t \to \infty} P(\sup_{V \leq v \leq t}(X_t - X_{t-v} - cv) > 0) = 0$,

where $\{B_{-t}^H, t \in R_+\}$ is a centered Gaussian process with autocorrelation function:

$$E_H[B_{-s}^H B_{-t}^H] \triangleq \frac{V_H}{2}(|s|^{2H} + |t|^{2H} - |t - s|^{2H}) = R_H(s, t).$$

Consequently,

$$\{B_t^H, t \in R_+\} \stackrel{(d)}{=} \{B_{-t}^H, t \in R_+\}. \tag{3.5}$$

Lemma 3.1. Let $(b_i)_{i=1}^n$ be K-Lipschitz continuous with respect to their second variable. Assume that there exist $(g_i)_{i=1}^n$ in $L^2(R)$ such that for all s in R_+ and $i = 1, ...n$, $|b_i(s,x)| \leq g_i(s)$. Then i), ii) and iii) are satisfied and,

$$W = \sup_{t\geq 0}(X_t - ct). \tag{3.6}$$

Proof: i) :
Consider first the process $\{X_t - ct, t \geq 0\}$. Using the Cauchy-Schwarz inequality, we have :

$$\begin{aligned} X_t - ct &= \sum_{i=1}^{n} \int_0^t K_{H_i}(t,s) b(s, X_s) ds - ct \\ &\leq t^H \sum_{i=1}^{n} V_{H_i}^{1/2} (\int_0^\infty g_i^2(s) ds)^{1/2} - ct, \end{aligned}$$

where $H = \sup_{i=1,...n} H_i$.
Since, as t grows to infinity, $t^H \sum_{i=1}^n (V_{H_i})^{1/2} (\int_0^\infty g_i^2(s)ds)^{1/2} \longrightarrow 0$,

$$X_t - ct \longrightarrow -\infty \text{ a.s..}$$

Now consider the Law of the Iterated Logarithm, applied to the fractional Brownian motion (see Arcones [1]). Hence, for all $i = 1, ..., n$, almost surely,

$$P(\limsup_{u\to 0+} \frac{B_u^{H_i}}{u^{H_i}\sqrt{\ln\ln(u^{-1})}} = \sqrt{V_{H_i}}) = 1.$$

Hence, the self-similarity of the fractional Brownian motion leads to

$$P(\limsup_{u\to 0+} \frac{u^{H_i} B_{1/u}^{H_i}}{\sqrt{\ln\ln(u^{-1})}} = \sqrt{V_{H_i}}) = 1.$$

Hence, setting $t = 1/u$,

$$P(\limsup_{t\to +\infty} \frac{B_t^{H_i}}{t^{H_i}\sqrt{\ln\ln(t)}} = \sqrt{V_{H_i}}) = 1.$$

Therefore, as t grows to infinity, $\sum_{i=1}^n B_t^{H_i} = O(1/(t^H\sqrt{\ln\ln(t)}))$ dP-a.s. and

$$\lim_{t\to\infty} \sum_{i=1}^{n} \frac{B_t^{H_i}}{t} = 0 \text{ a.s..} \tag{3.7}$$

ii) :
Using the results (3.5) and (3.7), we have:

$$\lim_{t\to\infty} \sum_{i=1}^{n} \frac{B_{-t}^{H_i}}{t} = 0 \text{ a.s.,}$$

and property ii)follows.

iii) :

The proof is based on the Girsanov theorem. In fact, consider Q defined by

$$\left.\frac{dQ}{dP}\right|_{\mathcal{F}_t}(\omega) \triangleq \Lambda_t \triangleq E[\exp(\sum_{i=1}^{n}\int_0^t b_i(s,X_s)\delta_{H_i}B_s^{H_i} - \frac{1}{2}\sum_{i=1}^{n}\int_0^t b_i^2(s,X_s)ds)].$$

By Theorem 2.2,

$$Q(\sup_{V\le v\le t}(X_t - X_{t-v} - cv) > 0) = P(\sup_{V\le v\le t}(\sum_{i=1}^{n}(B_t^{H_i} - B_{t-v}^{H_i}) - cv) > 0).$$

Since for all $i = 1,...,n$, for all x in R, $|b_i(.,x)| \le g_i(.)$, where g_i belongs to $L^2(R_+)$, then $E[\Lambda_\infty] = 1$. Hence, using equality (3.5), the Girsanov theorem, the Hölder inequality with $1 < p < 2$ and the stationarity of the increments of the fBm, we have:

$$\begin{aligned}
&P(\sup_{V\le v\le t}(X_t - X_{t-v} - cv) > 0)\\
&\le \left(Q(\sup_{V\le v\le t}(X_t - X_{t-v} - cv) > 0)\right)^{1/p}\\
&\quad\left(E_Q[\exp(-q\sum_{i=1}^{n}\int_0^t b_i(s,X_s)\delta_{H_i}B_s^{H_i} + \frac{q}{2}\int_0^t b_i^2(s,X_s)ds)]\right)^{1/q}\\
&\le \left(P(\sup_{V\le v\le t}(\sum_{i=1}^{n}(B_t^{H_i} - B_{t-v}^{H_i}) - cv) > 0)\right)^{1/p}\\
&\quad\left(E[\exp(-(q-1)\sum_{i=1}^{n}\int_0^t b_i(s,X_s)\delta_{H_i}B_s^{H_i} + \frac{q-1}{2}\int_0^t b_i^2(s,X_s)ds)]\right)^{1/q}\\
&\le \left(P(\sup_{V\le v\le t}(\sum_{i=1}^{n}B_{-v}^{H_i} - cv) > 0)\right)^{1/p}\\
&\quad\left(E[\exp(-2(q-1)\sum_{i=1}^{n}\int_0^t b_i(s,X_s)\delta_{H_i}B_s^{H_i} + 2(q-1)^2\int_0^t b_i^2(s,X_s)ds)]\right)^{1/2q}\\
&\quad\left(E[\exp(-(q-1)(q-2)\sum_{i=1}^{n}\int_0^t b_i^2(s,X_s)ds)]\right)^{1/2q}\\
&\le \left(P(\sup_{V\le v\le t}(\sum_{i=1}^{n}B_v^{H_i} - cv) > 0)\right)^{1/p}.
\end{aligned}$$

Now, using the self–similarity of the fractional Brownian motions, we get:

$$P(\sup_{V\le v\le t}(\sum_{i=1}^{n}B_v^{H_i} - cv) > 0) \;\le\; \sum_{i=1}^{n}P(\sup_{V\le v\le t}(B_v^{H_i} - \frac{cv}{n}) > 0)$$

$$\leq \sum_{i=1}^{n} P(\sup_{V\leq v\leq t} v^{H_i}(B_1^{H_i} - \frac{cv^{1-H_i}}{n}) > 0)$$
$$\leq \sum_{i=1}^{n} P(B_1^{H_i} > \frac{cV^{1-H_i}}{n}).$$

Since for all $i = 1, ..., n$, $B_1^{H_i}$ is a centered Gaussian random variable of variance $\sigma_i^2 = V_{H_i}$, then,

$$\sum_{i=1}^{n} P(B_1^{H_i} > \frac{cV^{1-H_i}}{n}) \leq \sum_{i=1}^{n} \frac{2\sigma_i}{\sqrt{2\pi}cV^{1-H_i}/n} \exp\Big(-\frac{(cV^{1-H_i}/n)^2}{2\sigma_i^2}\Big).$$

Therefore,

$$P(\sup_{V\leq v\leq t}(X_t - X_{t-v} - cv) > 0)$$
$$\leq \Big(\sum_{i=1}^{n} \frac{2\sigma_i}{\sqrt{2\pi}cV^{1-H_i}/n} \exp(-\frac{(cV^{1-H_i}/n)^2}{2\sigma_i^2})^p,$$

and,

$$\lim_{V\to\infty}\lim_{t\to\infty} P(\sup_{V\leq v\leq t}(X_t - X_{t-v} - cv) > 0) = 0.$$

Since the queue is stable, we have the following theorem:

Theorem 3.2. Assume that the hypothesis of Theorem 3.1 and Lemma 3.1 are satisfied. When the input process is the solution of the differential equation (3.4), then the workload process satisfies

$$\lim_{x\to\infty} x^{2H-2} \ln P(W \geq x\} = -\frac{1}{2V_H}\frac{c^{2H}}{\kappa(H)^2}, \tag{3.8}$$

where $H = \sup_{1\leq i\leq n} H_i$.

Proof: The proof is based on the expression of W given by (3.6):

$$W = \sup_{t\geq 0}(X_t - ct)$$

and the Girsanov Theorem. In fact, consider Q defined by (3.8). Thus,

$$\frac{dP}{dQ}\Big|_{\mathcal{F}_t}(\omega) \triangleq \tilde{\Lambda}_t \triangleq E[\exp(-\sum_{i=1}^{n}\int_0^t b_i(s, X_s)\delta_{H_i} B_s^{H_i} + \frac{1}{2}\sum_{i=1}^{n}\int_0^t b_i^2(s, X_s)ds)].$$

Then, by Theorem 2.2,

$$Q(\sup_{t\geq 0}(X_t - ct) \geq x) = P(\sup_{t\geq 0}(\sum_{i=1}^{n} B_t^{H_i} - ct) \geq x).$$

On the other hand, by the very definition of Q,

$$Q(\sup_{t\geq 0}(X_t - ct) \geq x) = E_P[1_{\sup_{t\geq 0}(X_t - ct)\geq x}\Lambda_\infty].$$

Hence, by standard calculations, for all $p > 1$, $\frac{1}{p} + \frac{1}{q} = 1$, we have:

$$\begin{aligned}
P(\sup_{t\geq 0}(X_t - ct) \geq x) &= E_Q[1_{\sup_{t\geq 0}(X_t - ct)\geq x}\tilde{\Lambda}_\infty] \\
&\leq \left(Q(\sup_{t\geq 0}(X_t - ct) \geq x)\right)^{1/p} \\
&\quad \left(E_Q[\exp(-q\sum_{i=1}^n \int_0^\infty b_i(s, X_s)\delta_{H_i} B_s^{H_i} + \frac{q}{2}\sum_{i=1}^n \int_0^\infty b_i^2(s, X_s)ds)]\right)^{1/q} \\
&\leq \left(P(\sup_{t\geq 0}(\sum_{i=1}^n B_t^{H_i} - ct) \geq x)\right)^{1/p} \\
&\quad \left(E_Q[\exp(-2q\sum_{i=1}^n \int_0^\infty b_i(s, X_s)\delta_{H_i} B_s^{H_i} + 2q^2\sum_{i=1}^n \int_0^\infty b_i^2(s, X_s)ds)]\right)^{1/2q} \\
&\quad \left(E_Q[\exp(q(1-2q)\sum_{i=1}^n \int_0^\infty b_i^2(s, X_s)ds)]\right)^{1/2q} \\
&\leq \left(P(\sup_{t\geq 0}(\sum_{i=1}^n B_t^{H_i} - ct) \geq x)\right)^{1/p}.
\end{aligned} \tag{3.9}$$

On the other hand, with $d > 1$ and $\frac{1}{d} + \frac{1}{d_*} = 1$,

$$\begin{aligned}
P(\sup_{t\geq 0}(\sum_{i=1}^n B_t^{H_i} - ct) \geq x) &= E_P[1_{\sup_{t\geq 0}(X_t - ct)\geq x}\Lambda_\infty] \\
&\leq \left(P(\sup_{t\geq 0}(X_t - ct) \geq x)\right)^{1/d} \\
&\quad \left(E_P[\exp(d_*\sum_{i=1}^n \int_0^\infty b_i(s, X_s)\delta_{H_i} B_s^{H_i} - \frac{d_*}{2}\sum_{i=1}^n \int_0^\infty b_i^2(s, X_s)ds)]\right)^{1/d_*} \\
&\leq \left(P(\sup_{t\geq 0}(X_t - ct) \geq x)\right)^{1/d} \\
&\quad \left(E_P[\exp(2d_*\sum_{i=1}^n \int_0^\infty b_i(s, X_s)\delta_{H_i} B_s^{H_i} - 2d_*^2\sum_{i=1}^n \int_0^\infty b_i^2(s, X_s)ds)]\right)^{1/2d_*} \\
&\quad \left(E_Q[\exp(d_*(2d_* - 1)\sum_{i=1}^n \int_0^\infty b_i^2(s, X_s)ds)]\right)^{1/2d_*} \\
&\leq \left(P(\sup_{t\geq 0}(X_t - ct) \geq x)\right)^{1/d} \exp(\frac{1}{2}(2d_* - 1)\sum_{i=1}^n \int_0^\infty g_i^2(s)ds).
\end{aligned}$$

Hence, for all $d > 1$

$$P(\sup_{t\geq 0}(X_t - ct) \geq x) \geq$$
$$(P(\sup_{t\geq 0}(\sum_{i=1}^n B_t^{H_i} - ct) \geq x))^d \exp(-\frac{d}{2}\frac{d+1}{d-1}\sum_{i=1}^n \int_0^\infty g_i^2(s)ds). \quad (3.10)$$

Using the inequalities (3.9) and (3.10), for all $p > 1$ and $d > 1$,

$$-V_H x^{2H-2} \frac{d}{2}\frac{d+1}{d-1}\sum_{i=1}^n \int_0^\infty g_i^2(s)ds + dV_H x^{2H-2} \ln P(\sup_{t\geq 0}(\sum_{i=1}^n B_t^{H_i} - ct) \geq x)$$
$$\leq V_H x^{2H-2} \ln P(\sup_{t\geq 0}(X_t - ct) \geq x) \leq pV_H x^{2H-2} \ln P(\sup_{t\geq 0}(\sum_{i=1}^n B_t^{H_i} - ct) \geq x) \quad (3.11)$$

The precise conclusion of the theorem is obtained using a result of Duffield and O'Connell (see [7]) which stands that

$$\lim_{x\to\infty} V_H x^{2H-2} \ln P(\sup_{t\geq 0}(\sum_{i=1}^n B_t^{H_i} - ct) \geq x) = -\inf_{c>0} \frac{(c+\mu)^2}{2c^{2-2H}} = -\frac{1}{2}\frac{\mu^{2H}}{\kappa(H)^2}.$$

Taking the limit in the inequality (3.11), we get:

$$-d\frac{1}{2}\frac{\mu^{2H}}{\kappa(H)^2} \leq \lim_{x\to\infty} V_H x^{2H-2} \ln P(\sup_{t\geq 0}(X_t - ct) \geq x) \leq -p\frac{1}{2}\frac{\mu^{2H}}{\kappa(H)^2},$$

and the result is obtained by letting $p \searrow 1$ and $d \searrow 1$.

4. When is the law of the workload absolutely continuous with respect to the Lebesgue measure?

Except for the very particular case of an input flow with independent increments, very little is known about the load process. Hence, it is of some interest to know whether its law is absolutely continuous with respect to the Lebesgue measure.
For the sake of simplicity, we assume now that $n = 1$ and $H > \frac{1}{2}$. Thus, the input process $\{X_t, t \in [0, T]\}$ is the solution of the equation :

$$X_t = \int_0^t K_H(t,s)b(s, X_s)ds + B_t^H.$$

Theorem 4.1. Let b be K-Lipschitz continuous with respect to its second variable. We assume also that there is g in $L^2([0,T])$ such that for all t in $[0,T], x$ in $R, |b(t,x)| \leq g(t)$. Moreover, suppose that b is continuously differentiable with bounded, K-Lipschitz continuous and positive derivative. Then, for all t in R_+, the law of $W_t = \sup_{s\in[0,t]}(X_s - cs)$ is absolutely continuous with respect to the Lebesgue measure..

The rather technical proof is based on the result of Nualart and Vives (see [16] and Decreusefond-Üstünel ([6])), which states that: If there is a random variable ζ in $L^2(\Omega;\mathcal{H})$ such that,

$$\inf_{t,X_t=M}(\nabla_H X_t,\zeta)_{\mathcal{H}} > 0 \quad dP-\text{a.s.},$$

Then, for all t in R_+, the law of the workload at time $t, W_t = \sup_{s\in[0,t]}(X_s - cs)$ is absolutely continuous with respect to the Lebesgue measure.

References

1. Arcones, M.A. (1995): On the Law of the Iterated Logarithm for Gaussian Processes. *Journal of Theorical Probability*, 8(4), 877–904.
2. Borovkhov, A.A. (1976): *Stochastic Processes in Queuing Theory*. Springer-Verlag.
3. Crovella, M., Bestavros, A. (1996): Self-similarity in World Wide Web Traffic: Evidence and Causes. In *ACM Sigmetrics Conference on Measurement and Modeling of Computer Systems.*
4. Decreusefond, L., Lavaud, N. (1996): Simulation of the Fractional Brownian Motion and Application to the Fluid Queue. In *Proceedings of the ATNAC'96 Conference.*
5. Decreusefond, L., Üstünel, A.S.: Stochastic Analysis of the Fractional Brownian Motion. To appear in Potential Analysis.
6. Decreusefond, L., Üstünel, S. (June 1995): The Benes Equation and Stochastic Calculus of Variations. *Stochastic Processes and Their Applications*, 57(2), 273–284.
7. Duffield, N.G., O'Connell, N. (1995): Large Deviations and Oerflow Probabilities for the General Single-Server Queue, with Applications. *Mathematical Proceedings of the Cambridge Philosophical Society*, 118:363–374.
8. Lebedev, N.N. (1975): *Special Functions and their Applications*. Dover Publication, INC..
9. Leland, W.E., Taqqu, M.S., Willinger, W.W., Wilson, D.V. (1994): Ethernet Traffic is Self-similar : Stochastic Modeling of Packet Traffic Data. *IEEE/ACM Trans.Networking*, 2(1), 1–15.
10. Mandelbrot, B., Van Ness, J. (1968): Fractional Brownian Motions, Fractional Noises and Applications. *SIAM Review*, 10(4), 422–437.
11. Nikiforov, A.F., Uvarov, V.B. (1988): *Special Functions of Mathematical Physics*. Birkhäuser.
12. Norros, I. (1994): A Storage Model with Self-similar Input. *Queuing systems*, 16(3,4), 387–396.

13. Norros, I. (1995): On the Use of the Fractional Brownian Motion in the Theory of Connectionless Networks. *IEEE Journal on Sel. Areas in Telecommunication*, 13, 953–963.
14. Nualart, D. (Preprint 1995): Analysis on Wiener Space and Anticipating Stochastic Calculus.
15. Nualart, D. (1995): *The Malliavin Calculus and Related Topics*. Probability and its Application. Springer-Verlag.
16. Nualart, D., Vives, J. (1988): Continuité absolue de la loi du maximum d'un processus continu. *C. R. Acad. Sci. Paris*, 307, 349–354.
17. Üstünel, A.S. (1995): *An Introduction to Analysis on Wiener Space*, volume 1610 of *Lect. Notes in Math.* Springer-Verlag.

Fractional Brownian motion and data traffic modeling: The other end of the spectrum

J. Lévy Véhel and R. Riedi

INRIA - Groupe Fractales,
B.P. 105, Domaine de Voluceau,
78153 Le Chesnay Cedex, FRANCE
{Jacques.Levy_Vehel,Rolf.Riedi}@inria.fr

Abstract We analyze the fractal behavior of the high frequency part of the Fourier spectrum of fBm using multifractal analysis and show that it is not consistent with what is measured on real traffic traces. We propose two extensions of fBm which come closer to actual traffic traces multifractal properties.

Keywords: fractional Brownian motion, Multifractal analysis, TCP

1. Introduction

Fractal analysis of computer traffic has received considerable attention since the seminal work of Leland and al. [11] who provided experimental evidence that some traces of data traffic exhibit long range dependence (LRD). This is a typical fractal feature which is not found with the classical Poisson models. An important issue since then has been to propose "physical" models that lead to such fractal behavior. A popular model [27] is based on the superposition of simple i.i.d ON/OFF sources which ON and/or OFF periods follow a heavy tailed law ($Pr(X > \lambda) \sim c\lambda^{-\alpha}, 1 < \alpha < 2$). When properly normalized, the resulting traffic is a fractional Brownian motion (fBm) of LRD exponent $H = (3 - \alpha)/2$. Several practical implications of LRD traffic have consequently been investigated, e.g. the queuing behavior [15] (see also the two previous papers in this volume). Experimental justifications of the ON/OFF models have also been searched for, one of them being found in the heavy tailed laws of computer file sizes in the context of WWW traffic [6].

In [20], we performed an experimental study of traffic traces to describe fractal features different from LRD and based on Multifractal analysis (MA). Roughly speaking, MA is concerned with "the other end" of the Fourier spectrum, i.e. it looks at the high frequency content of the signal instead of the low frequency one. For some simple fractal models such as fBm, the Hurst exponent H ruling the LRD and, thus, the low frequencies governs also the local singular behavior which is related to the high frequencies. For general processes, however, the local singular behavior depends on time and is best described by the multifractal spectrum of singularity exponents. At this point it is worthwhile noting that studies of the distributions of the sizes of files provide a solid explanation of the LRD behavior of data traffic on large time scales. They do not account, however, for the short time behavior which results from the fragmentation and aggregation in protocol networks and from waiting queues. In some applications such as traffic control, understanding the structure of short term variations, and in particular their maximum amplitude is important. This information is comprised in the multifractal spectrum. Furthermore, MA may enable to evidence differences between traffic types (e.g. incoming/outgoing, LAN/WAN, ftp/WWW, etc...) that remain hidden to a second order analysis such as LRD characterization. Finally, multifractals offer a simple way of building realistic synthetic traces of traffic. MA may thus help to assess the adequacy of previously proposed models by looking at properties different from LRD.

It is our aim in this work to check whether the results of the MA of traffic traces as performed in [20] and [14] is compatible with a modeling based on fBm, and if not, to propose some simple extensions that allow a better account of the multifractal properties. In that view, and after some recalls on MA in Section 2., we compute in Section 3. the multifractal spectrum of fBm, both theoretically and estimated numerically on synthetic traces of fBm.

The comparison of this spectrum with the ones obtained on real traffic traces shows that more refined models are needed. As a first step in this direction, we study in Section 4. generalizations that involve 1) lumpings of fBm-s with different exponents, and 2) multifractional Brownian motion (mBm), which is basically an fBm with continuously varying H [18]. The motivation for considering such models is to account for a possible non stationarity in the data while loosely keeping some of the physical explanation provided by ON/OFF type models. Again, theoretical and numerically estimated spectra are produced and compared to spectra of traces of real traffic.

2. Multifractal analysis

MA deals with the description of the singularity structure of "signals" (which can be measures [3, 4, 12, 13, 19], functions [9] or capacities [26]), both in a local and a global way. The local information is given by the *Hölder exponent* at each point, while the global information is captured through a characterization of the geometrical or statistical distribution of the occurring Hölder exponents, called *multifractal spectrum*. Such an analysis is useful when one deals with very irregular signals (such as traffic traces), and when the singularity structure has practical consequences [24].

There are several ways of measuring the local irregularity, and different global characterizations may be considered. We briefly recall in this section some of the most classical approaches.

2.1 Local irregularity

Among several ways of measuring the local irregularity we will mention two. For simplicity we consider a signal $X(t)$ defined on [0,1] which is nowhere differentiable[1] and define the analysis with respect to the dyadic intervals:

$$I_n^k = [k2^{-n}, (k+1)2^{-n}[, \ k = 0, \dots 2^n - 1, n \in \mathbb{N}$$

Define the *coarse Hölder exponents* through

$$\alpha_n^k = -\frac{1}{n}\log\left|X((k+1)2^{-n}) - X(k2^{-n})\right|$$

where all logarithms are taken to the base 2 and where $\log 0 := -\infty$. For a fixed t in [0,1] let (k_n) be such that $I_n(t) := I_n^{k_n}$ contains t. Then, $k_n 2^{-n} \to t$ as $n \to \infty$. The limiting exponent at t

$$\alpha(t) := \liminf_{n\to\infty} \alpha_n^{k_n}$$

is called the *local* Hölder exponent of X at t. We mention in passing that the exponent considered usually is

$$\tilde{\alpha}(t) := \liminf_{\varepsilon\to 0} \tilde{\alpha}_\varepsilon(t) \quad \text{where} \quad \tilde{\alpha}_\varepsilon(t) = \frac{1}{\log \varepsilon}\log \sup_{s,s'\in B(t,\varepsilon)} |X(s) - X(s')|.$$

Since it is better adapted to numerical computations we will consider mainly α.

Multifractal analysis consists in giving a compact representation of the local singularity structure of a signal as measured through $\alpha(t)$. In that view, two approaches arise naturally: either use a *geometrical* description, or use a *statistical* one. The former leads to the Hausdorff spectrum f_h which we will not address further, while the latter leads to the large deviation spectrum f_g. A third spectrum is also considered at the end of this section, namely the Legendre spectrum, the interest of which is to provide a simple way to compute f_g or f_h when some conditions are met.

2.2 Large deviation spectrum

The large deviation spectrum f_g measures, loosely speaking, how "fast" the probability of observing a coarse Hölder exponent different from the expected value tends to zero as the resolution tends to ∞. More precisely, f_g is related to the rate function appearing in the large deviation analysis of such quantities. A heuristic explanation is the following: assume we want to assess how much the traffic T_n may vary in a (small) time interval I_n^k of duration 2^{-n}, with respect to n. The strength of the variation is measured by α_n^k, defined as: $T_n \propto |I_n^k|^{\alpha_n^k} = 2^{-n\alpha_n^k}$. If $\alpha_n^k = 1$, the traffic varies smoothly with respect to the scale of measurement ε. The smaller α_n^k is, the more intense the traffic is, while values of α_n^k greater than one correspond to sparse traffic.

It is thus of interest to evaluate the distribution of the α_n^k's when k is picked randomly from $0, \dots, 2^{n-1}$ and n is large. This will allow to characterize the traffic in terms of:

[1] This way we will not have to deal with polynomial trends.

- the highest degree of burstiness (resp. sparseness) occurring in the signal,
- the probability of hitting a given burstiness when measuring the traffic in a "small" time interval picked randomly.

In this view, we set:

$$f_g(\alpha) = \lim_{\varepsilon \to 0} \limsup_{n \to \infty} \frac{\log N_n^{\varepsilon}(\alpha)}{n}, \tag{2.1}$$

where

$$N_n^{\varepsilon}(\alpha) = \#\{\alpha_n^k / |\alpha_n^k - \alpha| < \varepsilon\}.$$

The large deviation spectrum f_g describes the distribution of the local singularities, since the number of dyadic intervals of size 2^{-n} with coarse Hölder exponent $\simeq \alpha$ varies roughly as $2^{nf_g(\alpha)}$ for large n. Equivalently:

$$P_n(\alpha_n^k \approx \alpha) \propto 2^{-n(1-f_g(\alpha))}$$

where the probability is related to a random choice of k uniformly in $\{0, \ldots, 2^n - 1\}$, i.e. P_n is the uniform distribution on the set of all dyadic intervals I_n^k of size 2^{-n}.

2.3 Legendre spectrum

It is natural to interpret the spectrum f_g as a rate function in a *large deviation principle* (LDP). The general theory on large deviations provides conditions under which such rate functions may be calculated as the Legendre transform of a limiting moment generating function. As we are about to show this procedure provides a much more robust estimation of f_g than a direct computation via (2.1), provided that the data satisfy the necessary conditions.

From a multifractal point of view it is most convenient to define for $q \in \mathbb{R}$:

$$S_n(q) = \sum_{k=0}^{2^n-1} |X((k+1)2^{-n}) - X(k2^{-n})|^q \tag{2.2}$$

with the convention $0^q := 0$ for all $q \in \mathbb{R}$. Next, define the 'structure function'

$$\tau(q) = \liminf_{n \to \infty} \frac{\log S_n(q)}{-n}$$

and the so-called *Legendre spectrum* of X

$$f_l(\alpha) = \tau^*(\alpha) := \inf_{q \in \mathbf{R}} (\alpha q - \tau(q)).$$

Returning to the large deviation principle, let us follow [19] and define the sequence of random variables $Z_n := \log Y_n^k$ where k is distributed with P_n as before. Consider the corresponding moment generating functions

$$c_n(q) := -\frac{1}{n}\log \mathbb{E}_n[\exp(qZ_n)] = -\frac{1}{n}\log\left(2^{-n}S_n(q)\right).$$

A version of the Gärtner-Ellis theorem on LDP [7] allows then to conclude that if $\lim c_n(q)$ exists (in which case it equals $1+\tau(q)$), and is differentiable, then $c^* = f_g - 1$. In other words, the so-called *weak multifractal formalism*:

$$f_l(\alpha) = f_g(\alpha)$$

holds then. This case is favorable since $\tau(q)$, being computed by averaging and involving a single limit, is easier to estimate than f_g which requires the evaluation of local quantities and of a double limit.

In general, however, these functions may differ, and one has:

Lemma 2.1 ([19, 21]). *f_l is the concave hull of f_g, i.e. $\tau(q) = f_l^*(q)$ and $f_l(\alpha) = f_g^{**}(\alpha)$. Consequently, $\tau(q)$ is concave and $f_g(\alpha) \leq f_l(\alpha)$. Furthermore,*

$$\begin{array}{rcll} f_g(\alpha^+) & = & q\alpha^+ - \tau(q) & (q>0) \\ f_g(\alpha^-) & = & q\alpha^- - \tau(q) & (q<0) \end{array} \tag{2.3}$$

where $\alpha^+ := \tau'(q+)$ and $\alpha^- := \tau'(q-)$ denote the one-sided derivatives of $\tau(q)$. Finally, due to the special form of P_n f_g cannot assume negative, finite values. Consequently, $\tau(q)$ is finite either in all of $\mathbb{R}$ or exactly in $q \geq 0$.

This shows that f_g contains in general more information than f_l.

3. Multifractal Analysis of fBm

Let $X(t)$ be an fBm of exponent H. As is well known [23] $X(t)$ is a zero mean Gaussian process with stationary increments and is statistically self-similar in the sense of finite-dimensional distributions. It follows that $X(0) = 0$ a.s., and that

$$Y_n^k := X((k+1)2^{-n}) - X(k2^{-n}) \stackrel{d}{=} X(2^{-n}) \stackrel{d}{=} N(0, 2^{-nH}) \tag{3.1}$$

where $\stackrel{d}{=}$ means equality in distribution and where $N(m,\sigma)$ stands for the normal distribution with mean m and variance σ.

The LRD property of fBm manifests itself in the asymptotic behavior of the Fourier spectrum $S(\lambda) \sim |\lambda|^{1-2H}$ as the frequency λ tends to zero. The high frequency content of X is also ruled by H: one can in particular show that with probability one $\tilde{\alpha}(t) = H$ for all t [2]. fBm displays, thus, a very peculiar fractal behavior, where the same parameter H allows to describe both parts of the Fourier spectrum.

We continue this section by computing the large deviation and the Legendre spectrum of fBm, elaborating especially on the complications induced by the randomness. Indeed, measured along paths, f_g and f_l are random functions and hard to compute. As a first step towards this goal, we take expectations before going to the limits (Subsection 3.1). The spectra F_g and F_l obtained in this way are deterministic and represent the rate functions of a LDP in a higher probability space. Using estimates between f_l and F_l we are then able to treat the pathwise spectra in Subsection 3.2. It will not surprise that all spectra are essentially determined by H.

3.1 Deterministic spectra

For the calculation of the multifractal spectra one would like to exploit this self-similarity in distribution. Therefore, we take expectation in (2.2) and set

$$\overline{S}_n(q) := \mathbb{E}_\omega S_n(q,\omega) = \mathbb{E}_\omega \sum_{k=0}^{2^n-1} |Y_n^k|^q = 2^n \mathbb{E}_\omega |Y_n^1|^q, \tag{3.2}$$

where we used the stationarity of increments. Letting

$$c_q := \mathbb{E}_\omega |Y_1^1|^q = \mathbb{E}_\omega |X(1)|^q = \frac{2}{\sqrt{2\pi}} \int_0^\infty y^q e^{-y^2/2} dy,$$

which is finite exactly for $q > -1$, we find $\mathbb{E}_\omega |Y_n^1|^q = 2^{-nqH} c_q$. Thus, one finds the scaling law

$$T(q) := \lim_{n\to\infty} \frac{\log \overline{S}_n(q)}{-n} = \begin{cases} -\infty & \text{for } q \le -1 \\ qH - 1 & \text{for } q > -1, \end{cases}$$

which is — up to the constant 1 — the structure function of an LDP on the sequence of probability spaces $(\Omega \times \{1, \ldots, 2^n\}, P_\omega \times P_n)$. Note in particular the built-in averaging on the dyadic intervals through S_n. The Legendre transform of T is

$$F_l(\alpha) = \inf_q (q\alpha - T(q)) = \inf_{q>-1} (q\alpha - T(q)) = \begin{cases} -\infty & \text{for } \alpha < H \\ 1 + H - \alpha & \text{for } \alpha \ge H. \end{cases}$$

We will address F_l as the 'deterministic Legendre spectrum'. To define the corresponding 'deterministic spectrum of large deviations' set

$$\pi_n^\varepsilon(\alpha) := P_\omega \times P_n[\alpha_n^k \in (\alpha - \varepsilon, \alpha + \varepsilon)]$$

and define

$$F_g(\alpha) = 1 + \lim_{\varepsilon\to 0} \limsup_{n\to\infty} \frac{\log \pi_n^\varepsilon(\alpha)}{n}.$$

Note, first of all, that F_g may assume non-trivial negative values, whence the last conclusion of Lemma 2.1 is not valid. The other claims, though, continue to hold for this more general 'random procedure' of choosing α_n^k's.

We argue that $F_g(\alpha)$ should be a good estimator of $f_g(\alpha, \omega)$, the usual spectrum measured along paths. Indeed, equality holds a.s. at least for $H = 1/2$ (Lemma 3.2). In general, the increments of fBm with constant lag form a stationary process which is ergodic under time-shift [8]. This implies that $N_n^\varepsilon(\alpha)$ divided by the total number 2^n of dyadic intervals considered should be well approximated by $\pi_n^\varepsilon(\alpha)$. Also,

$$\mathbb{E}_\omega[N_n^\varepsilon(\alpha)/2^n] = \frac{1}{2^n}\mathbb{E}_\omega \sum_{k=0}^{2^n-1} \mathbf{1}_{(\alpha-\varepsilon,\alpha+\varepsilon)}(\alpha_n^k) = \pi_n^\varepsilon(\alpha),$$

which closes the circle started with (3.2). For a computation of F_g note that

$$\pi_n^\varepsilon(\alpha) = \frac{2^{nH+1}}{\sqrt{2\pi}} \int_{2^{-(\alpha+\varepsilon)n}}^{2^{-(\alpha-\varepsilon)n}} \exp\left(-\frac{x^2}{2^{-2nH+1}}\right) dx.$$

A crude estimate of the monotonous integrand gives [25]

$$F_g(\alpha) = \begin{cases} -\infty & \text{for } \alpha < H \\ 1+H-\alpha & \text{for } \alpha \geq H. \end{cases} \tag{3.3}$$

Note that F_g assumes negative values, which is not possible for f_g. Consequently, F_g may be expected to be a good estimator of f_g only where it is positive. Negative $F_g(\alpha)$ values correspond to probabilities of observing a coarse Hölder exponent α which decay faster than the number of 'samples' α_n^k considered. Oversampling the process, i.e. analyzing several independent realizations will, thus, result in larger observed α_n^k. In fact, in $\exp(-n\ln(2)F_g(\alpha))$ independent traces one has a fair chance to see at least one increment $|Y_k^n|$ with coarse Hölder exponent α.

Let us note that

$$F_l = F_g$$

which is a weak form of the so-called multifractal formalism.

3.2 Path-wise spectra of fBm

Let us return now to the spectra calculated along a path of fBm. Though $2^{nqH-n}S_n$ is in general not equally distributed to

$$\frac{1}{2^n}\sum_{k=0}^{2^n-1} |Y_1^k|^q. \tag{3.4}$$

it is at least equal in expectation. As we will show now in a general argument the study of $\overline{S}$ leads to the right normalization term. For any q with finite $T(q)$ and $\varepsilon > 0$

$$\mathbb{E}\limsup_{n\to\infty} 2^{n(T(q)-\varepsilon)}S_n(q,\omega) \leq \mathbb{E}\sum_{n\in\mathbb{N}} 2^{n(T(q)-\varepsilon)}S_n(q,\omega)$$

$$\leq \sum_{n\in\mathbb{N}} 2^{n(T(q)-\varepsilon)}\mathbb{E}S_n(q,\omega) \leq \frac{b(q,\varepsilon)}{1-2^{-\varepsilon/2}} < \infty,$$

by definition of T. Here b is a suitable constant depending on q and ε only. This allows to conclude that almost surely $\limsup_{n\to\infty} 2^{n(T(q)-\varepsilon)} S_n(q,\omega) < \infty$. Whence, $\tau(q) \geq T(q) - \varepsilon$, which is trivial if $T(q) = -\infty$. It is clear that this estimate holds with probability one for all $\varepsilon = 1/m$ ($m \in \mathbb{N}$) and some countable, dense set of q values with $T(q) < \infty$ simultaneously. Using that $\tau(q)$ is always continuous on open sets due to Lemma 2.1 we find the following result valid for any process:

Lemma 3.1. *With probability one*

$$\tau(q,\omega) \geq T(q) \qquad \textit{for all } q \textit{ with } T(q) < \infty.$$

As an immediate consequence we obtain with Lemma 2.1 that with probability one

$$f_g(\cdot,\omega) \leq f_l(\cdot,\omega) \leq F_l \tag{3.5}$$

As a matter of fact, we have equality at least for $H = 1/2$, the ordinary Brownian motion due to

Lemma 3.2 ([25]). *For any process with independent and stationary increments and any α*

$$f_g(\alpha,\omega) = F_g(\alpha) \qquad a.s.$$

Due to the linearity of T the bound provided by lemma 3.1 is all we need. Indeed, for each n, $2^n S_n(0,\omega)$ counts the number of non-vanishing increments which is 2^n almost surely. Thus, $\tau(0) = -1$ a.s. and since $\tau(q)$ must be concave the obtained bound is necessarily sharp.

In conclusion, with probability one:

$$\tau(q,\omega) = qH - 1 \qquad \text{for } q > -1, \tag{3.6}$$

and

$$f_g(\alpha,\omega) \begin{cases} = f_l(\alpha,\omega) & = -\infty & \text{for } \alpha < H \\ = f_l(\alpha,\omega) & = 1 & \text{for } \alpha = H \\ \leq f_l(\alpha,\omega) & \leq 1 + H - \alpha & \text{for } \alpha \geq H. \end{cases}$$

Finally, the implicit restrictions on $\tau(q)$ imply bounds for $q \leq -1$: with probability one.

$$qH + q \leq \tau(q,\omega) \leq qH - 1 \qquad \text{for } q \leq -1. \tag{3.7}$$

The upper bound follows from concavity combined with (3.6), while the lower bound reflects the fact that $f_l(\alpha,\omega)$ is either positive or $-\infty$ (Lemma 2.1). This together with $f_l(\alpha,\omega) \leq 1 + H - \alpha$ implies that the slopes of τ have to be smaller than $1 + H$.

Let us end the section with a remark on the scaling behavior for $q \leq -1$. Following the intuitive approximation (3.4) one may argue with Aaronson [10, p 15] that there is no power γ which makes $2^{n\gamma} S_n$ converge a.s. to a positive, finite number, due to the infinite expectation of $|Y_1^1|^q$. For $H = 1/2$ we have independence which allows to apply the theory of infinitely divisible laws.

With $|Y_1^1|^q$ lying in the domain of attraction of an α-stable law with $\alpha = -1/q$ this suggests that a renormalization of the type $2^{n(qH+q)}S_n(q,\omega) - A_n$ might provide convergence in distribution to a stable law. In this result, we recognize the limiting value $Hq + q$ for $\tau(q)$. However, the weak sense of 'convergence in distribution' and the divergence of $A_n \simeq 2^{n(qH+q-(q+1)/q)}$ hint to a poor scaling behavior for $q \leq -1$. Indeed, in Subsection 3.3 we provide numerical evidence for unreliable estimates starting at $q \approx -1/2$ which corresponds to infinite variance of $|Y_1^1|^q$.

3.3 Experimental results

We show in this section estimated large deviation and Legendre spectra computed on synthetic fBm traces. The traces consist of 2^{14} points and were generated using the Choleski decomposition method with $H = 0.2, 0.6, 0.8$. The estimation procedures for both f_g and f_l are the ones described in [20]. We first compare log-log plots of the partition sum $S_n(q)$ for an fBm and a real traffic trace on Fig. 3.1. To illustrate more clearly whether the assumption of linearity of these log-log plots holds, Fig. 3.2 displays their increments $\log S_{2m}(q) - \log S_m(q)$. If constant over a considerable range of m they provide a reliable estimate of $\tau(q)$.

From these figures it becomes apparent that fBm displays a good scaling behavior for q larger $\simeq -1/2$ while $\tau(q)$ does not seem to converge for lower values of q, a result which is consistent with (3.7). In contrast, the computation on real traces reveals a good scaling even for $q = -2$ and sufficient level of aggregation. This is a first point where the fBm model departs significantly from the data.[2]

Finally, Fig. 3.3 shows the estimated spectra. Both f_g and f_l almost perfectly match the theoretical expressions obtained in the previous sections, supporting in particular the fact that the positive part of F_l is indeed a good approximation of f_g. Again, the shapes differ clearly from the typical ones obtained on real traces [20].

3.4 Discussion

The preceding results show that neither spectrum of fBm, f_g nor f_l, displays the shape estimated on real traffic traces as presented in [20] and [14]. This calls for fractal traffic models which are more refined than fBm. The fact that an estimation of H alone is not sensitive to certain features observed with a multifractal analysis [20] further supports this point of view.

---

[2] One may argue that the bad scaling for negative q comes from the fact that fBm is a centered process. But adding a constant to fBm in order to obtain a positive trace worsens the situation by resulting in the trivial multifractal behavior $\tau(q) = q - 1$. Furthermore, the cut-off value for the divergence of τ_n is not 0 but lies between $-1/2$ and -1.

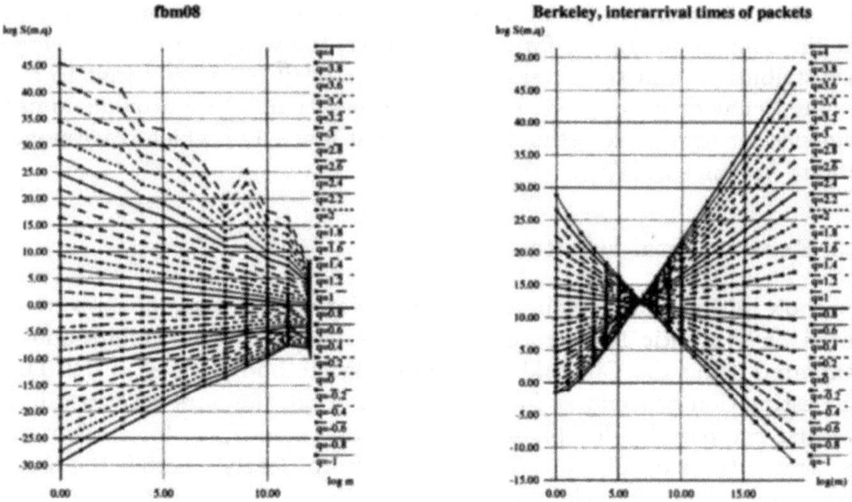

Fig. 3.1. Log-log plots of the partition sum $S_m(q)$ for fractional Brownian motion with $H = 0.8$ (left) and for the interarrival times of packets as observed at the gateway of Berkeley [20] (right).

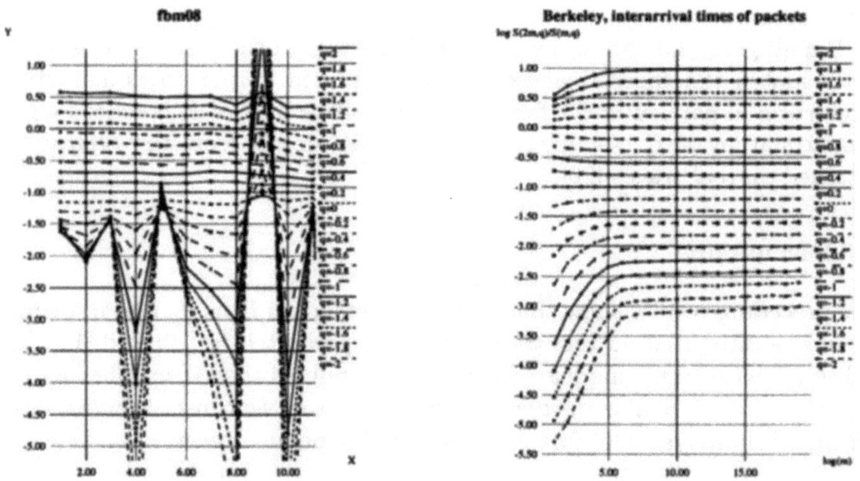

Fig. 3.2. Increments of the log-log plots of the partition sum $S_m(q)$ for fractional Brownian motion with $H = 0.8$ (left) and for the interarrival times of packets as observed at the gateway of Berkeley [20] (right).

The most simple multifractal processes bearing some analogies with real traffic traces are *random multiplicative cascades*. For a detailed introduction see [20]. The striking resemblance of such cascades with actual traces of real data traffic makes them a natural candidate for data traffic modeling.

If random cascades are to be invoked for justifying the multifractal behavior of real traffic, it is important to find "physical" reasons why such a multiplicative structure should occur. Loose equivalents of the multiplicative structure can be found in the fact that Telecommunication protocols are organized in a hierarchy and that these protocols work by splitting wholes into pieces. For instance, in the case of file transfer, we find at the highest level the decision to send a file. At the next level, the TCP control comes into play, and time sharing with other processes results in a random variability. The maximal window size and the speed regulation (e.g. slow start) still emphasizes this effect. Further levels are the maximal IP packet size, and finally, if the network is an ATM one, the breaking of IP packets into cells, which depends on the rate of the ATM virtual connection ([16]).

However, such explanations remain somewhat vague and much work is needed before one can come up with a physical model for multiplicative cascades as neat as the ON/OFF source model for fBm. Here, we propose a different approach which exhibits multifractal properties similar to real traffic traces, while retaining some of the physical explanation of the ON/OFF model.

The basic idea is simply to relax the hypothesis of stationarity. Although some experimental studies [1] conclude that the LRD parameter seems to remain constant over large time intervals, we argue that the exponent ruling the heavy tail of each individual ON/OFF source may undergo some changes in time, due to important intra-day variations (lunch time, busy hours, night-time). In addition, while H, as a parameter of LRD, should not depend much on the protocol used, it is conceivable that it varies with the content of the communication. Finally, it is likely that even minor intra-day variations can cause modifications in the local singular behavior, leading to a Hölder exponent that would depend on time.

4. Extensions of fBm

4.1 Lumpings

The most simple way of obtaining a non stationary process is to assume that, depending on the time of the day, we are observing different fBm processes with various exponents H. Formally, if $X_H(t)$ denotes an fBm with exponent H, the traffic model X is given by:

$$X(t) = X_{H_k}(t) \qquad \text{for } t \in [t_{k-1}, t_k[$$

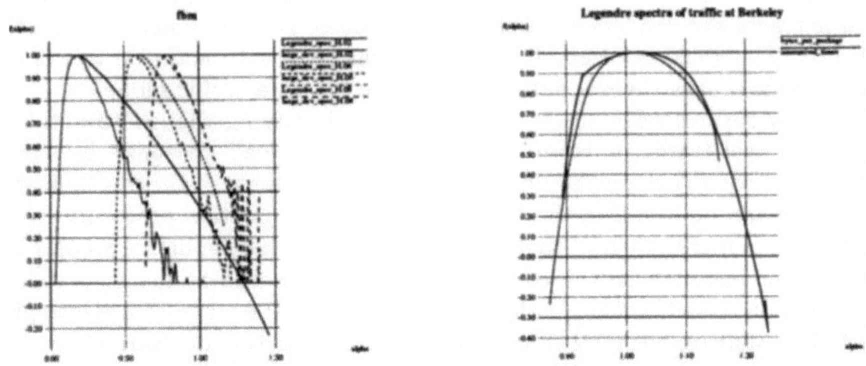

Fig. 3.3. Estimated large deviation spectrum and Legendre spectrum for fractional Brownian motion with $H = 0.2$, 0.6 and 0.8 (left) and for the interarrival times of packets as observed at the gateway of Berkeley [20] (right).

fro $k = 1, \ldots, m$, where m is a fixed number and $t_0 := 0 < t_1 < t_2 \ldots < t_{m-1} < t_m := 1$. For simplicity let us assume that $t_k 2^n$ is integer for all k from some large n on. Since m is fixed and finite, and due to the built-in averaging over dyadic intervals in their definitions, $\overline{S}_n(q)$ and $\pi_n^\varepsilon(\alpha)$ will be composed of m terms corresponding to the various $X_{H_k}(t)$; e.g.

$$\pi_n^\varepsilon(\alpha) = \sum_{k=1}^{m} \frac{t_k - t_{k-1}}{2^n} \pi_n^\varepsilon[X_{H_k}](\alpha)$$

where we used stationarity in the intervals $[t_{k-1}, t_k[$. It is clear that largest scaling exponent will govern the asymptotic behavior of the overall sum. In short, $T(q)$ will be the minimum of the corresponding rate functions of the $X_{H_k}(t)$, while F_g will be the corresponding maximum.

Thus, F_g assumes a zig-zag shape, consisting of m lines of slope -1 which start in $(H_k, 1)$ and continue to $(H_{k+1}, 1 + H_k - H_{k+1})$ (with the convention $H_{m+1} = \infty$). Letting $\underline{H} = \min_{1 \le i \le m} H_i, \overline{H} = \max_{1 \le i \le m} H_i$, we have

$$F_l(\alpha) = \begin{cases} -\infty & \text{for } \alpha < \underline{H} \\ 1 & \text{for } \underline{H} \le \alpha \le \overline{H} \\ 1 + \overline{H} - \alpha & \text{for } \alpha > \overline{H}. \end{cases}$$

In Section 3. we argued that the spectra numerically measured on a path of an fBm with Hurst parameter H_k should be well approximated by (3.3). In any case, (3.5) provides an upper bound. This is indeed confirmed by our experiments (see Fig. 4.2). In particular, f_g is not concave and differs from f_l, comprising more information than the latter.

4.2 Cantor superpositions of fBm-s

With a whole range of observable Hölder exponents the previous simple model shows indeed a non trivial multifractal behavior, although different from the one observed on real traffic. Its peculiar shape stems from the fact that the different values of the self-similarity parameter H occur in entire intervals (which are of dimension 1). In order to obtain the approximate $\cap$ shape corresponding to actual traces, one needs to build a superposition of two fBm processes with exponents $H_1 < H_2$ such that the 'size' of the support of X_{H_1} is 'essentially smaller' than the one of X_{H_2}.

A somewhat artificial way to achieve this is to set:

$$X(t) = Z_{H_1}(t) + X_{H_2}(t)$$

where $X_{H_2}(t)$ is an fBm of exponent H_2 and $Z_{H_1}(t)$ is a 'Cantor Brownian motion' obtained as follows: let $X_{H_1}(t)$ be an fBm of exponent $H_1 < H_2$ and independent of $X_{H_2}(t)$. Let C be a Cantor-like set of dimension $D_1 < 1$ constructed with dyadic intervals. Set $Z_{H_1}(t) = X_{H_1}(t)$ for $t \in C$ and interpolate Z_{H_1} linearly elsewhere. Since C is a closed set without isolated points, Z_{H_1} is well defined and continuous.

The computation of the deterministic Legendre spectrum follows the lines of Subsections 3.1 and 4.1, observing that $\overline{S}_n$ can again be split into two parts, one coming from dyadic intervals used in the construction of C and one from the remaining intervals, i.e. those contained in a gap (the components of the closure of the complement of C). Bearing this in mind, a lengthy but straightforward calculation gives [25]

$$F_l(\alpha) = \begin{cases} -\infty & \text{for } \alpha < H_1 \\ (1-D_1)/(H_2-H_1)(\alpha - H_1) + D_1 & \text{for } H_1 \le \alpha \le H_2 \\ 1 + H_2 - \alpha & \text{for } \alpha > H_2. \end{cases} \quad (4.1)$$

Adding more than two fBm with well chosen pairs (H_i, D_i) would allow to design spectra that would more or less look like the ones obtained on real traffic. A loose "physical" interpretation of this model could be that different processes corresponding to various kinds of applications and/or traffic types and/or users (machines) coexist on the network. Some, more regular ones, being active most of the time (the ones with the largest values of H and D), and the others appearing intermittently and dominating the traffic during these periods (low values of H and D). Such an interpretation is also supported by other types of analysis [5]. Of course, this model is very unstable because any shift in the origin of the analysis would affect the obtained spectrum. Although it would be possible to refine the construction, we prefer to turn to another generalization which possesses nicer analytical properties.

4.3 Multifractional Brownian motion

While the above extensions remain quite simple, they are not fully satisfactory from a theoretical point of view. A step towards a better model is *multifractional Brownian motion* (mBm), a process introduced in [18]. Roughly, mBm is an fBm where the parameter H is a function of time. The precise definition is as follows:

Let $H : (0, \infty) \to (0, 1)$ be a Hölder function with exponent $\beta > 0$. We will write H_t instead of $H(t)$ for the ease of notation. Multifractional Brownian motion (mBm) is the Gaussian process $Y_t = X_{H_t}(t)$ defined on $(0, \infty)$ by:

$$Y_t = \frac{1}{\Gamma(H_t + \frac{1}{2})} \int_{-\infty}^{t} [(t-s)_+^{H_t - 1/2} - (-s)_+^{H_t - 1/2}] dW(s),$$

where $(x)_+$ equals x if positive and 0 otherwise.

The main properties of mBm are the following:

1. $(Y_t)_{t>0}$ is an a. s. continuous process.
2. Under the technical assumption $H_t < \beta$ for all t, the graph of mBm has the 'intuitively correct dimension', i.e. with probability one,

$$\dim_H\{(t, Y_t) : t \in [a, b]\} = 2 - \min\{H_t, t \in [a, b]\}.$$

3. With probability one, the pointwise Hölder exponent of Y_t at t is H_t for all t provided $H_t < \beta$ for all t.

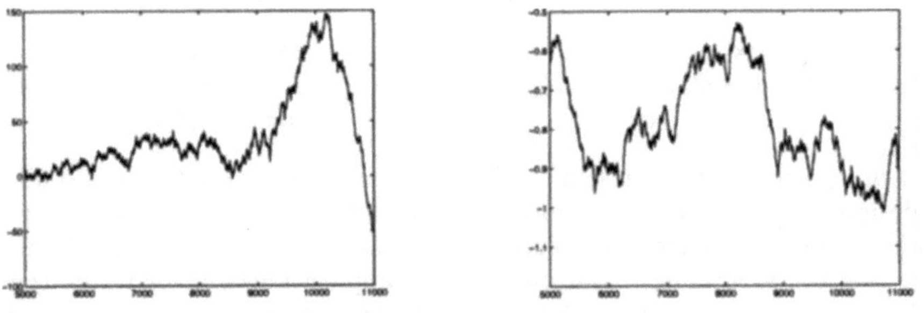

Fig. 4.1. Normalized mBm with $H(t) = t$ for $t \in [0.4, 0.8]$ (left) and fractional Brownian motion with $H = 0.6$ (right).

Multifractional Brownian motion allows, thus, to take into account a possible non stationarity of the exponent H as long as its evolution is smooth ($H(t)$ is a Hölder function). Let us now move to the computation of the deterministic Legendre spectrum of mBm. We present a heuristic computation to make the basic idea clear. A rigorous argument is contained in [25].

Recall the definition (3.2) of $\overline{S}_n$ in the random case. For n large, using property 3 above, we see that $Y_n^k = X((k+1)2^{-n}) - X(k2^{-n})$ is approximately equal to an $N(0, 2^{-nH(k2^{-n})})$ random variable. As earlier

$$\mathbb{E}(Y_n^k)^q \approx \begin{cases} \infty & \text{for } q \leq -1 \\ c_q 2^{-nqH(k2^{-n})} & \text{for } -1 < q. \end{cases}$$

In the special case $H(t) = t$ for $t \in (0,1)$ we obtain

$$\overline{S}_n(q) \simeq c_q \frac{1 - 2^{-nq}}{1 - 2^{-nq2^{-n}}} = c_q \frac{1 - 2^{-nq}}{nq2^{-n}\xi_n} \qquad (q \neq 0),$$

where $\xi_n \in [1/2, 2]$ for large n. With $\overline{S}_n(0) = c_q 2^n$ this leads to

$$T(q) = \begin{cases} -\infty & \text{for } q \leq -1 \\ q - 1 & \text{for } -1 < q < 0 \\ -1 & \text{for } 0 \leq q. \end{cases}$$

In general, for any non constant C^1 function H: $\mathbb{R}^{+*} \to [a, b] \subset]0,1[$ one

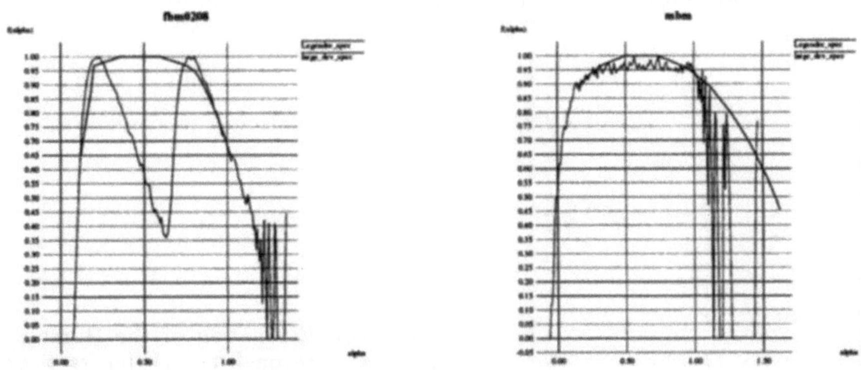

Fig. 4.2. Estimated large deviation spectrum and Legendre spectrum for the lumping of fBm $H = 0.2$ and $H = 0.8$ (left) as well as for the same mBm as before (right).

gets:

$$F_l(\alpha) = \begin{cases} -\infty & \text{for } \alpha < a \\ 1 & \text{for } a \leq \alpha \leq b \\ 1 + b - \alpha & \text{for } b < \alpha. \end{cases}$$

One finds, thus, the same spectrum F_l as the one of the lumping of two fBm with exponents corresponding to the minimum and maximum of the function $H(t)$.

Instead of $H(t) = t$, let us now consider a Hölder continuous function H which assumes the value H_1 on a Cantor set C as before and the value

$H_2 > H_1$ in points of distance larger than 2^{-N} from C. Any numerical computation of the spectra restricted to resolutions coarser than 2^{-N} would yield the same spectrum as in (4.1).

4.4 Experimental results

Fig. 4.1 displays a sample path of a normalized mBm (for all t, $X(t)$ is divided by the instantaneous variance) for which $H(t) = t$ for $t \in [0.4, 0.8]$, and, for comparison, a trace of an fBm with $H = 0.6$. Fig. 4.2 displays the estimated spectra on a lumping of fBm-s and on an mBm. As in Subsection 3.3, a good match with the theoretical expression for F_l is found. As for f_g, the estimated result support the heuristic argument presented in Subsection 4.1. Comparing with Fig. 3.3, one sees that these spectra indeed bear some analogies with the ones obtained on real traces (approximate $\cap$ shape, asymmetry), and could explain the findings reported in [20] and [14].

Finally, Fig. 4.3 displays estimates of the parameter H, based on the simple relation: $H = 1 + \tau(1, \omega)$ a.s. (3.6). For 'pure' fBm-s, the obtained value is quite good. For a lumping, the lowest H is estimated as predicted by theory. Lastly, no meaningful value is obtained in the case of mBm, which is again consistent with the non stationarity of H in this situation.

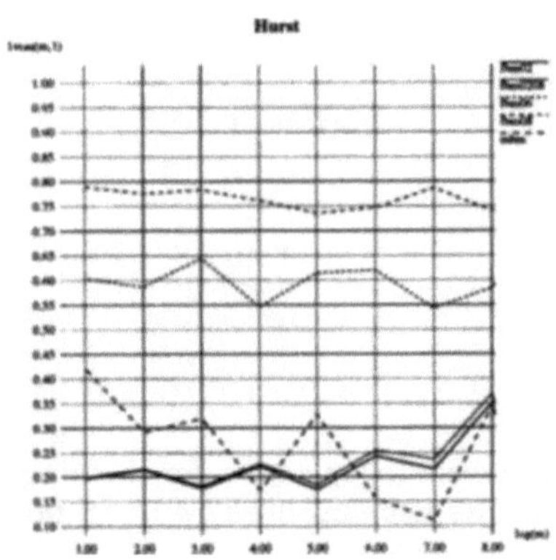

Fig. 4.3. Estimation of the Hurst parameter H through $1+\tau(1)$ for fractional Brownian motion with $H = 0.2$, 0.6 and 0.8 as well as for mBm and a lumping of fBm-s.

Conclusion

LRD analysis and associated fBm models allow to describe "robust" features of the traffic, i.e. features that are not sensitive to local variations in time or space. Multifractal analysis, on the other hand, is well fitted to the investigation of the fine structure of the traffic, that reflects some particular aspects of the network at a given location and hour of the day. Such a description is important because it allows to obtain more precise models and to characterize different types of traffic. A fine scale analysis is generally difficult for various reasons, and it seems that multifractal analysis may be a good tool for this task.

In as much as one is interested in the high frequency content of the data, we have shown that fBm is not a good model for computer traffic: The multifractal spectra of fBm, which we computed theoretically and estimated on synthetic traces, do not coincide with the ones of real traffic data. We have proposed to refine this model by taking into account a possible non stationarity in the singularity exponent.

Two situations have been studied. In the first one, several consecutive regimes are considered by lumping together plain fBm processes. The second model allows for a continuous variation of H. An adapted choice of the various parameters associated to these models allows to obtain spectra that look like real ones, at least when estimated on a finite number of resolutions. This result along with the fact that some 'physical' mechanism can be invoked which lead to such processes makes them possible candidates for modeling. A further argument is that a non stationary model is probably better suited to the high intra-day variability of the data.

One important discrepancy with real traces remains, namely that a non multifractal behavior is predicted for $q \leq -1$ in both models, the lumping of fBm-s and the mBm, while traces of real data traffic exhibit good scaling even for q lower than -4 [20]. In this respect, a radically different modeling based on multinomial cascades, which still lacks precise physical interpretation, would be more adapted. We should add, however, that such multinomial processes exhibit an everywhere discontinuous H function.

Despite of its drawbacks, it would be worthwhile to pursue the mBm type modeling and to study, for instance, the associated queueing behavior.

Acknowledgement. The authors gratefully acknowledge the financial support of CNET, France, under contract # 95 8B 069 and express their thanks to Fabrice Clerot, Ilkka Norros, Murad Taqqu and Marie-Claude Viano for helpful comments.

References

1. Abry, P., Veitch, D. (Preprint 1996): Wavelet analysis of lange range dependent traffic.
2. Adler, R. (1981): *The Geometry of Random Fields*. John Wiley & Sons, New York.
3. Arbeiter, M., Patzschke, N. (1996): Self-similar random multifractals. *Math. Nachr.*, 181, 5–42.
4. Brown, G., Michon, G., Peyrière, J. (1992): On the multifractal analysis of measures. *J. Stat. Phys.*, 66, 775–790.
5. Clerot, F. (1997). Private communication.
6. Crovella, M., Bestavros, A. (May 1996): Self-similarity in world wide web traffic. evidence and possible causes. In *Proceedings of SIGMETRICS '96*.

7. Ellis, R. (1984): Large deviations for a general class of random vectors. *Ann. Prob.*, 12, 1–12.
8. Itô, K. (1944): On the ergodicity of a certain stationary process. *Proc. Imp. Acad. Tokyo*, 20, 54–55.
9. Jaffard, S. (1993): Multifractal formalism for functions. *CRAS*, 317, 745–750.
10. Krengel, U. (1985): *Ergodic Theorems*. Walter de Gruyter, Berlin.
11. Leland, W., Taqqu, M., Willinger, W., Wilson, D. (1994): On the self-similar nature of ethernet traffic (extended version). *IEEE/ACM Transactions on Networking*, 1–15.
12. Mandelbrot, B.B. (1974): Intermittent turbulence in self similar cascades : divergence of high moments and dimension of the carrier. *J. Fluid. Mech.*, 62, 331.
13. Mandelbrot, B.B., Riedi, R.H. (1997): Inverse measures, the inversion formula and discontinuous multifractals. *Adv. Appl. Math.*, 18, 50–58.
14. Mannersalo, P., Norros, I. (1997): Multifractal analysis of real ATM traffic: a first look. *COST257TD*.
15. Norros, I. (1994): A storage model with self-similar input. *Queueing Systems*, 16, 387–396.
16. Norros, I. (1997). Private communication.
17. Paxson, V. (August 1994): Emperically-derived analytic models of wide-area TCP connections. *IEEE/ACM Transactions on Networking*, 2(4), 316–326.
18. Peltier, R., Lévy Véhel, J. (1995): Multifractional Brownian motion: Definition and preliminary results. *INRIA Research Report, No 2645*.
19. Riedi, R.H. (1995): An improved multifractal formalism and self-similar measures. *J. Math. Anal. Appl.*, 189, 462–490.
20. Riedi, R., Lévy Véhel, J. (1997): TCP traffic is multifractal: a numerical study. *INRIA Research Report No. 3129*,1997 *http://www-syntim.inria.fr/fractales/*.
21. Riedi, R.H. (1997): Seven definitions of multifractal scaling exponents. *In preparation*.
22. Roberts, J., Mocci, U., Virtamo, J. (eds.) (1996): Braodband network teletraffic. In *Lecture Notes in Computer Science, No 1155*. Springer.
23. Samorodnitsky, G., Taqqu, M. (1994): *Stable non-Gaussian random processes*. Chapman and Hall, New York ISBN 0-412-05171-0.
24. Lévy Véhel, J. (1996): Fractal approaches in signal processing. *Fractal Geometry and Analysis, The Mandelbrot Festschrift, Curacao 1995, C.J.G. Evertsz, H.-O. Peitgen, R.F. Voss (eds.), World Scientific*.
25. J. Lévy Véhel and R. Riedi. Fractional Brownian motion and data traffic modeling: The other end of the spectrum. *Technical Report, INRIA Rocquencourt, France*, 1997. http://www-syntim.inria.fr/fractales/ Extended version to be published as 'Large deviation spectra of some multifractal processes'.
26. Lévy Véhel, J., Vojak, R. (1995): Multifractal analysis of Choquet capacities: Preliminary results. *Adv. Appl. Math.* (to appear 1997).
27. Willinger, W., Taqqu, M., Sherman, R., Wilson, D. (1995): Self-similarity through high-variability: Statistical analysis of ethernet lan traffic at the source level. *IEEE/ACM Transactions on Networking (Extended Version)*, (to appear).

FRACTALS AND WAVES

Fractal Electrodynamics: From Super Antennas to Superlattices

D. L. Jaggard

Complex Media Laboratory
Moore School of Electrical Engineering
University of Pennsylvania
Philadelphia, PA 19104-6390, USA
jaggard@seas.upenn.edu

Abstract We provide a selected overview of *fractal electrodynamics* by examining the properties of fractal antennas, arrays and apertures; and investigating the nature of electromagnetic wave scattering from fractal surfaces and superlattices.

Fractals offer an ideal blend of order and disorder that can be used in the design of robust antenna arrays with low or moderate side lobes. More recently, fractal geometry has also been used to design individual antenna elements suitable for multiband operation. We also investigate and review the diffraction of electromagnetic waves by fractal apertures.

Finally, we examine and review the scattering of electromagnetic waves from fractal surfaces and fractal superlattices. In each case fractals imprint their distinctive symmetry on interrogating waves. In this way fractal descriptors of an object are embedded in the scattered wave in such a way that they can often be extracted.

1. Introduction

1.1 Background

The research of the 1980's on wave interactions with fractal structures led to the development of *fractal electrodynamics* [1, 2, 3]. This was part of an effort to understand the physical processes and mathematical analysis of electromagnetic wave propagation in fractal media and to investigate the scattering and radiation of electromagnetic waves from fractal objects. As a result, this and related research formed the basis for the use of fractals in device synthesis and modeling in the microwave, millimeter wave, and optical regimes; and the use of fractals in remote sensing, characterization and classification. Work in fractal electrodynamics builds on the initial literature concerning wave scattering and diffraction from fractal geometries [4, 5, 6, 7, 8, 9, 10]; and x-ray, photon, and neutron scattering from fractal aggregates, colloids and porous media [11, 12, 13, 14, 15, 16, 17, 18, 19, 20, 21, 22]. In addition, a number of books on fractals, too numerous to list here, provide a wealth of supporting material.

We provide here a selective overview of the research in three areas of fractal electrodynamics with an emphasis on research closely related to our own work. First, we consider use of fractals in antenna arrays [23, 24, 25, 26, 27], multiband and broadband antenna element design [28, 29, 30, 31, 32, 33], and in aperture antennas [34, 35, 36, 37, 38, 39, 40, 41]. The results of this work demonstrate the powerful application of fractals for the engineering design of robust sparse antenna arrays, in broad-band or frequency-independent arrays and elements, and in some cases for the synthesis of compact antennas for telecommunications. Second, we examine electromagnetic and optical waves scattering from fractally rough surfaces [6, 7, 8, 42, 43, 44, 45, 46, 47]. Here we clearly find the methods by which fractals imprint their distinctive geometry on interrogating waves. We envision applications that range from the radar remote sensing of sea state to the remote characterization of both natural and man-made surfaces. Third, we investigate the reflection of electromagnetic and optical waves from fractal superlattices or multi-layers [48, 49, 50, 51, 52, 53]. Again we find that fractal superlattices impress their fractal characteristics on the reflected waves in quantifiable ways. Applications include the modeling and synthesis of resonant cavities and mirrors for microwaves, millimeter waves, and optical waves and the analysis and modeling of wave propagation in finely divided layered media.

1.2 Fractal Descriptors and Electromagnetic Yardsticks

Physical fractal objects such as coastlines are characterized by their fractal dimension D which is calculated by using variable length physical yardsticks [54, 55]. The idea is to cover the fractal with disks or circular yardsticks of size δ. The number $N(\delta)$ of disks needed is related to the dimension by $D = \lim_{\delta\to 0} \frac{\ln[N(\delta)]}{\ln[1/\delta]}$. For deterministic mathematical fractals in which N copies of the original shape are made at each iteration, each scaled by the similarity factor or ratio r, a common relation for fractal dimension becomes $D = \frac{\ln[N]}{\ln[1/r]}$. This is often denoted the *similarity dimension.*

The ***disk covering method*** of finding the fractal dimension given by the relations above for finding the fractal dimension of objects from physical measurements has its analog for wave propagation. Instead of using a variable length ***physical yardstick,*** one can use a variable length ***electromagnetic yardstick*** [1, 2, 3] as demonstrated pictorially in Fig. 1.1. Appropriate electromagnetic yardstick candidates are the wavelength (λ) or the pulse width (σ) for frequency-domain and time-domain problems, respectively. The frequency-domain version will be demonstrated in the sections on scattering from fractal surfaces and superlattices.

While dimension is frequently used to characterize a fractal structure, it does not provide a complete description. For example, the fractal dimension D indicates the roughness or amount of real estate a fractal occupies. However, it does not quantify its texture. To remedy this, the term ***lacunarity*** Λ was coined by Mandelbrot [54] to describe the degree of translational invariance or homogeneity of a fractal. Structures which are inhomogeneous, with gaps at all size scales, or those which show little translational invariance, are said to possess ***high lacunarity.*** Conversely, fractals which are very homogeneous

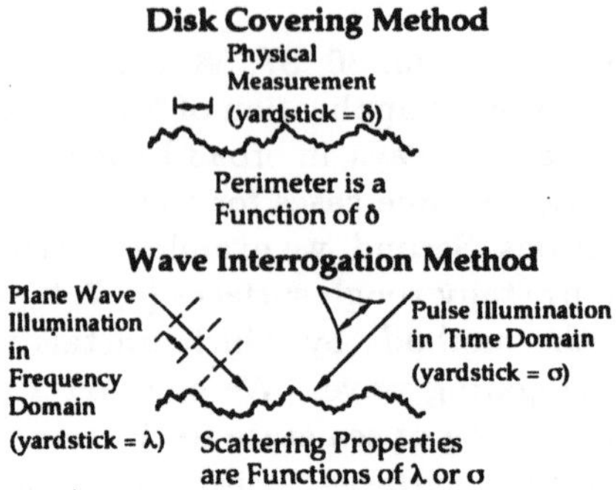

Fig. 1.1. Correspondence between physical yardsticks (upper) used for finding the fractal dimension from geometry and electromagnetic yardsticks (lower) used for extracting fractal descriptors from scattered waves.
(Figure adapted from D. L. Jaggard [1, 2, 3].)

and approach translational invariance are said to possess *low lacunarity*. An example of two fractals with the same dimension D but differing lacunarities Λ is shown in Fig. 1.2. The fractal on the left side is more *lumpy* or *gappy* and hence more lacunar than the one on the right.

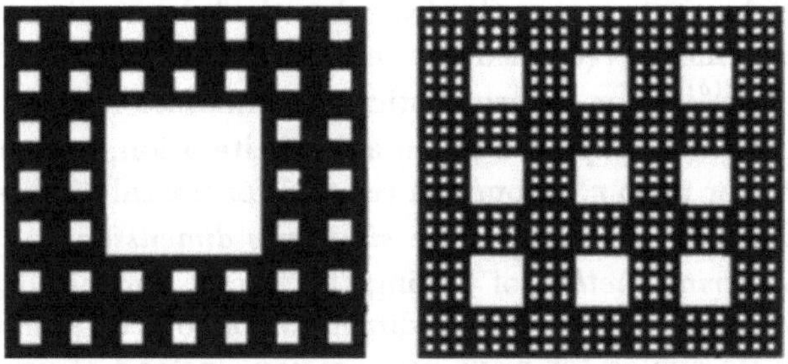

Fig. 1.2. Two Sierpiński carpets of equal similarity fractal dimension $D = ln(40)/ln(7)$ but different lacunarity. The less lacunar fractal on the right has greater translational invariance than the one on the left characterized by greater lacunarity.
(Figure adapted from B. B. Mandelbrot [54].)

Although there is not a universal mathematical definition of lacunarity, a number of candidate measures for the lacunarity Λ have been proposed [56, 57, 58, 59, 60, 61] following the initial discussion of Mandelbrot [54]. In many of these, $\Lambda = \Lambda(\delta)$ is not a single number but rather a function of a characteristic disk size δ used in its evaluation. As a result, some researchers have proposed that this lacunarity function $\Lambda(\delta)$ is a characteristic of texture useful for classification and discrimination of images.

The use of lacunarity in fractal electrodynamics will be demonstrated in the section on scattering from fractal superlattices [52, 53]. However, it clearly has an additional, and largely unexplored, role in the synthesis of new antenna arrays [41] and in wave scattering from fractally rough surfaces as well as in many other areas of engineering and applied physics. For example, in the case of rough surface scattering, lacunarity provides a more complete measure or characterization of surface texture. Therefore, we propose it as an additional means to quantify and classify sea state in the remote sensing of ocean surfaces and to further characterize other rough surfaces.

2. Fractal Antenna Arrays and Elements

There are three separate but related concepts that are relevant in a discussion of fractal arrays and elements. The first concept is concerned with the balance of long-range order and short-range disorder inherent in fractals to design fractal arrays. Here favorable characteristics of ordered arrays (relatively low sidelobes) are blended with like characteristics of disordered or random arrays (robustness with respect to element failure). The second concept is concerned with the folding large arrays into small geometric regions using fractal geometry so that multiple resonances corresponding to each scale-length in the antenna will overlap and provide wide bandwidth or multi-band operation. In select cases, it appears this method may be used to design more compact antennas suitable for certain ranges of frequencies. This has obvious applications in communications devices and telecommunications systems. The third concept concerns the relation between the symmetry of the fractal antenna or array and the symmetry of its radiation pattern. Here self-similar arrays produce self-similar array factors or radiation patterns in accordance with the symmetry of functions and the theory of Fourier transforms.

2.1 Linear Fractal Antenna Arrays

Our original motivation to study fractal antennas, circa 1985, was to blend the attractive characteristics of both periodic and random arrays and apply them to the design of a new class of sparse linear arrays we denote *fractal arrays*. The problem with traditional sparse arrays is that for thinned periodic arrays, grating lobes appear and ruin the sidelobe characteristics needed in many applications. Thinned random arrays typically exhibit large grating-like sidelobes. They are also robust and tend to degrade gracefully with element failure.

The fractal idea for antenna arrays was triggered by the work of Schelkunoff [62] who mentioned *arrays of arrays* in the 1940's in the design of tapered arrays. The connection to fractals is obvious. The details are given elsewhere

[23] but the idea is to use a fractal initiator and a subarray as a generator to combine the best of ordered and disordered systems. The result, shown in Fig. 2.1, displays the power radiation pattern or array factor for three sparse arrays, all with average element spacing of two wavelengths.

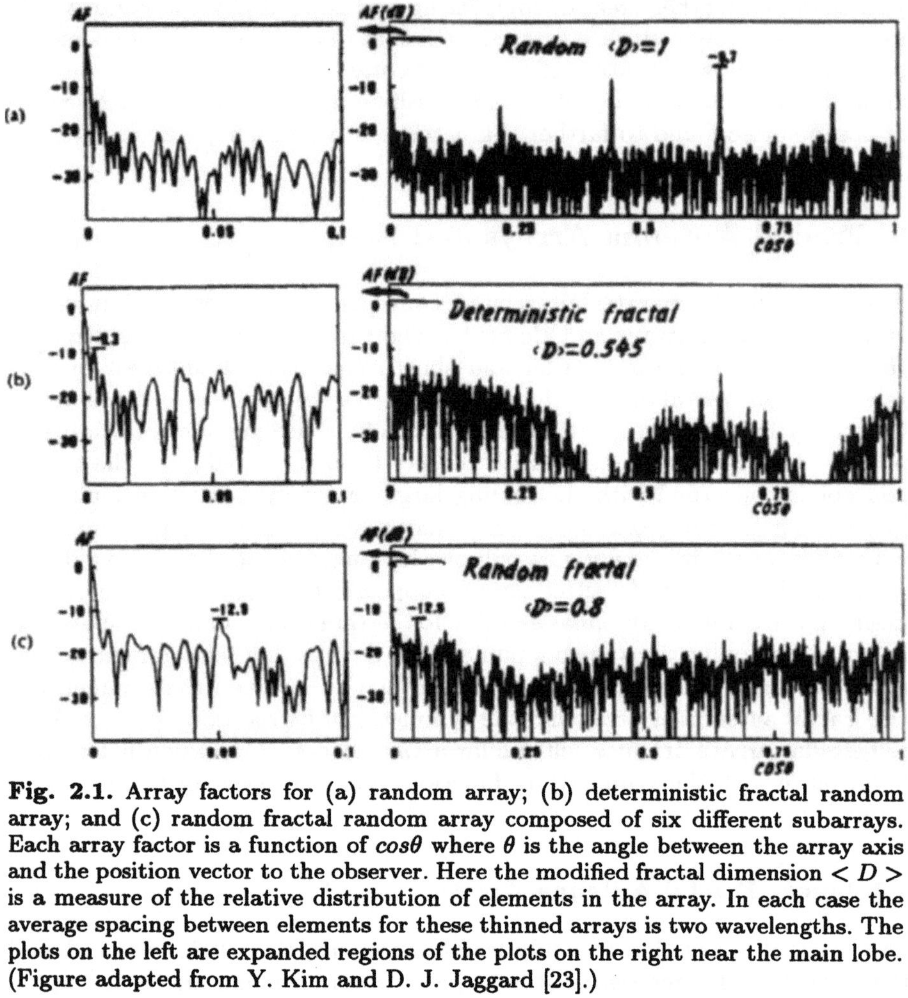

Fig. 2.1. Array factors for (a) random array; (b) deterministic fractal random array; and (c) random fractal random array composed of six different subarrays. Each array factor is a function of $cos\theta$ where θ is the angle between the array axis and the position vector to the observer. Here the modified fractal dimension $< D >$ is a measure of the relative distribution of elements in the array. In each case the average spacing between elements for these thinned arrays is two wavelengths. The plots on the left are expanded regions of the plots on the right near the main lobe. (Figure adapted from Y. Kim and D. J. Jaggard [23].)

The random array (top) yields the usual large sidelobes. For the example displayed here, the largest is $-5.7db$. A deterministic fractal array (middle) formed by a periodic initiator and generator yields excellent results away from the mainlobe but exhibits a large sidelobe of strength $-9.3db$ adjacent to the mainlobe that effectively increases the mainlobe width. The random fractal array (bottom), even though thinned, has a maximum sidelobe level of $-12.5db$. This is comparable to the sidelobes of non-thinned uniform arrays. It is anticipated that with tapering, further sidelobe reduction could be accomplished. Generalizations to two-dimensional arrays are possible using similar concepts.

Alternative approaches to fractal antenna arrays are given in the inventive work shown in Ref. [26, 27]. For example, Ref. [26] describes the use of Cantor sets and van Koch curves for the design of multiband fractal arrays. Likewise in Ref. [27] there is a description of multiband antennas and arrays and a review of frequency-independent antennas from the perspective of fractal geometry. Additional fractal approaches to the creation of antenna arrays have also appeared [24].

2.2 Fractal Antenna Elements

Fractal geometry has also been proposed to design individual antenna elements, to increase the bandwidth of antennas, and to provide candidate designs for efficient but compact antennas. Two recent creative examples are given by the fractal monopoles of Puente et. al. [28, 32, 33] and the fractal wire antennas of Cohen [29, 30, 31].

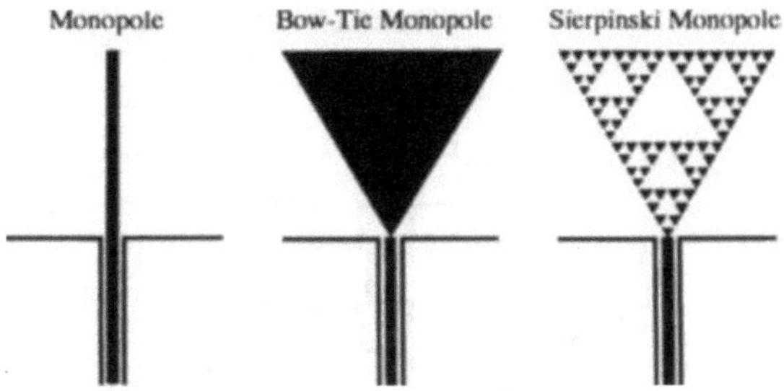

Fig. 2.2. Evolution of fractal antenna element for broadband applications starting with narrowband monopole (left), a moderate bandwidth bow-tie antenna (middle), and proceeding to the multiband Sierpiński monopole (right). In each case the monopole is placed over a ground plane and is driven by a coaxial line from below. (Figure on the right adapted from C. Puente, et. al. [32].)

In Fig. 2.2 is shown a typical monopole (left), a bow-tie monopole (middle), and the fractal Sierpiński monopole (right). The monopole is highly frequency sensitive due to its single characteristic length while the Sierpiński monopole has multiband characteristics that correspond to the various scale lengths present in the structure. Details on the frequency response, induced currents, and principles of operation of this and related monopoles and dipoles are provided elsewhere in the literature [32, 33] and in this book.

In Fig. 2.3 is displayed two fractal wire antennas of Cohen, a fractal dipole (upper) and a fractal loop (lower). The idea for the fractal dipole is to reconfigure a wire structure with a single characteristic length to one having many characteristic lengths and many characteristic resonances in order to

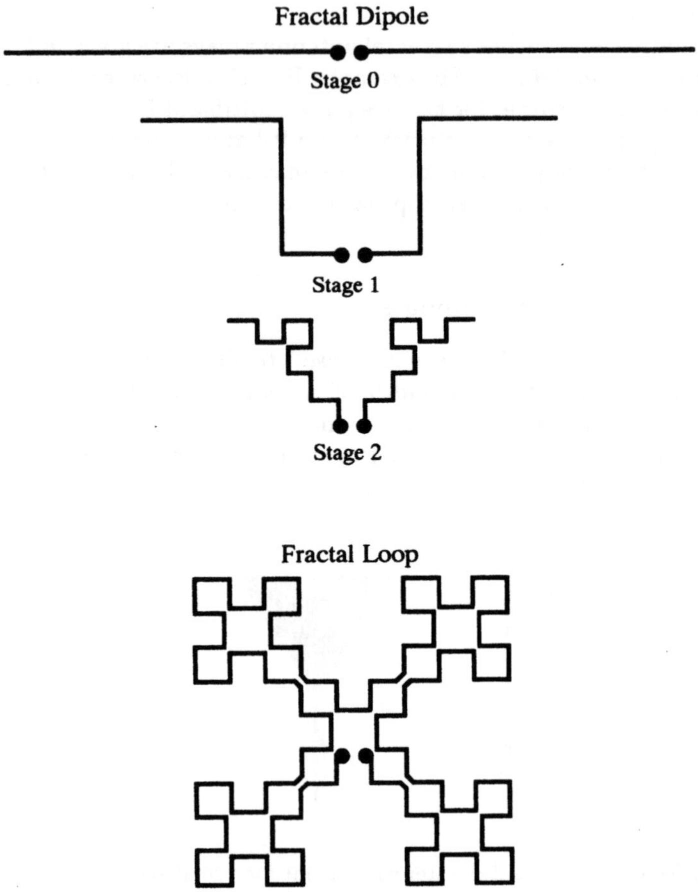

Fig. 2.3. Evolution of fractal dipole antenna (top) to the second stage of growth and the fractal loop antenna (bottom) at the second stage of growth. In each case the driving terminals are indicated at the center of the antenna.
(Figure adapted from N. Cohen [29, 30, 31].)

increase bandwidth and in selected cases, reduce size. Details can be found in the literature [29, 30, 31]

2.3 Fractal Aperture Diffraction and Antennas

Diffraction by fractal apertures is the third example of fractal antennas to be reviewed here. Aperture diffraction is one of several possible two-dimensional extensions of linear fractal array theory. In addition, this topic also provides a natural transition from fractal antennas and arrays to scattering from fractal structures. Understanding of this phenomena is assisted by the simple Fourier relation that exists between an aperture field or an aperture transmission function and its far-zone diffracted field for both the scalar and vector case of aperture diffraction in the physical optics approximation.

The fractal arrays considered here demonstrate one possible route to the design of more general planar fractal arrays. However, here the emphasis will be on the way in which these structures transfer fractal information concerning their geometry to their diffracted fields. In particular it is evident that self-similar structures provide self-similar radiation patterns. Furthermore, the self-similarity of these apertures provides a recursive method by which their radiation patterns can be calculated. Finally, information related to the fractal dimension is embedded in the radiation patterns due to the Fourier transform relation between aperture function and diffracted field under appropriate approximations.

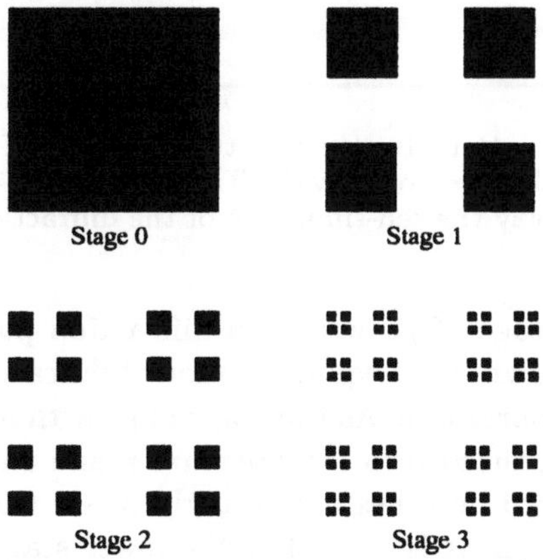

Fig. 2.4. Several stages of growth of the Cantor Square used as an aperture transmission function for fractal aperture diffraction. The fractal dimension is given by $D = ln(4)/ln(3)$.

The physical configuration is described by the normal incidence of a plane wave on a planar aperture function described by a Cantor square [38]. The

Cantor square used here is displayed in Fig. 2.4 while the resulting diffraction pattern for different stages of growth is shown in Fig. 2.5. This relative diffraction pattern is shown as a function of normalized coordinate (denoted *Fourier Distance* in the figure) along a line coincident with the symmetry axis of the fractal aperture function.

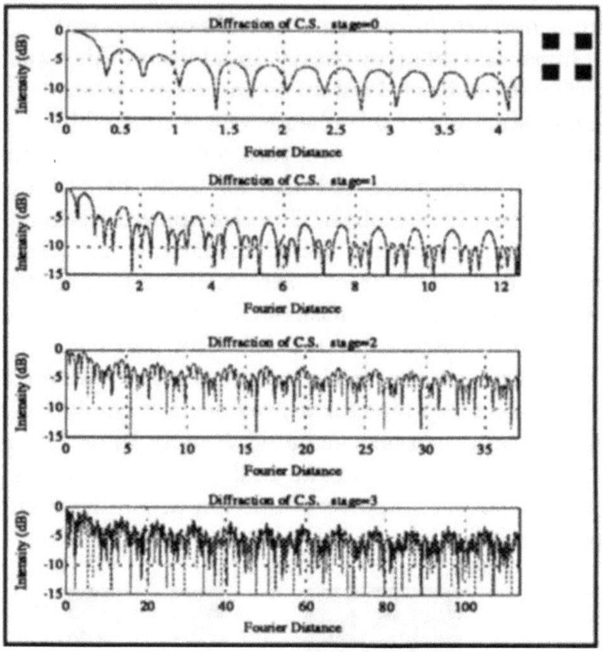

Fig. 2.5. Diffraction pattern for the aperture function of Fig. 2.4 along the symmetry axis for several stages of growth. The axes are re-scaled for each stage of growth to clearly display the self-similarity of the diffraction pattern.

For the zeroth stage of growth (top diffraction pattern in Fig. 2.5), the result is the well-known "sinc-squared" plot of diffracted intensity as a function of normalized coordinate. As the stage of growth increases, higher spatial frequencies are introduced into the aperture plane and the fine structure of the diffraction pattern becomes increasingly dense and the energy is spread further from the origin. It is seen that the lower stages of growth provide a diffraction pattern that is the envelope for higher stages of growth when the latter is appropriately scaled. This displays the self-similarity of the diffraction pattern for these two dimensional fractal apertures. Not surprisingly, these results are very similar to those of the circularly symmetric Cantor target we examined previously [38, 39]. The angularly averaged distribution of the diffraction pattern is related to the fractal dimension of the aperture transmission function as noted previously [36, 65, 36].

Other research related to fractal aperture diffraction includes work on the theoretical and experimental Fourier transforms of fractals [63, 34, 64, 65] and the diffraction of waves by fractally serrated apertures [37] in which a circular aperture edge is perturbed by a fractal function of variable dimension and the diffracted field is found both analytically and experimentally.

3. Scattering from Fractal Surfaces

3.1 Problem Formulation

We next consider the scattering of electromagnetic waves from fractally corrugated surfaces [42]. The largest scattered waves will satisfy the Bragg's Law or the conservation of momentum condition.

As with the scattering of optical waves, x-rays and neutrons from fractal surfaces and volumes, the fractal dimension is embedded in the scattered wave in a simple manner. Details of the calculations for the scattering cross-section can be found in the literature [42]. We turn now to the results for one set of calculations.

3.2 Examples and Results

In Fig. 3.1 we display the distribution of scattered energy, or the scattering cross-section, as a function of scattering angle θ for surfaces of four roughnesses for four fractal dimensions D using the Kirchhoff method. To avoid shadowing and other limitations of this method, we examine the surface scattering with waves incident at 30° degrees from the normal to the plane containing the surface when the corrugations become infinitesimal and we keep the corrugations small compared to the radius of curvature of the surface irregularities. For this reason the *rms* height and the fractal dimension of the corrugations are modest.

In the top half of Fig. 3.1 is shown the polar plots of scattering for fractal dimension D ranging from 1.05 to 1.70. As the fractal dimension increases, the surface roughness increases and more of the scattered energy is spread away from the specular reflection direction indicated by the major scattering lobe. This is expected since the higher spatial frequency plays an increasingly important role as the fractal dimension increases [42, 43, 44, 46, 47].

The role of fractal dimension is brought into sharp focus from the plots of the scattered energy shown in the bottom half of Fig. 3.1. These figures display a version of of the data shown in the top half of the figure, replotted on a log-log scale. Two envelopes are indicated in these plots. The background envelope (superposed thin line with constant slope of -1) is due to the finite patch size and consists of the specularly reflected main beam and its associated sidelobes. The coupling lobe envelope (superposed wide

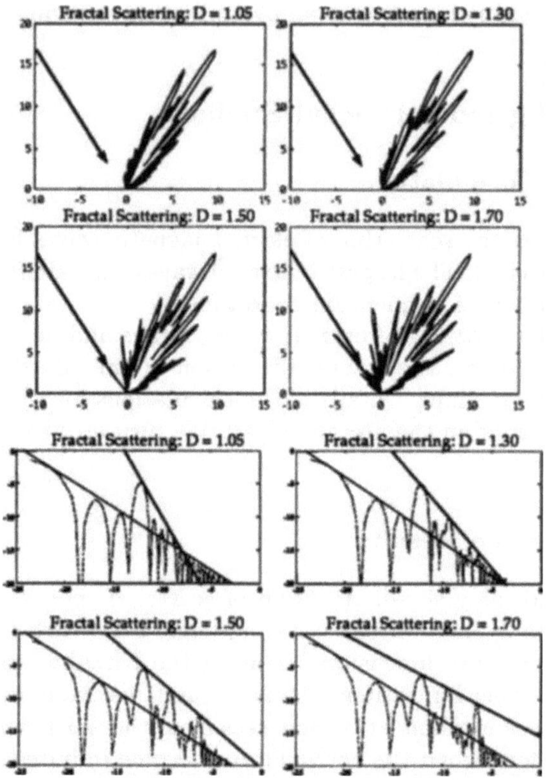

Fig. 3.1. The differential scattering cross-sections as a function of polar angle θ (upper set of four) and as a function of $sin[(\theta - 30°)/2]$, both in dB scales (lower set of four), for fractal dimensions $D = 1.05, 1.30, 1.50$, and 1.70 with waves incident at 30° degrees. In the lower set of plots, the envelope slopes of coupling sidelobes (superposed wide line with variable slope) vary monotonically with the fractal dimension while the background slope (superposed thin line with slope of -1) is constant for varying D and represents the scattering sidelobes due to the patch size of the scattering area.
(Figure adapted from D. L. Jaggard [1, 3, 42].)

line with variable slope) is due to surface coupling and consists of the first-order Bragg coupling of the surface harmonics. These coupling lobe envelope slopes are found to be $-2.50, -1.92, -1.22$, and -0.72 for fractal dimensions $D = 1.05, 1.30, 1.50$, and 1.70, respectively. The slope of the coupling lobe envelope is indicative of surface roughness and is a monotonic function of fractal dimension. Thus we have a method of remotely characterizing the fractal dimension from the scattered waves. The magnitude of slope for these bandlimited fractal surfaces varies approximately as $2.7(2 - D)$ for the *rms* heights considered in this example. This result is similar to the scattering of optical waves from fractal aggregates.

Alternative investigations of wave scattering from fractal surfaces using both generalized Rayleigh methods [43, 44] and a perturbational approach [45] have followed the original work displayed in Fig. 3.1. We also note that recently applied exact calculations using the extended boundary condition method (EBCM) or T-matrix approach [46, 47] yields quantitatively similar results but can be applied to surfaces which violate some of the restrictions (e.g., small radii of curvature, small wave heights, no shadowing) inherent in the Kirchhoff, Rayleigh or perturbational methods. In addition, these exact methods demonstrate the effect of polarization on the scattered wave and so provide a more sensitive diagnostic tool for the remote sensing of rough surfaces.

4. Scattering from Fractal Superlattices

4.1 Problem Formulation

In this section we consider the scattering of an optical or electromagnetic wave that is normally incident on a fractal superlattice composed of alternating layers of refractive indices $\hat{n}_0$ and $\hat{n}_1$ [48, 49, 50, 51, 52, 53] (not shown here). The variation in index is governed by a family of generalized polyadic Cantor sets on the interval $[0, L]$ at finite stages of growth. For the case governed by the set E, the refractive index at the coordinate z is $\hat{n}_1$ for $z \in E$ and $\hat{n}_0$ otherwise.

For the symmetric Cantor bars in which we are interested, we characterize the lacunarity by ϵ (denoted *Epsilon* in the next figure), the normalized width of the outermost gaps in the first stage of construction. This has physical significance for the reflection problem under consideration and can be immediately identified in the scattering data displayed.

The problem of transmission and reflection from a number of layers of arbitrary width and refractive index can by treated by well-known chain-matrix approach. However, this method does not recognize the inherent order of the fractal superlattices examined here and involves an exponentially increasing number of matrix multiplications $\{4(N_{gaps} + 1)^S - 1\}$ where S is the stage of growth and N_{gaps} is the number of gaps. Alternatively, one can use approximate methods [51] to provide qualitative information on scattered fields. However, this technique tends to become inaccurate when the scattering becomes large. Exact solutions using the dilation symmetry of the fractal structure and a single recursion were first introduced for this problem in the early 1990's [48, 50]. Here we follow more recent research in which a doubly recursive method is developed and applied [52, 53].

In the first recursion a generating function is used to create a general reflection (transmission) function through recursion on the number of gaps. Noting that at stage S the bar contains N scaled copies of the interval at stage $S-1$, we use these functions to find reflection (transmission) coefficients through a recursion on the stage of growth. This forms the second recursion. The details of these calculations can be found in the literature [52, 53] and the results for normal incidence will be shown next. The analytical results are valid for any stage of growth, refractive index, number of gaps, fractal dimension, and spacing of the scaled copies within $[0, L]$ for bilaterally symmetric polyadic Cantor superlattices. The case of oblique incidence is immediately found by appropriately normalizing the refractive index by the cosine of the propagation angle in each layer [53].

4.2 Results and Examples

A clear picture of the effect of lacunarity is given in the *twist plots* of Fig. 4.1 in which the reflection coefficient amplitudes are displayed as a function of normalized frequency kL and continuous variations of ϵ (denoted *Epsilon* in the figure) for two stages of growth. The nulls in Fig. 4.1 are especially informative. There are three sets of these characteristic nulls in the upper twist plot. The first set of nulls are the vertical dark lines at $kL = \frac{14\pi m}{3} (m = 1, 2, 3, \ldots)$. We expect their independence of lacunarity as they are caused by destructive interference from the front and back surface reflections of individual layers of index $\hat{n}_1$. The second set of nulls are broad curves which are almost vertical for small normalized wavenumber but appear as arcs for larger normalized wavenumbers. These are due to the changed spacing between regions of index $\hat{n}_1$ in the superlattice. Therefore, these are sensitive to variations in both frequency and lacunarity. The final set of nulls which provide "barber pole" striations are due to the gap width at the center of each superlattice. These last two sets of nulls are indicators of the degree of lacunarity and demonstrate that in addition to information on stage of growth and fractal dimension lacunarity is also distinctively imprinted on scattering data [52, 53]. The nulls in the lower plot of Fig. 4.1 are explained in a similar way. However, as expected, there are additional fine striations here due to the finer structure of the second stage of growth. Therefore, just as in the aperture diffraction case, the reflection pattern displays the characteristic dilation symmetry of the fractal structure.

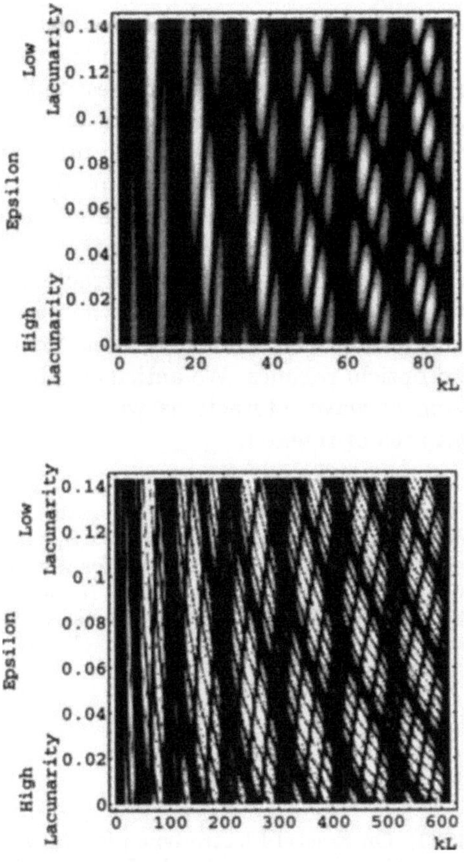

Fig. 4.1. The amplitude of the reflection coefficient R shown as a density plot as a function of the normalized frequency kL (horizontal axis) and continuous variations *Epsilon*($= \epsilon$) (vertical axis) for first (upper plot) and second (lower plot) stages of growth. These ***twist plots*** also demonstrate the features that are independent of lacunarity (vertical nulls) and those which are descriptors of lacunarity (diagonal and curved nulls). The second stage of growth plot (lower) is similar to a scaled version (note change of horizontal scale) of the first with an overlaid fine structure. This indicates the self-similar or fractal nature of these superlattices. The null structure, shown in black, is directly related to the lacunarity, fractal dimension and stage of growth.
(Figure adapted from D. L. Jaggard and A. D. Jaggard [52, 53].)

5. Conclusions

Here we have covered several aspects of fractal electrodynamics—the study of the interaction of electromagnetic waves with fractal structures. Our emphasis has been on antennas, arrays and apertures; on the scattering of waves from fractally rough surfaces; and the reflection and transmission of waves from fractal superlattices or multilayers. In the first case we examined the usefulness of fractal geometry in enhancing the performance of arrays or antenna elements. In the last two cases we investigated the way in which fractal structures imprinted their characteristic geometry on scattered waves and methods by which fractal descriptors can be extracted.

Although relatively young, fractal electrodynamics has already yielded useful and intriguing results in a number of areas covering the microwave, millimeter wave and optical regimes. We anticipate that we can reach a still deeper understanding of wave interactions with fractals and discover new devices through this area of research.

Acknowledgement. The author acknowledges the enjoyable collaborations with Yun Jin Kim, Panayiotis Frangos, Xiaoguang Sun, Terrence Spielman, Michael Dempsey, and Aaron D. Jaggard in fractal electrodynamics for almost fifteen years. These interactions led directly or indirectly to the results reported here. This work was supported in part through a NATO Collaborative Research Grant No. 920 702 and the Complex Media Laboratory (University of Pennsylvania).

References

1. Jaggard, D.L. (1990): On Fractal Electrodynamics. In *Recent Advances in Electromagnetic Theory*, H.N. Kritikos and D.L. Jaggard, editors, Springer-Verlag, New York, 183–224.
2. Jaggard, D.L. (1991): Fractal Electrodynamics and Modeling. In *Directions in Electromagnetic Wave Modeling*, H.L. Bertoni and L.B. Felsen, editors, Plenum Publishing Co., New York, 435–446.
3. Jaggard, D.L. (1995): Fractal Electrodynamics: Wave Interactions with Discretely Self-Similar Structures. In *Electromagnetic Symmetry*, C. Baum and H. Kritikos, eds., Taylor and Francis Publishers, Washington, D.C., 231–281.
4. Berry, M.V. (1979): Diffractals. *J. Phys. A: Math. Gen. 12*, 781–797.
5. Berry, M.V., Blackwell, T.M. (1981): Diffractal Echoes. *J. Phys. A: Math. Gen. 14*, 3101–3110.
6. Jakeman, E. (1982): Scattering by a Corrugated Random Surface with Fractal Slope. *J. Phys. A 15*, L55–L59..
7. Jakeman, E. (1982): Fresnel Scattering by a Corrugated Random Surface with Fractal Slope. *J. Opt. Soc. Am. 72*, 1034–1041.
8. Jakeman, E. (1983): Fraunhofer Scattering by a Sub-Fractal Diffuser. *Opt. Acta 30*, 1207–1212.
9. Jakeman, E. (1986): Scattering by Fractals. In *Fractals in Physics*, L.Pietronero and E. Tosatti, editors, Elsevier Science Publishers B.V., New York, 55–60.

10. Allain, C., Cloitre, M. (1986): Optical Fourier Transforms of Fractals. In *Fractals in Physics*, L. Pietronero and E. Tosatti, editors, Elsevier Science Publishers B.V., New York, 61–64.
11. Bale, H.D., Schmidt, P.W. (1984): Small-Angle X-Ray-Scattering Investigation of Submicroscopic Porosity with Fractal Properties. *Phys. Rev. Lett. 53*, 596–599.
12. Schaefer, D.W.,Martin, J.E., Wiltzius, P., Cannell, D.S. (1984): Fractal Geometry of Colloidal Aggregates. *Phys. Rev. Lett. 52*, 2371–2374.
13. Kaye, B.H., Le Blanc, J.E., Abbott, P. (1985): Fractal Description of the Structure Fresh and Eroded Aluminum Shot Fineparticles. *Part. Charact. 2*, 56–61.
14. Schmidt, P.W., Dacai, X. (1986): Calculation of the Small-Angle X-Ray and Neutron Scattering From Nonrandom (Regular) Fractals. *Phys. Rev. A 33*, 560–566.
15. Rojanski, D., Huppert, D., Bale, H.D., Dacai, X., Schmidt, P.W., Farin, D., Seri-Levy, A., Avnir, D. (1986): Integrated Fractal Analysis of Silica: Adsorption, Electronic Energy Transfer, and Small-Angle X-Ray Scattering. *Phys. Rev. Lett. 56*, 2505–2508.
16. Teixeira, T. (1986): Experimental Methods for Studying Fractal Aggregates. In *On Growth and Form: Fractal and Non-Fractal Patterns in Physics*, H.E. Stanley and N. Ostrowsky, editors, M. Nijhoff, Boston, MA, 145–162.
17. Rarity, J.G., Pusey, P.N. (1986): Light Scattering from Aggregating Systems: Static, Dynamic (QELS) and Number Fluctuations. In H. E. Stanley and N. Ostrowsky, editors, *On Growth and Form: Fractal and Non-Fractal Patterns in Physics*, Martinus Nijhoff Publishers, Boston, MA, 218–221.
18. Kaye, B.H. (1986): Fractal Dimension and Signature Waveform Characterization of Fine Particle Shape. *American Laboratory*, 55–63.
19. Chen, Z., Sheng, P., Weitz, D.A., Lindsay, H.M., Lin, M.Y., Meakin, P. (1988): Optical Properties of Aggregate Clusters. *Phys. Rev. B37*, 5232–5235.
20. Rarity, J.G., Seabrook, R.N., Carr, R.J.G. (1989): Light-Scattering Studies of Aggregation. In *Fractals in the Natural Sciences*, M. Fleischmann, D.J. Tildesley, and R.C. Ball, editors, Princeton University Press, Princeton, NJ, 89-101.
21. Lin, M.Y., Lindsay, H.M., Weitz, D.A., Ball, R.C., Klein, R., Meakin, P. (1989): Universality of Fractal Aggregates as Probed by Light Scattering. In *Fractals in the Natural Sciences*, M. Fleischmann, D. J. Tildesley, and R. C. Ball, editors, Princeton University Press, Princeton, NJ, 71–87.
22. Sinha, S.K. (1988): Scattering from Fractal Structures. *Physica D 38*, 310–314.
23. Kim, Y., Jaggard, D.L. (1986): The Fractal Random Array. *Proc. IEEE. 74*, 1278–280.
24. Lakhtakia, A., Varadan, V.K., Varadan, V.V. (1987): Time Harmonic and Time-Dependent Radiation by Bifractal Dipole Arrays. *Int. J. Electronics 63*, 819-824.
25. Werner, D.H., Werner, P.L. (1995): On the Synthesis of Fractal Radiation Patterns. *Radio Science 30*, 29–45.
26. Puente, C., Pous, R. (1996): Fractal Design of Multiband and Low Side-Lobe Arrays. *IEEE Trans. Antennas Propagat. 44*, 730–739.
27. Werner, D.H., Werner, P.L. (1996): Frequency-Independent Featues of Self-Similar Fractal Antennas. *Radio Science 31*, 1331–1343.
28. Puente, C. (Dec. 1993): Fractal Design of Multiband Antenna Arrays. Departmenet of Electrical and Computer Engineering, University of Illinois, Urbana-Champaign, *ECE 477 Term Project.*
29. Cohen, N. (Summer 1995): Fractal Antennas Part 1. *Comm. Quart.*, 7–22.
30. Cohen, N. (Summer 1996): Fractal Antennas Part 2. *Comm. Quart.*, 53–66.

31. Cohen, N., Hohfeld, R.G. (Winter 1996): Fractal Loops and the Small Loop Approximation. *Comm. Quart.* 77–81.
32. Puente, C., Romeau, J., Pous, R., Garcia, X., Benitez, F. (1996): Fractal Multiband Antenna Based on the Sierpinski Gasket. *IEE Elec. Lett. 32*, 1–2.
33. Puente, C., Romeau, J., Pous, R., Garcia, X., Benitez, F. (1996): Perturbation of the Sierpinski Antenna to Allocate Operating Bands. *IEE Elec. Lett. 32*, 2186–2187.
34. Allain, C., Cloitre, M. (1987): Spatial Spectrum of a General Family of Self-Similar Arrays. *Phys. Rev. A 36*, 5751–5757.
35. Beal, M.M., George, N. (1989): Features in the Optical Transforms of Serrated Apertures and Disks. *J. Opt. Soc. Am A6*, 1815–1826.
36. Jaggard, D.L., Spielman, T., Sun, X¿ (June 24–27, 1991): Diffraction by Cantor Targets," presented at the 1991 AP-S/URSI Meeting, London, Ontario.
37. Kim, Y., Grebel, H., Jaggard, D.L. (1991): Diffraction by Fractally Serrated Apertures. *J. Opt. Soc. A8*, 20–26.
38. Spielman, T., Jaggard, D.L. (July 20–25, 1992): Diffraction by Cantor Targets: Theory and Experiment. presented at the *1992 AP-S/URSI Meeting*, Chicago, IL.
39. Jaggard, D.L., Spielman, T. (1992): Diffraction from Triadic Cantor Target Diffraction. *Microwave and Optical Technology Letters 5*, 460–466.
40. Jaggard, D.L., Spielman, T., Dempsey, M. (June 28–July 2, 1993): Diffraction by Two-Dimensional Cantor Apertures. Presented at the *1993 AP-S/URSI Meeting*, Ann Arbor, MI.
41. Jaggard, D.L., Jaggard, A.D. (July 13–18, 1997): Fractal Apertures: The Effect of Lacunarity. to be presented at the *1997 AP-S/URSI Meeting*, Montréal.
42. Jaggard, D.L., Sun, X. (1990): Scattering by Fractally Corrugated Surfaces. *J. Opt. Soc. A7*, 1131–1139.
43. Sun, X., Jaggard, D.L. (1990): Wave Scattering from Non-Random Fractal Surfaces. *Opt. Comm. 78*, 20–24.
44. Jaggard, D.L., Sun, X. (1990): Rough Surface Scattering: A Generalized Rayleigh Solution. *J. Appl. Phy. 68*, 5456–5462.
45. McScharry, P., Cullen, P.J. (1995): Wave Scattering by a One-Dimensional Band-Limited Fractal Surface Based on a Perturbation of the Green's Function.*J. Appl. Phy. 78*, 6940–6948.
46. Savaidis, S., Frangos, P., Jaggard, D.L., Hizanidis, K. (Dec. 1995): Scattering from Fractally Corrugated Surfaces: An Exact Approach. *Optics Lett. 20*, 2357–2359.
47. Savaidis, S., Frangos, P., Jaggard, D.L., Hizanidis, K. (1997): Scattering from Fractally Corrugated Surfaces Using the Extended Boundary Condition Method. *J. Opt. Soc. Am. A14*, 475–485.
48. Jaggard, D.L., Sun, X. (1990): Reflection from Fractal-Layers. *Opt. Lett. 15*, 1428–1430.
49. Konotop, V.V., Yordanov, O.I., Yurkevich, I.V. (1990): Wave Transmission Through a One-Dimensional Cantor-Like Fractal Medium. *Europhys. Lett.12*, 481–485.
50. Sun, X., Jaggard, D.L. (1991): Wave Interactions with Generalized Cantor Bar Fractal Multi-Layers. *J. Appl. Phy. 70*, 2500–2507.
51. Megademini, T., Pardo, B., Jullien, R. (1991): Fourier Transform and Theory of Fractal Multilayer Mirrors. *Opt. Comm. 80*, 312–316.
52. Jaggard, D.L., Jaggard, A.D. (1997): Polyadic Cantor Superlattices with Variable Lacunarity. *Opt. Lett. 22*, 145–147.

53. Jaggard, A.D., Jaggard, D.L. (July 13–18, 1997): Fractal Superlattices and Scattering: Lacunarity, Fractal Dimension, and Stage of Growth. To be presented at the *1997 AP-S/URSI Meeting*, Montréal; also Jaggard, A.D., Jaggard, D.L. (1997): Scattering from Fractal Superlattices with Variable Lacunarity. In preparation.
54. Mandelbrot, B.B. (1983): *The Fractal Geometry of Nature*. W. H. Freeman and Comapny, New York.
55. Barnsley, M. (1988): *Fractals Everywhere*. Academic Press, New York.
56. Voss, R. (1985): Random Fractals: Characterization and Measurement. In *Scaling Phenomena in Disordered Systems*, NATO ASI Series, R. Pynn and A. Skjeltorp, editors, Plenum Press, New York, 1–11.
57. Mandelbrot, B.B. (1994): A Fractal's Lacunarity, and How it can be Tuned and Measured. In *Fractals in Biology and Medicine*, T. F. Nonnenmacher, G. A. Losa, and E. R. Weibel, editors, Birkhäuser Verlag, Basel, 8–21.
58. Mandelbrot, B.B., Stauffer, D. (1994): Antipodal Correlations and the Texture (Fractal Lacunarity). In Critical Percolation Clusters. *J. Phys. A: Math. Gen.* *27*, L237–L242.
59. Fábio, D.A., Reis, A., Riera, R. (1994): Lacunarity Calculation in the True Fractal Limit. *J. Phys. A: Math. Gen 27*, 1827–1835.
60. Mandelbrot, B.B., Vespignani, A., Kaufman, H. (1995): Crosscut Analysis of Large Radial DLA: Departures from Self-Similarity and Lacunarity Effects. *Europhys. Lett 33*, 199–204.
61. Hovi, J.-P., Aharony, A., Stauffer, D., Mandelbrot, B.B. (1996): Gap Independence and Lacuanrity in Percolation Clusters. *Phys. Rev. Lett. 77*, 877–880.
62. Schelkunoff, S.A. (1943): A Mathematical Theory of Linear Arrays. *Bell Syst. Tech. J. 22*, 80–107.
63. Lakhtakia, A., Holter, N.S., Messier, R., Varadan, V.K., Varadan, V.V. (1986): On the Spatial Fourier Transforms of the Pascal-Sierpinski Gaskets. *J. Phys. A: Math. Gen. 19*, 3147–3152.
64. Uozumi, J., Kimura, H., Asakura, T. (1990): Fraunhofer Diffraction by Koch Fractals. *J. Mod. Optics 37*, 1011–1031.
65. Uozumi, J., Kimura, H., Asakura, T. (1991): Fraunhofer Diffraction by Koch Fractals: The Dimensionality. *J. Mod. Optics 38*, 1335–1347.

Multiband Fractal Antennas and Arrays

C. Puente, J. Romeu, R. Pous, and A. Cardama

D3-Electromagnetics and Photonics Engineering group
Universitat Politèctina de Catalunya
Modul D3, Campus Nord, c/Gran Capita s/n
08034 Barcelona, Spain.
Fax: 34-3-4017232

Abstract Antennas and antenna arrays are essentially narrowband devices. Their characteristic size determines their operating wavelength. Fractal structures, having no characteristic size and a multiscale self-similar shape, are suggested for the design of multifrequency antennas and arrays. The multiband properties of some fractal arrays and array factors are presented here. Experimental results on multiband fractal antennas are described as well. Fractal antennas are shown to display a multiband behavior from both the return-loss and radiation pattern points of view.

1. Introduction

Fractal Electrodynamics is a quite novel research area which relates fractals [1, 2] and electromagnetics [3]. The term was coined by D.L. Jaggard et al who recently described [3]-[5] the scattering and diffraction properties of some fractal bodies. It has been already shown in [3]-[5] and other related works [6], that electromagnetic waves acquire some fractal properties when interacting with fractal bodies.

Fractal structures present some special features, not shared by common Euclidean shapes, which can be of interest when designing antennas. First of all they have a complex, convoluted form that can properly enhance radiation [7, 8]. For instance, some fractal small loops, enclosing a small area with an arbitrarily large perimeter, might have a radiation resistance larger than the one predicted by the Small Loop Approximation (SLA) as suggested by N.Cohen in [7]. Similarly, Landstorfer et al had already shown in 1985 how some weird, highly irregular wire structures improved some features of common simple antennas.

Second of all, fractals are usually self-similar structures, which means that several copies of the main shape are found at several scales within the fractal body. This property can be used to design multifrequency antennas [9, 10] and arrays as it was first suggested by Puente et al in [11, 12]. The main problem with the design of wide-band or frequency-independent radiating systems is that once the system is designed to match its characteristic size to the operating wavelength, its parameters will change when operating at other different wavelengths. For single antennas, such a characteristic size is the length or area of the antenna. For arrays, the spacing among elements is also important.

Angles, cones and spirals are examples of some of the few shapes that have successfully been used to design frequency-independent antennas [9, 10]. All of them present a common feature: their shape is invariant under a scaling transformation such that their form relative to all wavelengths is held constant. Similarly, it appears natural to assume that fractal shapes, having no characteristic size or at least, a multiscale structure, should be appropriate candidates to become multifrequency antennas[1].

The Sierpinski antenna was, at the extent known by the authors, the first reported example of a fractal multiband antenna from both the input impedance and radiation pattern points of view [13]. In the present work, the authors outline some of the main results regarding multiband fractal antennas and arrays.

2. Fractal Arrays and Array Factors

Array theory is a rather broad discipline which concerns several branches of science. Some papers relating fractals and arrays have been already published [4, 6, 12, 14, 15]. In general, it was shown in those works how some fractal properties of the arrays were impinged on their array factors. Mainly, some relationship between fractal dimension and the side-lobe ratio (SLR) of the array factor was described [4, 15]. Also, some similarity features of the radiated fields were also introduced [6, 14].

The multifrequency operation of fractal arrays and array factors was explored in [11, 12]. The Cantor set was taken as the basic fractal structure where the array elements were located. Similarly, the Koch curve was there properly modified to design self-similar array factors.

2.1 Fractal Arrays

A Cantor antenna linear array $c(z)$ aligned along the z direction can be conveniently described by a multiple convolution succession of several delta function sets $f(z)$, each one scaled by an arbitrary scale factor d, such that

$$c(z) = ...f(z/\delta) * f(z) * f(z \cdot \delta)... = \underset{n=-\infty}{\overset{\infty}{X}} f(z \cdot \delta^n) \tag{2.1}$$

where δ is an arbitrary scale factor, X indicates the convolution operator, and $f(z)$ is the set of equally spaced N delta functions

$$f(z) = \sum_{n=0}^{N-1} \delta\left(z - nd + \frac{N-1}{2}d\right) \tag{2.2}$$

[1] Fractal and Multifractal Antennas are patent pending.

being d the spacing between array elements for each set. For the classical Cantor set array, the number of elements in $f(z)$ is $N = 2$, and the characteristic scale factor is $\delta = 3$ (see Figure 2.1). The Cantor set generating algorithm has been also generalized here to allow growth to both infinitely small and infinitely large scales.

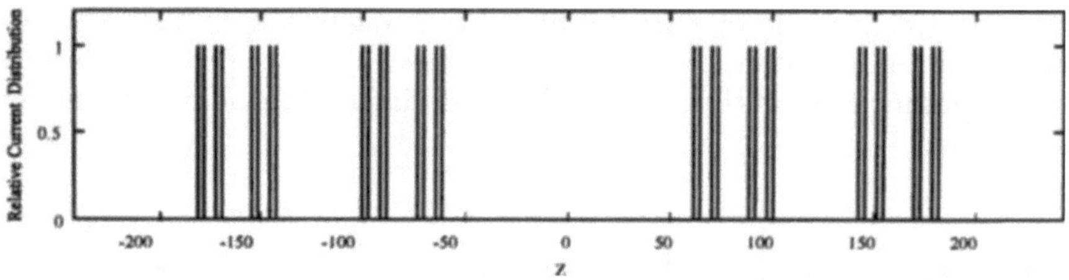

Fig. 2.1. The classical Cantor array constructed after 7 iterations ($N = 2$, $\delta = 3$).

The corresponding array factor $C(\Psi)$ can be now written as an infinite succession of products

$$C(\Psi) = \prod_{n=-\infty}^{\infty} \frac{1}{\delta^n} F\left(\frac{\Psi}{\delta^n}\right) \tag{2.3}$$

where $\Psi = kd.\cos\theta + \beta$ as usual in array theory. The array factor is self-scalable when scaled down by a factor d, i.e.,

$$C(\Psi/\delta) = \prod_{n=-\infty}^{\infty} \frac{1}{\delta^{n-1}} F\left(\frac{\Psi}{\delta^{n-1}}\right) = C(\Psi) \tag{2.4}$$

which means that the array factor is held constant for any set of frequencies $f_n = f_o.\delta^n$ since Ψ is proportional to frequency. Obviously, such a property is only strictly held by the infinite array constructed by an infinite succession of convolutions (i.e., fractal growth stages). However, some similarities among patterns are still observed in a feasible, truncated version of the array. Figure 2.2 displays the array factor at five different wavelengths of a Cantor array constructed after 7 iterations ($N = 2$, $\delta = 3$).

Mainly, the array factors keep a similar lobe structure at each band, although each lobe is broken in d lobes when frequency is increased by the same factor d. Such a result is analogous to the pattern modulation at different growth stages described in [5]. It was also shown in [12] that the SLR was highly dependent on the array lacunarity; in particular, a safe value for the lacunarity ratio $\delta/N < 1.2$ was recommended to avoid an increase of the secondary lobe amplitude beyond the uniform array SLR level.

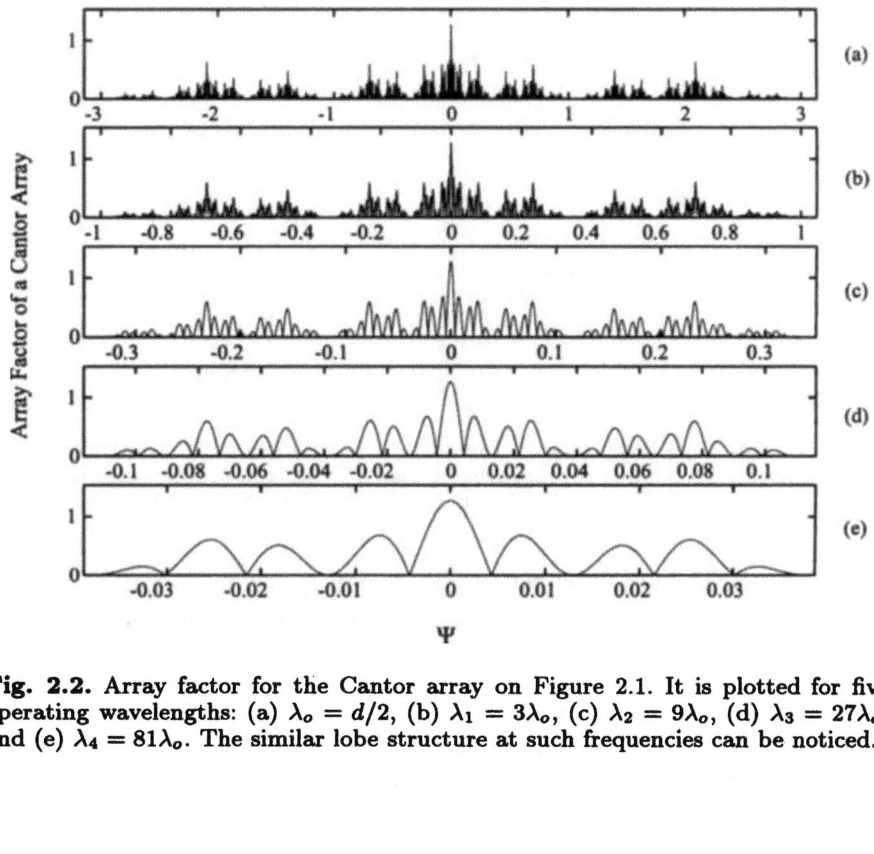

Fig. 2.2. Array factor for the Cantor array on Figure 2.1. It is plotted for five operating wavelengths: (a) $\lambda_o = d/2$, (b) $\lambda_1 = 3\lambda_o$, (c) $\lambda_2 = 9\lambda_o$, (d) $\lambda_3 = 27\lambda_o$, and (e) $\lambda_4 = 81\lambda_o$. The similar lobe structure at such frequencies can be noticed.

2.2 Fractal Array Factors

Although the patterns of the Cantor arrays exhibit certain multiwavelength features, they do not keep the same number of lobes and main beam width at each band. Since one looks for similarities among patterns instead of similarities among the array itself, a fractal design of array factors is suggested instead [12]. A self-similar curve such as the fractal Koch curve can be properly modified to introduce a new set of self-similar patterns (see Figure 2.3).

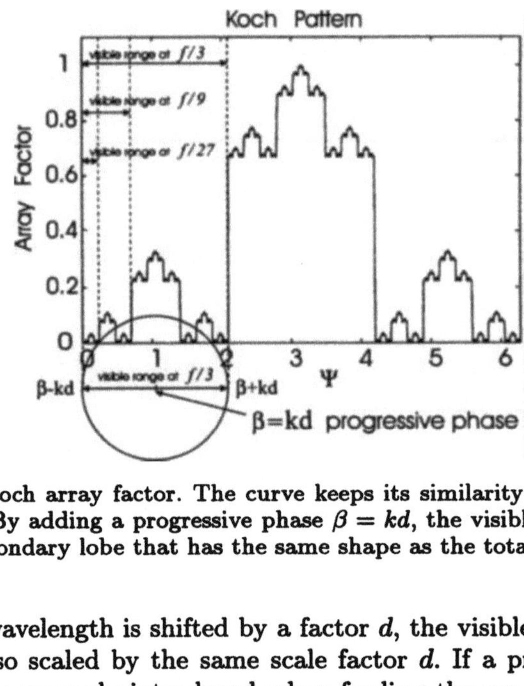

Fig. 2.3. The Koch array factor. The curve keeps its similarity properties at six different scales. By adding a progressive phase $\beta = kd$, the visible range is always centered at a secondary lobe that has the same shape as the total pattern.

Each time wavelength is shifted by a factor d, the visible range (i.e., the Ψ range) is also scaled by the same scale factor d. If a progressive phase factor $\beta = kd$ is properly introduced when feeding the array elements, the array factor becomes set around a secondary lobe which has exactly the same shape as the overall array factor. This way, the radiation pattern is held constant at each band except for a slight decrease of resolution due to a feasible finite fractal shape implementation.

A smoother array factor named the Blackman-Koch array factor was also introduced in [12]. Constructed by using the same scheme as in the Koch array factor construction of Figure 2.3, but using a Blackman window instead of a rectangular window as the fractal generator, the Blackman-Koch array factor presents some convenient advantages with respect to the Koch design. Basically, the element weight distribution displays a spiky shape, which suggests that some elements have a stronger influence than the others on the final pattern shape. This allows a proper reduction on the number of elements without significantly distorting its performance. It has been shown [11] that the array can be truncated by keeping only the 75 most significant elements (as opposed to the original 729 elements), and still keep a similar behavior at five bands through a whole 81:1 frequency range.

3. Fractal Antennas

Any practical multiband antenna array should be composed by multiband antenna elements as well. Therefore, it appears natural to explore the application of fractal structures to the design of multiband single antennas. This section describes the multiband behavior of some fractal antennas through both numerical and experimental results. Numerical results have been procured by means of an FDTD algorithm running in a connection machine CM-5, while experimental results on the radiation pattern and input return loss have been obtained in an anechoic chamber and an HP8510B network analyzer at the UPC.

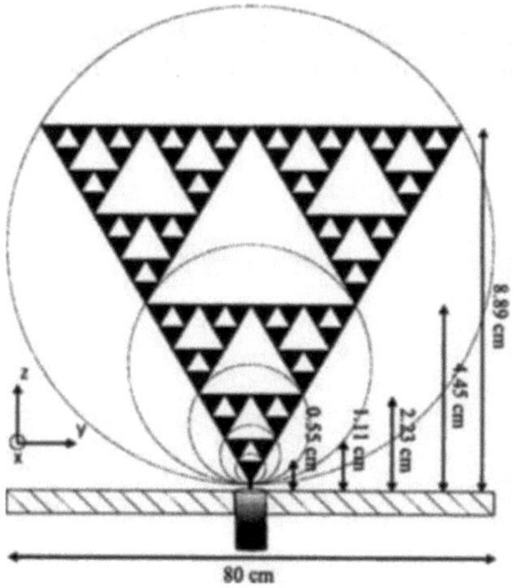

Fig. 3.1. The Sierpinski monopole.

3.1 The Sierpinski Antenna

A five-iteration Sierpinski monopole was printed using standard printed circuit techniques over a Cuclad 250 dielectric substrate (h=1.588 mm, ε_r=2.5). The structure was mounted over an 80 x 80 cm^2 aluminum ground plane and fed through a 50Ω coaxial probe. The antenna structure is plotted in Fig. 3.1 while its input reflection coefficient for a 0.05 GHz - 16 GHz frequency range is plotted in Fig. 3.2. The antenna clearly evidences five operating bands at frequencies 0.44, 1.75, 3.51, 7.01 and 13.89 GHz. The bands are log-periodically spaced by a factor of 2, which is precisely the scale factor between the gasket and sub-gaskets of the Sierpinski structure. It must be noted that the first band is shifted with respect to its expected position. This phenomenon can be related to the truncation effect, namely, the lack of larger fractal iterations within the antenna body which breaks the consonance with the remaining bands.

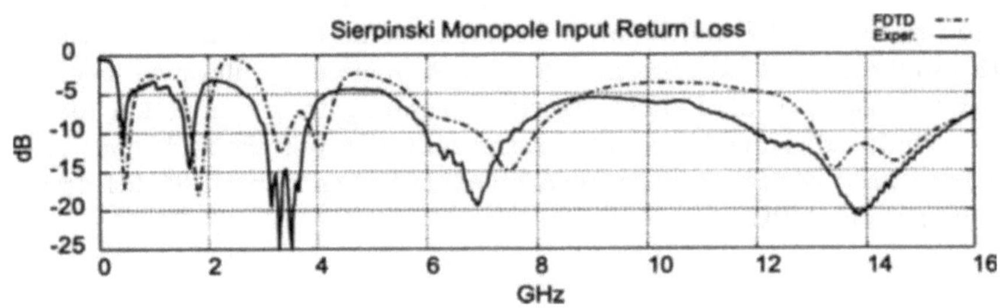

Fig. 3.2. Input reflection coefficient of the Sierpinski monopole. The antenna is well matched at five bands. Bands are spaced by a factor of 2.

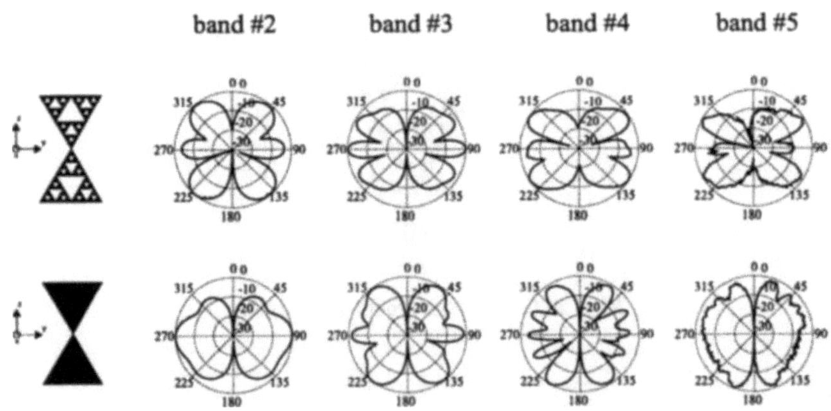

Fig. 3.3. Sierpinski dipole front to back ($\varphi = 0°$) radiation pattern (E_θ component) at the four upper bands (up). The same patterns for a standard triangular (bow-tie) antenna of the same size are plotted for comparison (bottom).

Figure 3.3 shows the measured front to back radiation patterns ($\varphi = 0°$ cut, $\overrightarrow{E_\theta}$ component) of a Sierpinski dipole at the four upper bands. The per-

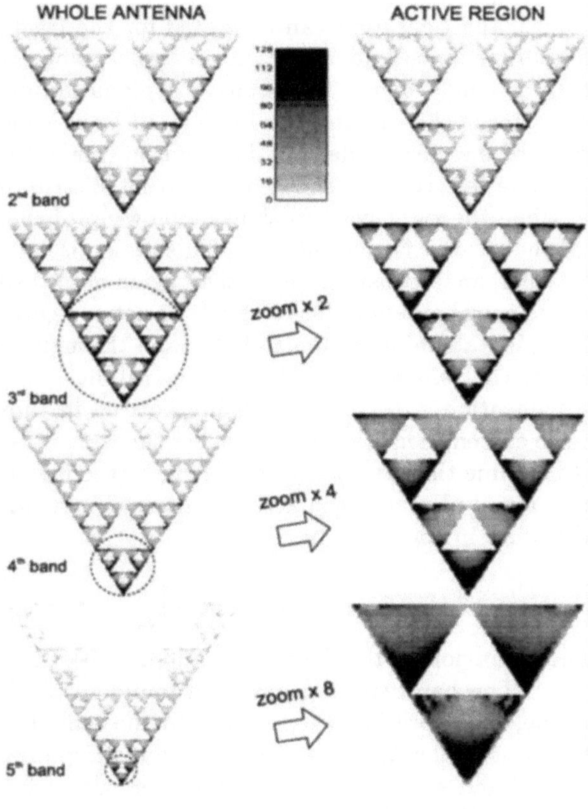

Fig. 3.4. Computed (FDTD) surface current densities over the Sierpinski dipole. The active region is compressed around a smaller similar sub-gasket each time the wavelength is reduced.

formance of the dipole is basically the same as the monopole but the radiation pattern extends to the whole space instead of only the upper half-space covered by the monopole. The patterns corresponding to a classical triangular (bow-tie) antenna are plotted for comparison as well. Both antennas were fed by a fussed coaxial to two-wire transmission line balun transformer.

It is clear that the basic pattern shape of the fractal antenna is kept almost constant throughout the bands. On the contrary, the Euclidean triangular antenna features a pattern with a growing number of lobes when frequency is increased.

While a multiband operation can be stated for the fractal antenna from both the input impedance and radiation patterns points of view, the Euclidean antenna is single band. An intuitive explanation on the multiband behavior of the Sierpinski antenna can be derived from Figure 3.4. It shows a reconstruction of the current density distribution (magnitude) over the antenna surface. The left column shows the current distribution at the four upper bands, while the right column shows a blow-up of the antenna region where most of the current is concentrated. This region is called the active region of the antenna, analogously to the active region described in log-periodic dipole arrays and spiral antennas [9, 10]. Such an active region self-scales with wavelength, keeping a similar proportional shape through the bands, which is the key feature that allows the multiband operation.

Also, the truncation effect can be readily noticed in those plots. At the second band the current density almost reaches the top of the antenna. Thus, it is normal to assume that at a larger wavelength (the first band) symmetry will be broken, mainly because the active region will not have enough room to expand as it does in the other bands.

3.2 Perturbation of the Sierpinski Antenna

The multiband behavior of the Sierpinski antenna has been stated in the previous section. The band spacing has been found to be tightly related to the fractal shape, more precisely, it is equal to the scale factor ($\delta = 2$) between the iterations of the fractal shape. An interesting question arises now, i.e., whether the band spacing can be tailored or not to fulfill the requirements of several specific applications. Some insight on this issue is gained on Figures 3.5 and 3.6.

Assuming that each resonant frequency of the Sierpinski structure is proportionally related to the height of the five subgaskets located near the feeding point, one can modify the antenna shape to match the size of each gasket to a specific wavelength. The so called P3/5SPK and P2/3SPK antennas [17] are designed following such a rule to keep an spacing of $\delta = 3/5$ and $\delta = 2/3$ ($1/\delta = 1.66$ and $1/\delta = 1.5$) among bands. The resulting shapes are self-affine rather than self-similar, which does not seem to be a major relevant issue (except for the scale factor) regarding the antenna performance. Both antennas again present a multiband behavior upon the radiation patterns and

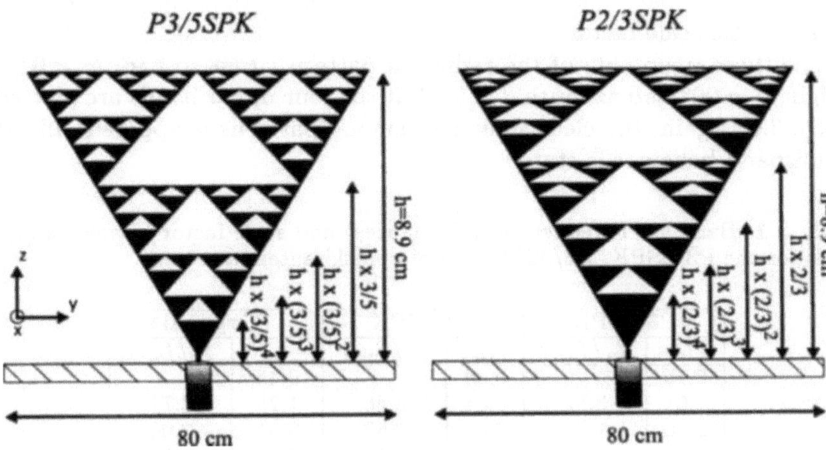

Fig. 3.5. Perturbation of the Sierpinski antenna to allocate the operating bands.

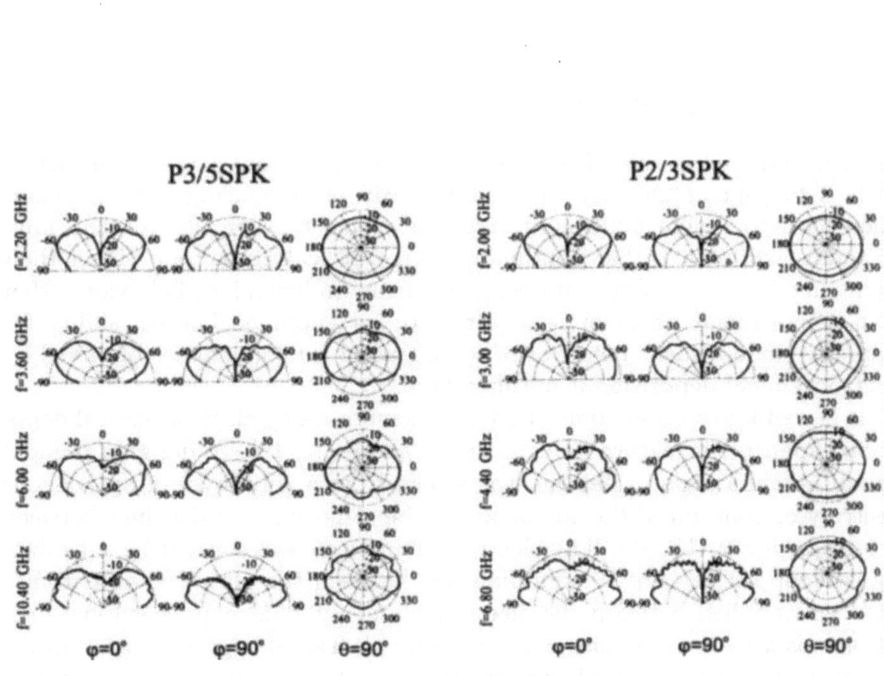

Fig. 3.6. Main radiation pattern cuts for the fractal P3/5SPK and P2/3SPK antennas.

input impedance [17]. Table 3.1 relates the experimentally measured resonant frequencies of the P3/5SPK, P2/3SPK and Sierpinski antennas respectively. Also, the band spacing is calculated to show its relationship with the fractal characteristic scale factor.

The three main cuts of the radiation pattern (front to back $\varphi = 0°$, left to right $\varphi = 90°$ and azimuth $\theta = 90°$) at the four upper bands are plotted in Figure 3.6. Again, the clear similarity among patterns is apparent and thus a multiband behavior is stated.

Table 3.1. Band start frequencies (resonances) and scale factor between adjacent bands for the P3/5SPK, P2/3SPK and Sierpinski antennas.

P3/5SPK		P2/3SPK		Sierpinski	
f_n	f_n/f_{n-1}	f_n	f_n/f_{n-1}	f_n	f_n/f_{n-1}
0.45	-	0.44	-	0.46	-
1.61	3.57	1.54	3.46	1.75	3.77
2.76	1.71	2.43	1.57	3.51	2.00
4.54	1.65	3.69	1.51	7.01	2.00
7.52	1.66	5.43	1.47	13.89	1.98

3.3 Random Fractal Antennas

Deterministic self-similar fractal shapes usually present large scale factors ($\delta > 2$ for planar shapes) and consequently large spacings among bands when used to design multiband antennas. Some random fractals and multifractals exhibit a rather continuous range of scaling factors within their main shape. In some applications, a frequency independent behavior rather than a multiband behavior is required for an antenna. The suitability for frequency independent behavior of some fractal random shapes obtained by electrochemical deposition is outlined here.

It is well known that some aggregates generated by electrochemical deposition experiments feature random fractal shapes [2, 19]. To design a random fractal antenna form, a deposit was grown from a thin layer of an aqueous electrolyte, containing the metal ion to be deposited sandwiched between two glass plates [18, 19]. The electrochemical cell was formed by introducing two copper wires of 100mm diameter between the plates. A potential of 12 V was applied between the electrodes and a 0.05M copper sulphate solution was taken as the aqueous electrolyte. A fractal aggregate was formed this way, and its photographed shape was used to print the tree-like antenna of Figure 3.7 over a dielectric substrate using the same technique as in the Sierpinski antenna.

Although random, the antenna still keeps a similar shape at several scales around the feeding point. To explore the multifrequency behavior of the an-

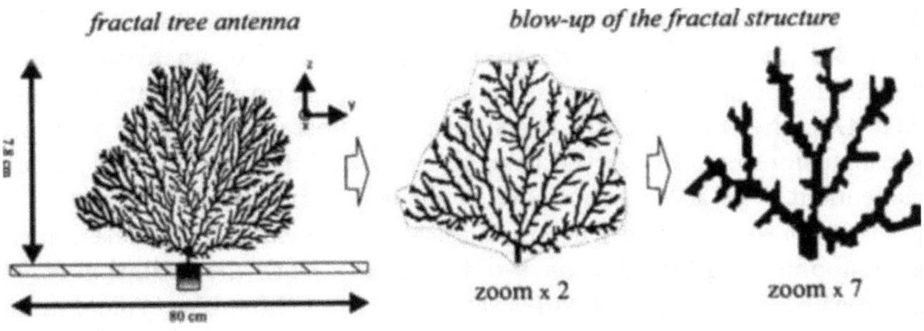

Fig. 3.7. Fractal tree antenna generated by electrochemical deposition. Although random, its shape displays self-similarity around the feeding point.

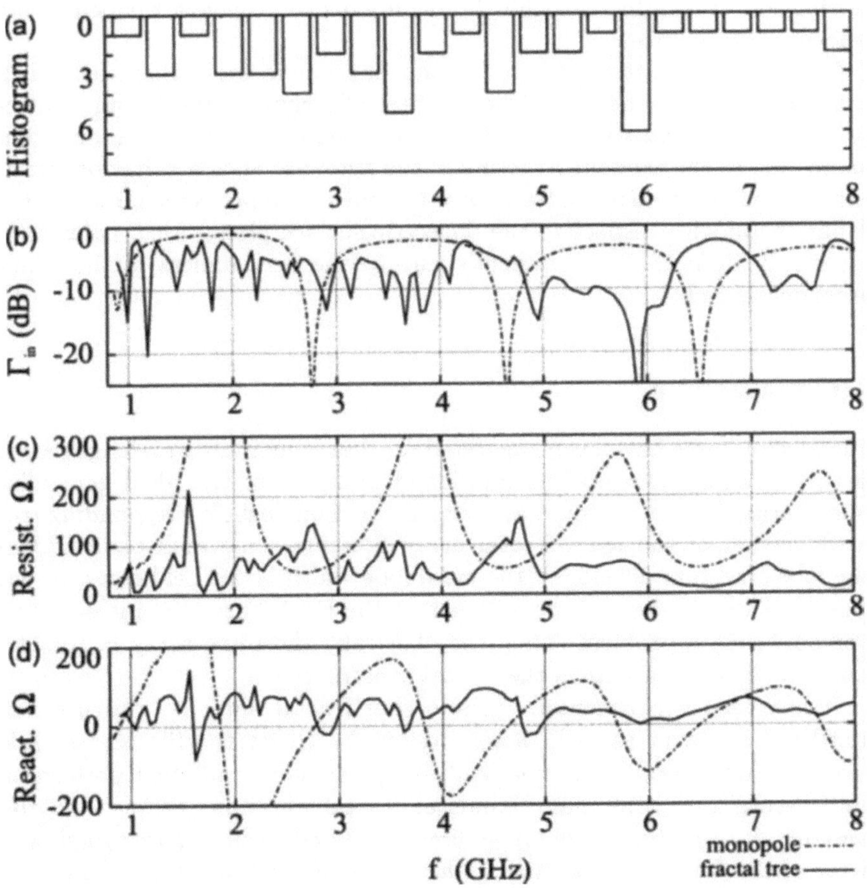

Fig. 3.8. Multiwavelength behavior of the fractal tree antenna. The length distribution histogram denotes certain agreement with the frequency distribution.

tenna, both the input reflection coefficient and antenna patterns were measured. The reflection coefficient, input resistance and reactance are shown in Figure 3.8 for a 0.9 to 8 GHz frequency range. Also, an histogram of the branch length distribution is plotted for comparison (associated frequencies $f = c/4\lambda$ rather than lengths are shown for a better likening). Some interesting conclusions can be derived from this result. Although not truly frequency independent, the antenna exhibits a denser distribution of matched bands with respect to a typical Euclidean antenna (the same parameters for a linear monopole are plotted in dash-dotted lines). Both the radiation resistance and input reactance show smaller variations than those of a typical monopole of the same height, and thus better approach the ideal characteristic of a frequency independent antenna. On the other hand, the band structure seems to hold some relationship with the length distribution within the fractal shape. Therefore, it might suggest that other fractal shapes with a more uniform distribution of lengths could fill the band gaps to achieve a frequency independent performance.

The radiation patterns of the fractal tree antenna were also measured [18] at the 15 operating bands. Although basically similar, the left-to-right cut ($\varphi = 90°$) featured some random fluctuations on the pointing directions, mainly due to the random structure of the antenna. Such a behavior was not held by the front-to-back pattern ($\varphi = 0°$) due to the perfect symmetry of the planar shape with respect to the dielectric surface plane. Also, polarization was highly unpredictable, basically due to the randomness of the branch directions as well.

4. Conclusions

Fractal structures have been suggested for designing multiwavelength antennas and antenna arrays. The multiband behavior of the ideal generalized Cantor array has been proven analytically. Although the corresponding truncated version of the ideal fractal array presents some multiband features (the lobe structure envelope), the main lobe width is not held constant with frequency. Fractal array factors based on the Koch curve are introduced to overcome such a problem. They keep basically the same array factor at as many bands as iterations in the fractal construction.

It has been experimentally proven that fractal antennas can present a multiband behavior as well. Some designs based on the Sierpinski gasket have been introduced. They basically keep the same radiation pattern and input return-loss at as many bands as fractal iterations are used in their construction. The band spacing is directly related to the characteristic scale factor of the fractal shape. Also, it has been shown that such a band spacing can be changed by properly perturbing the fractal shape. Some results regarding the multifrequency behavior of a fractal tree antenna generated by electrochemical deposition have been outlined as well.

Given the particular behavior of multiband fractal antennas, one can conclude that they might be specially suitable where the operation of several communication systems (GSM, GPS, etc.) in a single antenna is required. Similarly, it becomes apparent that they might fit multiwavelength needs of many spread spectrum and frequency diversity systems as well.

Acknowledgement. The authors would like to thank P. Mayes from the University of Illinois and D. L. Jaggard from the University of Pennsylvania for encouraging this work. They also acknowledge J. Claret, F. Sagus, X. Garcia, F. Benítez, Ll. Milà, X. Fernández, A. Seguí, A. Hijazo, R. Bartolomé, A. Medina, J. Ramis, C. Borja and M. Navarro for many contributions in the antenna measurements and simulations. This work has been partially supported by The Spanish Commission of Science and Technology" under grants MAT95-1038-C02-02, TIC-93-0055-C03-03 and TIC-96-0724-C06-04.

References

1. Mandelbrot, B.B. (1983): *The Fractal Geometry of Nature*. W.H. Freeman and Company.
2. Peitgen, H.O., Jürgens, H., Saupe, D. (1990): *Chaos and Fractals, New Frontiers of Science*. Springer-Verlag.
3. Jaggard, D.L. (1990): On Fractal Electrodynamics. In *Recent Advances in Electromagnetic Theory*, H.N.Kritikos, 183–224, D.L.Jaggard Editors; Springer Verlag.
4. Kim, Y., Jaggard, D.L. (1986): The Fractal Random Array. *Proc. of the IEEE, Vol. 74, No. 9*, Ref375809862, 1278–1280.
5. Jaggard, D.L., Spielman, T. (1992): Triadic Cantor Target Diffraction. *Microwave and Optical Technology Letters*, Vol. 5, No. 9, Ref375809859, 460–466.
6. Lakhtakia, A., Varadan, V.K., Varadan, V.V. (1987): Time-harmonic and time-dependent radiation by bifractal dipole arrays. *Int. J. Electronics, Vol. 63, No. 6*, 819–824.
7. Cohen, N., Hohlfeld, R.G. (1996): Fractal Loops And The Small Loop Approximation. *Communications Quarterly*, 77–81.
8. Cohen, N. (1996): Fractal and Shaped Dipoles. *Communications Quarterly*, 25–36.
9. Mayes, P.E. (1992): Frequency-Independent Antennas and Broad-Band Derivatives Thereof. *Proc. of the IEEE, Vol. 80, No. 1.*
10. Rumsey, V.H. (1966): *Frequency Independent Antennas*. Academic Press.
11. Puente, C. (1993): Fractal Design of Multiband Antenna Arrays. *Elec. Eng. Dept. Univ. Illinois*, Urbana-Champaign, ECE 477 term project.
12. Puente, C., Pous, R. (1996): Fractal Design of Multiband and Low Side-Lobe Arrays. *IEEE Trans. Antennas and Propagation*, Vol. 44, No. 5, 1–10.
13. Puente, C., Romeu, J., Pous, R., Garcia, X., Benitez, F. (1996): Fractal multiband antenna based on the Sierpinski gasket. *IEE Electronics Letters, Vol. 32, No.1*, 1–2.
14. Allain, C., Cloitre, M. (1987): Spatial spectrum of a general family of self-similar arrays. *Physical Review A, Vol. 36, No. 12*, 5751–5757.

15. Goutelard, C. (1992): Fractal theory of large arrays of lacunar antennas. *Electromagnetic Wave Propagation Panel Symposium* (AGARD-CP-528), 35, 1–15, (in french).
16. Puente, C., Romeu, J., Bartolomé, R., Pous, R. (1996): Perturbation of the Sierpinski antenna to allocate operating bands. *IEE Electronics Letters, Vol. 32*, No. 24, 2186–2187.
17. Puente, C., Romeu, J., Pous, R., Cardama, A. (1996): On the Behavior of the Sierpinski multiband fractal antenna. *IEEE Trans. Antennas and Propag*, (Submitted).
18. Puente, C., Romeu, J., Claret, J., Sagués, F., López-Salvans, M.Q., Pous, R. (1996): Multiband properties of a fractal tree generated by electrochemical deposition. *IEE Electronics Letters, Vol. 32., No. 25*, 2298–2299.
19. Trigueros, P.P., Claret, J., Mas, F., Sagués, F. (1991): Pattern morphologies in zinc electrodeposition. *Journal of Electroanalytical Chemistry, Vol. 312, No. 1-2*, 219–235.

Fractal-Surface-Enhanced Optical Responses

R. Botet[1], E.Y. Poliakov[2], V.M. Shalaev[2], and V.A. Markel[2]

[1] Laboratoire de Physique des Solides, Universit Paris-Sud, Centre d'Orsay, 91405 Orsay Cedex, FRANCE
[2] Department of Physics, New Mexico State University, Box 30001/Dept.3D Las Cruces, New Mexico 88003, USA

Abstract Local filed distribution and non-linear optical responses of self-affine metal films are calculated numerically in the quasi-static approximation. It is shown that the distribution of local field on the metal-air interface is highly inhomogeneous. In the areas of high local field intensity ("hot zones"), the latter can exceed that of the applied field by three orders of magnitude. Spatial location of "hot zones" inside the sample depends strongly on the frequency and the polarization of the incident light. As a result, non-linear optical responses of self-affine metal films are dramatically enhanced compared to those of smooth interfaces. We calculated enhancement factors for surface-enhanced Raman scattering (SERS) and second harmonic generation (SHG). We predict that, under certain circumstances, these non-linear effects can be enhanced by the factor of 10^7 due to the non-trivial geometrical structure of self-affine films.

1. Introduction

A number of optical phenomena are strongly enhanced on a rough metal film [1]. This enhancement is commonly associated with excitation of surface plasmon oscillations. Two varieties of surface plasmons are typically recognized. They are surface plasmon waves (SPW) and localized surface plasmons (LSP). SPW propagate laterally along the metal surface; LSP are confined to metal particles that are much smaller in size than the wavelength of the incident light.

Recent evidence suggests that cold-deposited metal films are self-affine fractal structures [2, 3, 4]. Fractal structures (fractals) do not possess translational invariance and cannot transmit ordinary running waves (such as SPW, for example). Mathematically, this is a consequence of the fact that running waves are not eigenmodes of the operator of dilation symmetry that, in general, characterizes self-affine surfaces and fractal clusters. The efficiency of dumping running waves by fractals may lead to detailed understanding of "self-stabilization" of many of the fractals found in nature [5, 6].

Dipolar eigenmodes in fractal composites are substantially different from those in other media. Fractal aggregates possess a variety of dipolar eigenmodes (fractons), distributed over a wide spectral range. For the case of a conventional three-dimensional collection of particles (such as a random close-packed sphere of particles or a random gas of particles), the absorption spectra are peaked near a relatively narrow resonance of individual particles. Dipolar eigenmodes of fractals can be excited by a homogeneous electric field, whereas only one dipolar mode can be excited by the same type of field in a dielectrical sphere[1]. In the continuous media as well as in the majority of microscopically disordered, but homogeneous on average, media, dipolar modes (polaritons) are usually delocalized over large areas. All the monomers absorb light energy with approximately equal rate in the regions whose linear dimensions significantly exceed the incident field wavelength. Contrary to that, fractons in nanocomposite materials, such as fractal clusters and self-affine thin films, are short range and tend to be strongly localized [7, 8, 9, 10, 11]. These composites have eigenstates that very often belong to the sub-wavelength resolution. Such irregular media are not homogeneous on average, and stay disordered on all possible scales; therefore, they do not possess translational symmetry. Instead, they are symmetrical with respect to the scale transformation (i.e. scale-invariant).

Light absorption by a collection of monomers confined in sub-wavelength regions can be much higher than that by another monomer ensemble simply because of density variation in different regions of an object (where less or more voids are present). Therefore, one can expect strong fluctuations of local fields that are associated with such a localization of optical eigenmodes. Spatial fluctuations of local fields ultimately lead to huge enhancements in a number of optical effects for nanocomposites [9, 10, 11, 12, 17].

In self-affine films, we anticipate that plasmon oscillations associated with different roughness features strongly interact with each other via dipolar or, more generally, multipolar forces. Thus, plasmon oscillations on a self-affine surface are neither conventional SPW nor LSP. They should be treated as collective eigenmodes which are formed by the interaction between different features of a self-affine film.

Unlike in a random surface with small spatial inhomogeneities, the contribution of highest spatial harmonics (with the amplitude larger than the harmonic wavelength) does play an important role in the Fourier decomposition of a self-affine surface. This means that neither the Rayleigh perturbation approximation[2] nor the Kirchoff (geometrical optics) approach can be applied to describe optical properties of a self-affine structure [18]. Apart from the two basic approaches, there exists a phase perturbation approach [19, 20], which is, in some sense, intermediate between former two methods and also cannot be applied for a self-affine surface.

2. Discrete Dipole Approximation

We base our approach on the microscopic "discrete-dipole approximation" (DDA) which was initially suggested by Purcell and Pennypacker [21] and

[1] The are three (non-degenerate) resonances with non-zero total dipole moment for a spheroid.

[2] The complete set of formulas for the case of electromagnetic waves scattering from smooth random surfaces can be found in [14, 15, 16].

lately developed by Draine [22] to calculate an optical response from the object of an arbitrary shape.

We treat a self-affine film as a collection of N linear polarizable particles (monomers) under assumption that all the particles are identical and possess a scalar polarizability, α_0. Being irradiated by a plane monochromatic incident wave of the form

$$E^{inc}(r,t) = E^{(0)} \exp(ik \cdot r - i\omega t), \tag{2.1}$$

the monomers interact with each other through the dipole-dipole interactions. Total local electric field at monomer's position r_i is given by the sum of the incident wave and the scattered (secondary) one $E_i \equiv E_i{}^{loc} = E^{inc}(r_i,t) + E^{sc}(r_i,t)$. Dipole moment d_i at the i−th monomer site is determined as

$$d_i = \alpha_0 E_i \tag{2.2}$$

The field $E^{sc}(r_i)$, scattered from all other dipoles, generally contains both the near-zone and the far-zone terms. In this paper we restrict our consideration to the quasi-static limit, i.e., the linear size of a system l is assumed to be much smaller than wavelength $\lambda = 2\pi c/\omega$. In this approximation, the factor $\exp(ik{\cdot}r_i)$ is always close to unity. The incident wave (2.1) is also monochromatic, so that the whole exponential factor can be omitted. We note that the lowest order part of the near-zone term is only important in $E_i{}^{sc}$ (see (2.3) below).

A solution for an induced dipole at i−th site is obtained from a set of coupled-dipole equations (CDE) [7, 8, 9, 10, 11]:

$$\begin{aligned} d_{i,\alpha} &= \alpha_0 \left(E_\alpha^{(0)} + \textstyle\sum_j W_{ij,\alpha\beta} d_{j,\beta} \right) \\ W_{ij,\alpha\beta} &= -(\delta_{\alpha\beta} r_{ij}^2 - 3 r_{ij,\alpha} r_{ij,\beta})/r_{ij}^5, \end{aligned} \tag{2.3}$$

where $W_{ij,\alpha\beta}$ is the interaction operator between two dipoles, with radius-vectors r_i and r_j; $r_{ij} = r_j - r_i$. Greek indices denote Cartesian components and should not be confused with the polarizability, α_0. Summation over repeated Greek indices is implied.

Polarizability α_0 can be expressed in terms of the bulk dielectric permittivity, $\epsilon \equiv \epsilon'(\lambda) + i\epsilon''(\lambda)$, and dipole density n by using the Clausius-Mossotti relation [12, 21, 22]:

$$\alpha_0 = (3/4\pi n)[(\epsilon - 1)/(\epsilon + 2)] \tag{2.4}$$

Point dipoles located on a cubic lattice with period $a_0 = n^{-1/3}$ provide an equivalent continuous representation for such an irregularly shaped object, as a self-affine surface, if the lattice period is much smaller than a characteristic size of a field variation (which is the skin depth).

Expression (2.4) is much similar to that for the polarizability for an isolated dielectric sphere (see, for example, [23]). A characteristic monomer size, $R_m = (3/4\pi)^{1/3} a_0$, provides equality of the total volume occupied by the film and the volume represented by individual spheres [13, 21, 22]. Thus, in the DDA, the equivalence of a point dipole consideration and a sphere representation follows as long as R_m is defined correctly.

Since $W_{ij,\alpha\beta}$ is independent of frequency ω in the quasi-static approximation, the spectral dependence of solutions (2.3) is manifested only through $\alpha_0(\omega)$. For convenience, we introduce the variable $Z_0 = \alpha_0^{-1} = -(X + i\delta)$, where

$$X \equiv -Re[\alpha_0^{-1}] = -R_m^{-3}[1 + 3(\epsilon' - 1)/|\epsilon - 1|^2], \qquad \delta \equiv -Im[\alpha_0^{-1}] = 3R_m^{-3}\epsilon''/|\epsilon - 1|^2. \tag{2.5}$$

The variable X describes the proximity to resonance of an individual particle and, thus, plays the role of a frequency parameter; δ characterizes dielectric losses. The smaller δ, the higher the resonance quality-factor, $q \sim \delta^{-1}$. For a specific material, using the spectral dependence, $\epsilon(\lambda)$, and formulas (2.5), one can find $X(\lambda)$ and $\delta(\lambda)$.

We notice that interaction operator W in (2.3) is a real Hermitian operator, therefore its eigenfunctions, $|n\rangle$, as well as the eigenvalues, w_n, are also real. Following [7, 8, 9, 10, 11], we introduce a $3N$-dimensional vector space R^{3N} and an orthonormal basis $|i\alpha\rangle$. Vectors of dipole moments $|d\rangle$ and incident field $|E^{inc}\rangle \in R^{3N}$; their components in this basis are $\langle i\alpha|d\rangle = d_{i,\alpha}$ and $\langle i\alpha|E^{inc}\rangle = E^{inc}_{i,\alpha} = E^{(0)}_{\alpha}$, respectively. Matrix element $W_{ij,\alpha\beta} = \langle i\alpha|W|j\beta\rangle$. Then Eq.(2.3) takes the form

$$(Z_0 - W)|d\rangle = |E^{inc}\rangle. \tag{2.6}$$

By diagonalizing interaction matrix W with $W|n\rangle = w_n|n\rangle$ and expanding $3N$- dipole vectors in terms of eigenstates $|n\rangle$ (as $|d\rangle = \sum_n C_n|n\rangle$), we obtain a relation between local fields and the amplitudes of linear dipoles induced by wave (2.1) [7, 8, 9, 10, 11]:

$$E_{i,\alpha} = \alpha_0^{-1} d_{i,\alpha} = Z_0 \alpha_{i,\alpha\beta} E^{(0)}_{\beta}, \qquad \alpha_{i,\alpha\beta} \equiv \alpha_{i,\alpha\beta}(\omega) = \sum_{j,n} \frac{\langle i\alpha|n\rangle\langle n|j\beta\rangle}{Z_0 - w_n}, \tag{2.7}$$

where $\alpha_{i,\alpha\beta}$ is the total polarizability matrix of the i-th dipole at the fundamental frequency ω.

We note here that the general solutions (2.7), expressed in terms of parameters X and δ (see (2.5)), depend only on the object's morphology; they are independent of material properties. Substituting a given dielectric function $\epsilon = \epsilon(\lambda)$, we can explicitly express parameter $Z_0 = Z_0(\lambda)$ in terms of the wavelength, i.e. the general solutions can be specified for any material [9, 10, 11].

3. Geometrical Properties and Numerical Algorithms

In this section we summarize the geometrical properties of cold deposited metal surfaces and describe algorithms to simulate such structures.

3.1 General properties of self-affine films

In many cases surfaces formed by condensing atomic beams onto a low temperature substrate can be characterized by microscopic surface roughness [2, 3]. Rough surface profiles obtained during the period of film growth (deposition time) possess the properties of self-affine fractal structures [4]. The obtained self-affine surfaces belong to the Kardar-Parisi-Zhang universality class [24].

Unlike fractal assembles of particles, which are prevalent, for example, in nanocomposites, self-affine structures have two different scaling factors; one is applied in the surface plane, the other is in the normal (growth) direction, $\hat{z}$. Contrary to the case of "usual" roughness, there is no correlation length for self-affine surfaces, which implies that inhomogeneities of all scales are present (within a certain size interval) according to a power-law distribution. A self-affine surface contains roughness features of very small (asymptotically zero) radii of curvature, i.e., the profile's derivatives can be very large. However, this kind of divergence is only formal, because the scale-invariance is valid in the intermediate asymptote region, i.e. on the scales between the characteristic atomic lengths and the size of a sample.

3.2 Numerical models of cold-deposited metal films

To generate a self-affine film (see Figure 3.1), we used the restricted solid-on-solid (RSS) model [25, 26].

Fig. 3.1. The self-affine film obtained in the restricted solid-on-solid model. The scaling exponent $D_H = 0.4$ and the fractal dimension $D = 3 - D_H = 2.6$.

In this model a particle is incorporated in a growing aggregate only if the newly created interface does not have steps which are higher than the lattice period a_0. Such an algorithm provides that the surface structure of metal deposits does not have overhangs, and true scaling behavior pronounces clear, even for small dimensions. Initially, we generated two different random ensembles of self-affine samples with $N \sim 10^3$ and $N \sim 10^4$ monomers per cluster. These relatively big numbers allowed us to achieve the scaling condition

$$\langle [h(R_i) - h(0)]^2 \rangle \approx R^{2D_H}, \tag{3.1}$$

where R_i is the vector in the x-y plane of growth, and scaling exponent D_H is related to the fractal dimension, $D = 2.6$, by expression $D_H = 3 - D$. Formula (3.1) is known as a height-height correlation function. As function (3.1) is valid for large values of the average height (proportional to the characteristic deposition time), it becomes well pronounced in the large time regime of the self-affine surface realization.

The analysis performed in section 2. requires the knowledge of the eigenstates for matrix $W_{ij,\alpha\beta}$ (see, for example, (2.3)). Diagonalization of $W_{ij,\alpha\beta}$ can be done in several different ways in quasi-static approximation [7, 8, 27]. We used Jacobi algorithm to diagonalize $W_{ij,\alpha\beta}$. For calculations of linear optical effects, we did not need to know the local fields explicitly, and we used both cluster ensembles. However, for non-linear effects, it is important to know the local field distributions. Local polarizabilities must be calculated from the average optical response of a whole cluster set (see definition (2.7)). Apparently, the storage of a matrix with $\sim 10^9$ elements is not possible and has to be reestablished on each stage of time extensive calculations. That is why we limited our model to an average number of $N \sim 10^3$ monomers per cluster.

The actual characteristics turned out to be slightly different, namely, total number of monomers, $N_t = 8271$, providing $N = 690$ and standard deviation $\sigma_N = 110$, the total number of monomers occupying the surface, $N_S = 2620$. A ratio of $N_S/N_t \sim 1:3$, gave us, in fact, a very good condition of "thickness", and indeed none of the clusters was composed of more than five subsequent monomer layers.

To make the reduction of particles, we cut a "parent" cluster until the total number of monomers left was placed in the given boundaries. The "excessive" monomers were removed from the "bottom" of a cluster, a regular part of the generated film, so that the resultant sample had, at least, one hole. Note that such a procedure does not violate the scaling condition (3.1).

As an example of a cold deposited film, we used a silver aggregate and data of [28] to tabulate the dependence $\epsilon(\lambda)$. For RSS silver samples, the inequality $l < \lambda$, was preserved, so that the quasi-static approximation was valid. The bottom layer of a film was about 14x15 lattice units with $a_0 \sim 5$nm. The latter quantity well reflected the geometrical properties of silver nanocomposites.

4. Optical Properties of Cold Deposited Metal Films

In this section we consider some linear optical properties of self-affine structures.

4.1 Mode Localization and Field Distribution

First, we investigate the absorption by rough films by considering the parallel and perpendicular components of the mean polarizability per particle, defined as $\alpha_{\parallel} \equiv (1/2)\langle \alpha_{i,xx} + \alpha_{i,yy} \rangle$ and $\alpha_{\perp} \equiv \langle \alpha_{i,zz} \rangle$ (see Eq.(2.7)). The parallel component, $\alpha_{\parallel}$, characterizes the polarizability of a self-affine film in the (x, y)-plane, whereas the perpendicular component, $\alpha_{\perp}$, gives the polarizability in the normal, z, direction, provided the external field has the same polarization.

We observed a strong dichroism expressed in the difference between the two spectra, $\alpha_{\parallel}(X)$ and $\alpha_{\perp}(X)$ (the data are not shown). The modes contributing most to $\alpha_{\parallel}$ (the "longitudinal" modes) are located in the long wavelength part of the spectrum (negative X), whereas the "transverse" modes tend to occupy the short wavelength part of the spectrum (positive X). To some extent, this can be understood by considering a film as a prolate spheroid, where the longitudinal and transverse modes are shifted to the red and blue part of a spectrum, respectively, in comparison with the eigenmode of a sphere. However, in contrast to the case of a spheroid, there is a large variety of eigenmodes in self-affine films. The widths of absorption spectra are much larger than the width of an individual resonance, δ; this indicates a strong inhomogeneous broadening associated with a variety of the dipolar eigenmodes on a self-affine surface. Thus, the dipole-dipole interactions of constituent monomers in a self-affine film "generate" a wide spectral range of resonant modes.

We also make an important conclusion that in the quasi-static approximation the linear optical properties of a self-affine film do not depend on the number of monomers, N, and, therefore, on the linear size, l, of the film. The calculations that were performed for the ensembles of samples with different numbers of particles and linear sizes (see section 3.2) gave similar results.

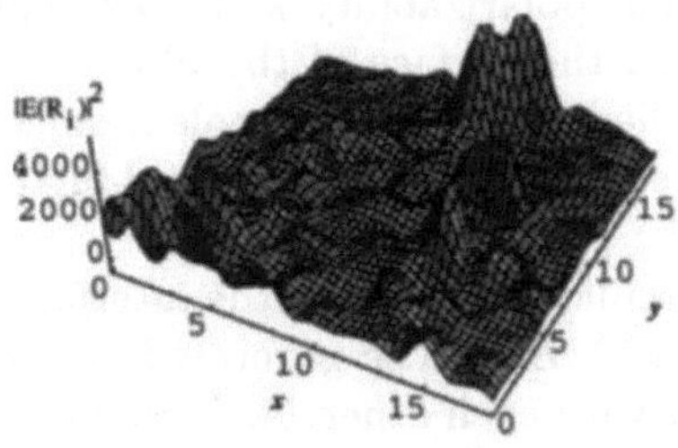

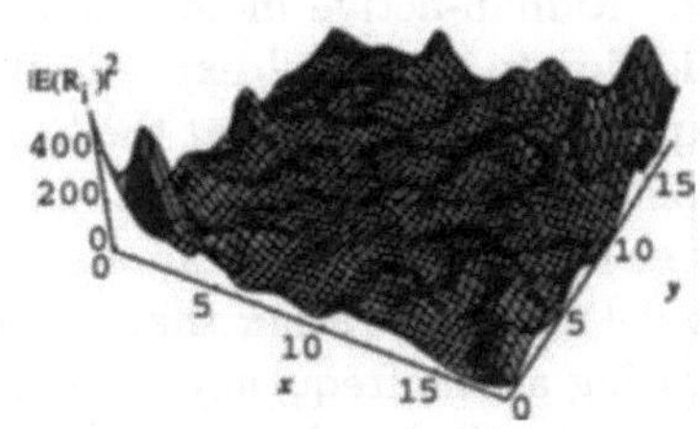

Fig. 4.1. **Fig. 4.2.**

Spatial distributions of the local field intensities, $|E(R_i)|^2$, on the self-affine surface for different values of frequency parameter, X. Figure 4.1: $X = -3$, Figure 4.2: $X = -2$. The decay parameter $\delta = 0.03$ in both cases. The applied field is polarized in the (x, y)-plane, $E^{(0)} = (2)^{-1/2}(1, 1, 0)$.

The field distributions of eigenmodes on a self-affine surface are extremely inhomogeneous. There are "hot spots" associated with areas of high local fields, and "cold zones" with small local fields. (Similar patchwork-like picture of the field distribution is observed in fractal clusters [7, 8, 9, 10, 11, 29, 30, 31].) Spatial locations of the modes are very sensitive to both frequency and polarization of the applied field.

To demonstrate this, in Figures 4.1, 4.2, we show the intensity distributions for the local fields, $|E(R_i)|^2$, on the film-air interface, calculated from Eq.(2.7). The results are shown for different values of frequency parameter, X. Note that the local field distributions, $|E(R_i)|^2$, can be measured with the use of a near-field scanning optical microscopy. provided the probe is passive [32].

As seen in Figures 4.1, 4.2, for a value of $\delta = 0.03$, which is typical for metals in the visible and near-infrared parts of the spectrum, the local field intensities in the "hot zones" can significantly exceed the intensity of the applied field, up to three orders of magnitude. For smaller values of δ, the enhancements can be even larger. The high frequency and polarization sensitivity of the field distributions is also obvious from the figures.

Strongly inhomogeneous distributions of local fields on a self-affine surface bring about large spatial fluctuations of local fields and strong enhancements in various optical processes. These enhancements are especially large for non-linear optical phenomena that are proportional to the local fields raised to some integer power.

Below we consider two typical examples of optical phenomena, surface-enhanced Raman scattering (SERS) and second harmonic generation (SHG), occurring in rough thin films.

4.2 Enhanced Raman Scattering from Thin Surfaces

Suppose that Raman-active molecules cover homogeneously the surface of a self-affine film, so that each site at the film-air interface, apart from the linear polarizability $\alpha_0(\omega)$, possesses also a Raman polarizability κ, for example, due to Raman-active molecules adsorbed on the surface of the film. Then, in addition to dipoles $d_i \exp(-i\omega t)$, the applied field induces dipole moments $d_i^s \exp(-i\omega_s t)$ oscillating at the Stokes-shifted frequency, ω_s. To avoid further complications we assume that κ is a scalar.

The field acting on a Raman-active molecule at i-th site of the interface is the local field, E_i, rather than the applied field, $E^{(0)}$. The generated dipoles, oscillating at the frequency ω_s, can interact with each other. As a result, the CDE for the Stokes dipoles have the form [9, 11, 17]:

$$d_{i,\alpha}^s = \kappa_i E_{i,\alpha} + \alpha_0^s \sum_j W_{ij,\alpha\beta} d_{j,\beta}^s, \tag{4.1}$$

where $\alpha_0^s = \alpha_0(\omega_s)$ is the linear polarizability at the frequency ω_s, and the local fields, $E_{i,\alpha}$, at the fundamental frequency, ω, are given by Eq. (2.7).

The enhancement of Raman scattering, G_{RS}, is defined as

$$G_{RS} = \langle |\sum_i d_i^s|^2\rangle/(N_0|\kappa|^2|E^{(0)}|^2), \tag{4.2}$$

where N_0 is the number of Raman-active molecules. Because of an anisotropical structure of the self-affine film, the enhancement greatly depends on the polarization of the incident light, *s*- or *p*-polarization. Therefore, we introduce two different quantities, $G_{RS,\parallel} = (G_{RS,x} + G_{RS,y})/2$ and $G_{RS,\perp} = G_{RS,z}$, describing the enhancements for the applied field polarized in the plane of the surface and perpendicular to it.

For spontaneous Raman scattering, the Raman polarizabilities κ_i at different surface sites contain uncorrelated phases, $\langle \kappa_i \kappa_j \rangle = |\kappa|^2 \delta_{ij}$, where the angle brackets stand for averaging over the ensemble of random samples (not to be confused with the *bra* and *ket* vectors).

The formal solution to CDE in (4.1) for the Stokes dipoles is very similar to that to Eqs. (2.3) (see details in section 2.). Using this technique and the above properties of the Raman polarizabilities, we obtain the following representation of (4.2):

$$G_{RS} = \frac{|Z_s Z_0|^2}{|E^{(0)}|^2} \langle \alpha^s_{j,\beta\alpha}[\alpha_{j,\beta\gamma} E^{(0)}_\gamma]\alpha^{s*}_{j,\delta\alpha}[\alpha^*_{j,\delta\delta'} E^{(0)*}_{\delta'}]\rangle_S, \tag{4.3}$$

where $Z_s = Z(\omega_s)$, $Z_0 = Z(\omega)$, and $\alpha_i^s \equiv \alpha_i(\omega_s)$. The terms in the square brackets in (4.3) give the local fields at the fundamental frequency ω acting on j-th site. The notation $\langle ... \rangle$ stands for an average over positions of dipoles on the surface of the film only.

If the difference $\omega - \omega_s$ is small (small Stokes shift), we can put $Z_s \approx Z_0$, and $\alpha_j^s \approx \alpha_j$. Then for light linearly polarized along the η-axis, the enhancement is given by

$$G_{RS,\eta} = |Z_0|^4 \langle \sum_\alpha |(\alpha_j^2)_{\alpha\eta}|^2\rangle_S, \tag{4.4}$$

where $(\alpha_j^2)_{\alpha\beta} = \alpha_{j,\gamma\alpha}\alpha_{j,\gamma\beta}$.

In Figure 4.3, we show the average enhancement of Raman scattering for small Stokes shift on self-affine films generated in the RSS model. For the dielectric function of silver, $\epsilon(\lambda)$, we used the data of [28]. As seen in the figure, the enhancement increases toward the long-wavelength part of the spectrum and reaches very large values, $\sim 10^7$; this agrees well with the experimental observations of SERS on cold-deposited thin films [1]. Note that the present SERS-theory contains no adjustable parameters.

In Figures 4.4, we plot the field spatial distributions at the fundamental and Stokes frequencies for $|Z_s\kappa|^2 = 0.25$. As seen in the figure, the distributions contain "hot spots", where the fields are very high. The spatial positions of these spots are strong functions of the frequency and polarization of the applied field [12]. Although the Stokes signal is proportional to the local field at the fundamental frequency, ω, the generated Stokes field, with frequency ω_s, excites, in general, other eigenmodes. Hence the field spatial distributions produced by the applied field and by the Raman signal can be different. (As seen in Figures 4.4, the "hot" spots at the left side of Figure 4.4b and 4.4c are spatially correlated, whereas the largest Stokes peak is positioned at the right, where the ω-field is relatively small.)

This picture is expected to be typical for various optical processes in strongly disordered fractal systems, such as self-affine thin films. Specifically, "hot spots" associated with fields at different frequencies and polarizations can be localized in spatially separated nm-sized areas. Recent near-filed optical microscopy observations of optical excitations on a silver self-affine film [33] verify basic predictions of the developed theory.

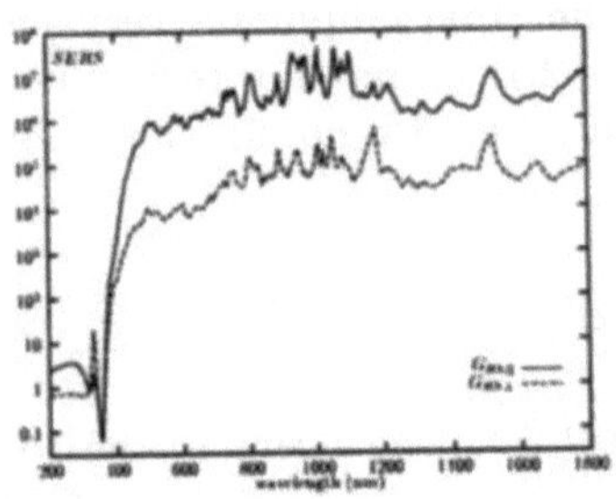

Fig. 4.3. The enhancements of Raman scattering, $G_{RS,\parallel}$ and $G_{RS,\perp}$, on silver self-affine films.

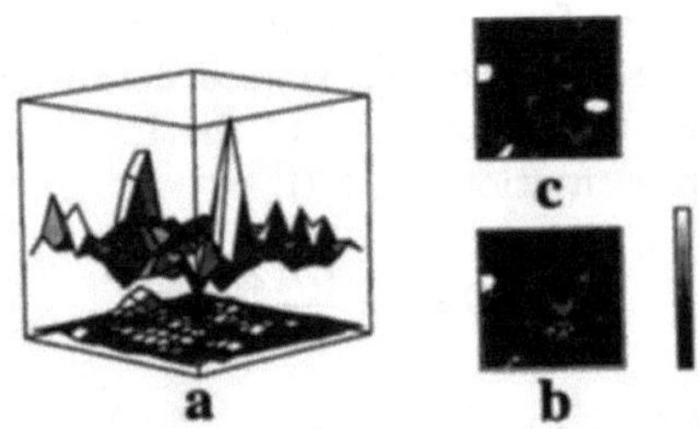

Fig. 4.4. a): The spatial distributions for the local fields at the fundamental wavelength, $\lambda = 550nm$, (bottom: the field distribution is magnified by 3) and for the Stokes fields, $\lambda_s = 600nm$, (top). The applied field is linearly polarized in the plane of the film. b) and c) : The contour-plots for the field distributions shown on a).

5. Nonlinear Optical Processes

In this section we consider second harmonic generation (SHG) [34, 35] as a most suitable example of a nonlinear optical phenomena of the second order. As SHG is extremely sensitive to the surface roughness conditions (which made it a widely used technique for numerous structural and electronic properties of surfaces and interfaces [35, 36, 37]), this process is of fundamental interest for studying the nonlinear interaction of light with self-affine films.

There were extensive studies of different mechanisms of SHG from surface [38, 39] and in bulk [40, 41] for jellium models since they were first proposed by Rudnick and Stern [42]. Here we consider contributions to SHG from adsorbent molecules on the film surface. Thus, non-local effects (connected with the spatial dispersion) are left out of the analysis.

For SHG, the nonlinear polarization $P^{(2)}(2\omega)$ is commonly introduced through the definition of a nonlinear polarizability tensor $\hat{\chi}(2\omega;\omega,\omega)$ of a third rank as

$$P^{(2)}(2\omega) = \hat{\chi}(2\omega;\omega,\omega) : E(\omega) : E(\omega). \tag{5.1}$$

Specifically, we imply the following non-centrosymmetrical structure of impurities. These molecules are assumed to have a "preferred" direction, $\hat{n}$, which coincides with the normal vector to the (x, y)–plane of film, so that the film anisotropy is completely reproduced by its impurities[3]. Such a molecule can be thought of as a spiral having the counter-clockwise direction of loops; thus, providing rotation in the positive $\hat{z}$- direction.

Due to the geometrical constraints, imposed on the impurity structure, we construct the vector $P^{(2)}(2\omega)$ from two independent combinations of the triplet $(\hat{n}, E, E)$ in Eq.(5.1)[4]:

$$P^{(2)}(2\omega) = A(E \cdot E)\hat{n} + B(\hat{n} \cdot E)E, \tag{5.2}$$

where A and B are two independent complex scalars determined only by the inherent structure of impurities. They are not related to parameters a and b introduced in the classical paper [42] and frequently used in the literature describing SHG.

Because induced nonlinear dipoles can interact with each other, CDE (2.3) transforms to its most general matrix form (which is also very similar to (4.1))

$$\begin{aligned} |d^{NL}\rangle &= |\hat{\chi} : E^2\rangle + \alpha_0 W |d^{NL}\rangle, \\ \langle i\alpha|\hat{\chi} : E^2\rangle &= a n_\alpha E_{i\beta} E_{i\beta} + b E_{i\alpha} n_\gamma E_{i\gamma}, \end{aligned} \tag{5.3}$$

where $a = A/v_0$, $b = B/v_0$ with v_0 being a volume of one particle ($v_0 = 4\pi R_m^3/3$ for a spherical representation of DDA). In (5.3) we included the interaction of nonlinear dipoles given through a scalar polarizability at $\omega_g = 2\omega$, apart from the mechanism responsible for generation of nonlinear dipoles (given by the impurity surface tensor of the second order - see (5.2)).

For an optical process under consideration, the enhancement of local fields occurs mainly due to the dipole-dipole interaction given by the scattered fields E_i^{sc}. An incident field induces nonlinear dipoles having the different relative phase shifts. Because we are interested in the average intensity generated from the system, we sum amplitudes of the generated signal over all the monomers belonging to the film. Also, because of (2.2), the enhancement of the local fields is equivalent to the following dipole enhancement:

---

[3] Note that in this simplified model, the orientation of $\hat{n}$ does not change from site to site.

[4] Similar analysis for a spherically symmetrical medium in the case of degenerate four-wave mixing can be found in [43].

$$G^{(K)} = \frac{|\langle D_{int}^{(K)}(\omega_g)\rangle|^2}{|\langle D_0^{(K)}(\omega_g)\rangle|^2}, \tag{5.4}$$

where $D_{int} = \sum_i d_i^{NL}(\omega_g)$ is the total non-linear dipole moment, calculated under assumption that induced nonlinear dipoles interact with each other at the generated frequency ω_g; D_0 is the total nonlinear dipole moment when the local field is completely determined by the incident wave. For SHG, $K = 2$ and $\omega_g = 2\omega$.

$P^{(2)}(2\omega)$ is extremely sensitive to the incident polarization (see section 4.), and for the rest of this paper, we assume that incident wave (2.1) is linearly polarized and given by

$$E^{(0)} = E_M^{(0)} e^{i\phi} \hat{M}, \tag{5.5}$$

By neglecting interaction term $\alpha_0 W|d^{NL}\rangle$ in CDE (5.3), we find the integral enhancements (5.4) for p- and s-polarizations, respectively:

$$\begin{aligned} G_z^{(2)} &= |Z_0|^4 |\langle d_z^2\rangle + \Gamma^{-1}\langle d_{||}^2\rangle|^2 / |E_z^{(0)}|^4 \\ G_\tau^{(2)} &= |Z_0|^4 |\Gamma\langle d_z^2\rangle + \langle d_{||}^2\rangle|^2 / |E_\tau^{(0)}|^4 \end{aligned} \tag{5.6}$$

with the abbreviation $\langle d_{||}^2\rangle = \langle d_x^2\rangle + \langle d_y^2\rangle$. $E_\tau^{(0)} \equiv E_{x,y}^{(0)}$, depending on what s-polarization is maintained ($\hat{x}$- or $\hat{y}$-linear polarization). Components of linear dipoles (2.7) depend on the incident polarization (5.5) as

$$d_{i,\alpha} = \sum_{j,n} \frac{\langle i\alpha|n\rangle\langle n|jM\rangle}{Z_0 - w_n} E_M^{(0)}. \tag{5.7}$$

Oblique coefficient Γ is defined as

$$\Gamma \equiv \frac{\chi_{zzz}}{\chi_{zss}} = 1 + b/a. \tag{5.8}$$

The absolute value of Γ is always finite, $1 < |\Gamma| < \infty$. The case when either $a \to 0$ or $b \to 0$ forbids the generation of the second harmonic as such conditions represent a centrosymmetric medium. The latter is clearly seen from (5.2).

To take into account the linear coupling of nonlinear dipoles at the frequency 2ω, we have to use (5.3). The formal solution to this equation was described in section 2. for its linear analog (2.6), and it can be easily obtained as:

$$d_{i,\alpha}^{NL} = Z_{2\omega} \sum_{j,n} \frac{\langle i\alpha|n\rangle\langle n|j\beta\rangle}{Z_0 - w_n} \langle j\beta|\hat{\chi} : E^2\rangle \tag{5.9}$$

where $Z_{2\omega} \equiv \alpha_0^{-1}(2\omega)$, and $\alpha_0(2\omega)$ is a scalar polarizability at double frequency.

Using definition (5.4), we write the corresponding enhancement for SHG process as

$$G_{SHG} = \frac{\langle D_\alpha \rangle \langle D_\alpha \rangle^*}{|\langle D_0 \rangle|^2}, \tag{5.10}$$

where $\langle D_\alpha \rangle = \sum_i d_{i,\alpha}^{NL}(2\omega)/N_s$.

In Figure 5.1 we show the values of integral enhancements for SHG from a self-affine surface. We obtained the family of curves corresponding to variation of $b/a \in [10^{-2}, 10^1]$ (see Eq.(5.8)). We note that real values of Γ indicate the absence of a non-linear absorption from the surface impurities[5]. We observed that all the curves appeared to be similar in terms of numerical values of G as well as in spectral positions of their maximums and minimums (the graphs are not shown). This feature is the consequence of the same spectral dependence of linear dipoles in (5.7), which does not effectively correlate with the value of Γ chosen to be independent of the incident wavelength.

We see that expected inequality $G_\parallel \gg G_\perp$ holds, as the components of a linear dipole (5.7) are the largest on average for $s-$polarized incident field. Figure 5.1 also demonstrates the additional enhancement of the local fields if interactions of non-linear dipoles are included.

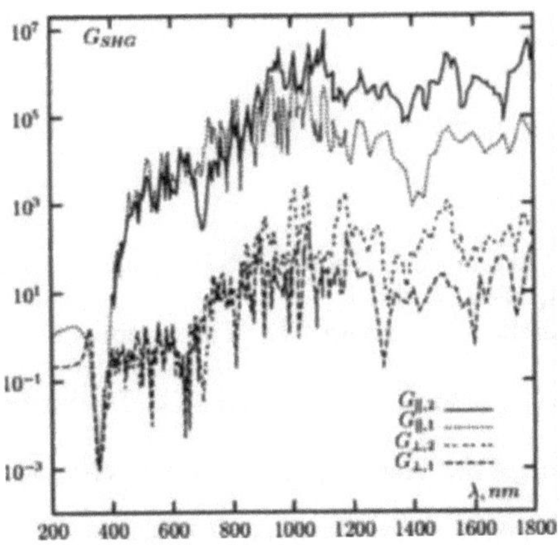

Fig. 5.1. The enhancement of SHG from self-affine silver surface. G_1 are calculated based on Eqs.(5.6), G_2 are from Eqs.(5.10). $\Gamma = 2$; $G_\parallel = \sum_\tau G_\tau/2$, $G_\perp = G_z$.

At last, we note that Eqs.(5.6) can be well estimated as $|\left\langle (E_i/E^{(0)})^2 \right\rangle|^2$.

[5] It also turned out that complex values of Γ did not significantly change the enhancement profile.

6. Concluding Remarks

To summarize, we showed that for a self-affine surface, the spatial distributions of local fields are extremely inhomogeneous and consist of "hot" and "cold" zones, with large and small local fields, respectively. Some of the dipolar eigenmodes on a self-affine film are strongly localized (whereas others involve into excitation a significant fraction of the film). On average, the modes are sublocalized. The local field distributions are very sensitive to the frequency and polarization of the applied field. Large spatial fluctuations of local fields bring about strong enhancements of optical processes on a self-affine film. The enhancement parameter for surface-enhanced Raman scattering (SERS) can be high as $\sim 10^7$ for s-polarization and $\sim 10^5$ for p-polarization (compared to that on a smooth interface). For the second harmonic generation (SHG), the values are $\sim 10^7$ and $\sim 10^3$, respectively.

Acknowledgement. Work at NMSU was supported in part by NSF under grant DMR-9500258 and by NATO under grant CRG 950097; work at Université Paris-Sud was supported by NATO under grant CRG 950097; work at Sandia National Laboratories was supported by the US Department of Energy under contract DE-AC04-94AL85000.

References

1. In: Chang, R.K., Furtak, T.E. (eds) (1982): *Surface Enhanced Raman Scattering.* Plenum Press, New York.
2. Chiarello, R., Panella, V., Krim, J., Thompson, C. (1991). *Phys Rev Lett* 67:3408.
3. Krim, J., Heyvaert, I., Van Haesendonck, C., Bruynseraede, Y. (1993). *Phys Rev Lett* 70:57.
4. Douketis, C., Wang, Z., Haslett, T.L., Moskovits, M. (1995). *Phys Rev* B 51:11022.
5. Mandelbrot, B.B. (1982): *The Fractal Geometry of Nature.* Freeman, San-Fransisco.
6. Sapoval, B. (1990): *Fractals.* Aditech, Paris.
7. Markel, V.A., Muratov, L.S., Stockman, M.I., George, T.F. (1991). *Phys Rev* B 43:8183.
8. Stockman, M.I., Pandey, L.N., George, T.F. (1996). *Phys Rev* B 53:2183.
9. Shalaev, V.M. (1996): *Physics Reports.* 272:61.
10. Markel, V.A., Shalaev, V.M., Stechel, E.B., Kim, W., Armstrong, R.L. (1996). *Phys Rev* B 53:2425.
11. Shalaev, V.M., Poliakov, E.Y., Markel, V.A. (1996). *Phys Rev* B 53:2437.
12. Shalaev, V.M., Botet, R., Mercer, J., Stechel, E.B. (1996). *Phys Rev* B 54:8235.
13. Singham, S.B., Bohren, C.F. (1988). *J Opt Soc Am* A 5:1867.
14. Marvin, A., Toigo, T., Celli, V. (1975). *Phys Rev* B 11:2777.
15. Toigo, F., Marvin, A., Celli, V., Hill, N. (1977). *Phys Rev* B 15:5618.
16. Maradudin, A.A., Mills, D.L. (1975). *Phys Rev* B 11:1392.

17. Poliakov, E.Y., Shalaev, V.M., Markel, V.A., Botet, R. (1996): *Opt Lett.* 21:1628.
18. Beckmann, P., Spizzichino, A. (1987): *The Scattering of Electromagnetic Waves from Rough Surfaces.* Artech, Norwood, Mass.
19. Ishimaru, A. In: Nieto-Vesperinas, M., Dainity, J.C. (eds) (1990): *Scattering in Volumes Surfaces.* Elsever Science Publishers, New York.
20. Fitzgerland, R.M., Maradudin, A.A. (1994): *Waves in Random Media.* 4:275.
21. Purcell, E.M., Pennypacker, C.R. (1973). *Astrophys J* 186:705.
22. Draine, B.T. (1988). *Astrophys J* 333:848.
23. Jackson, J.J. (1975): *Classical Electrodynamics.* Jonh Willey & Sons, Inc, New York.
24. Kardar, M., Parisi, G., Zhang, Y.C. (1986). *Phys Rev Lett* 56:889.
25. Meakin, P., Ramanlal, P., Sander, L.M., Ball, R.C. (1986). *Phys Rev* A 34:5091.
26. Kim, J.M., Kosterlitz, J.M. (1989). *Phys Rev Lett* 62:2289.
27. Shalaev, V.M., Botet, R., Jullien, R. (1991). *Phys Rev* B 44:115.
28. Johnson, P.B., Christy, W. (1972). *Phys Rev* B 6:4370.
29. Shalaev, V.M., Botet, R., Butenko, A.V. (1993). *Phys Rev* B 48:6662.
30. Tsai, D.P. et al. (1994). *Phys Rev Lett* 72:4149.
31. Shalaev, V.M., Moskovits, M. (1995). *Phys Rev Lett* 75:2451.
32. Betzig, E. et al. (1992): *Science* 257:189.
33. Zhang, P., Haslett, T., Douketis, C., Moscovits, M. (1997). Submitted to *Phys Rev* B.
34. Bloembergen, N., Pershan, P.S .(1962). *Phys Rev* 128:606.
35. Shen, Y.R. (1984): *The Principles of Nonlinear Optics.* Wiley, New York.
36. Richmond, G.L., Robinson, J.M., Shanon, V.L. (1988): *Prog Surf Sci* 28:1.
37. Janz, S., van Driel, H.M. (1993): *Int J Nonlinear Opt Phys.* 2:1.
38. Sipe, J.E., So, V.C.Y., Fukui, M., Stegeman, G.I. (1980). *Phys Rev* B 21:4389.
39. Guyot-Sionnest, P., Chen, W., Shen, Y.R. (1986). *Phys Rev* B 33:8254.
40. Pershan, P.S. (1963). *Phys Rev Lett* 130:919.
41. Guyot-Sionnest, P., Shen, Y.R. (1988). *Phys Rev* B 38:7985.
42. Rudnick, J., Stern, E.A. (1971). *Phys Rev* B 4:4274.
43. Boyd, R.W. (1992): *Nonlinear Optics.* Academic Press, Inc, New York.

Fractional Fourier analysis of objects with scaling symmetry

T. Alieva[1,2] and A.M. Barbé[1]

[1] Katholieke Universiteit Leuven, Department of Electrical Engineering ESAT, Kardinaal Mercierlaan 94, B-3001 Leuven, BELGIUM

[2] A. L. Mintz Radiotechnical Institute, Moscow, RUSSIA

Abstract The fractional Fourier analysis is used for investigation of fractal structures. As is shown, it reveals the scaling features of deterministic as well as of the random fractals, and allows to determine their main characteristics. As an example we present the fractional Fourier analysis of some self-affine functions. The map of the Radon-Wigner transform of the triadic Cantor set is considered numerically. The application of the fractional Fourier analysis in optical engineering is discussed.

Keywords: fractional Fourier transform, Radon-Wigner transform, Cantor set.

1. Introduction

In recent years, different physical phenomena have been analysed within a fractal framework [1]. There are various methods of investigation of fractal objects. One of them is Fourier analysis [2, 3, 4, 5]. The wavelet concept has also been widely used in the study of the scaling properties of fractal structures [6, 7]. The purpose of this article is to consider the possibility of using fractional Fourier analysis, which successfully unites the advantages of wavelet and Fourier methods, for investigation of structures with scaling properties.

The fractional Fourier transform (fractional FT) R^α [8] is an extension of the ordinary Fourier transform and depends on a parameter α that can be interpreted as a sort of rotation by the angle α in the position-frequency plane. Here under position and frequency we will understand any Fourier conjugated pair of variables. The fractional FT is linear and additive in this parameter: $R^\alpha R^\beta = R^{\alpha+\beta}$. For $\alpha = \pi/2$ it reduces to the ordinary Fourier transform (FT). Thus changing α from 0 to $\pi/2$, we have a continuous transformation of a function to its Fourier image. Due to the uncertainty principle it is not possible to obtain arbitrary fine localisation in both position and frequency. Then the localisation in position is loosening with increasing of α, while the localisation in frequency becomes better. From this point of view the fractional FT is similar to the wavelet transform.

The kernel of the fractional FT is, except for a phase factor, equal to the propagator of the non-stationary Schrödinger equation for the harmonic oscillator [9]. This transform is also used in optics. The fractional Fourier transform property of quadratic refractive index media (lenses, optical fibers, spherical mirrors) has been widely discussed by the optical community (see, for example [9, 10, 11]). Furthermore, the fractional FT arises in signal processing through chirp modulation [12] and describes the pulse distortion in dispersive media [9]. So the fractional FT is realised in many physical systems and these can be used for the experimental analysis of different objects.

In this paper the features of the fractional FT of deterministic and random objects with scaling symmetry are investigated. We pay more attention to the analysis of the squared moduli of the fractional FT (often named Radon-Wigner transform (RWT) [13]), because they relate to the easily measured physical parameters such as probability in quantum mechanics and intensity distribution in optics and signal processing. The RWT for $\alpha = \pi/2$ corresponds to the ordinary Fourier spectrum, which is usually used for the analysis of fractals [2, 3, 4, 5]. Thus, in the context of this paper, fractional Fourier analysis of some object (signal, function ...) refers to the map of its Radon-Wigner transformations for different angles α ranging from 0 to $\pi/2$. It will be shown that the scaling properties of a fractal object are reflected on the map of RWTs. The RWTs of a self-affine fractal at the different angles are similar to each other. The analysis of the RWT map gives information about the hierarchical structure of a fractal and allows to determine its scaling exponent and to localise the point around which the scaling behaviour is observed. For simplicity, we will consider only the one-dimensional case.

The outline of this paper is as follows: Section 2. gives a brief overview of the fractional FT and RWT. Section 3. discusses the features of the fractional FT and RWT of some deterministic fractals. In Section 4. the method of the fractional FT analysis is generalized to the case of random fractals. Section 5. briefly considers the application of the fractional FT analysis in optical engineering.

2. Definitions of fractional Fourier transform and Radon-Wigner transform

The one-dimensional fractional FT with parameter α of $f(x)$, denoted by $R^\alpha f(x)$, performs a linear operation, given by the integral transform

$$[R^\alpha f(x)]\,(u) = F_\alpha(u) = \int_{-\infty}^{\infty} K_\alpha(x,u) f(x)dx \tag{2.1}$$

where the kernel $K_\alpha(x,u)$

$$K_\alpha(x,u) = (2\pi i \sin\alpha)^{-1/2} \exp(i\frac{\alpha}{2}) \exp\left(\frac{i}{2\sin\alpha}\left((x^2+u^2)\cos\alpha - 2xu\right)\right) \tag{2.2}$$

is the propagator of the non-stationary Schrödinger equation for a harmonic oscillator which is well known in quantum mechanics (where $\alpha = \omega t$ relates to time t and classical frequency ω , and u is a position at the moment t)[14]. This transformation is associated with rotation mapping in phase space

$$\begin{pmatrix} u \\ k_u \end{pmatrix} = \begin{pmatrix} \cos\alpha & \sin\alpha \\ -\sin\alpha & \cos\alpha \end{pmatrix} \begin{pmatrix} x \\ k_x \end{pmatrix} \tag{2.3}$$

Here x, u and k_x, k_u are the position and frequency variables. Changing gradually the angle α the fractional FT permits to look at the continuous transformation of an input function $f(x)$ to its Fourier image $F_{\pi/2}(u)$ for $\alpha = \pi/2$, then to $f(-x)$ for $\alpha = \pi$ and to $F_{\pi/2}(-u)$ for $\alpha = 3\pi/2$. The fractional FT at angles equal to $2\pi n$ (n is an integer) corresponds to the identity operator. The main properties of fractional FT have been considered in [9, 12]. One of them is that the fractional FT produces the rotation of the Wigner function in the phase space. The Wigner function [15] belongs to a class of bilinear distributions and gives a representation of a function $f(x)$ in the phase space. It is a generalized convolution at non-zero frequency

$$W_f(x, k_x) = \int_{-\infty}^{\infty} f\left(x + \frac{x_0}{2}\right) f^*\left(x - \frac{x_0}{2}\right) \exp(ix_0k_x)dx_0 \tag{2.4}$$

where f^* means the complex conjugate of f. The inverse transform can be written as

$$f(x) = \frac{1}{2\pi f^*(0)} \int_{-\infty}^{\infty} W_f(x/2, k_x) \exp(ixk_x)dk_x \tag{2.5}$$

It has been shown (see, for example, [12]) that the Wigner function of the fractional FT for the parameter α of $f(x)$ is the Wigner function of $f(x)$ rotated at the angle α in the phase space

$$W_{F_\alpha}(u, k_u) = W_f(u\cos\alpha - k_u\sin\alpha, u\sin\alpha + k_u\cos\alpha) \tag{2.6}$$

Moreover, the projection of W_f upon the direction rotated at angle α is the squared modulus of the corresponding fractional FT, named the Radon-Wigner transform (RWT) [13]

$$|F_\alpha(u)|^2 = \int_{-\infty}^{\infty} W_f(u\cos\alpha - k_u\sin\alpha, u\sin\alpha + k_u\cos\alpha)dk_u \tag{2.7}$$

In particular, $|F_0(u)|^2 = |f(u)|^2$ and $\left|F_{\pi/2}(u)\right|^2$ is the power spectrum of f. The inverse Radon transform allows to reconstruct the Wigner function W_f (and then f) from the set of its projections upon different angles from 0 to π (phase-space tomography)[16]. Note that the RWT relates with the easily measured physical parameters like probability in quantum mechanics and intensity distribution in optics and signal processing. Besides that, the concept of the Wigner function has been generalized to the case of random functions $f(x)$ [17]

$$W_f(x,k_x) = \int_{-\infty}^{\infty} G_f(x,x_0)\exp(ix_0k_x)dx_0 \quad (2.8)$$

where $G_f(x,x_0) = \langle f\left(x+\frac{x_0}{2}\right) f^*\left(x-\frac{x_0}{2}\right)\rangle$ is a two-point correlation function of f . This allows to apply the phase-space tomography for the reconstruction of the correlation functions from the set of $\left\langle |F_\alpha(u)|^2 \right\rangle$ where $\alpha \in [0,\pi]$.

3. Fractional Fourier analysis of deterministic fractals

The typical property of fractal objects is that they keep their own shapes under a change of scale [18]. We consider fractals described by a self-affine (or homogeneous) function $f(x)$, so that the following relation holds around the point x_0

$$f(\lambda x + x_0) = \lambda^H f(x+x_0) \quad (3.1)$$

For some deterministic functions this relation can be satisfied exactly. For random functions it should be valid in distribution. The structure of fractals is usually described by some iterative procedure and relation (3.1) holds not for all λ but for some sequence $\lambda = \lambda_0^k$, where $k = 1, ..., N$. In a mathematically strict sense the number of iterations is $N \to \infty$.

In physical situations, no exact fractals exist, because of finite N and because of the finite size of the structures. We will call such structures real-world fractals of level N. Then for a fractal function of level N in some finite region one has approximately

$$f_N(\lambda_0^k x + x_0) \approx \lambda_0^{kH} f_{N-k}(x+x_0) \quad (3.2)$$

where $\lambda_0 > 1$ (see, for example the different steps leading to the Devil's staircase). In this case the parameter λ_0 and the generator $f_0(x)$ become important in the characterisation of a fractal. Below we will stress the difference in structure of the RWT maps of real-world and exact fractals.

Let us consider the features of the fractional FT of scale-symmetric or fractal objects. We suppose that for the function $f(x)$ relation (3.1) holds or/and its measure $\mu(x)$ $(d\mu(x) = f(x)dx)$ scales around the point x_0 with scaling exponent $H+1$ [19]. One can write, using the shift and scaling theorems for the fractional FT [9], the following relation

$$\int_{-\infty}^{\infty} K_\alpha(x,u) f(\lambda x + x_0)dx = \lambda^{-1}\int K_\alpha(x,u)d\mu(\lambda x + x_0) =$$

$$\sqrt{\frac{\cos\beta}{\cos\alpha}}\exp(i\varphi)\int K_\beta\left(\xi, \sqrt{\left|\frac{\sin 2\beta}{\sin 2\alpha}\right|}u + x_0\cos\beta\right)d\mu(\xi)$$

where $\varphi = (\alpha - \beta)/2 - u^2 \tan(\alpha)/2 + (u \sin\beta/\sqrt{|\sin 2\alpha|} + x_0\sqrt{|\sin 2\beta|}/2)^2$. Here, $\lambda > 0$ and the angles α and β are connected by

$$\tan\beta = \lambda^2 \tan\alpha \tag{3.3}$$

The linearity of the transform and the scaling behaviour of the fractal function (or/and its measure) lead to an important relation for the RWT of $f(x)$ at angles α and β

$$|F_\alpha(u + x_0 \cos\alpha)|^2 = \left(\frac{\cot\beta}{\cot\alpha}\right)^H \frac{\cos\beta}{\cos\alpha} \left| F_\beta\left(\sqrt{\left|\frac{\sin 2\beta}{\sin 2\alpha}\right|} u + x_0 \cos\beta\right)\right|^2 \tag{3.4}$$

Then, if relation (3.1) holds for any λ, we can deduce that the RWTs of a homogeneous function at different angles not equal to $\pi n/2$, are affine to each other. Thus the RWT at the angle $\alpha = \arctan\lambda^{-2}$ is represented through the RWT at the angle $\pi/4$

$$|F_\alpha(u)|^2 = \frac{(\tan\alpha)^H}{\sqrt{2}\cos\alpha} \left| F_{\pi/4}\left(u\,|\sin 2\alpha|^{-1/2} + x_0 \frac{1-\sqrt{\cot\alpha}}{\sqrt{2}}\right)\right|^2 \tag{3.5}$$

Let us consider the RWTs at angles α and $\pi/2 - \alpha$, which are symmetric with respect to $\pi/4$. Using the results in (3.3) and (3.4) for the RWT at the angles $\alpha = \arctan\lambda^{-2}$ and $\pi/2 - \alpha$ we can write

$$\left|F_{\pi/2-\alpha}(u + x_0 \cos\alpha)\right|^2 = (\cot\alpha)^{2H+1} |F_\alpha(u + x_0 \sin\alpha)|^2 \tag{3.6}$$

Thus both profiles can be superposed by suitable dilation and translation of one at $x_0(\cos\alpha - \sin\alpha)$. For $x_0 = 0$, the RWTs at angles symmetric with respect to $\pi/4$, are the same except for a normalisation factor. Then comparing the similar profiles of the RWT at angles α and $\pi/2 - \alpha$ one can find the point x_0 where the scaling symmetry is observed and determine the scaling exponent from the ratio $\xi(\alpha) = \left|F_{\pi/2-\alpha}(u + x_0 \cos\alpha)\right|^2 / |F_\alpha(u + x_0 \sin\alpha)|^2$ as $H = (\log_{|\cot\alpha|} \xi(\alpha) - 1)/2$. The local box dimension of the self-affine fractal is defined as $D = 2 - H$ [18].

Let us demonstrate relations (3.4) and (3.6) on some homogeneous functions with scaling symmetry in the point $x_0 = 0$.

Example 1. The simplest homogeneous function is a constant everywhere $f(x) = 1$ ($H = 0$). Its fractional FT is
$F_\alpha(u) = \exp(i\alpha/2)(\cos\alpha)^{-1/2}\exp(-iu^2(\tan\alpha)/2)$ and$|F_\alpha(u)|^2 = |\cos\alpha|^{-1}$. Therefore $|F_\alpha(u)|^2 / |F_\beta(u)|^2 = |\cos\beta/\cos\alpha|$. It means that the RWT of this function is always a constant except at points $\alpha = \pi/2 + \pi n$, where it becomes infinite in the point $u = 0$.

Example 2. We now investigate the algebraic functions $f(x) = |x|^H$ ($-1 < H < 0$), which are often used in the description of critical phenomena [18] .

These functions are self-similar for any scale λ. The fractional FT at angle α is

$$\left[R^{\alpha}|x|^{H}\right](u)= \frac{H\Gamma(H)}{\sqrt{2\pi\sin\alpha}}\exp\left[i\left(\frac{\pi H}{4}+\frac{u^2\cot 2\alpha}{2}\right)\right](\tan\alpha)^{(H+1)/2}\times \left\{D_{-(H+1)}\left(\frac{\sqrt{2}u}{\sqrt{\sin 2\alpha}}e^{-i\pi/4}\right)-D_{-(H+1)}\left(\frac{\sqrt{2}u}{\sqrt{\sin 2\alpha}}e^{i\pi/4}\right)\right\} \tag{3.7}$$

where $D_\nu(z)$ is a parabolic cylinder function [20]. Hence, the relations (3.4), (3.5) and (3.6) hold for any $\alpha \neq \pi n/2$ and

$$\frac{\left[R^{\pi/2-\alpha}|x|^{H}\right](u)}{\left[R^{\alpha}|x|^{H}\right](u)}=(\cot\alpha)^{H+1/2}\exp\left(-iu^2\cot 2\alpha\right) \tag{3.8}$$

The FT of these functions are real

$$\left[R^{\pi/2}|x|^{H}\right](u)=-\frac{H\sqrt{2}\Gamma(H)\sin(H\pi/2)}{\sqrt{\pi}\,|u|^{1+H}} \tag{3.9}$$

From (3.6) follows that for homogeneous functions with scaling exponent $H=-1/2$ the RWTs at the angles α and $\pi/2-\alpha$ are equal. Thus, for the function $|x|^{-1/2}$ which is invariant under Fourier transform: $\left[R^{\pi/2}|x|^{-1/2}\right](u)=|u|^{-1/2}$, the fractional FT at angle α equals

$$\left[R^{\alpha}|x|^{-1/2}\right](u)=\frac{|u|^{-1/2}}{\sqrt{2\pi\sin 2\alpha}}\exp\left(\frac{i}{2}u^2\cot 2\alpha\right)K_{1/4}\left(\frac{-iu^2}{2\sin 2\alpha}\right) \tag{3.10}$$

where $K_\nu(z)$ is a modified Bessel function [20]. It means that the fractional FT of $|x|^{-1/2}$ at any angle $\alpha \neq \pi n/2$ differs from that at angle $\pi/2-\alpha$ only by $\exp\left(iu^2\cot 2\alpha\right)$. Hence its RWT possesses the following symmetry for any α : $|F_\alpha(u)|^2=\left|F_{\pi/2-\alpha}(u)\right|^2$. Note that this relation holds for any homogeneous functions with the scaling exponent $H=-1/2$.

In case a fractal is scale invariant for some sequence of scaling factors λ_0^k relation (3.4) holds only for angles α and β satisfying (3.3), where $\lambda=\lambda_0^k$. In particular relation (3.6) is true for the angles $\alpha_k=\arctan\lambda_0^{-2k}$. This has been already demonstrated for the Mandelbrot-Weierstrass cosine fractal [21]. The scaling factor λ_0 of the investigated fractal can be determined from the ratio $\tan\alpha_k/\tan\alpha_{k+1}$, where α_k and α_{k+1} are the nearest angles for which (3.6) holds.

Using (3.2) one can write for fractals of level N

$$\left|F_\beta^{(N)}\left(\sqrt{\left|\frac{\sin 2\beta}{\sin 2\alpha}\right|}u+x_0\cos\beta\right)\right|^2\approx\left(\frac{\cot\alpha}{\cot\beta}\right)^{H}\frac{\cos\alpha}{\cos\beta}\left|F_\alpha^{(N-k)}(u+x_0\cos\alpha)\right|^2 \tag{3.11}$$

where we suppose that $\lambda_0>1$, $k=1,\dots,N$ and

$$\tan\beta = \lambda_0^{2k} \tan\alpha \tag{3.12}$$

Note that while relation (3.4) for fractals with infinite support holds everywhere, except for angles $\pi n/2$, equation (3.11) for finite structures is true only in some region of angles $\beta \in [\delta, \pi/2 - \delta]$, where δ depends on the size of a real-world fractal (for similar reasons as this is the case for Fresnel integrals used in optics [22]). Then, at the RWT map of the fractal of level N we can observe that, at angles β_k, such that $\tan\beta_k = \lambda_0^{2k}\tan\delta$, the structure associated with the fractal of level $N-k+1$ transforms to the another one corresponding to the fractal of level $N-k$ ($k = 1, ..., N$). Thus, at the angle β_N the structure closely related to the generator of the investigated fractal f_0 is observed. It means that like a wavelet analysis the fractional Fourier one reveals the hierarchical structure of real-world fractal objects.

Example 3

Let us now consider the map of the RWTs of the triadic Cantor set of level 5, represented by its characteristic function $\xi(x)$ as shown in the lower part of Fig. 3.1 (for a Cantor set $\xi(x)$ coincides with RWT ($\alpha = 0$)). The function $\xi(x)$ is not self-affine but its measure scales around point $x_0 = -0.5$ with scaling factor $\lambda_0 = 3$.

In Figure 3.1 the Cantor set (lower pattern) , its RWT at the angle $\pi/4$ (middle pattern) and its Fourier spectrum (upper pattern), as obtained numerically, are shown. It can be seen that the RWT at the angle $\pi/4$ contains a mixture of features of the direct and Fourier images.

Figure 3.2 shows the continuous evolution of $|\xi(x)|^2$ to the Fourier spectrum of the Cantor set (RWT map). The levels of intensity are marked by different scales of grey. One can see the hierarchical structure (or iterative procedure of the construction) of the Cantor set: the RWT patterns at the angles satisfying (3.12) are similar to each other, in agreement with (3.11). The angles of the branching of RWT, in agreement with the above analysis, are given by $\alpha_i \approx \arctan(\delta 3^{-2i})$ (marked with arrows). It means that the scaling factor of the fractal is $\lambda_0 \approx 3$.

The similarity dimension of the triadic Cantor set can be directly obtained from the branching ratio $r = 2$ (the number of copies of fractal of level $k-1$ presenting in the fractal of level k) and the scale factor λ_0 according with the formula [18] $D = \ln r / \ln \lambda_0 \approx 0.63$.

In Figure 3.3 the RWT at the angles $\alpha = \arctan 3^{-1}$ and $\pi/2 - \alpha$ are shown. It can be appreciated that, indeed, both profiles can be superposed in some region of u (where boundary effects are not important) by some translation and dilation of one, in agreement with (3.11). The point of the scaling symmetry x_0 can be determined from the position shift $x_0(\cos\alpha - \sin\alpha)$ and it is approximately equal to -0.5. The ratio of average RWT in the similarity region gives approximately the value of scaling exponent equal to D in agreement with [23].

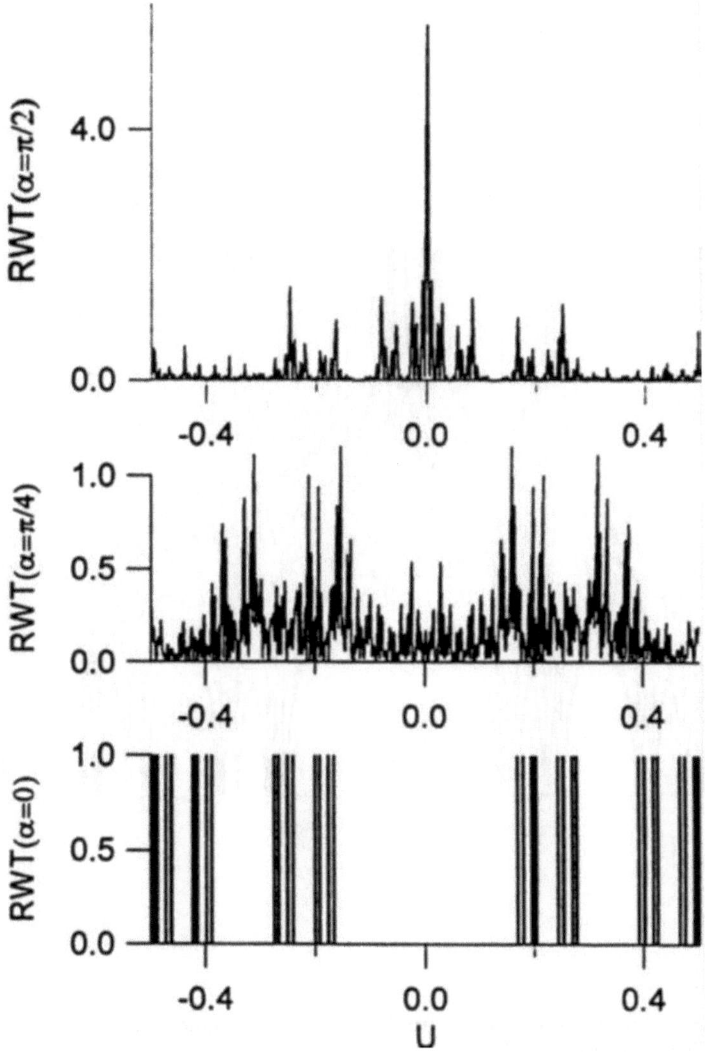

Fig. 3.1. Profiles of RWT of a Cantor set at angles $\alpha = 0$ (lower pattern), $\alpha = \pi/4$ (middle pattern) and $\alpha = \pi/2$ (upper pattern) versus a normalized coordinate U.

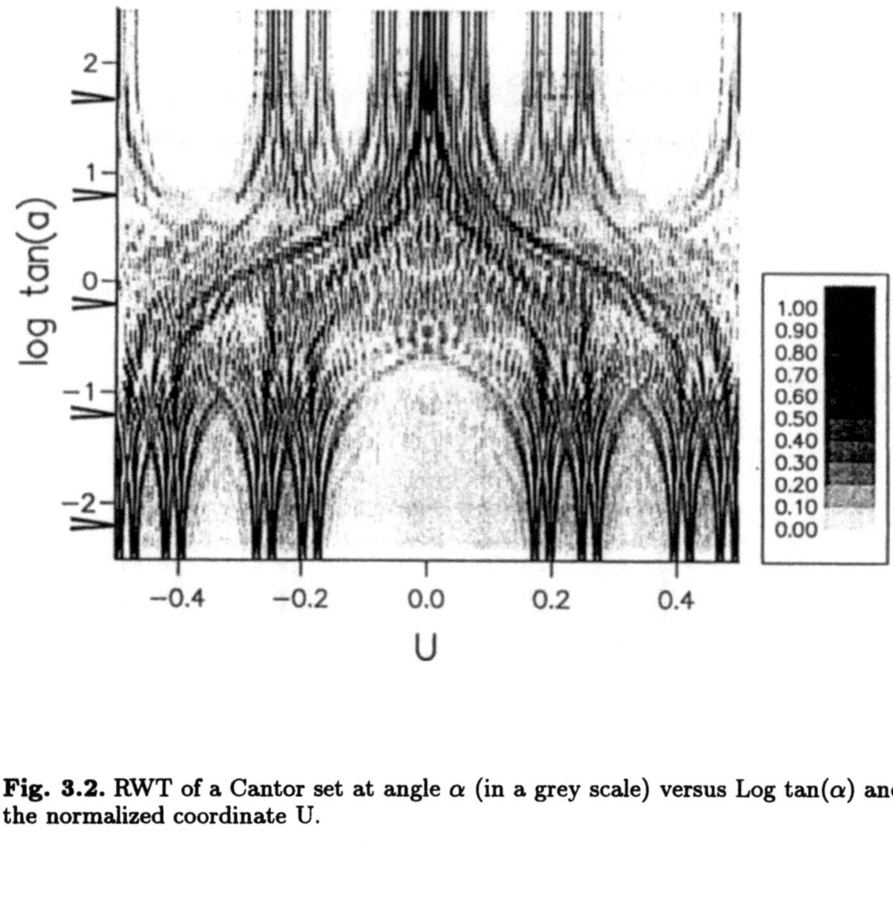

Fig. 3.2. RWT of a Cantor set at angle α (in a grey scale) versus Log tan(α) and the normalized coordinate U.

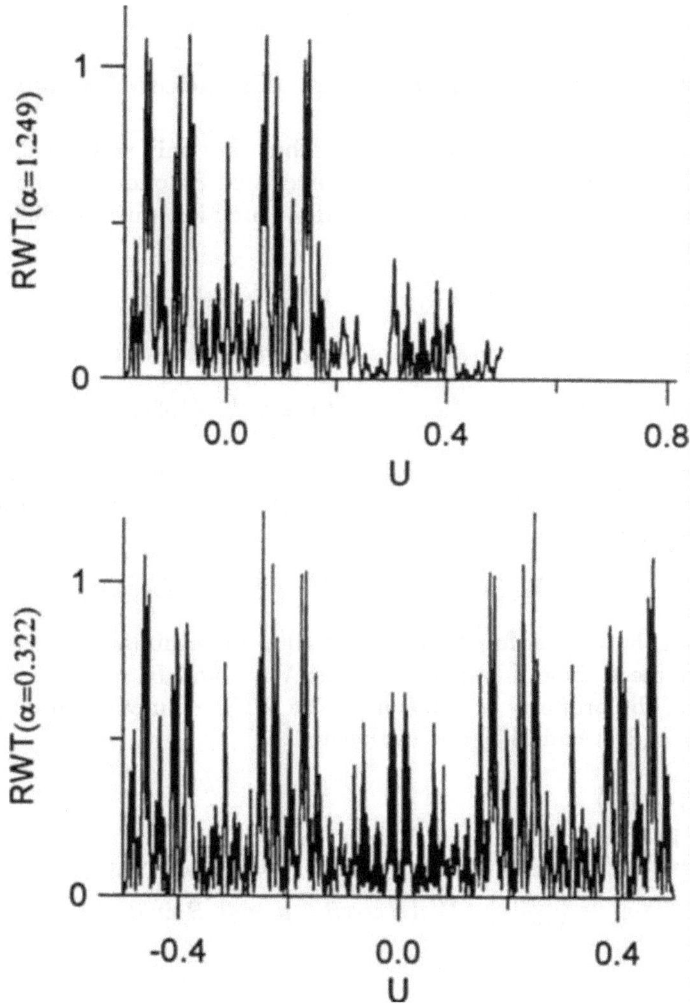

Fig. 3.3. Profiles of RWT of a Cantor set versus a normalized coordinate U for angles α =arctan 3^{-1} and $\pi/2 - \alpha$.

Then the fractional Fourier analysis of a fractal allows us to look at its hierarchical structure, to determine the scaling factor λ_0, the scaling exponent H and the point x_0 around which the scaling symmetry is observed.

4. Fractional Fourier analysis of random fractals

Let us consider now random fractals, which are only statistically self-affine. We will assume that the two-point correlation function of a non-stationary (in time or in space) process is a self-affine function with the different scaling of coordinates, so that

$$G_0(\lambda X, \sigma x) = \lambda^{H_\lambda}\sigma^{H_\sigma}G_0(X, x) \tag{4.1}$$

In particular, for the average intensity distribution we have that

$$\left\langle |F_0(X)|^2 \right\rangle = G_0(X, 0) = \lambda^{-H_\lambda}\left\langle |F_0(\lambda X)|^2 \right\rangle \tag{4.2}$$

In the case when $G_0(X, x)$ is self-affine only for one of the coordinates, we can put λ or σ equal to 1 in (4.1). As it follows from (2.8) and (4.1) the Wigner function is also self-affine

$$W_0(\lambda X, \sigma^{-1}\omega) = \lambda^{H_\lambda}\sigma^{H_\sigma+1}W_0(X, \omega) \tag{4.3}$$

Note that the Wigner function of a self-affine deterministic object is self-affine with inverse scaling of two coordinates: $\lambda = \sigma$ and $H_\sigma = H_\lambda$.

Using the property of rotation of the Wigner function through the fractional FT (2.7) and (4.3) one can write

$$\begin{aligned}
&\left\langle |F_\alpha(X)|^2 \right\rangle \\
&= \int W_0(X\cos\alpha - \omega\sin\alpha, X\sin\alpha + \omega\cos\alpha)d\omega \\
&= \lambda^{-H_\lambda}\sigma^{-H_\sigma-1} \times \int W_0(\lambda X\cos\alpha - \lambda\omega\sin\alpha, \sigma^{-1}X\sin\alpha + \sigma^{-1}\omega\cos\alpha)d\omega \\
&= \lambda^{-H_\lambda}\sigma^{-H_\sigma}\left(\frac{\cos\beta}{\cos\alpha}\right)\int W_0(R\cos\beta - p\sin\beta, R\sin\beta + p\cos\beta)dp
\end{aligned} \tag{4.4}$$

where

$$\tan\beta = \lambda\sigma\tan\alpha \tag{4.5}$$

and

$$R = X\sigma^{-1}\frac{\sin\beta}{\sin\alpha} \tag{4.6}$$

Then we have that the Radon-Wigner transforms of a random fractal with the correlation function (4.1) at the angles α and β connected by (4.5) are related by

$$\left\langle |F_\alpha(X)|^2 \right\rangle = \lambda^{-H_\lambda}\sigma^{-H_\sigma}\left(\frac{\cos\beta}{\cos\alpha}\right)\left\langle \left|F_\beta\left(X\sigma^{-1}\frac{\sin\beta}{\sin\alpha}\right)\right|^2 \right\rangle \tag{4.7}$$

In particular, for $\lambda = \sigma$ we have the same result as for deterministic fractals (3.4). But otherwise the set of Radon-Wigner transformations in the angle region from 0 to $\pi/2$ for random fractal will differ from the deterministic one. If relation (4.1) holds for any λ and σ then we can always choose $\sigma = \sin\beta/\sin\alpha$ and $\lambda = \cos\alpha/\cos\beta$ that yields

$$\left\langle |F_\alpha(X)|^2 \right\rangle = \left(\frac{\cos\alpha}{\cos\beta}\right)^{-H_\lambda - 1} \left(\frac{\sin\beta}{\sin\alpha}\right)^{-H_\sigma} \left\langle |F_\beta(X)|^2 \right\rangle \tag{4.8}$$

It means that comparing the intensity distributions of the Radon-Wigner transformations for the different angles we can find the scaling exponents of the correlation function.

If the correlation function is self-affine only for some sequence $\sigma = \sigma_0^k$, then relation (4.8) holds for some set of angles α and β such that $\sigma_0^k = \sin\beta/\sin\alpha$. In particular, for stationary processes ($G_0(X, x) = G_0(x)$ is independent of X , $H_\lambda = 0$) we have

$$\left\langle |F_\alpha(X)|^2 \right\rangle = \left(\frac{\cos\alpha}{\cos\beta}\right)^{-1} \left(\frac{\sin\beta}{\sin\alpha}\right)^{-H_\sigma} \left\langle |F_\beta(X)|^2 \right\rangle \tag{4.9}$$

and for $\sigma_0^k = \cot\alpha$

$$\left\langle |F_\alpha(X)|^2 \right\rangle = (\cot\alpha)^{-H_\sigma - 1} \left\langle |F_{\pi/2-\alpha}(X)|^2 \right\rangle \tag{4.10}$$

If the scaling relation for random fractals holds only for some sequences of both parameters $\lambda = \lambda_0^k$ and $\sigma = \sigma_0^k$, $k = 0, 1, ...$ the symmetry in the position scaling of the Radon-Wigner transformations for the angles α and $\pi/2 - \alpha$ observed for deterministic fractals (see (3.6)) is broken. Thus, for $\cot^2\alpha = \lambda_0^k \sigma_0^m$ one can write (4.7) as

$$\left\langle |F_\alpha(X)|^2 \right\rangle = \lambda_0^{-k(H_\lambda + 1/2)} \sigma_0^{-m(H_\sigma + 1/2)} \left\langle \left| F_{\pi/2-\alpha}\left(\sqrt{\lambda_0^k \sigma_0^{-m}}\, X\right) \right|^2 \right\rangle \tag{4.11}$$

In particular, for a non-stationary process whose average intensity distribution is not self-affine ($\lambda = 1$) for $\sigma_0^m = \cot^2\alpha$, we have

$$\left\langle |F_\alpha(X)|^2 \right\rangle = (\tan\alpha)^{2H_\sigma + 1} \left\langle \left| F_{\pi/2-\alpha}(X\tan\alpha) \right|^2 \right\rangle \tag{4.12}$$

Note that, as is the case for deterministic fractals, the scaling of intensity distributions in (4.11) is absent for the scaling exponents $H_\lambda = H_\sigma = -1/2$.

Like for the deterministic case, the map of RWT reveals the hierarchical structure of real-world random fractals.

5. Application to optical engineering

Below we briefly discuss how the fractional FT analysis of fractal objects can be experimentally realized in optical engineering. Note that the above results formulated for one-dimensional fractional FT are successfully generalized to the n-dimensional case [19, 21].

We suppose that the complex field amplitude of the planar monochromatic wave $\Psi(x_0, y_0, 0)$ at $z = 0$ just after passing through a fractal screen (which can be a random one) has the same scale-invariant properties as the corresponding fractal structure [2, 3, 4].

The two-dimensional fractional FT describes in scalar paraxial approximation of the diffraction theory the evolution of the complex field amplitude $\Psi(x, y, z)$ along the optical axis z during wave propagation through a medium with a quadratic refractive index profile (optical fiber, lenses) : $\Psi(x, y, z) = \left[R^{\alpha_x(z)\alpha_y(z)}\Psi(x_0, y_0, 0)\right](x, y) = F_{\alpha_x\alpha_y}(x, y)$. Thus the intensity distribution at plane $z = const$ is the RWT of the input complex field amplitude $\Psi(x_0, y_0, 0)$ at angles $\alpha_x(z)$ and $\alpha_y(z)$. Varying the lengths of optical fibers or alternatively playing with the distances from input and output observation planes to the lens or with its focal length one can scan the rotation angle from 0 to $\pi/2$. The fractional FT at the different angles α_x and α_y for two orthogonal coordinates is realized by the using cylindrical lenses or relating optical fibers. A detailed analysis of the optical implementation of the fractional FT can be found, for example, in [9, 10, 11, 13, 16]. Measuring the average intensity distributions at the different planes corresponding to the rotation angles from 0 to $\pi/2$ one can get the RWTs map and apply the results described in the previous sections.

In summary, we can conclude that the fractional Fourier analysis can be successfully realized in optics, and can be used for the experimental investigation of the scaling properties of deterministic as well as random fractal objects.

Acknowledgement. This work has been partially supported from a Concerted Action Project of the Flemish Community entitled *Model-based Information Systems* GOA-MIPS.

References

1. Mandelbrot, B.B. (1982). *The Fractal Geometry of Nature.* Freeman, New York.
2. Allain, C., Cloitre, M. (1986). *Phys Rev E* 33, 3566–3569.
3. Lakhtakia, A., Holter, N.S., Varadan, V.K., Varadan, V.V. (1987). IEEE Trans Ant Prop 35, 236–239.
4. Uozumi, J., Sakurada, Y., Asakura, T. (1995). *J Mod Optics* 42, 2309–2322.

5. Giona, M., Patiero, O. (1997): Integral transforms of multifractal measures. In: this volume.
6. Argoul, F., Arneodo, A., Elezgaray, J., Grasseau, G., Murenzi, R. (1990). *Phys Rev A* 41, 5537–5560.
7. Muzy, J.F., Bacry, E., Arneodo, A. (1991). *Phys Rev Lett* 67, 3515–3518.
8. Namias, V. (1980). *J. Inst. Math. Appl.* 25, 241–265.
9. Alieva, T., Lopez, V., Agullo-Lopez, F., Almeida, L.B. (1994). *J Mod Optics* 41, 1037–1044.
10. Mendlovic, D., Ozatkas, H.M. (1993). *J Opt Soc Am A* 10, 1875–1881.
11. Lohmann, A.W. (1993). *J Opt Soc Am A* 10, 2181–2186.
12. Almeida, L.B. (1994). *IEEE Trans Signal Process* 42, 3084–3091.
13. Lohmann, A.W., Soffer, B.H. (1994). *J Opt Soc Am A* 11, 1798–1801.
14. Feynman, R.P., Hibbs, R.P. (1965): *Quantum Mechanics and Path Integral.* McGraw-Hill, New York.
15. Wigner, E. (1932). *Phys Rev* 40, 749–759.
16. Raymer, M.G., Beck, M., McAlister, D.F. (1994). *Phys Rev Lett* 72, 1137–1140.
17. Bastiaans, M.J. (1978). *Optics Comm* 25, 26–30.
18. Feder, J. (1988). *Fractals.* Plenum Press, New York.
19. Alieva, T., Agullo-Lopez, F. (1996). *Optics Comm* 125, 267–274.
20. Gradshteyn, I.S., Ryzhik, I.M. (1994): *Table of Integrals, Series, and Products.* Academic Press, New York.
21. Alieva, T. (1996). *J Opt Soc Am A* 13, 1189–1192.
22. Born, M., Wolf, E. (1965): *Principles of Optics.* Pergamon Press, Oxford.
23. Halsey, T.C., Jensen, M.H., Kadanoff, L.P., Procaccia, I., Shraiman, B.I. (1986). *Phys Rev A* 33, 1141–1151.

Potential Scattering on Fractals in One Dimension

C.-A. Guerin and M. Holschneider

Centre de Physique Théorique, Unité Propre de Recherche 7061
CNRS - Luminy, Case 907, F-13288 Marseille Cedex 9, FRANCE
{guerin,hols}@cpt.univ-mrs.fr

1. Introduction

Scattering is known to be a powerful method of analysing fractal systems. A classical heuristic result in small-angle X-ray or neutron scattering relates the scattered intensity $I(q)$ in the high-frequency regime to some power-law of the momentum transfer q

$$I(q) \sim |q|^{-D},$$

where D is some fractal dimension of the scatterer. In the last decade, the case of one-dimensional systems has been studied more rigorously, in the framework of optical and quantum scattering ([1], [2],[13],[19],[14]). For deterministic fractals such as Cantor-like features, the heuristic argument needs to be refined. As was first pointed out in ([1]), some scaling law actually exists if one performs frequency averages on the scattering data. Indeed as figure 1.1 shows, the reflection coefficient for quantum scattering on a triadic Cantor measure gives a "wrong" scaling. Only after some suitable averages have been performed, the right scaling behavior is recovered (figure 1.2). Recently, a general method was proposed ([7]) to retrieve a fractal dimension, namely the correlation dimension, of a potential barrier from the reflection amplitude. In this paper, we wish to give further developments to this method. In particular, we show how the same kind of results can be extended to the scattering phase for half-line scattering. Moreover we shall show how to recover the self-similarity (if there is any) of the potential through a large scale renormalization of the scattering amplitude and scattering phase, respectively.

2. Scattering on the line

2.1 Scattering formalism in one dimension.

Consider the one-dimensional Helmholtz equation

$$\left(\frac{d^2}{dx^2} + p^2\right) \Psi_{V,p} = V\Psi_{V,p}, \; p > 0, \tag{2.1}$$

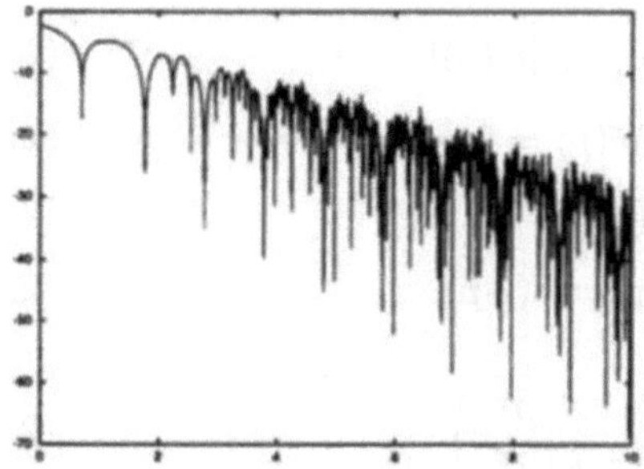

Fig. 1.1. $\log_3 |r(p)|^2$ against $log_3(p)$ for the standard triadic Cantor measure ($p_1 = p_2 = 1/2$, $l_1 = l_2 = 1/3$), which correlation dimension is $D = \log 2/\log 3 \simeq 0.63$. Although the curve seems to follow a general drift, the linear regression is very bad ($\rho = 0.531$) and gives anyway a wrong dimension $A \simeq -3.72 << -2 - D$.

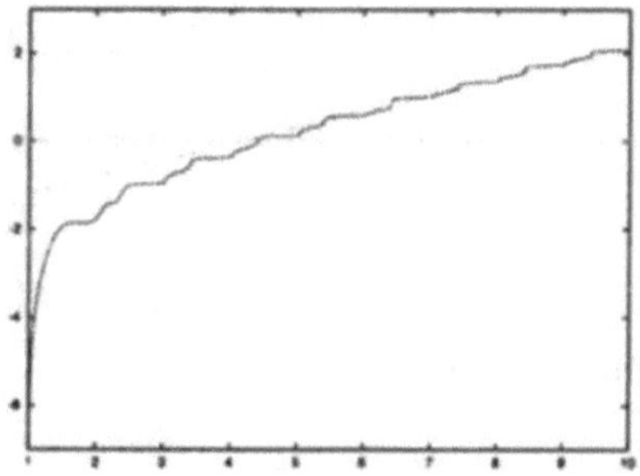

Fig. 1.2. $\log_3 \left(\int_1^p dp'\, p'^2\, |r(p')|^2\right)$ against $log_3(p)$. Once $|r(p)|^2$ has been integrated, it gives the right correlation dimension of the Cantor measure $A \simeq 0.372 \simeq 1 - D$, with $\rho = 0.992$.

where V is a multiplication operator by some smooth real-valued function $V(x)$ with compact support and p is a positive real. By 'smooth' we mean that $V(x)$ is of regularity C^k for some $k \geq 0$, that is k times continuously differentiable on $\mathbb{R}$. Left of the support of V, the solution $\Psi_{V,p}$ of (2.1) has the form $\Psi_{V,p}(x) = A\ e^{ipx} + B\ e^{-ipx}$. Right of the support, it has the same form with other coefficients, say $\Psi_{V,p}(x) = A'\ e^{ipx} + B'\ e^{-ipx}$. The complex coefficients A, B, A' and B' are related by the so-called transfer matrix $\mathcal{M}[V,p]$

$$\begin{pmatrix} A \\ B \end{pmatrix} = \mathcal{M}[V,p] \begin{pmatrix} A' \\ B' \end{pmatrix}.$$

For each p, there are two pairs of complex numbers $(t_V(p), r_V(p))$ and $(t'_V(p), r'_V(p))$ uniquely defined by

$$\begin{pmatrix} 1 \\ r_V(p) \end{pmatrix} = \mathcal{M}[V,p] \begin{pmatrix} t_V(p) \\ 0 \end{pmatrix} \quad \text{and} \quad \begin{pmatrix} t'_V(p) \\ 0 \end{pmatrix} = \mathcal{M}[V,p] \begin{pmatrix} r'_V(p)) \\ 1 \end{pmatrix}.$$

The coefficients $t_V(p)$ and $r_V(p)$ are known respectively as the transmission and reflection amplitude from the left. The associated solution $\Psi_{V,p}$ corresponds (in the stationary scheme) to the physical situation of an ongoing plane wave e^{ipx} traveling to the right, encountering an obstacle V and then splitting into a transmitted part $t(p)e^{ipx}$ and a reflected part $r(p)e^{-ipx}$. Analogously, one defines $t'_V(p)$ and $r'_V(p)$ as the transmission and reflection amplitude from the right. When no confusion is possible, we shall omit in the following the dependence in V and simply write $t(p)$, $t'(p)$, $r'(p)$ or $r(p)$. The transmission and reflexion amplitudes satisfy the following relations (for an exhaustive presentation we refer to [4])

$$|t(p)|^2 + |r(p)|^2 = |t'(p)|^2 + |r'(p)|^2 = 1, \tag{2.2}$$

$$\bar{t}(p)r'(p) + t'(p)\bar{r}(p) = 0, \tag{2.3}$$

$$t'(p) = t(p), \tag{2.4}$$

$$\bar{r}(p) = r(-p),\ \bar{t}(p) = t(-p). \tag{2.5}$$

The matrix

$$\mathcal{S}[V,p] = \begin{pmatrix} t(p) & r'(p) \\ r(p) & t'(p) \end{pmatrix} \tag{2.6}$$

is the *scattering matrix* in the basis (e^{ipx}, e^{-ipx}). The first relation (eq. (2.2) can be interpreted as the conservation of energy during the scattering process, and with (2.3) it expresses the unitarity of the scattering matrix. Equation (2.4) is a consequence of the time reversal invariance and equation (2.5) comes from the realness of the potential. Thanks to theses relations, the transfer matrix can be easily expressed in terms of transmission and reflections amplitudes from the left

$$\mathcal{M}[V,p] = \begin{pmatrix} \frac{1}{t(p)} & \frac{\bar{r}(p)}{\bar{t}(p)} \\ \frac{r(p)}{t(p)} & \frac{1}{\bar{t}(p)} \end{pmatrix}. \tag{2.7}$$

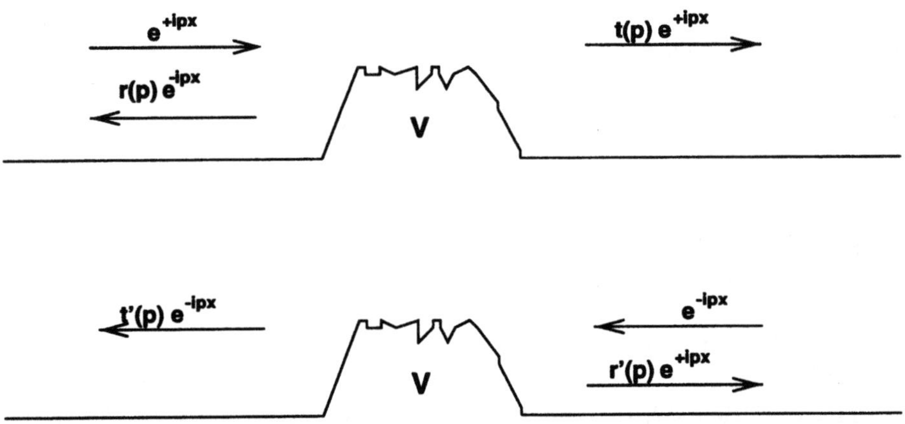

Fig. 2.1. Shematic representation of the transmission and reflection amplitudes.

2.2 Existence of the scattering states

We now wish to show that the transmission and reflection coefficients $t(p)$ and $r(p)$ remain well defined, at least for p large enough, if the potential is a finite measure, possibly singular. Consider the solution $\Psi^+_{V,p}$ of the Helmholtz equation (2.1) satisfying the boundary conditions

$$\Psi^+_{V,p}(x) = \begin{cases} e^{ipx} + r(p)\, e^{-ipx}, & x \to -\infty, \\ t(p)\, e^{ipx}, & x \to +\infty. \end{cases} \tag{2.8}$$

Analogously, we define $\Psi^-_{V,p}$ the solution with boundary conditions

$$\Psi^-_{V,p}(x) = \begin{cases} t'(p) e^{-ipx}, & x \to -\infty, \\ e^{-ipx} + r'(p)\, e^{ipx}, & x \to +\infty, \end{cases} \tag{2.9}$$

For smooth potentials, it can be shown that these solutions are given by the Lippmann-Schwinger equations

$$\Psi^\pm_{V,p}(x) = e^{\pm ipx} + \int dy\, \frac{e^{ip|x-y|}}{2ip} V(y) \Psi^\pm_{V,p}(y) = \Phi_{\pm p} + G_p * (V \Psi^\pm_{V,p}), \tag{2.10}$$

where we have set $G_p(x) = \frac{1}{2ip} e^{ip|x|}$ and $\Phi_p(x) = e^{ipx}$, $*$ standing for the convolution. Here and everywhere, the subscripts $\pm$ refer to two different equations. Now, the Lippmann-Schwinger equations still make sense if V is a measure, in which case $V(y)$ is to be understood in the distributional sense. Moreover, the operator

$$\mathcal{A}_{V,p} : \ \phi \mapsto \ G_p * (V\phi)$$

is a bounded operator in $C^0(\mathbb{R})$ (the space of continuous functions) with norm less than $\|V\|/2$. Thus, $(1 - \mathcal{A}_{V,p})$ is invertible as soon as $p > \frac{\|V\|}{2}$ and the solutions $\Psi^\pm_{V,p}$ are then given by their Neumann series

$$\Psi^\pm_{V,p} = \sum_{n=0}^{\infty} \mathcal{A}^n_{V,p} \Phi_{\pm p} \tag{2.11}$$

which converge uniformly in $C^0(\mathbb{R})$. If we let $x \to \pm\infty$ in (2.10) and identify with (2.8) and (2.9), respectively, we obtain integral expressions for the reflection and transmission amplitudes from the left and from the right

$$r(p) = \tfrac{1}{2ip} \int dy\, e^{ipy} \Psi^+_{V,p}(y) V(y), \qquad r'(p) = \tfrac{1}{2ip} \int dy\, e^{-ipy} \Psi^-_{V,p}(y) V(y),$$

$$t(p) = 1 + \tfrac{1}{2ip} \int dy\, e^{-ipy} \Psi^+_{V,p}(y) V(y), \quad t'(p) = 1 + \tfrac{1}{2ip} \int dy\, e^{ipy} \Psi^-_{V,p}(y) V(y). \tag{2.12}$$

Therefore, for a compactly supported measures V the scattering problem has a well defined meaning since there exist continuous functions $\Psi^+_{V,p}, \Psi^-_{V,p}$, $r_V(p)$, $t_V(p)$, $r'_V(p)$ and $t'_V(p)$ as soon as $p > \frac{\|V\|}{2}$. Moreover, the scattering problem on irregular potentials can be obtained as a limiting case of scattering on smooth potentials. Precisely, we have the following

Proposition 2.1. *If $V_n \to V$ in the weak topology of $\mathcal{M}$, then there is some p_0 such that for all $p \geq p_0$,*

$$r_{V_n}(p) \to r_V(p),\ t_{V_n}(p) \to t_V(p),\ r'_{V_n}(p) \to r'_V(p),\ t'_{V_n}(p) \to t'_V(p),\ n \to \infty.$$

For the proof, we refer to [7]. An important consequence is that relations (2.2), (2.3), (2.4) and (2.5) still hold for potentials defined as measures. Indeed, these identity hold for smooth potentials and by the proposition we may go to the limit. Finally, let us estimate the reflection and transmission amplitudes. In view of (2.11), we have

$$\Psi^{\pm}_{V,p} = \Phi_{\pm p} + O\left(\frac{1}{p^2}\right),\ p \to \infty.$$

Inserting this last expression in (2.12) we obtain

$$r(p) = \frac{\widehat{V}(-2p)}{2ip} + O(\frac{1}{p^2}),\ r'(p) = \frac{\widehat{V}(2p)}{2ip} + O(\frac{1}{p^2}),\ p \to +\infty, \tag{2.13}$$

and

$$t(p) = t'(p) = 1 + \frac{\widehat{V}(0)}{2ip} + O(\frac{1}{p^2}),\ p \to +\infty, \tag{2.14}$$

where $\widehat{V}$ is the Fourier transform of V

$$\widehat{V}(\xi) = \int dx\ e^{-i\xi x} V(x).$$

Recall that the notation $f = O(g)$ means $\forall x,\ f(x) \leq Cg(x)$ for some positive constant C.

2.3 Recovering some fractal dimensions.

As we have seen, the reflection and transmission amplitude behave, up to a trivial factor, like the Fourier transform of the potential. In the general case, this is contradictory with the expectation that the reflected intensity $|r(p)|^2$ should follow a power-law. However, this happens to be true in an average sense. Indeed, it has been shown that the quadratics means of the Fourier transform are asymptotically governed by some fractal dimension. Consider first the case of positive measures. Then we have the following result, to be found in [8].

Lemma 2.1. *For any finite positive measure V on $\mathbb{R}$,*

$$\lim_{p\to\infty} \begin{matrix}\sup\\ \inf\end{matrix} \log\left(\int_{0\leq\xi\leq p} d\xi\ \left|\widehat{V}(\xi)\right|^2\right) / \log p = 1 - D^{\mp}[V],$$

where $D^+[V]$ and $D^-[V]$ are the upper and lower correlation dimension of V, respectively.

One of the usual definitions of these last dimensions is the following [16]. If we set

$$\Omega_V(r) = \int dV(x)\; V\left(B(x,r)\right),$$

where $B(x,r)$ is the ball of radius r around the point x, then the upper and lower correlation dimension are given by

$$D^{\pm}(V) = \lim_{r \to 0} \begin{matrix}\sup \\ \inf\end{matrix} \frac{\log \Omega_V(r)}{\log r}.$$

When $D^-[V]$ and $D^+[V]$ coincide, we write $D(V)$ for their common value and call it simply correlation dimension. The terminology comes from the fact that the function Ω_V can be interpreted as a correlation function according to the probability measure μ. Note that if $V \in \mathcal{L}^2(\mathbb{R})$, then $D^{\pm}[V] = 1$.

For signed measures, the correlation dimensions are not sufficient to characterise the Fourier asymptotic. The relevant dimensions in this case are the upper and lower *2-wavelet dimensions* $\kappa_2^{\pm}$, also called *wavelet correlation dimensions*, introduced in [9]. They are defined via the following procedure. Let $\mathcal{S}^+$ be the subspace of functions in the Schwartz class which Fourier transform is supported by positive frequencies only. A function g in $\mathcal{S}^+$ has all its moments vanishing or, what amounts to the same, has a flat Fourier transform in zero

$$\int dx\; x^n g(x) = 0, \quad \text{for all } n \Leftrightarrow \widehat{g}(\omega) \le O(\omega^n),\ \omega \to 0 \text{ for all } n$$

Now take some function g in $\mathcal{S}^+$, and consider the wavelet transform (see e.g [5], [3], [10]) of V with respect to g

$$\mathcal{W}_g V(b,a) = (\tilde{g}_a * V)(b) = < g_a(\cdot - b), V >,$$

where $g_a(x) = a^{-1}g(a^{-1}x)$ and $\tilde{g}_a(x) = a^{-1}\overline{g}(-a^{-1}x)$ are essentially dilatations of g. Now set

$$G_{g,q}(a) = \int db\,|\mathcal{W}_g V(b,a)|^q,$$

and

$$\Gamma_{g,q}(a) = \min\left\{\int_a^1 \frac{d\alpha}{\alpha} G_{g,q}(\alpha), \int_0^a \frac{d\alpha}{\alpha} G_{g,q}(\alpha)\right\}.$$

Then the q-wavelet dimensions $\kappa_q^{\pm}(V)$ are defined as

$$\kappa_q^{\pm}[V] = \lim_{a \to 0} \begin{matrix}\sup \\ \inf\end{matrix}\ \log \Gamma_{g,q}(a)/\log a$$

and do not depend on the choosen function g in $\mathcal{S}^+$ for $q \ge 1$, provided $g \ne 0$. In particular $\kappa_2^{\pm}[V]$ is called the wavelet correlation dimension.

Remark. A function in $\mathcal{S}^+$ cannot have compact support. However, admitting an arbitrary small error on $\kappa_2^{\pm}$, we may suppose that the wavelet is compactly supported (see [9]).

The wavelet correlation dimensions satisfy the following

Theorem 2.1. *If V is a finite complex measure, then either $V \notin \mathcal{L}^2(\mathbb{R})$ and*

$$\lim_{p\to\infty} \begin{matrix}\sup\\ \inf\end{matrix} \log\left(\int_0^p dp' \left|\widehat{V}(p')\right|^2\right) / \log p = -\kappa_2^{\mp}[V]$$

or $V \in \mathcal{L}^2(\mathbb{R})$, in which case the above limits are zero. Instead we have

$$\lim_{p\to\infty} \begin{matrix}\sup\\ \inf\end{matrix} \log\left(\int_p^\infty dp' \left|\widehat{V}(p')\right|^2\right) / \log p = -\kappa_2^{\mp}[V].$$

Let us make some remarks on this last theorem. First note that $\kappa_2^-[V] \leq 0$ if $V \notin \mathcal{L}^2(\mathbb{R})$ and $\kappa_2^-[V] \geq 0$ if $V \in \mathcal{L}^2(\mathbb{R})$. More generally, the faster $\widehat{V}$ decreases, the higher are the wavelet dimensions. In fact, κ_2^- can be seen as an index of regularity. Secondly, verifying $\kappa_2^+[V] = \kappa_2^-[V] = \kappa_2[V]$ is not sufficient for having a scaling

$$\int_0^p dp' \left|\widehat{V}(p')\right|^2 \sim p^{-\kappa_2[V]} \text{ or } \int_p^\infty dp' \left|\widehat{V}(p')\right|^2 \sim p^{-\kappa_2[V]}.$$

This last statement is false in general and can only be proven in some specific cases (e.g self-similar measures, see further). Finally, we have chosen a wavelet in $\mathcal{S}^+$. However, if the measure is real, the wavelet dimensions $\kappa_2^{\mp}$ can also be computed with a wavelet g in $\mathcal{S}^-$, the subspace of Schwartz functions having negative frequencies only, or a wavelet g in $\mathcal{S}^0 = \mathcal{S}^+ \oplus \mathcal{S}^-$. The procedure is the same and theorem (2.1) still holds.

Now the above theorem can be applied to the scattering problem when the potential is not too regular (but only "fractal" and therefore irregular potentials are of interest). Using (2.13), it is straightforward to prove the next

Theorem 2.2. *Let V be a finite, real measure with compact support and let $p_o \geq \|V\|/2$. Define*

$$E(p) = \min\left\{\int_{p_0}^p dp'\, p'^2 \left|r(p')\right|^2,\ \int_p^\infty dp'\, p'^2 \left|r(p')\right|^2\right\}.$$

If $\kappa_2^-[V] \leq \kappa_2^+[V] < 1$, then

$$\lim_{p\to\infty} \begin{matrix}\sup\\ \inf\end{matrix} \frac{\log E(p)}{\log p} = -\kappa_2^{\mp}[V]. \tag{2.15}$$

3. Scattering on the half-line

We wish here to adapt the result of the last section to the scattering of the half-line. The scattering states are now given by the Helmholtz equation on the positive semi-axis

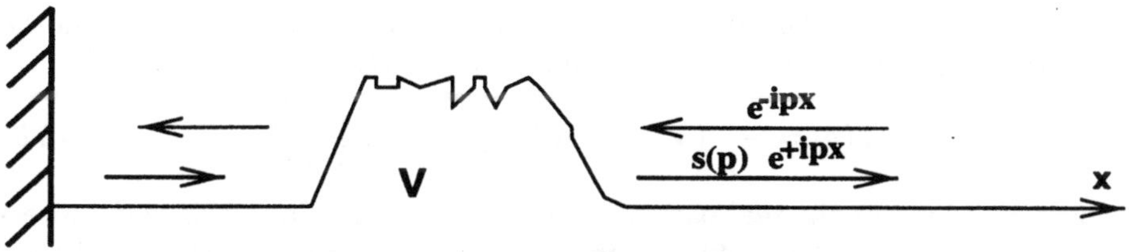

Fig. 3.1. Scattering on the half-line

$$\left(\frac{d^2}{dx^2}+p^2\right)\phi_{V,p}=V\phi_{V,p},\ x>0, \tag{3.1}$$

where $V(x)$ is a measure supported by some compact in $]0,+\infty[$.

Consider the solution $\phi_{V,p}$ which looks asymptotically like $e^{-ipx}+s(p)e^{ipx}$, with homogeneous Dirichlet, Neumann or mixt condition at zero. The complex number $s(p)$ is the *scattering amplitude* (or the scattering matrix, which reduces to a scalar of modulus one in the half-line case). Using the transfer matrix $\mathcal{M}[V,p]$ introduced in the first section, we easily derive the following expression for $s(p)$

$$\phi_{V,p}(0)=0\ \text{(Dirichlet)}\Rightarrow s_D(p)=-\frac{t(p)}{\bar{t}(p)}\frac{1+\bar{r}(p)}{1+r(p)}, \tag{3.2}$$

$$\phi'_{V,p}(0)=0\ \text{(Neumann)}\Rightarrow s_N(p)=-s_D(p), \tag{3.3}$$

$$\phi'_{V,p}(0)=\alpha\phi_{V,p}(0)\ \text{(Mixt)}\Rightarrow s_\alpha(p)=\frac{\alpha+ip}{\alpha-ip}s_D(p). \tag{3.4}$$

Since the scattering amplitude is unitary, we can rewrite it as

$$s(p)=e^{i(\delta_0(p)+\delta(p))}, \tag{3.5}$$

where we have introduced δ_0, the change of phase due to the wall in the absence of potential ($V=0$), and δ, the specific disturbance caused by the potential. The term δ is called *scattering phase shift*. It carries the information about the potential. Note the difference with the case of scattering on the line, where the interesting quantity to study was the reflected energy $|r(p)|^2$. Now, identifying with (3.2), (3.3) and (3.4), we obtain

$$\delta_0(p)=\begin{cases}-\pi & \text{(Dirichlet)}\\ 0 & \text{(Neumann)}\\ 2\arctan(p/\alpha)-\pi & \text{(mixt)}\end{cases}, \tag{3.6}$$

and

$$\delta(p)=\arg\left(\frac{t(p)}{\bar{t}(p)}\frac{1+\bar{r}(p)}{1+r(p)}\right) \tag{3.7}$$

in every case. Using the Born approximation (2.13) and (2.14), we obtain the following asymptotics for the scattering phase shift, as $p \to \infty$

$$\delta(p) = -\frac{1}{p}\left(\hat{V}(0) - \Re[\hat{V}(2p)]\right) + O\left(\frac{1}{p^2}\right), \tag{3.8}$$

where $\Re[\cdot]$ denotes the real part of a complex number. From this last expression we can derive a Weyl-like asymptotic formula for the scattering phase.

Theorem 3.1. *Let V be a finite, real measure with compact support in $]0,+\infty[$ and let $p_o \geq \|V\|/2$. Then if $V \notin \mathcal{L}^2(\mathbb{R})$ we have*

$$\int_{p_0}^{p} dp' p'^2 \delta^2(p') = C_1 p + \xi(p) + C_2 + O(p^{-1}), \tag{3.9}$$

else if $V \in \mathcal{L}^2(\mathbb{R})$ and $\kappa_2^{\pm}[V] < 1$,

$$\int_{p}^{\infty} dp' (p'^2 \delta^2(p') - C_1) = \xi(p) + O(p^{-1}) \tag{3.10}$$

where, in both cases,

$$C_1 = \left|\hat{V}(0)\right|^2 = \lim_{p\to\infty} \frac{1}{p} \int_{p_0}^{p} dp' p'^2 \delta^2(p') \tag{3.11}$$

and

$$\lim_{p\to\infty} \begin{array}{c}\sup\\ \inf\end{array} \log \xi(p)/\log(p) = -\kappa_2^{\mp}[V] \tag{3.12}$$

Proof: We limit ourselves to the first part statement, the second one being completely analogous. By (3.8) we have

$$p^2\delta^2(p) = \left|\hat{V}(0)\right|^2 + \left|\Re[\hat{V}(2p)]\right|^2 - 2\Re[\hat{V}(0)\hat{V}(2p)] + O(p^{-3}),$$

whence

$$\begin{aligned}\int_{p_0}^{p} dp' p'^2 \delta^2(p') &= \left|\hat{V}(0)\right|^2 p + \int_{p_0}^{p} dp' \left|\Re[\hat{V}(2p')]\right|^2 - \\ &\quad 2\hat{V}(0)\Re\left[\int_{p_0}^{p} dp' \hat{V}(2p')\right] + p_0 \left|\hat{V}(0)\right|^2 + O(p^{-1}).\end{aligned}$$

Now since V is supported on $]0,+\infty[$ we have

$$\lim_{p\to+\infty} \int_{0}^{p} dp' \ \hat{V}(2p') = V(0)/2 = 0,$$

and thus the term $2\hat{V}(0)\Re\left[\int_{p_0}^{p} dp' \hat{V}(2p')\right]$ converges toward some constant as $p \to +\infty$. We may therefore write

$$\int_{p_0}^{p} dp' p'^2 \delta^2(p') = C_1 p + \int_{p_0}^{p} dp' \left|\Re(\hat{V}(2p'))\right|^2 + C_2 + O(p^{-1}),$$

with $C_1 = \left|\hat{V}(0)\right|^2$. The conclusion follows from the following lemma

Lemma 3.1. *If V is a real finite measure supported in $]0,+\infty[$, then*

$$\lim_{p\to\infty} \begin{matrix}\sup\\ \inf\end{matrix} \frac{\log\left(\int_0^p dp' \; \left|\Re[\widehat{V}(p')]\right|^2\right)}{\log p} = \lim_{p\to\infty} \begin{matrix}\sup\\ \inf\end{matrix} \frac{\log\left(\int_0^p dp' \; \left|\widehat{V}(p')\right|^2\right)}{\log p}$$

and

$$\lim_{p\to\infty} \begin{matrix}\sup\\ \inf\end{matrix} \frac{\log\left(\int_p^\infty dp' \; \left|\Re[\widehat{V}(p')]\right|^2\right)}{\log p} = \lim_{p\to\infty} \begin{matrix}\sup\\ \inf\end{matrix} \frac{\log\left(\int_p^\infty dp' \; \left|\widehat{V}(p')\right|^2\right)}{\log p}.$$

Proof: We shall here only sketch the proof. Define $\tilde{V}(x) = \underline{V(-x)}$ and $V_{sym} = V + \tilde{V}$ and take some symmetric wavelet $g(x) = \tilde{g}(x) = g(-x)$. The wavelet transform of V_{sym} with respect to g is then given by

$$\mathcal{W}[\tilde{V}, g](b,a) = \overline{\mathcal{W}[V,\tilde{g}](-b,a)} = \overline{\mathcal{W}[V,g](-b,a)}$$

and thus

$$\begin{aligned} |\mathcal{W}[V_{sym},g](b,a)|^2 &= |\mathcal{W}[V,g](b,a)|^2 + \left|\mathcal{W}[\tilde{V},g](-b,a)\right|^2 \\ &\quad + 2\Re\left[\mathcal{W}[V,g](b,a)\mathcal{W}[V,g](-b,a)\right]. \quad (3.13) \end{aligned}$$

As was explained above, we may suppose that g is compactly supported. Then for a small enough, the functions $\mathcal{W}[V,g](b,a)$ and $\mathcal{W}[V,g](-b,a)$ have disjoint supports in b, so that the cross term in (3.13) vanishes. Thus, on integrating, we have

$$\|\mathcal{W}[V_{sym},g](\cdot,a)\|_2 = 2\,\|\mathcal{W}[V,g](\cdot,a)\|_2\,,\; a\to 0,$$

and therefore $\kappa_2^{\pm}[V_{sym}] = \kappa_2^{\pm}[V]$. Now we may apply theorem (2.1) to V_{sym} and V. Since $\widehat{V_{sym}} = \Re[\widehat{V}]$, the conclusion follows.

4. Scattering on self-similar potentials.

We now want to consider the case of potentials represented by self-similar measures. A self-similar measure on $\mathbb{R}$ was defined by Hutchinson ([12]) to be a probability measure V satisfying

Condition 4.1. (self-similarity equation) $\quad V = \sum_{i=1}^N p_i \, V \circ S_i^{-1}$,

where the p_i are a set of weights ($0 < p_i < 1$, $\sum_{i=1}^N p_i = 1$) and the S_i a set of similarities $S_i(x) = l_i x + b_i$, with $0 < l_i < 1$. A natural question might be: where do we recover in the scattering data the fact that the potential is self-similar. As we shall see, self-similar of the scatterer is mirrored by the existence of a weak renormalisation of the scattering data.

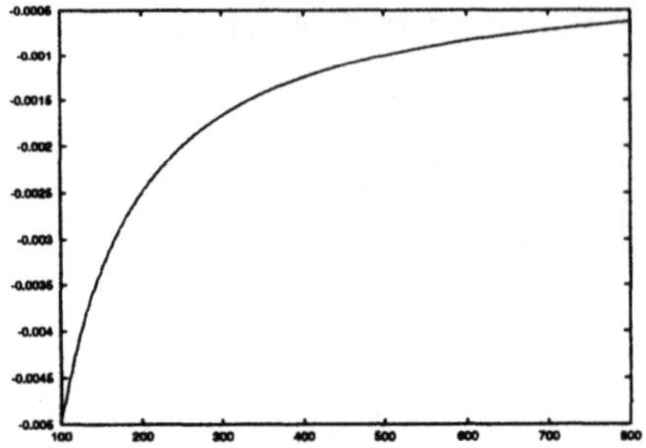

Fig. 4.1. Scattering phase shift $\delta(p)$ against p for the triadic Cantor measure. On a large frequency range, only the preponderant term p^{-1} is visible.

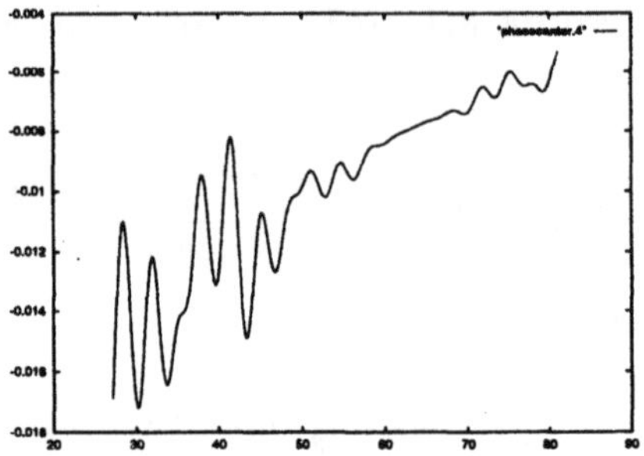

Fig. 4.2. Detail of the scattering phase.

The support of a self-similar V is the unique non empty compact K globally invariant under these contractions, i.e $K = \bigcup_{i=1}^{N} S_i(K)$. This support may be of zero Lebesgue measure. This occurs when the following condition is satisfied

Condition 4.2. (strong open set condition [18]) There exist a closed bounded set U such that $S_i(U) \cap S_j(U) = \emptyset$ for $i \neq j$.

It can be shown that the set U at least contains K. For measures satisfying condition (4.2), the upper and lower correlation dimension coincide ($D^+ = D^- = D$) and are given by the implicit equation (see e.g [6])

$$\sum_{i=1}^{N} p_i^2 l_i^{-D} = 1. \tag{4.1}$$

This implies in particular that $0 < D < 1$. If furthermore

Condition 4.3. (equicontractivity condition), all the contraction ratio are equal, that is $l_i = l,\ i = 1, ..N$,

then D is given explicitly by

$$D = \frac{\log \sum_{i=1}^{N} p_i^2}{\log l}$$

The case $p_1 = p_2 = 1/2$ and $l_1 = l_2 = 1/3$ reduces to the well-known triadic Cantor measure which dimension is $D = \log\ 2/\log\ 3$.

The $\mathcal{L}^2$-Fourier asymptotic of self-similar measures has been studied in [17],[18] and [15]. The last authors stated the most general result. They showed (among other things) that if the strong open set condition holds (the needed condition is actually more general), then

$$\int_0^R dp' \left|\widehat{V}(p')\right|^2 \sim R^{1-D}, \tag{4.2}$$

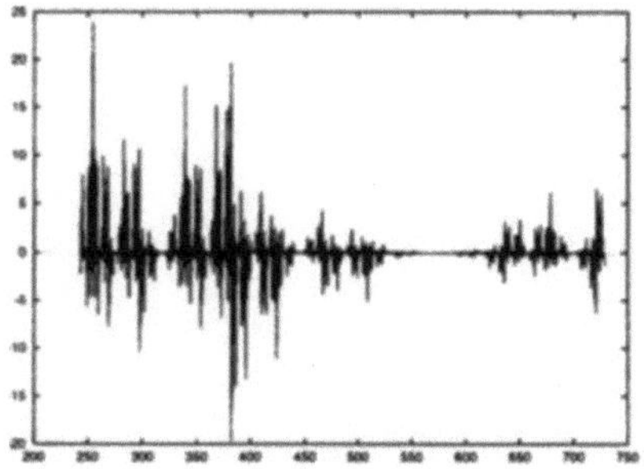

Fig. 4.3. $2^6(1 + 3^6 p\delta(3^6 p))\cos(2\ 3^6 p)$ against p.

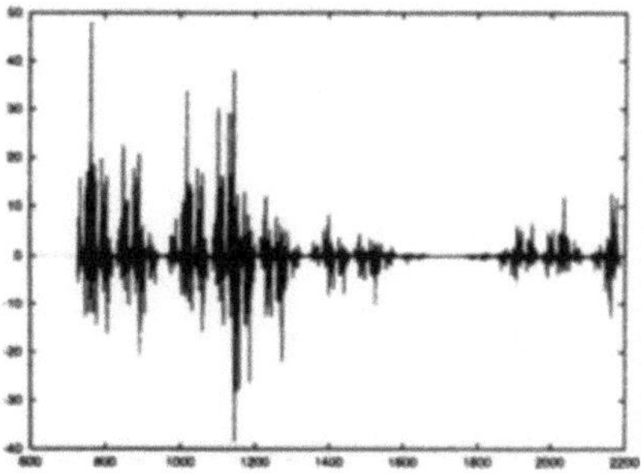

Fig. 4.4. $2^7(1 + 3^7 p\delta(3^7 p))\cos(2\ 3^7 p)$ against p.

where the notation $f \sim g$ means $Cf \leq g \leq C'f$ for some $C' \geq C > 0$. Immediate consequences are

Proposition 4.1. *If V satisfies conditions 4.1, 4.2 and 4.3, then*

$$\int_{p_0}^{p} dp'\ p'^2 \left|r(p')\right|^2 \sim p^{1-D}, \tag{4.3}$$

and

$$\int_{p_0}^{p} dp'\ p'^2 \delta(p')^2 = C_1 p^{-1} + \xi(p),\ \xi(p) \sim p^{1-D}, \tag{4.4}$$

where D is given by (4.1).

Note that this statement is stronger than the general result (2.1) for positive measures, since the scaling does not always exist. A numerical application has been performed on the triadic Cantor measure ($p_1 = p_2 = 1/2$, $l_1 = l_2 = 1/3$) We have approached the Cantor measure by a finite approximation and computed the transfer matrix by a recursive procedure using the self-similarity of the measure (for more details see [7]). This kind of method was already used in [14], where the authors computed the transmission amplitude on a Cantor-like potential by exhibiting a recurrent (non linear) system on the modulus and phase. On figure 1.1, we have plotted the reflected energy $|r(p)|^2$ in a logarithmic diagram. Then on figure 1.2, we have re-plotted the same quantities after an integration. On each graph, we have made a linear regression $y = Ax + B$ and indicated the coefficient of correlation ρ. This clearly illustrates our main result: the fractal dimension of the scatterer can only appear after averages have been performed on the scattering data.

We now wish to show how the self-similar nature of the potential reappears in the scattering data. Let us begin with some intuitive considerations. On figure 4.1, we have drawn the scattering phase $\delta(p)$ against p and then magnified in figure 4.2. At first sight, the phase exhibits some chaotic behavior. The same remark holds for the reflection amplitude (see figure 1.1). Hovever there is some hidden system behind these oscillations since one can show that a

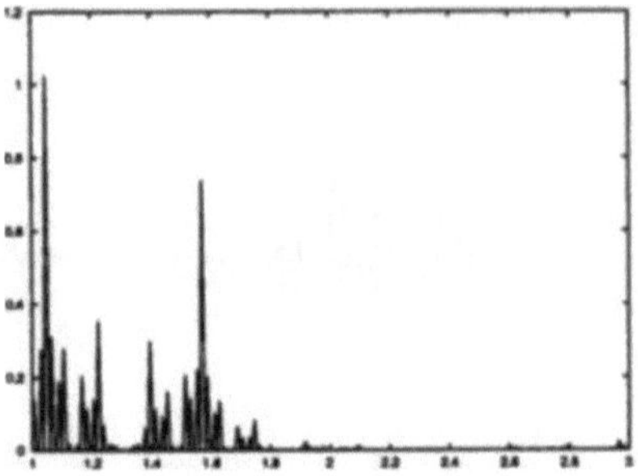

Fig. 4.5. $18^5 \left|r(3^5 p)\right|^2$ against p

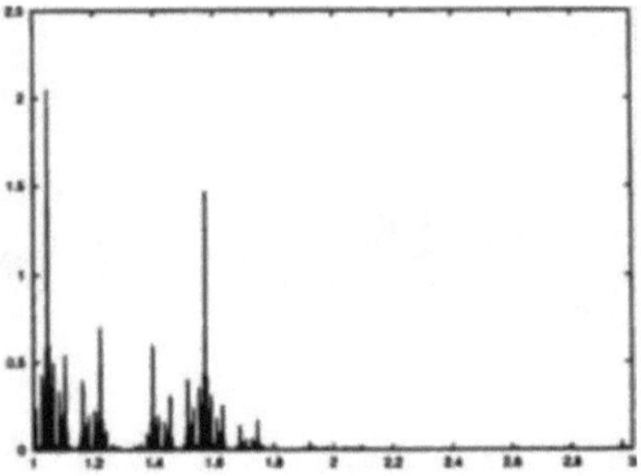

Fig. 4.6. $18^6 \left|r(3^6 p)\right|^2$ against p

renormalisation procedure on the frequencies exists in some suitable topology (see [7]). This can be summarized in the following theorem. We denote by Σ the set of dilation generated by the dilation factors l_i, $\Sigma = \{l_{i_1} l_{i_2} \dots l_{i_N}\}$. We consider simultaneously the case of scattering on the line and on the half-line.

Theorem 4.1. *Let V be given by a self-similar measure verifying conditions 4.1,4.2. Then for all $\beta \in \Sigma$ there are numbers $\alpha \in \mathbb{C}$ and $\gamma \in \mathbb{R}$ such that the sequence of functions*

$$f_n(p) = \alpha^n \rho(p/\beta^n), \quad \rho(p) = e^{i\gamma p} r(p),$$

converges in the sense of distributions to some $f^* \neq 0$. If in addition V satisfies 4.3 and if the auto-convolution $V * \widetilde{V}$ satisfies the strong open set condition, where as usual $\widetilde{V}(x) = V(-x)$, then $f(p) = |r(p)|^2$ has a large-scale renormalisation too in that

$$f_n(p) = \left(l^2 \sum_{i=1}^{N} p_i^2 \right)^{-n} f(l^{-n} p),$$

converges toward some positive measure μ. In the case of half-line scattering on a self-similar potential satisfying 4.1,4.2 we have the following renormalization of the scattering phase $\delta(p)$. For all $\beta \in \Sigma$ there is $\alpha \geq 1$ and $\gamma \in \mathbb{R}$ such that

$$f_n(p) = \alpha^n g(p/\beta^n), \quad g(p) = \cos(\gamma p)(p\delta(p) + \widehat{V}(0))$$

converges in the sense of distributions towards a limit $\eta \neq 0$. If in addition V satisfies 4.3 and if the auto-convolution $V * \widetilde{V}$ satisfies the strong open set condition, then $f(p) = (p\delta(p) + \widehat{V}(0))^2$ has a large-scale renormalisation too in the above sense.

To illustrate this theorem, with have plotted $18^n \left|r(3^n p)\right|^2$ against p for the triadic Cantor measure supported on the interval $[1, 2]$, with $n = 5, 6$ (figure 4.5 and figure 4.6). One can actually show that the limit measure is absolutely continuous with respect to the Riesz measure [11]. On figure 4.3 and figure 4.4 we have plotted $2^n(1 + 3^n p\delta(3^n p)) \cos(2\ 3^n p)$ against p for $n = 6$ and 7.

References

1. Allain, C., Cloitre, M. (1985): Optical diffraction on fractals. *Phys. Rev. B*, 33(5), 3566–3569.
2. Allain, C., Cloitre, M. (1987): Spatial spectrum of a general family of self-similar arrays. *Physical Review A*, 36(12), 5751–5757.
3. Daubechies, I. (1992): *Ten Lectures on Wavelets.* SIAM.
4. Deift, P., Trubovitz, E. (1979): Inverse scattering on the line. *Comm. Pure Appl. Math.*, 32, 121–251.
5. Grossmann, A., Morlet, J. (1984): Decomposition of Hardy functions into square integrable wavelets of constant shape. *SIAM J. Math. Anal.*, 15(2).
6. Guerin, C.A., Holschneider, M. (1995): l^2-Fourier asymptotic of self-similar measures. *Preprint CPT*, (P.3266).
7. Guerin, C.A., Holschneider, M. (1996): Scattering on fractal measures. *J. Phys. A: Math. Gen.*, (29), 7651–7667.
8. Guerin, C.A., Holschneider, M. (1997): On equivalent definitions of the correlation dimension for a probability measure. *Journ. Stat. Phys.*.
9. Holschneider, M. (1994): Fractal wavelet dimension and localization. *Comm. Math. Phys.*, 160, 457–473.
10. Holschneider, M. (1995): *Wavelets, an Analytical Tool.* Oxford University Press.
11. Holschneider, M. (1996): Large scale renormalisation of fourier transforms of self-similar measures and self-similarity of Riesz measures. *J. of Math. Anal and Appl.*, (200), 307–314.
12. Hutchinson, J.E. (1981): Fractals and self similarity. *Ind. Univ. Math. Jour.*, 30(5), 713–746.
13. Jaggard, D.L., Sun, X. (1990): Reflection from fractal multilayers. *Opt. Lett.*, 15, 1428–1430.
14. Konotop, V.V., Yordanov, O.I., Yurkevitch, I.V. (1990): Wave transmission through a one-dimensional Cantor-like fractal medium. *EuroPhys.Lett*, 12(6), 481–485.
15. Lau, K.S., Wang, J. (1993): Mean quadratic variations and Fourier asymptotics of self-similar measures. *Monatsh. Math.*, 115, 99–132.
16. Pesin, Y.B. (1993): On rigorous mathematical definition of the correlation dimension and generalized spectrum for dimension. *Journ. Stat. Phys.*, 71(3/4), 529–547.
17. Strichartz, R.S. (1990): Self-similar measures and their Fourier transform 1. *Indiana Univ. Math.Journ.*, 39(3), 797–817.
18. Strichartz, R.S. (1993): Self-similar measures and their Fourier transform 2. *Trans. Am. Math. Soc.*, 336(1), 335–361.
19. Sun, X., Jaggard, D.L. (1991): Wave interactions with generalized Cantor bar fractal multilayers. *J. Appl. Phys.*, 70, 2500–2507.

IMAGE COMPRESSION

Wavelet and Fractal Transforms for Image Compression

F.C. Cesbron and F.J. Malassenet

Georgia Tech Lorraine
The European Platform of the Georgia Institute of Technology
2-3, rue Marconi
57070 Metz, FRANCE
Tel: +33 3 87 20 39 39   Fax: +33 3 87 20 39 40
fcesbron@georgiatech-metz.fr

1. Introduction

With the development of communication technology and digital image processing, the demand for more and better images is increasing. New compression techniques are needed for storage and transmission. Fractal coding is one of the promising new coding techniques to increase compression ratios, that is beginning to be adopted worldwide by the way of the Internet. It is based on the work of Barnsley on fractals and iterated functions systems to describe them [1]. Jacquin was the first to derive an automatic fractal coding algorithm for still images [2], rapidly followed by many others [3].

The aim of this article is to improve the conventional fractal coding scheme by paying a particular attention to the high frequency components of images. Indeed, fractal techniques distort the spectral allocation of coded images, resulting in a poor coding of edges and globally blurred images.

2. Fractal Coding and Wavelet Transform

All fractal algorithms for image compression are based on the theory of Iterated Function Systems (IFSs). This theory is derived from contractive transformations applied to images. The coding part consists in finding a contractive transformation such that the distance between the original image and its transformed is as small as possible. The decoded image is the fixed point or attractor of the contractive transformation. The Collage theorem and the Shadowing theorem guarantee that the coded image is within a given bound of the original image, even when quantization error is introduced [1].

A fractal coder/decoder is implemented from these ideas. The original image is decomposed in $Rbl \times Rbl$ non-overlapping range blocks and $2Rbl \times 2Rbl$ non-overlapping domain blocks. For every (range, domain) block pair, an affine transform is defined by its scaling factor and offset such that the mean

[0] Project co-funded with Région de Lorraine, France.

square error between the range block and the transformed domain block is minimized. To obtain contractive transformations, scaling factors are constrained to be less than one. For each range block, the matching domain block and affine transform correspond to the smallest mean square error. The fractal code is the set of domain block addresses and affine parameters associated to range blocks. The scaling factor distribution is quite noisy. Scaling factors are thus uniformly quantized, typically with four bits. The offset distribution is closed to a Laplacian one. The offset quantizer is a Lloyd-Max quantizer, optimized for a Laplacian distribution, typically with four bits. At the decoding part, the fractal code is iterated eight times from an initial black image. Since only low bit rates are studied here, this algorithm is kept quite simple with no quadtree partitions, no overlapping domain blocks, and no isometries of the domain blocks.

Through the discrete wavelet transform, signals are decomposed in lower resolution and detail signals [4]. The simplest extension of one dimensional wavelet to images is used here: separable wavelets. The low-pass and high-pass filters derived from the wavelet and scaling function are first applied on the image rows, then on its columns. Thus, each iteration of the discrete wavelet decomposition transforms one image into four subimages, each of them one fourth the original size. The usual image representation of image multiresolution decomposition is depicted Figure 2.1. Four filters are defined from the discrete wavelet transform. Their supports are represented in Figure 2.2 on the wavelet transform domain. The wavelet used here is the simplest one, the Haar basis: the low-pass filter computes the average of two rows or columns, the high-pass one computes their difference.

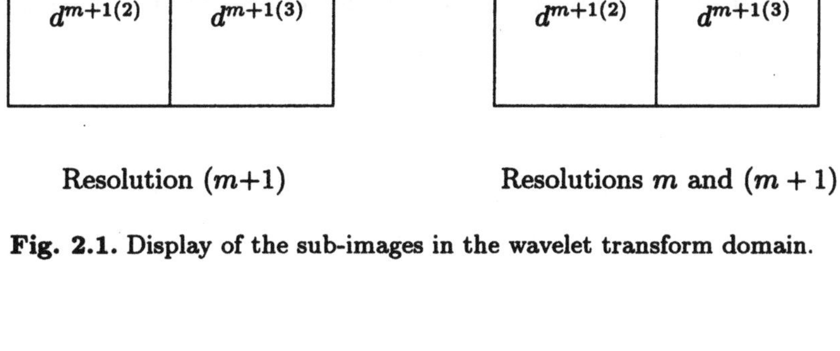

Fig. 2.1. Display of the sub-images in the wavelet transform domain.

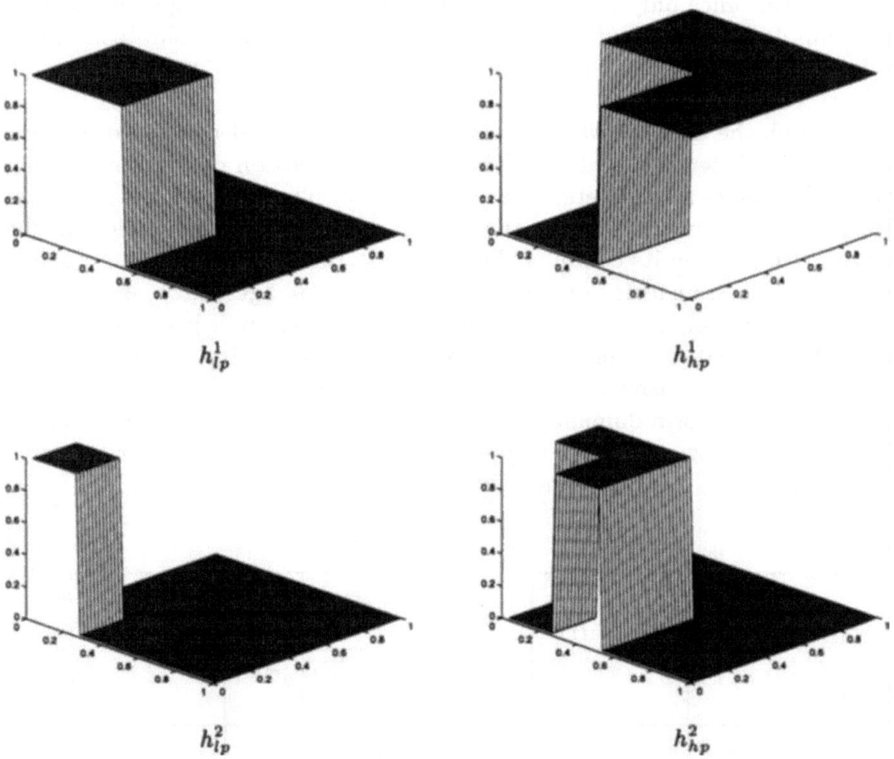

Fig. 2.2. Supports of the h_{lp}^1, h_{hp}^1, h_{lp}^2, and h_{hp}^2 filters in wavelet transform domain.

3. Fractal Transform in the Haar Wavelet Transform Domain

Many researchers have worked on the link between fractal coding and the wavelet transform. For example, Davis derived relationships between the scales of multiresolution decomposition of fractal images. He then used them to construct subtrees of decaying wavelet coefficients [5]. These subtrees were added to the zerotrees introduced by Shapiro [6]: fractal block coders were viewed as a form of Haar wavelet subtree quantization scheme. Krupnik et al. redefined both the fractal coding and decoding procedures in the Haar wavelet transform domain [7]. The fractal decoding became a non-iterative process where, starting from a coarse resolution, all the wavelet coefficients of a higher subband were computed from those of a lower subband.

In the proposed approach, explicit relationships are first derived between scales of fractal images. They translate the fractal property of self-similarities inside an image in the Haar wavelet transform domain. The conventional fractal algorithm is interpreted in the wavelet transform domain to obtain two frequential components. Modifications are then brought to this decomposition to improve the conventional coder and obtain multiresolution fractal coders.

Relationships between scales are first computed in the simple case of an image composed of only four range blocks, each of them associated with an affine transform of the whole image. These relationships state that each range block of any subimage at the resolution m, i.e., $c^m, d^{m(1)}, d^{m(2)}$ or $d^{m(3)}$, is a rescaled version of the whole corresponding subimage at the resolution $(m+1)$.

With some more notations and constraints, these relationships generalize to the images obtained at the output of a fractal coder/decoder. Direct applications include the reconstruction of the attractors, i.e., the decoded images, at any resolution, without any iteration but only one inverse wavelet transform.

The relationships in the multiresolution decomposition of fractal coded images can be seen as constraints between scales. The modifications introduced in the conventional fractal coder will loosen these constraints.

Before introducing the multiresolution fractal coder, the conventional algorithm is interpreted in the Haar wavelet transform domain. The fractal transform F, with fixed point f_0, is decomposed into a low-pass component and a high-pass one.

The low-pass component, F_{lp}, is the fractal transform F followed by the low-pass filter h^1_{lp}:

$$\forall f,\ F_{lp}(f) = F(f) * h^1_{lp},$$

where f is any image. It can be shown that F_{lp} is a contraction which fixed point is the low-pass part of f_0, $f_0 * h^1_{lp}$.

The high-pass component, F_{hp} is the sum of two terms:

$$\forall f,\ F_{hp}(f) = F(f) * h^1_{hp} + f_0 * h^2_{hp},$$

where f is any image. The first term is only the fractal transform F, high-pass filtered with h^1_{hp}. The second term is a condensation transform which condensation set, $c = f_0 * h^2_{hp}$, is a bandpass version of the attractor f_0. This high-pass component is a contraction which fixed point is a high-pass version of the attractor of F, $f_0 * (h^1_{hp} + h^2_{bp})$.

Finally, the fractal transform F may be expressed as the sum of F_{lp} and F_{hp} minus the condensation transform. The fixed point is thus written as the fixed point of F_{lp} plus the fixed point of F_{hp} minus the condensation set c.

4. Multiresolution Fractal Coder

The proposed new coder is based on the previous decomposition, except that two *different* fractal coders are used in the low-pass component and the high-pass one.

The low-pass component is a first fractal transform, F_1, followed by the low-pass filter h^1_{lp}. This can be implemented using the conventional fractal coder and keeping only the low frequencies of the result. In an alternative approach, the wavelet transform is first applied to the original image. Then, a fractal code is derived on the lower resolution image, one fourth the original size, with the conventional fractal scheme described above. This results in a significant decrease of the computation time.

The high-pass component is a second fractal coder, F_2, followed by the high-pass filter h^1_{hp}. In this second part, the conventional fractal scheme is slightly modified. The range blocks contain the high-frequencies of the original image. They are modeled by affine transforms of domain blocks containing the high frequencies of the attractor of the low-pass component. Since only high frequencies are considered, offsets are useless. Scaling factors are quantized with a Laplacian distribution, typically with four bits. Moreover, since only one scale is reconstructed, only one iteration will be needed at the decoder and no constraints are imposed on the scaling factors.

The new fractal coder is the combination of these two components [8].

The new coder is first tested on the 128 × 128 "Lena" image. No quantization is applied on the parameters so no bit rates are available. The PSNR obtained with the conventional fractal coder is 28.77 dB. With the new coder, the PSNR is 32.58 dB. The gain of 3.8 dB is clearly visible: the overall image is less blurred, high frequencies are better rendered. These results are depicted on Figure 4.1.

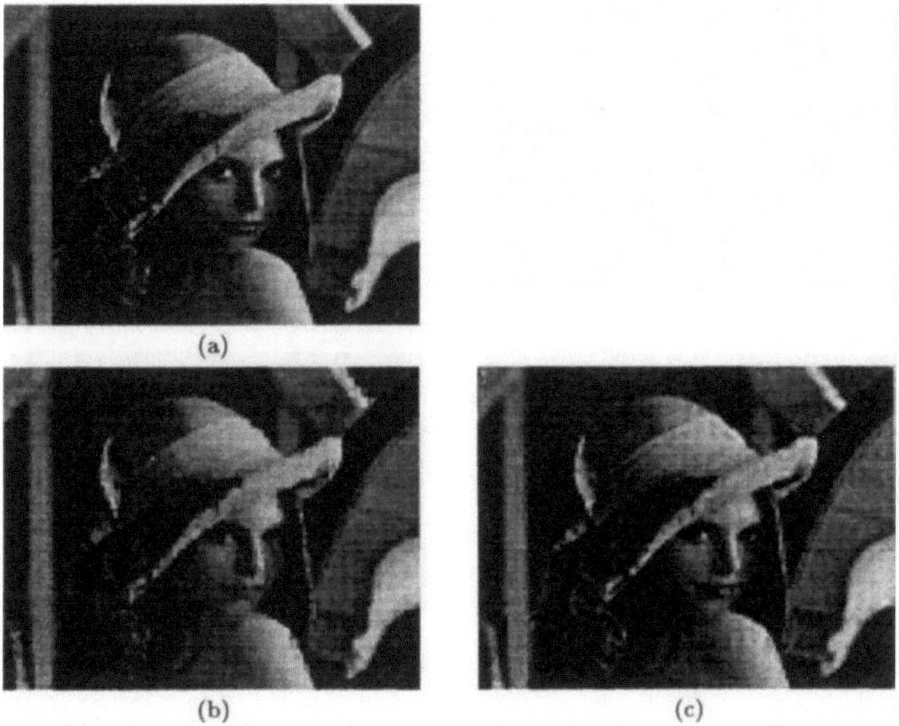

Fig. 4.1. "Lena" image (128 × 128): (a) original image,(b) conventional fractal algorithm (PSNR = 28.77 dB), (c) multiresolution fractal algorithm (PSNR = 32.58 dB).

For 512 × 512 images, the average gain with the new coder is 0.70 dB over the conventional coder. The bit rate increases from 0.07 bit per pixel to 0.11 bit per pixel. Moreover, the difference is clearly visible. Images coded with the multiresolution fractal algorithm are less blurred and more natural looking. Figure 4.2 represents the original 512 × 512 "baboon" image (a), and the coded images with the conventional coder (b) and the multiresolution one (c).

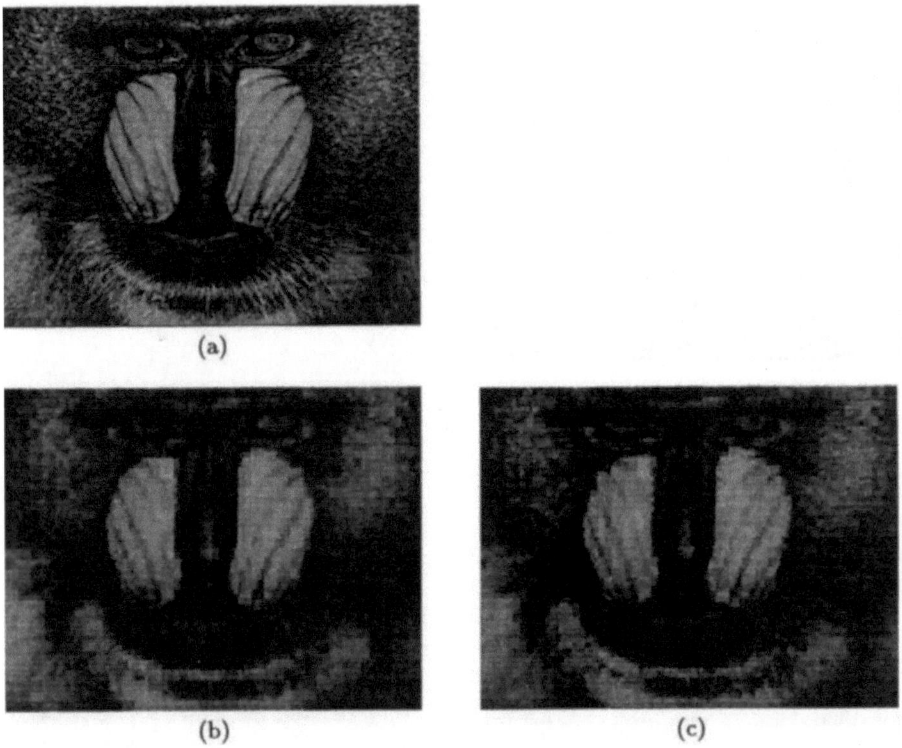

Fig. 4.2. "Baboon" image (512 × 512): (a) original image, (b) conventional fractal coder (PSNR = 18.89 dB, 0.07 bpp), (c) multiresolution fractal coder, Haar wavelet, cut-off frequency $\pi/2$ (PSNR = 19.19 dB, 0.11 bpp).

5. Extensions

In the first extension, the Haar basis used in the multiresolution decomposition is replaced by any wavelet basis. Ten different orthonormal bases are tested. In general, all bases perform better than the Haar basis, in particular by reducing blocking artifacts. However, the basis corresponding to the highest PSNR depends on the original image. The average difference between the best basis and the worst one is only 0.5 dB. Figure 5.1 (d) represents the "baboon" image coded with the multiresolution coder using the Daubechies wavelet with eight coefficients.

The second extension consists in another low-pass/high-pass decomposition. In the multiresolution coder previously described, the image used in the low-pass part is the one obtained after only one iteration of the wavelet transform. It corresponds to a filter with cut-off frequency $\pi/2$. Its size is one fourth the original size. In the secondary multiresolution coder, the wavelet transform is iterated twice. The image coded in the low-pass part corresponds to a low-pass filtered image with cut-off frequency $\pi/4$, and which size is one sixteenth the original size.

For eight 512 × 512 images, this new coder gives the best results, with an average gain of 1 dB over the conventional fractal algorithm. But, most important, the visual quality is clearly increased, see Figure 5.1 (e): edges are better coded (nose), textures are better rendered (hair), the overall image is less blurred than with the conventional fractal coder.

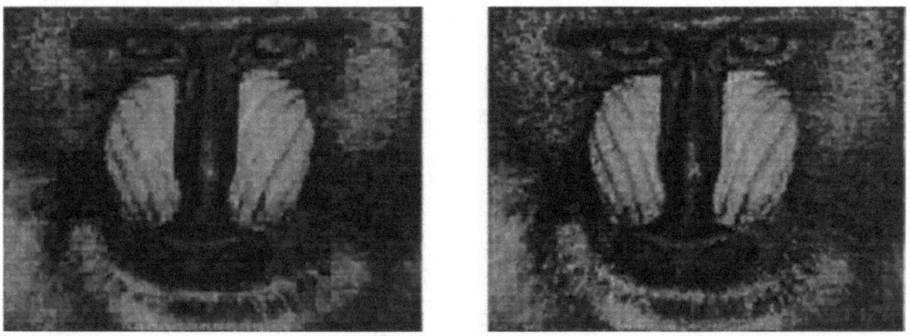

(d) (e)

Fig. 5.1. "Baboon" image (512×512): (d) multiresolution fractal coder, Daubechies wavelet, cut-off frequency $\pi/2$ (PSNR = 19.19 dB, 0.11 bpp), (e) multiresolution fractal coder, Daubechies wavelet, cut-off frequency $\pi/4$ (PSNR = 19.57 dB, 0.11 bpp).

Figure 5.2 represents the PSNRs obtained for the eight test images with four algorithms: the conventional fractal coder, the multiresolution fractal coder with cut-off frequency $\pi/2$ and the Haar wavelet, the multiresolution fractal coder with cut-off frequency $\pi/4$ and the Daubechies wavelet (8 coefficients), and the standard JPEG coder.

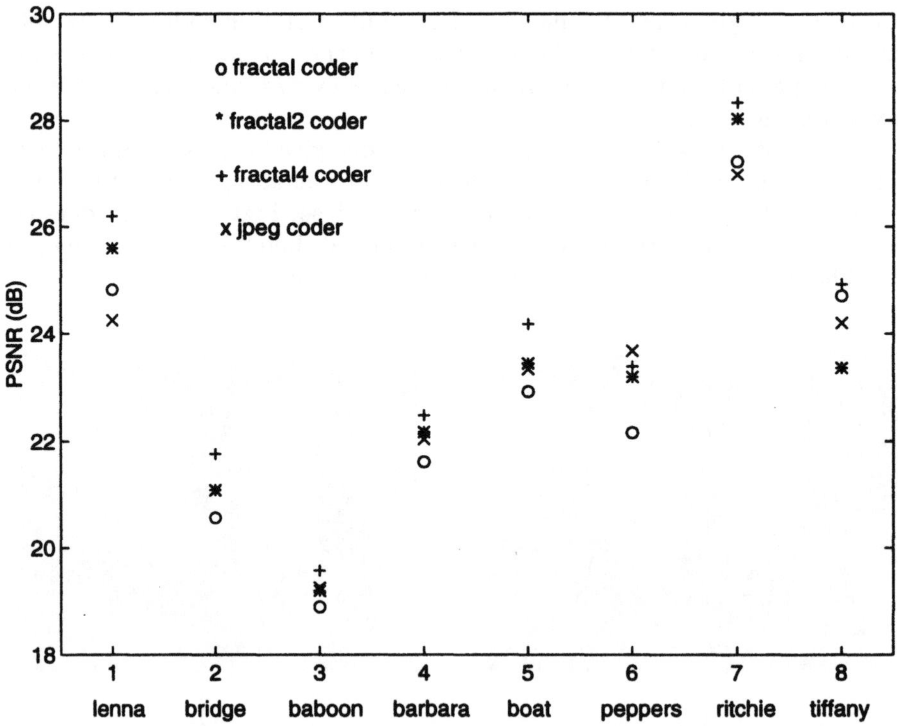

Fig. 5.2. Comparison of PSNRs for eight 512 × 512 images: *fractal coder* is the conventional fractal coder (0.07 bpp), *fractal2 coder* is the multiresolution fractal coder with cut-off frequency $\pi/2$ (0.11 bpp), *fractal4 coder* is the multiresolution fractal coder with cut-off frequency $\pi/4$ (0.11 bpp), *JPEG coder* is the standard JPEG algorithm (0.15 bpp).

6. Conclusions

The multiresolution fractal coder and its extensions have all the advantages of the conventional fractal coder. The quality of reconstructed images is good, even at low bit rates (< 0.1 bpp). The decoding part is fast, leading to an algorithm particularly suitable for applications such as database browsing or archiving. Moreover, the multiresolution coders decrease the two major distortions of fractal coded images: blocking artifacts are less annoying, high frequencies are better coded. A larger range of bit rates is now achievable. Finally, a hierarchical coder could be easily associated with multiresolution fractal algorithms: one can transmit only the low-pass part of the code to adapt to computational or display resources, or to time-varying channels.

The cost of these improvements is a slight increase of the computation time in the coding part since two different fractal coders are used. This increase is kept small since the low-pass part codes a smaller image than the original image and no offsets are computed in the high-pass part.

Further developments will generalize the multiresolution fractal scheme to any low-pass/high-pass decomposition and extend it to more than two different fractal coders inside a unique algorithm. They will also consider the study of such algorithms for a large range of bit rates.

References

1. Barnsley, M. (1988): *Fractals Everywhere*. Academic Press, New York.
2. Jacquin, A. (1989): A Fractal Theory of Iterated Markov Operators with Applications to Digital Image Coding. PhD thesis, Georgia Institute of Technology.
3. Fisher, Y. (editor) (1994): *Fractal Image Compression: Theory and Application*. Springer-Verlag.
4. Vetterli, M., Kovacevic, J. (1995): *Wavelets and Subband Coding*. Prentice Hall PTR.
5. Davis, G. (1995): Self-Quantized Wavelet Subtrees: a Wavelet-Based Theory for Fractal Image Compression. In *Data Compression Conference, DCC'95*.
6. Shapiro, J. (1996): Techniques for Fast Implementation of the Embedded Zerotree Wavelet Algorithm. In *International Conference on Acoustics, Speech, and Signal Processing, ICASSP'96*.
7. Krupnik, H., Malah, D., Karnin, E. (1995): Fractal Representation of Images via the Discrete Wavelet Transform. In *Eighteenth Convention of Electrical and Electronics Engineers in Israel*.
8. Cesbron, F., Malassenet, F. (1996): Multiresolution Fractal Coding of Still Images. In *International Conference on Signal Processing and Applications, ICSPAT'96*.

How to improve pixel-based fractal image coding with adaptive partitions

F. Davoine, G. Robert, and J.-M. Chassery

Laboratory TIMC-IMAG,
I.A.B., Domaine de la Merci,
38706 La Tronche Cedex, FRANCE

Abstract In this paper, we summarize adaptive partitioning methods traditionally used for pixel-based fractal image compression (pixel-based in opposition with hybrid methods). We discuss their properties in terms of flexibility, computing and encoding complexities, and we describe their use in a coding algorithm. In a second part, we propose a new polygonal partition which is fully flexible, and gives a better rate-distortion curve, in comparison with results based on adaptive triangulations. We describe a simple approach to compress images, based on unconstrained polygons and contractive affine mappings.

1. INTRODUCTION

The partitioning scheme is very important for fractal image compression in order to capture local self-similarities in natural images. Many researchers have already proposed different fractal coding algorithms making use of square blocks (quadtree) [6], rectangles (H-V) [6], triangles [4, 9], quadrilaterals [4] and polygons [12]. These partitions have different construction rules, which can be sorted from the most rigid to the most flexible ones. Moreover, authors have proposed to merge similar blocks in the partitions, before ([4]) or during ([13]) the fractal encoding process, in order to get better rate-distortion curves.
In this paper we show how to make use of the *Delaunay* and *Voronoï partitions*, from which we get triangles, quadrilaterals and polygons, in order to compute the fractal transform. The partitions satisfy the following properties:

- They are fully flexible, because they are computed on a set of points distributed on the image domain, in a grey level dependent way;
- The support of a block (triangle, quadrilateral or polygon) can have any orientation and shape;
- The partitions are computed in a graph structure, which permits to easily scan the blocks, and makes easier the neighborhood referencing. Moreover, we show that the storage of the adaptive partitions can be very efficient: during the computation of an adaptive triangulation, only one bit is used to code the addition or the suppression of a point in the data structure.

In the next section, we summarize the fractal image compression theory, already presented in [8] [1] [6]. We present traditional and more advanced partitioning coding schemes based on squares and rectangles in section 3.

A mixed quadrilateral and triangular partition is detailed in section 4., as well as its use in a fractal coding and decoding algorithm. In section 5., we show how to use the Voronoï partition for fractal image coding. We propose a simple algorithm making use of arbitrarily shaped polygons, and affine spatial transforms. Section 6. concludes the paper.

2. THE FRACTAL TRANSFORM

Let us consider an *eventually contractive operator* W defined on the metric space (X, d) of digital images to itself, where d is a given distance. W is eventually contractive if it is contractive at the nth iterate. With this property, to W is associated an unique *fixed image* $\mathcal{A}_t$ verifying $\mathcal{A}_t = W(\mathcal{A}_t)$, given by:

$$\mathcal{A}_t = \lim_{k \to \infty} W^{\circ k}(B) \qquad \forall B \in X, \tag{2.1}$$

where $W^{\circ k}$ is the kth iterate of W. In this case, the image $\mathcal{A}_t$ is completely defined by the operator W. The principle of fractal image compression is to construct an eventually contractive operator W for which the fixed image $\mathcal{A}_t$ is a good approximation of a given image A to be encoded. The operator W is computed so that it minimizes a distance between $W(A)$ and A, noted as $d(W(A), A)$. This requirement is given by the generalized collage theorem [6]:

$$d(A, \mathcal{A}_t) \leq \frac{1}{1-s} \, \frac{1-\sigma^n}{1-\sigma} \, d(A, W(A)) \tag{2.2}$$

where s is the contractivity of $W^{\circ n}$ and σ is the Lipschitz factor of W. Inequality 2.2 shows that the minimization of $d(A, W(A))$ is suboptimal, but easier to carry out than the minimization of $d(A, \mathcal{A}_t)$, because $\mathcal{A}_t$ is unknown during the encoding step. In order to find the operator W, Jacquin [8] proposed the use of a partition of the image A into N non-overlapping range squares R_i so that:

$$A = \bigcup_{i=1}^{N} R_i \quad \text{with} \quad R_i \cap R_j = \emptyset \quad \forall i, j \text{ and } i \neq j \tag{2.3}$$

Each original square R_i is then approximated by a *collage block* produced by a linear combination of a constant block and another domain block noted $D_{\alpha(i)}$, on average larger than R_i, and extracted from the original image itself. In the schemes we propose, the domains are selected from a regular partition D composed of M domain blocks (section 4.), or through a limited set of blocks centered on points issued from an irregular and adaptive sampling of the image support (section 5.). It is opposed to the classical selection from large librairies, composed of blocks positioned on lattices with different spacings. A domain block is used for the approximation of one or more different range blocks of the partition R. Thus α is an application from $[1, \ldots, N]$ to

$[1, \ldots, M]$. The collage block is noted $\omega_i(D_{\alpha(i)})$. In this case, the transformed image A under the operator W discussed above is expressed as:

$$W(A) = \bigcup_{i=1}^{N} \omega_i(D_{\alpha(i)}). \tag{2.4}$$

W returns a *collage image* $W(A)$ close to the original image A if this last one is sufficiently *piecewise self-similar* (large regions resemble smaller regions in the partition R). This kind of *scale-redundancy* makes very important to compute the partition R in a grey level dependent way.
Finally, the compressed representation of the image A typically contains the partition R construction rule, the coefficients of the N local contractive transforms w_i associated to each block R_i and the coding of the shapes and locations in the image of the corresponding blocks $D_{\alpha(i)}$.

3. SQUARES AND RECTANGLES

In order to build the fractal operator W, a simple rigid square partition has been firstly proposed to find local self-similarities in a natural image. However, this technique did not provide good visual quality decoded images because of the fixed size of the blocks. In this case, a compromise had to be found: in order to insure a significant compression ratio, the range blocks must be not too small (8×8 in [8]), but small enough to be accurately encoded. The first solution to increase the optimality of the partition was to split initial squares covering complex areas such as fine textures and edges, into four sub-squares [8]. But due to the fixed size of the partition, too small blocks still covered smoothly varying image regions.
The quadtree-based partition is a solution to this problem [6]. It provides better decoding results in terms of visual quality, but the method stays too rigid, because the partition returns a lot of unuseful squares near edges between homogeneous and complex regions of the images (Figure 3.1.a).
Other schemes based on square blocks have been proposed recently for fractal image coding [14, 13]. In [13], the authors introduce evolutionary computing in order to build a range partition with arbitrary shaped regions, each region being the union of edge-connected small *atomic* image squares (Figure 3.1.b). The algorithm starts with a uniform partition obtained by subdividing the image with 4×4 blocks. The evolutionary process makes use of populations composed of finite sets of configurations, each configuration consisting of a partition with its associated fractal code. At the end of each step of the evolutionary process, a new partition is constructed, which has one range less than its parent partition. The authors compare their method to the quadtree-based method [6] and observe PSNR-gains up to 2 dB.
The HV partition, more flexible than the quadtree-based partition, can also be constructed in an image-dependent way [6]. This partition is arranged

so that strong vertical or horizontal edges separate the blocks into homogeneous parts, or so that edges in the image tend to run diagonally through the blocks (Figure 3.1.c). Y. Fisher [6] compares his method to the quadtree-based method and observes PSNR-gains larger than 2 dB.

Fig. 3.1. a: Partition Quadtree: 5704 squares, b: Evolutionary computing: 500 regions (from [13]), c: Partition HV: 2910 rectangles (from [6]).

4. TRIANGLES AND QUADRILATERALS

4.1 Delaunay triangulation

Let S be a set of points in the plane. The Delaunay triangulation $DT(S)$ associated to S is the unique triangulation with *empty circles*. More formally, we can write:

$$DT(S) = \{(p,q,r) \in S^3, C(p,q,r) \cap S - (p,q,r) = \emptyset\} \tag{4.1}$$

where $C(p,q,r)$ is the circle circumscribed by the three points p, q, and r. An important property of the Delaunay triangulation for our compression

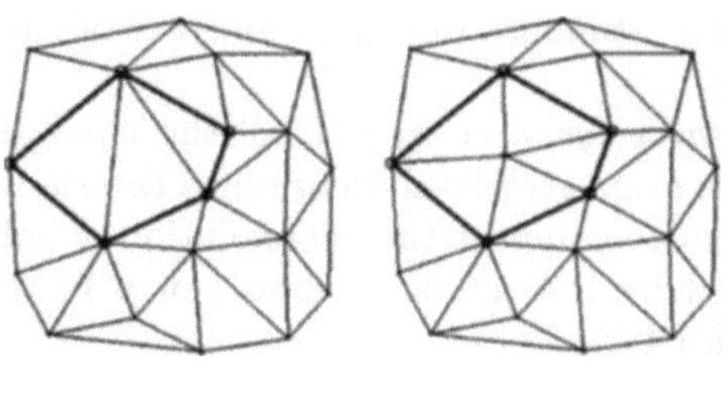

Fig. 4.1. Addition of a point in the Delaunay triangulation: Before and after the addition. The addition of a new point in the triangulation works locally, inside a *star* polygon.

application is the *local max-min angle* criterion which says that the smallest interior angle of the triangulation is maximized [10]. This property reduces the error in the localization of each pixel in the triangles. Moreover, we can note that the dual graph of the Delaunay triangulation is the well know Voronoï diagram. We will make use of this property in section 5. To compute the triangulation, the best known approaches are based on the "divide and conquer" [11] and the "incremental" algorithms. The incremental algorithm was formulated by Green and Sibson [7]. The idea is to compute the Delaunay diagram iteratively, by adding new points in the triangulation. We use for this purpose an adaptation of the algorithm initially proposed by Bowyer in [2]. In practice, 10000 triangles are computed in less than 1 second on a Silicon Graphics Indigo workstation with an incremental method.

4.2 Application to image representation

We present three approaches for the computation of the set of points adapted to the image content: the merge only, the split and merge and the edge constrained approaches.

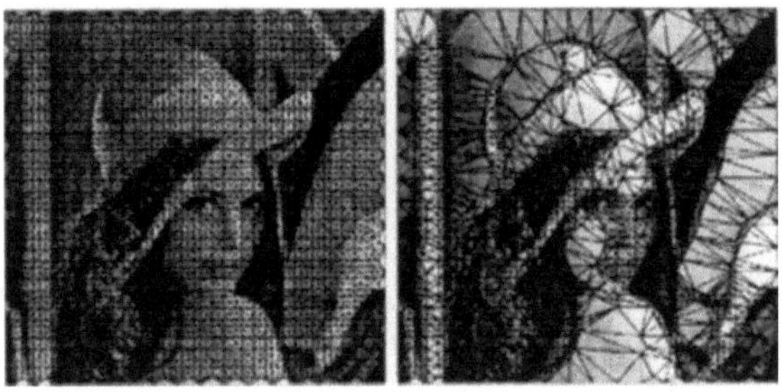

Fig. 4.2. Merge algorithm computed on the image Lena 512 × 512. a: Initialization, b: Merge step

The following algorithm shows the steps of the merge only approach:

1. Initialize the process with a dense lattice on the image support (Poisson process with intensity λ or regular lattice).
2. Compute the Delaunay triangulation.
3. Compute the grey level mean value, standard deviation and size of each triangle.
4. Merge the triangles under the *star polygon* criteria. If all the triangles arising from a top point p have the same grey level according a threshold, then p is deleted. This is equivalent to compute the triangulation of only the star polygon which is defined by the set of neighbors of p in the Delaunay triangulation (see Figure 4.1).

The step (4) consists in deleting "useless" triangles from the Delaunay triangulation. In practice, small triangles stay on the texture regions of the image and large triangles appear on the homogeneous regions. The result of such an algorithm is shown on Figure 4.2.

The second approach is the split and merge approach. The algorithm proceeds in the following three major steps:

1. Initialize the process with a lattice on the image support.
2. Repeat until convergence: homogeneity of all the triangles (standard deviation of each triangle is less than a fixed threshold E and triangle sizes are greater than a fixed threshold T).
   a) Compute the Delaunay triangulation,
   b) Compute the grey level mean value, standard deviation and size of each triangle,
   c) Split each non-homogeneous triangle (standard deviation is greater than E) by adding a point on its barycenter: this is a non-adaptive point addition to simplify the coding of the adaptive triangulation.
3. Merge the triangles under the *star polygon* criteria.

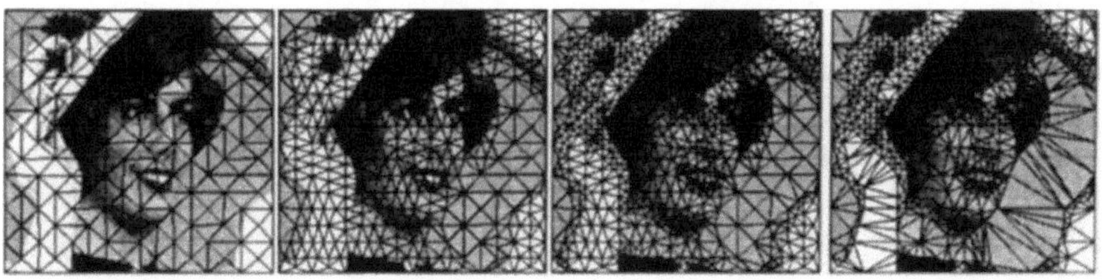

Fig. 4.3. 2 splits and 1 merge step, initialized on a lattice. Final partition: 893 triangles.

Step (3) is equal to step (4) of the merge only algorithm. With this algorithm, the image is adaptively divided into small triangles that cover texture and edge regions, and into larger triangles on homogeneous regions. However, the method gives better results than the merge only method near the edges. The triangles are better adapted in those regions during the split steps and are left unchanged during the merge step (see Figure 4.3).

The third approach is called edge-based approach:

1. Detect the edges in the image.
2. Operate a curve partitioning task by positioning:
   a) points p_i on extrema of local curvature along the contours,
   b) points p_j on the contours, regularly distributed between the points p_i, separated from an increment d_1,
   c) points p_k, on each side of the segments resulting from steps (a) and (b), at a distance $\frac{d_2}{2}$ to the contour. During this step, the condition $d_2 > d_1$ must be satisfied in order to have triangles along, and not above the contours.
3. Triangulate the set of points p_i, p_j and p_k (Figure 4.4.b),

4. If necessary, split the triangles larger than a fixed size T by adding a new point on their barycenter.

The result of this algorithm is given on Figure 4.4.c. The partition is com-

Fig. 4.4. Egde constrained triangulation: edges, points positioning and associated triangulation.

posed of homogeneous triangles along the main edges in the image. This can be compared to the H-V partition which returns homogeneous rectangles on each side of the horizontal and vertical edges.

4.3 Insertion of quadrilaterals

We proposed in a previous work to reduce the number of comparisons between a range triangle and the domain triangles. This was done by integrating a classification step of the blocks, based on a modified version of a classical vector quantization algorithm [3]. The reduction of the number of comparisons was obtained by keeping a set of best representative triangles in the domain blocks partition. In this section we propose a different optimization approach

Fig. 4.5. A mixed partition of the image Lena (512 × 512), composed of 1706 triangles and 715 quadrilaterals. The algorithm returns an adaptive partition in which many quadrilaterals are distributed in homogeneous and texture regions, and triangles are more densely distributed in regions containing interesting features such as corners and edges.

of the encoding step by extracting from the range partition R adjacent triangles that form convex quadrilaterals. The purpose is to significantly reduce the number of range blocks, and consequently the number of transformations ω_i to code. The first step consists in testing whether or not a quadrilateral is concave (or almost degenerated). Thereafter, we use a multiple test regarding the differences of mean values and standard deviations between the two triangles. If the values are less than given thresholds we merge the two triangles (see Figure 4.5).

4.4 Fractal encoding algorithm

The encoding algorithm is based on a two-level Delaunay triangulation of the image domain. The partition D, discussed in section 2., contains regular domain blocks and is not adapted to the image content. It provides a pool of domain blocks available for the encoding process. It should be noted that the partition D can provide simultaneously triangles and quadrilaterals. Moreover, the size of the pool is increased by considering the 6 (resp. 8) isometries (rotations and flips) applied on triangular (resp. quadrilateral) blocks. The partition R returns range blocks R_i of smaller or equal "mean area" and is computed with one of the algorithms proposed in section 3.. Neighboring triangles that form convex quadrilaterals are then merged; the other triangles are left unchanged.

Transformations ω_i composing the operator W determine the way to map a block $D_{\alpha(i)}$ onto a block R_i and is written as:

$$\begin{aligned}
\omega_i(D_{\alpha(i)}) &\equiv \omega_i(x_m, y_m, f(x_m, y_m)) \quad \forall\, (x_m, y_m) \in D_{\alpha(i)} \\
&= (x'_m, y'_m, f(x'_m, y'_m)) \\
&= (v_i(x_m, y_m), s_i\, f(x_m, y_m) + o_i),
\end{aligned}$$

where $f(x_m, y_m)$ is the luminance of the pixel at coordinates (x_m, y_m) in the block $D_{\alpha(i)}$, $f(x'_m, y'_m)$ is the luminance of the pixel at coordinates (x'_m, y'_m) in the block R_i, and s_i, o_i respectively control the contrast and brightness of the grey level transformation.

The contractivity of the transform W is controlled by the parameters s_i. If $s = Max_i\{s_i\} \;\; \forall \;\; i = 1 \dots N$ is less than 1, W is insured to be eventually contractive, considering the mean square error. Moreover, it has been experimentally verified [6] that W stays eventually contractive if $s < 1.5$. The

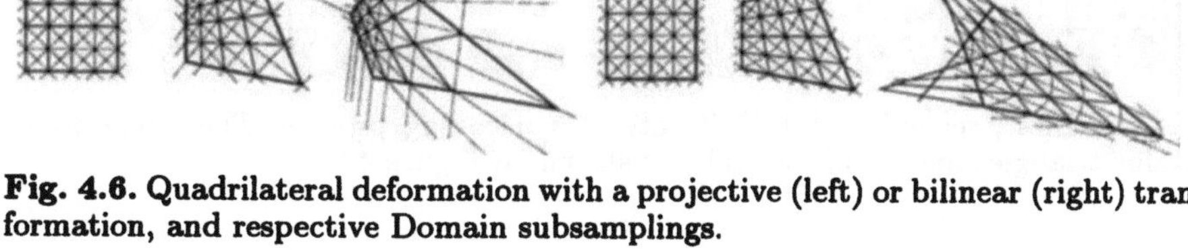

Fig. 4.6. Quadrilateral deformation with a projective (left) or bilinear (right) transformation, and respective Domain subsamplings.

spatial transformation v_i maps a block $D_{\alpha(i)}$ onto a block R_i ($v_i(D_{\alpha(i)}) = R_i$). The transformation used to map a triangle onto another is affine and is used in other fractal image compression schemes based on square and rectangular blocks [8] [1] [6]. Two planar quadrangles are generally not affine images of each other. For this purpose, we have tested two possible transformations: the *projective* and *bilinear* spatial transformations, and retained the first one for its computational efficiency (see Figure 4.6 and [5]).
The similarity (in terms of grey levels) between a block R_i and the transformed block $D_{\alpha(i)}$ is measured by the square error (SE) given by:

$$d(R_i, \omega_i(D_{\alpha(i)})) = \sum_{m=1}^{n_r} \left[f(x'_m, y'_m) - s_i f(v_i^{-1}(x'_m, y'_m)) - o_i\right]^2 \tag{4.2}$$

for any $(x'_m, y'_m) \in R_i$, where n_r is the number of pixels included in R_i. We use the inverse transformation v_i^{-1} in order to consider all "discrete" points in the block R_i. If the block R_i is smaller than $D_{\alpha(i)}$, the block $D_{\alpha(i)}$ is subsampled and averaged. However the subsampling is not uniform when two different shape quadrilaterals are compared, as it is shown in Figure 4.6.

4.5 Fractal decoding algorithm

The decoding step consists, after reconstruction of the two partitions R and D, in iterating the operator W, starting with any initial image B. The reconstructed image is given by $A \simeq \lim_{n\to\infty} W^{o_n}(B)$. One iteration consists in scanning each block R_i and in applying the transformation ω_i on their corresponding domain block $D_{\alpha(i)}$. The grey level $f_n(x'_m, y'_m)$ of the pixel m in the block R_i at the nth iteration of W is given by :

$$f_n(x'_m, y'_m) = s_i\, f_{n-1}(v_i^{-1}(x'_m, y'_m)) + o_i \quad \forall\, (x'_m, y'_m) \in R_i. \tag{4.3}$$

Algorithm needs 5 to 10 iterations to converge. We observe that the number

Fig. 4.7. First, second and 10th iterate of a decoding process. The partition R contains triangles and quadrilaterals, as shown on Figure 4.5.

of iterations is mainly dependent on the block size differences between the two partitions. The computation time for a 256 × 256 image is less than 3 seconds. Figure 4.7 illustrates three decoding iterations.

Fig. 4.8. The figure illustrates the decoding process of one pixel. Considering one pixel in a given range triangle, we show its antecedent in the associated domain triangle. Starting from this new pixel that is included itself in a range triangle, we show its antecedent, included in a new domain block. We continue this process 15 times. Experimentally, we observe that if we do not constrain the searches of similarities between blocks, these last ones are distributed randomly on all the image support. Many parts of the image participate to the iterative reconstruction of one given pixel.

4.6 Compression

To encode an image, we need to store the coefficients of the N mappings ω_i that compose the operator W. One mapping is needed for each block R_i in partition R. The Delaunay partition is coded in a graph environment. Thus, for one transformation ω_i, it is only necessary to code:

- the position of the domain triangle D_i in Delaunay graph with M_T bits.
- the collage process between triangles (rotation, flips) with 3 bits.
- the scale coefficient s_i quantized with 6 bits (uniform scalar quantization)
- the offset coefficient o_i quantized with 6 bits (uniform scalar quantization)

In addition, the image adaptive partition R must be coded (see [5]). For example, the final partitions in Figure 4.2 (resp. 4.3) is coded with 3936 (resp. 1370) bits, before entropy coding. The remaining coefficients $a_i \dots h_i$ are not stored because they can be computed by the decoder, given the decoded partition R and the known non-adaptive partition D. The total number of bits necessary to code the operator W is given by:

$$N_{tot} = N(6 + 6 + 3 + M_T) + X , \tag{4.4}$$

with N = total number of blocks R_i in the partition R and X is the number of bits to code the partition R. An analysis of our practical results shows that usually 5% (resp. 25%) of the bit-stream is devoted to the partition computed with the split and/or merge (resp. edge-based) approach.

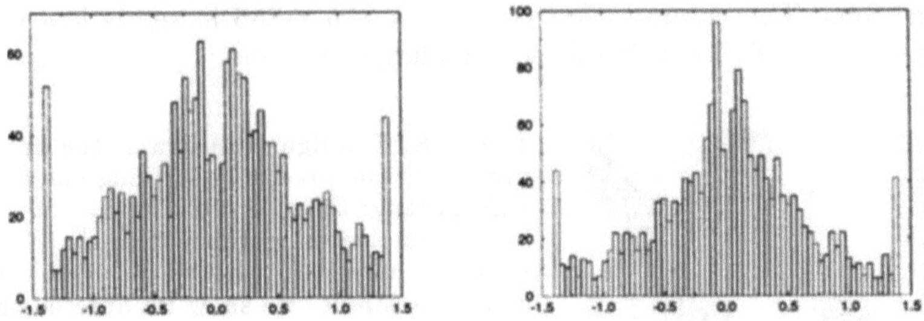

Fig. 4.9. Histogram of the scale coefficients s_i of two fractal codes computed on different triangulations composed 3450 triangles, considering the image *Lena* (512 × 512 pixels). Left: Split and merge triangulation. Right: edge-based triangulation.

4.7 Results

4.7.1 Split and merge or edge-based triangulation?. The edge-based triangulation was initially proposed in order to provide homogeneous (similar) triangles on homogeneous regions against detected image edges. Conversely, the split and merge approach provides non-homogeneous small triangles that cover the edges. Experimentally, we verify that the images are better reconstructed, in terms of PSNR, when most of the range blocks are non homogeneous. We show on Figure 4.9 that the scale coefficients s_i of a fractal code computed on an "edge-based" partition are mainly concentrated around zero. Moreover, we see that they are better distributed between s_{min} and s_{max} if the fractal code is computed on a "split and merge" partition that presents non homogeneous blocks. The influence of the scale coefficient on the higher bound of the distance between the original and the reconstructed image is shown in equation 2.2.

4.7.2 Triangles only, or triangles and quadrilaterals?. We show in this part that the use of quadrilaterals allows:

- to increase the compression ratio by reducing the number of transformations ω_i,
- or to increase the quality of the decoded image, while preserving a constant compression ratio, by computing smaller triangles on the image contours, before to extract quadrilaterals.

This triangular partitioning scheme is thus convenient for fractal image compression and could be further optimized by better adapting the partition on large homogeneous regions. However, today our PSNR and visual results stay inferior to the results given by the rectangular partition, initially proposed by Y. Fisher [6].

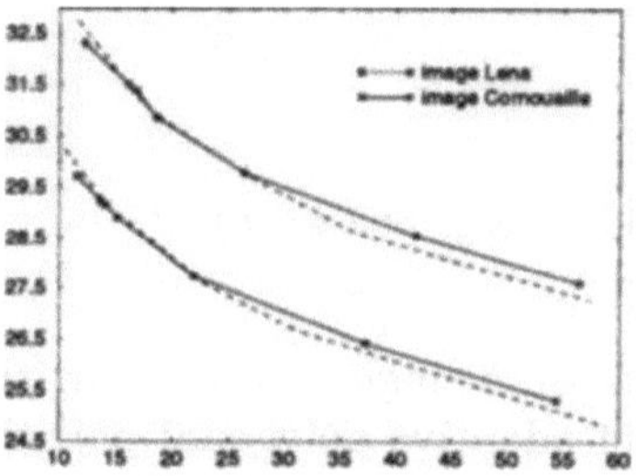

Fig. 4.10. *PSNR* for different compression ratios (T_c), considering the images Lena and Boat 512 × 512. Continuous lines: triangles and quadrilaterals. Dashed lines: triangles only. This figure shows that if the compression ratio increases, the decoded image quality stays higher with a mixed partition R than with a triangular partition. We verify experimentally that the reconstructed images contain fewer artifacts on the diagonal edges than methods based on the quadtree partitioning.

5. POLYGONS

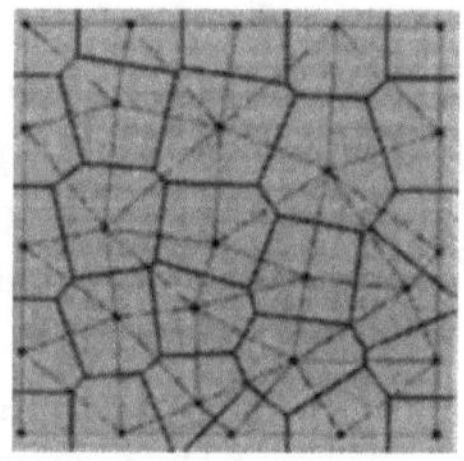

Computation of Voronoï diagram is similar to computation of Delaunay triangulation thanks to their duality. We use an incremental method, as described in section 4.1, which consists in addition or suppression of seeds, based on only local processing.

5.1 Application to image representation

A split and merge approach has been retained in order to adapt a Voronoï diagram to the image content. The split and merge algorithm used to build a Voronoï tessellation, is composed of three steps:

1. Initialize the process with some seeds on the image support (randomly distributed) and compute the Voronoï diagram. Compute grey level mean value and standard deviation on each polygon, considering the pixels included in each one.
2. Split the non homogeneous polygons (standard deviation greater than a fixed threshold), by adding new seeds on the middle of each edge of the polygon (Figure 5.1.a). Update the diagram.
3. Merge a polygon (suppress its seed) each time it is homogeneous and surrounded by other homogeneous polygons similar to it. The similarity criterion is based on the difference between mean grey level value of two polygons (Figure 5.1.b). Update the diagram.

The strategy consists in splitting the image until all polygons are homogeneous. Then, merge steps are applied until there is none useless seeds left.

5.2 Encoding algorithm

As we presented in section 4., the fractal encoding algorithm based on Delaunay triangulation uses two partitions, that return range and domain blocks.

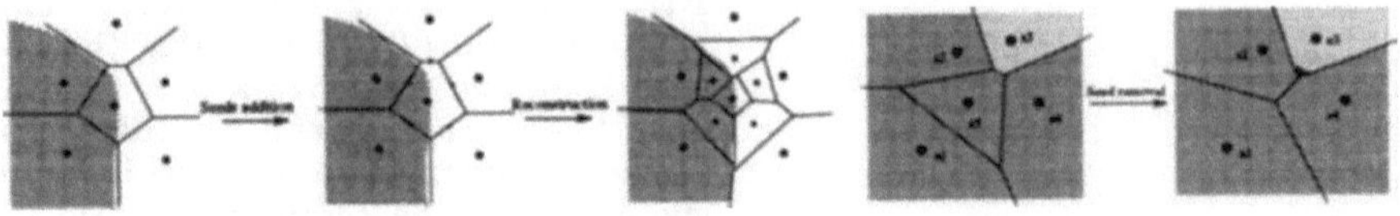

Fig. 5.1. Illustration of split (a) and merge phases (b)

In this part, we present a new approach based on a single Voronoï partition: only range blocks are extracted from a given adaptive polygonal partition. Domain blocks are directly evaluated from range blocks through affine spatial transformations. To each range block we associate a dilated polygon which is translated on various seeds (these seeds are issued from the set of points that define the polygonal partition). The goal of the encoding step is to determine and encode the best parameters θ_i (rotation), k_i (spatial contraction) and $(a_i, b_i)^t$ (translation) of each affine transform ω_i, in order to map as accurately as possible a polygonal part of the image on the range polygon R_i. Considering the range polygon R_i associated to a given seed of the adaptive Voronoï partition, the computation of ω_i can be summarized as follows :

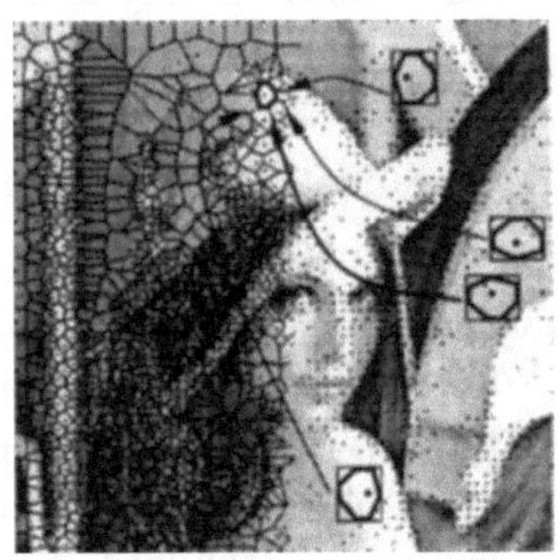

Look for the associated domain block D_i in the image, that minimizes the matching error $d(R_i, \omega_i(D_i))$ by using equation 4.2. The search is done by considering the set of seeds that define the partition. Each tested block D_i is an affine antecedent of R_i, centered on a seed. The size of the domain block depends on the size of the range block (the contraction parameter k_i is less than 4:1 and equal to 2:1 on average). Different rotation angles θ_i can be considered to improve the matching process.

5.3 Results

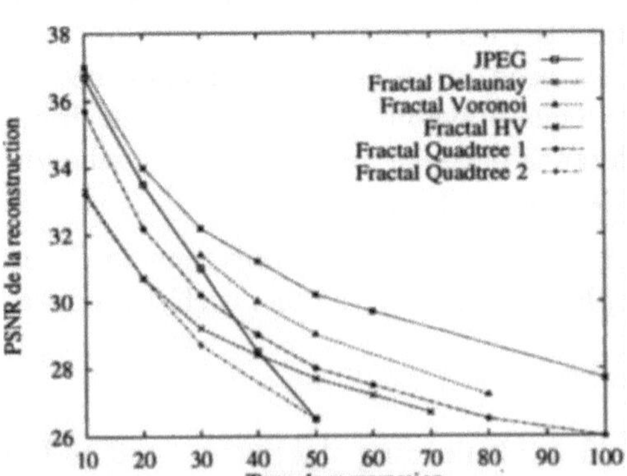

Fig. 5.2. PSNR versus compression ratio curves computed by using a single Voronoï destination partition. The compression ratio is directly linked to the number of polygons in the adaptive Voronoï partition.

Fig. 5.3. From top-left to bottom-right: Voronoï partition, 1st, 2nd, 3rd, 4th and 7th decoding iteration of the image Lena 512 × 512. (PSNR = 31.6 dB, $T_c = 30:1$)

Figure 5.2 presents the efficiency of the proposed method. We observe that polygonal adaptive partitions allow to improve previous results based on Delaunay triangulations [4] or square blocks (quadtrees) [6]. Moreover, the method can be optimized (in terms of computation times) by noticing that the set of seeds that define the adaptive polygonal partition are well distributed on the image support in a grey level dependent way. This non uniform distribution of points can be analysed in order to classify the content of each range polygon and cancel useless block comparisons during the encoding process.

6. CONCLUSION

In this paper, we have addressed the problem of fractal image compression on different adaptive partitions with variable block shapes. We have presented the main partitioning approaches based on squares, rectangles, triangles and quadrilaterals. The use of different triangulation construction rules gives us the possibility to analyse the influence of the partition on the decoded image quality. By this way, we compare the results obtained with two kind of triangulations: a triangulation with a lot of non homogeneous blocks, and another one maximizing the number of homogeneous block. In a second part, we present a new encoding algorithm based on a unique range adaptive polygonal partition, and affine spatial transforms. The polygonal partition is computed on a set of points randomly distributed on the image support, and adapted to the image in a grey level dependent way, by using a split and merge approach. Such a partition contains less blocks than its dual partition composed of triangles, and thus, we show that it is possible to improve the decoding results. We shall investigate in a future work how to make better use of the Voronoï partition properties (spatial relationships between polygons) in order to search domain blocks in the neighborhood of each range block. This should be useful in order to encode independent image regions (*objects* given by a segmentation process.

References

1. Barnsley, M.F., Hurd, L.P. (1993): *Fractal Image Compression.* AK Peters Ltd., Wellesley.
2. Bowyer, A. (1981): Computing Dirichlet tessellations. *The computer J.*, **24**(2), 162–166.
3. Davoine, F., Antonini, M., Chassery, J.-M., Barlaud, M. (1996): Fractal image compression based on Delaunay triangulation and vector quantization. *IEEE Transactions on Image Processing,* **5**(2), 338–346.
4. Davoine, F., Svensson, J., Chassery, J.-M. (1995): A mixed triangular and quadrilateral partition for fractal image coding. *Proc. of the International Conference on Image Processing,* **3**, 284–287, Washington, D.C..
5. Davoine, F. (1995): Compression d'images par fractales basée sur la triangulation de Delaunay. *Thèse de l'Institut National Polytechnique de Grenoble.*
6. Fisher, Y. (editor) (1995): *Fractal Image Compression: Theory and Application to Digital Images.* Springer Verlag, New York.
7. Green, P.J., Sibson, R. (1978): Computing Dirichlet tessellation in the plane. *The computer J.*, **21**, 168–173.
8. Jacquin, A.E. (1992): Image coding based on a fractal theory of iterated contractive image transformations. *IEEE Transactions on Image Processing,* **1**(1), 18–30.
9. Novak, M. (1993): Attractor coding of images. *Licentiate Dissertation, Dept. of Electrical Engineering, Linköping University.*
10. Okabe, A., Boots, B., Sugihara. K. (1992): *Spatial tesselations: concept and applications of the Voronoi diagram.* John Wiley and Sons, New York.
11. Preparata, J.P., Shamos, M.I.S. (1988): *Computational Geometry, an Introduction.* Springer Verlag, New York.
12. Reusens, E. (1994): Partitioning complexity issue for iterated function systems based image coding. *Proc. of the EUSIPCO*, Edinburg, U.K., **1**, 171–174.
13. Saupe, D., Ruhl, M. (1996): Evolutionary Fractal Image Compression. *Proc. of the International Conference on Image Processing,* **1**, 129–132, Lausanne.
14. Thomas, L., Deravi, F. (1995): Region-based fractal image compression using heuristic search. *IEEE Transactions on Image Processing,* **4**(6), 832–838.

CHEMICAL ENGINEERING

Transport Phenomena on Fractals and Green-Function Renormalization

M. Giona[1,3] and A. Adrover[2,3]

[1] Dipartimento di Ingegneria Chimica, Universitá di Cagliari piazza d'Armi, 09123 Cagliari, ITALY
[2] Dipartimento di Ingegneria Chimica, Universitá di Roma "La Sapienza" via Eudossiana 18, 00184 Roma, ITALY
[3] Centro Interuniversitario sui Sistemi Disordinati e sui Frattali nell'Ingegneria Chimica c/o Dip. Ingegneria Chimica Universitá di Roma "La Sapienza" via Eudossiana 18, 00184 Roma, ITALY

Abstract This article highlights the practical relevance of Green-function renormalization within the most common engineering applications of transport. Simple extensions of the method are discussed by considering Cayley trees and other loopless structures referred to as *ultrafractals*.

1. Introduction

Fractal structures are often encountered in chemical physics and transport phenomena involving solid materials (catalysts, adsorbents, porous media) [1, 2]. Many materials (such as aerogels and microporous adsorbents) possess fractal properties (related to surface structure, to distribution of active reacting/adsorbing sites, or to pore-network topology) over a given range of lengthscales. In engineering applications, fractal structures can be viewed as a limiting example of materials possessing complex geometry and topology. The study of transport and reaction phenomena on fractals permits us to highlight and extrapolate physical laws of general validity in engineering modeling and design. Furthermore, the development of a coherent and simple mathematical physics of dynamic phenomena on fractal structures, capable of solving problems which can be addressed in closed form for Euclidean continua, is of great scientific importance. In this perspective, one of the main objectives is to show that dynamic phenomena on fractals can be addressed by means of rigorous mathematical methods, and not only by means of computer simulations, random walks, or approximate models.

Over the last few years, a systematic study of transport models of engineering interest has been pursued by means of Green-function Renormalization (henceforth GFR) [3, 4, 5, 6] (a form of real-space renormalization). This method makes it possible to obtain within arbitrary precision on arbitrarily large lattice structures the relevant quantitative information found to be experimentally measurable in laboratory experiments, simply by iterating an extremely small number of exact recursions for the pivotal Green functions [7, 8]. The number of pivotal Green functions depends on the symmetries of the lattice structure and on the transport problem (diffusion, advection/diffusion etc.). Let us consider a discretized representation of the transport equations[1] on a graph $\mathcal{G}$:

$$\frac{dc(t)}{dt} = (A+B)c(t) + f(t) \,, \qquad c(t=0) = c_o \,, \tag{1.1}$$

where $c(t)$ is the concentration field defined on each lattice site, $f(t)$ a forcing term accounting for interactions with the surroundings, A the coefficient matrix expressing the transport dynamics on the internal nodes, and B the boundary-condition matrix defined on the external nodes of the structure. The solution of eq. (1.1) in the Laplace domain attains the form

$$\widehat{c}(s) = \widehat{G}^*(s)[c_o + \widehat{f}(s)] \,, \tag{1.2}$$

where $\widehat{G}^*(s) = (sI - A - B)^{-1}$ is the Green function matrix.

The lattice structure $\mathcal{G}$ can be viewed as the limit of a constructive approximation, i.e. as the limit of a sequence of graphs $\{\mathcal{G}^{(n)}\}$, such that the structure at iteration $n+1$, $\mathcal{G}^{(n+1)}$ is given by the union of ν copies of $\mathcal{G}^{(n)}$ connected only through the external nodes (this construction process defines finitely ramified structures). As a consequence of the construction principle, the relation between the coefficient matrices $A^{(n+1)}$ and $A^{(n)}$ for generic diffusion/reaction/adsorption processes[2] is given by $A^{(n+1)} = \overline{A}^{(n)} + V^{(n)}$, where $\overline{A}^{(n)}$ can be put in a block-diagonal form whose ν block equals $A^{(n)}$ (and corresponds to the coefficient matrix of the ν copies of $\mathcal{G}^{(n)}$ before connection), and $V^{(n)}$ is an extremely sparse matrix (referred to as the connection matrix) whose non-vanishing entries $V_{ij}^{(n)}$ occur for i, j representing the external sites of the ν copies of $\mathcal{G}^{(n)}$.

In the same way, the boundary matrix B admits non-zero entries B_{ij} for i, j corresponding to external nodes of $\mathcal{G}^{(n+1)}$.

Let us indicate with $\widehat{G}^{(n)}(s) = (sI - A^{(n)})^{-1}$ the *bare* Green-function matrix at iteration n [6]. From elementary matrix algebra it follows that

$$\widehat{G}^{(n+1)}(s) = \overline{G}^{(n)}(s) + \overline{G}^{(n)}(s)V^{(n)}\widehat{G}^{(n+1)}(s) \,, \tag{1.3}$$

where $\overline{G}^{(n)}(s)$ can be put in a block-diagonal form whose ν block equals $G^{(n)}(s)$, and

$$\widehat{G}^*(s) = \widehat{G}^{(n)}(s) + \widehat{G}^{(n)}(s)B\widehat{G}^*(s) \,. \tag{1.4}$$

Eq. (1.3) provides a self-consistent set of linear equations relating the pivotal Green functions (associated with the external nodes) at iteration $n+1$ to the pivotal Green functions at iteration n. The solution of this system of equations yields the renormalization recursion. In the same way, eq. (1.4) furnishes the boundary transformation for the specific boundary conditions under consideration.

[1] Analysis is limited here to the linear case.

[2] The matrix $A^{(n)}$ in pure diffusion equals the adjacency matrix $H^{(n)}$ of the graph $\mathcal{G}^{(n)}$ plus an isotropic term $-pI$, where p is the coordination number of the structure.

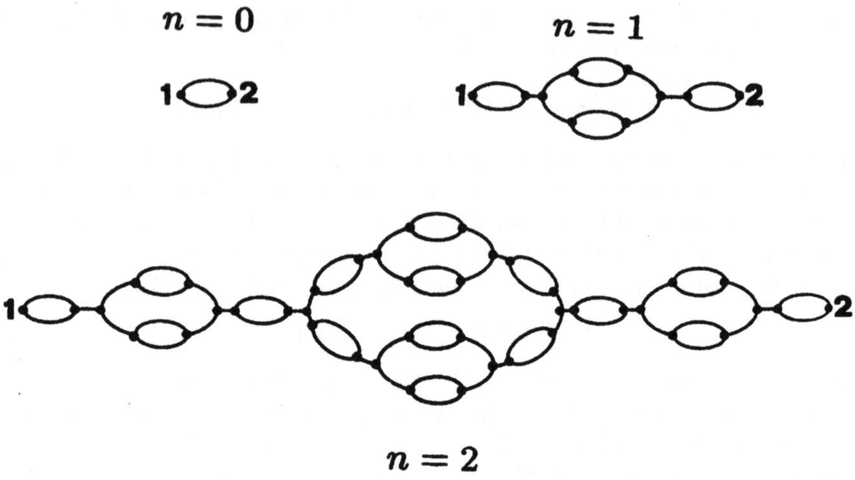

Fig. 1.1. Pictorial representation of the construction process of a finitely ramified fractal set. The structure is the ARC(3)-lattice. Nodes $i = 1, 2$ are the external nodes.

To give an example, Fig. 1.1 shows the construction of the ARC(3)-lattice [3, 9]. In the case of pure diffusion, there are two pivotal Green functions: $x = \widehat{G}_{11}$ and $y = \widehat{G}_{21}$. Nodes 1 and 2 are the external nodes of the structure. A detailed analysis of the renormalization method and of the renormalization recursion can be found elsewhere [3, 7, 8], and is not reported here so as to avoid pointless repletion.

It is important to stress that, while real-space renormalization is a fairly well-known technique, the systematic study of transport schemes (including convection, reaction and adsorption), the inclusion of proper boundary conditions, the analysis of different feed configurations, etc., represent the main novelty of this research project, which is mainly aimed at providing a rigorous framework for the study of transport properties on fractals.

The aim of this article is to show how the results obtained for a large class of fractal structures by means of GFR can be applied to address simple and practical engineering problems involving fractal structures and to derive quantitative results of general validity.

In addition, the final part of this article (section 4.) shows some new results on two classes of highly pathological lattices: Cayley trees and *ultrafractals*. While Cayley trees are extensively studied in the framework of percolation theory [10], ultrafractals are a new family of lattice structures, introduced for the first time in this article, which possess highly anomalous transport properties.

2. GFR: Exact Results and Connection with Experiments

In this section we briefly review some results obtained by means of GFR in the study of transport and reaction processes and show how these results may be connected to simple experiments.

2.1 Elementary Reaction Kinetics

In catalytic processes, chemical kinetics is coupled to diffusive transport inside porous particles [11]. For elementary kinetics, a rate equation of the form $-r(c) = kc^m$ [mol s^{-1} m^{-3}], $m > 0$, is assumed to describe the consumption rate of a reactant whose concentration is c.

In standard equipment, such as packed reactors, the effective reaction rate r_{eff} (which is, in most cases, the only quantity experimentally measurable from the value of the outlet conversion) can be expressed as

$$r_{eff} = \eta r(c_o) = -\eta k c_o^m \, , \tag{2.1}$$

where c_o is the reactant concentration in the fluid phase, and η the effectiveness factor,

$$\eta = \frac{1}{V_p r(c_o)} \int_{V_p} r(c) dV \, , \tag{2.2}$$

i.e. η is the ratio of the quantity reacted inside the pellet per unit time (V_p being the volume of the pellet) and the quantity that would have been reacted in the presence of uniform external concentration c_o.

The effectiveness factor for elementary reactions is a simple measurable quantity at steady state and characterizes the interplay between diffusion and reaction inside the porous particles. It is customary to plot it against the dimensionless parameter characterizing diffusion/reaction kinetics, i.e. the Thiele modulus ϕ, defined as $\phi^2 = kc_o^{m-1}L^2/D$, where D is the diffusion coefficient and L the characteristic linear size (e.g. the radius) of the porous particles.

The results obtained by applying GFR to first-order reactions on fractal particles (i.e $m = 1$) are summarized below. Before doing this, it is important to stress that in the definition of the transport scheme, the exchange manifold $\mathcal{B}$ should be specified. In physical terms, $\mathcal{B}$ represents the permeable surface of the catalyst, from which the reactant flows from the fluid phase inside the particle.

Let $d_f^{\mathcal{B}}$ be the fractal dimension of $\mathcal{B}$. For finitely ramified structures, the effectiveness factor scales with ϕ as [5, 6]

$$\eta \simeq \begin{cases} 1 & \phi < \phi^* \quad \text{reaction-controlled regime} \\ A\phi^{-\gamma} & \phi > \phi^* \quad \text{diffusion-controlled regime} \end{cases} \tag{2.3}$$

where A and $\phi^* \leq 1$ are two constants characterizing the structure (for model fractal structures, their values can be exactly predicted) and the exponent γ is given by

$$\gamma = \frac{2(d_f - d_f^{\mathcal{B}})}{d_w} , \tag{2.4}$$

where d_f is the fractal dimension of the structure and d_w the walk dimension. Moreover, it is possible to rescale the η-ϕ curve in a universal fashion by defining the quantities $\widetilde{\phi} = \phi/\phi^*$ and $\widetilde{\eta} = \eta^{1/\gamma}$. The resulting $\widetilde{\eta}$-$\widetilde{\gamma}$ graph is universal and equals $\widetilde{\eta} = \tanh\widetilde{\phi}/\widetilde{\phi}$ [5].

Indeed, the analysis developed for linear kinetics is of more general applicability. It is well known in reaction kinetic theory that the behavior of η vs ϕ is practically independent of m (i.e. of the order of nonlinearity in the elementary reaction) [11]. This observation holds true also for fractal catalysts, and implies that the scaling behavior, eqs. (2.3)-(2.4), holds for any reaction of the form $-r(c) = kc^m$.

2.2 Sorption Behavior

The study of sorption in batch conditions is probably the simplest and most common way of experimentally addressing the kinetics of adsorption processes in microporous materials, membranes and controlled-release systems [12, 13]. The basic quantity that can be obtained from such experiments is the fractional uptake curve $M(t)/M_\infty$ expressing the ratio of the solute adsorbed by the sample up to time t and its limiting value (for $t \to \infty$), M_∞.

The application of Fick's law in regular (Euclidean) structures gives at intermediate times (i.e. for $M(t)/M_\infty \leq 0.5$) the scaling behavior $M(t)/M_\infty \sim t^{1/2}$. The literature on controlled release (especially for pharmaceutical purposes) is full of articles showing how deviations from the Fickian result are observed in many experiments (i.e. $M(t)/M_\infty \sim t^\alpha$, with $\alpha > 1/2$), and are referred to as Case II diffusion [13]. This phenomenology depends upon the specific interactions between solute, solvent (e.g. in the case of swelling systems) and porous matrix.

In the case of model fractal structures, by applying GFR it follows that the Laplace transform of $M(t)/M_\infty$ is given by $\widehat{M}(s)/M_\infty = C\Psi(s)/s^2$, where C is a constant and $\Psi(s)$ is

$$\Psi(s) = 1 - \frac{1}{N_{\mathcal{B}}} \sum_{i,j \in \mathcal{B}} \widehat{G}_{ij}^{(d)}(s) , \tag{2.5}$$

where $N_{\mathcal{B}}$ is the number of sites of $\mathcal{B}$ and the superscript (d) indicates sorption Green functions, i.e. the Green functions obtained after inclusion of the proper boundary conditions characterizing batch-sorption experiments [3, 4]. The behavior of the function $\Psi(s)$ has been studied in detail for many structures and different feed configurations [4, 6]. The results obtained can be summarized by the scaling behavior,

$$\Psi(s) \sim s^{1-(d_f-d_f^{\mathcal{B}})/d_w} , \tag{2.6}$$

which implies for the uptake curve,

$$M(t)/M_\infty \sim t^{(d_f-d_f^{\mathcal{B}})/d_w} . \tag{2.7}$$

Eq. (2.7) can be applied as a design criterion for release systems tending towards a constant release rate i.e $M(t)/M_\infty \sim t$, which can be regarded as an optimal situation in many practical applications, e.g. drug delivery. According to eq. (2.7), a constant release rate can be obtained by decreasing the fractal dimension of the exchange manifold (in the limit, $d_f^{\mathcal{B}} = 0$) and by increasing the ratio d_f/d_w as much as possible.

2.3 Hierarchical Micro/Macropore Models

The structure of many porous materials possesses several lengthscales of porosity and micro/macroporous models are often considered in order to describe the presence of well-defined peaks in pore-size distibution [12].

This kind of model, and other more complex and hierarchical forms of pore distribution, can be easily studied by means of GFR by introducing the concept of coupling between two graphs.

Let $\mathcal{G}_c$ and $\mathcal{G}_q$ be two graphs. We indicate this coupling as $\mathcal{G}_c \leftarrow \mathcal{G}_q$, since for each node in the c-phase ("phase" being used here to indicate the macropores or c-phase, and the micropores or q-phase) a $\mathcal{G}_q$ graph is connected. This notation implies that each point in the $\mathcal{G}_c$ lattice is associated with or connected to the input site of a separate copy of $\mathcal{G}_q$.

In the case of feed from a single site, the balance equations in pure diffusion attains the form

$$\begin{aligned} \frac{dc_i}{dt} &= \overline{D}_c \sum_{j=1}^{N_c} H_{ij}^{(c)} c_j - c_i \overline{D}_c \left(\sum_{j=1}^{N_c} H_{ij}^{(c)} \right) + \delta_{i1} \overline{D}_c (c_o - c_1) - k_1 (c_i - q_{i,1}) \\ \frac{dq_{i,h}}{dt} &= \overline{D}_q \sum_{m=1}^{N_q} H_{hm}^{(q)} q_{i,m} - q_{i,h} \overline{D}_q \left(\sum_{m=1}^{N_q} H_{hm}^{(q)} \right) + \delta_{h1} k_2 (c_i - q_{i,1}) , \end{aligned} \tag{2.8}$$

where c_i is the concentration at the i-th node of $\mathcal{G}_c$, $q_{i,h}$ the concentration at the h-th node of the $\mathcal{G}_q$ graph connected to the i-th node of $\mathcal{G}_c$, k_1, k_2 the transfer coefficients and $\overline{D}_c = D/\Delta x_c^2$, $\overline{D}_q = D/\Delta x_q^2$ the diffusional rates in the two phases.

If we use $\widehat{G}^*(s)$ to indicate the Green-function matrix associated with eq. (2.9) for the c-phase, i.e. $\widehat{c}_i(s) = \widehat{G}_{i1}^*(s)\widehat{c}_o(s)$, it can be shown [5] that

$$\widehat{G}^*(s) = \widehat{G}^{(d,c)} \left(s + \frac{k_1 \overline{D}_q (1 - \widehat{G}_{11}^{(d,q)}(s))}{\overline{D}_q - (\overline{D}_q - k_2) \widehat{G}_{11}^{(d,q)}(s)} \right) . \tag{2.9}$$

Eq. (2.9) connects the Green-function matrix of the coupled structure to the sorption Green functions $\widehat{G}^{(d,c)}$, $\widehat{G}^{(d,q)}$ of the two lattices. This result makes it possible to analyze a unlimited variety of structures in which different fractal (or non-fractal) properties are associated with the macropore and micropore networks. An application of eq. (2.9) will be discussed in section 4..

As a result of the findings shown in this section, attention should also be drawn to the following points:
1) Although GFR has been applied to specific lattice structures, the results obtained are of general validity and can be used in a simple way to predict transport/reaction properties of fractals.
2) The results obtained are directly applicable to common engineering problems and provide useful predictions for the design of reaction/transport systems and devices.
3) In transport applications the structure of the transfer interface $\mathcal{B}$ plays a fundamental role, see e.g. eqs. (2.3)-(2.4) and (2.7). It is therefore quite meaningless to describe a transport phenomenon on fractals without specifying the properties of $\mathcal{B}$, i.e. without specifying the boundary conditions, or by considering extremely particular configurations, such as diffusional propagation on an infinitely extended system.

3. Infinitely Ramified Structures and Convolutions

Another point highlighting interesting mathematical properties of renormalization recursions concerns the spectral properties of product lattices.

Product lattices have been studied by Schwalm and Schwalm [14] in their seminal paper on extension theory, and further analyzed by Giona et al. [15] with regard to sorption properties.

Let $\mathcal{G}_1$, $\mathcal{G}_2$ be two graphs with N_1 and N_2 sites. The product graph $\mathcal{G}_\pi = \mathcal{G}_1 \otimes \mathcal{G}_2$ is a graph with $N_\pi = N_1 \times N_2$ sites labeled with two indices, and possessing an adjacency matrix $H^{(\pi)}$ given by

$$H^{(\pi)}_{(i,j)(h,k)} = H^{(1)}_{ih} \delta_{jk} + \delta_{ih} H^{(2)}_{jk} \ , \tag{3.1}$$

where δ_{ih} is the Kronecker symbol. Eq. (3.1) corresponds for graphs to the Cartesian product of two sets: the product graph of two one-dimensional lines is a two-dimensional square, etc.

Let us consider the case of pure diffusion. This implies that the matrix A in eq. (1.1) equals $H - pI$. By applying spectral decomposition [14], it is possible to show that the Green function for the product lattice can be expressed in terms of the Green functions of $\mathcal{G}_1$, $\mathcal{G}_2$ as

$$\widehat{G}^{(\pi)}_{(i,j)(h,k)}(s) = -\frac{1}{\pi} \int_{-\infty}^{\infty} \mathrm{Im}[\widehat{G}^{(1)}_{ih}(x^+)] \widehat{G}^{(2)}_{jk}(s-x) \, dx \ , \tag{3.2}$$

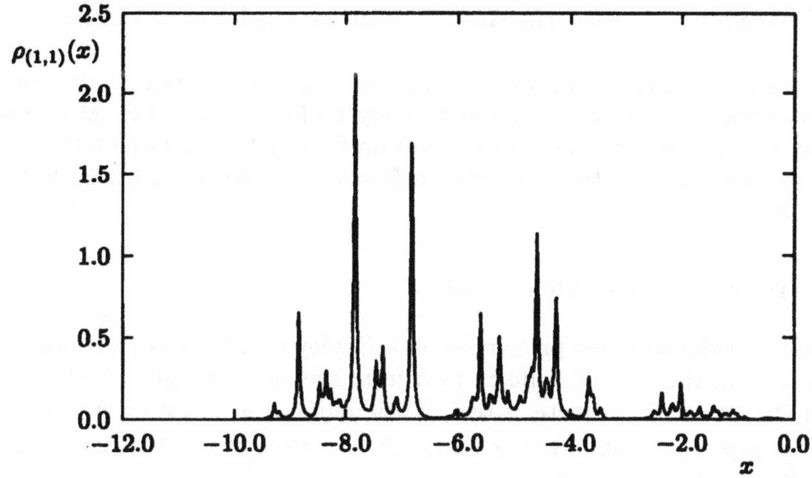

Fig. 3.1. Local density of states at site (1,1) for the product of two threefold Sierpinski gaskets.

where

$$\widehat{G}^{(1)}_{ih}(x^+) = \lim_{\eta \to 0^+} \widehat{G}^{(1)}_{ih}(x + I\eta) , \quad I = \sqrt{-1} . \tag{3.3}$$

The local density of states (spectral density) of the product lattice at site $(1,1)$ $\rho^{(\pi)}_{(1,1)}$ is a weighted distribution function for the eigenvalues of the discrete Laplacian operator on the lattice, and is defined as $\rho^{(\pi)}_{(1,1)}(x) = -\mathrm{Im}[\widehat{G}^{(\pi)}_{(1,1)(1,1)}(x^+)]/\pi$. From eq. 3.2), it follows that

$$\rho^{(\pi)}_{(1,1)}(x) = \int_{-\infty}^{\infty} \rho^{(1)}_1(x')\rho^{(2)}_1(x - x')\, dx' , \tag{3.4}$$

i.e. the spectral density of the product lattice is the convolution of the spectral densities associated with the graphs $\mathcal{G}_1$, $\mathcal{G}_2$. For example, Fig. 3.1 shows the spectral density at site (1,1) of the Cartesian product of the threefold Sierpinski gasket with itself.

In the case of fractal structure, $\rho_1(x)$ is defined on a Cantor-dust support of dimension less than 1 [16], and may possess multifractal features.

Apart from the intrinsic mathematical problems associated with the definition of densities, eq. (3.4) shows that the eigenvalue distribution (spectral measure) of $\mathcal{G}^{(\pi)}$ is the convolution of spectral measures. A deeper understanding of the spectral properties of product lattices can be achieved by considering the mathematical problem of the convolution of invariant measures associated with IFSP (particularly nonlinear IFSP). Some preliminary results on this issue have recently been obtained [17].

4. Further Developments: Pathological Structures

This section analyzes some other applications of GFR, focusing attention on pathological loopless structures possessing no finite dimension. More specifically, we consider transport properties on Cayley trees (Bethe lattices) and on a new class of structures called *ultrafractals*, possessing highly pathological features.

4.1 Transport on Cayley Trees

Let us consider sorption properties on a Cayley tree possessing coordination number p in the case of uniform feed from the external shell of sites Σ (i.e. from the external perimeter sites). Their number at iteration n is equal to $N_{\Sigma}^{(n)} = p(p-1)^n$, while the total number of sites of the structure is $N_n = 1 + p[(p-1)^n - 1]/(p-2)$.

In the case where all the perimeter sites are connected with an external reservoir at constant solute concentration c_o, the fractional uptake curve is given by

$$\frac{dM(t)/M_\infty}{dt} = \frac{h(p-1)}{c_o N} \sum_{h \in \Sigma} [c_o - c_h(t)] \, , \qquad (4.1)$$

where summation runs over the sites belonging to the external perimeter shell Σ, h is the mass-transfer rate, and the factor $p-1$ appears because each site on Σ should be saturated with $p-1$ connections with the external reservoir.

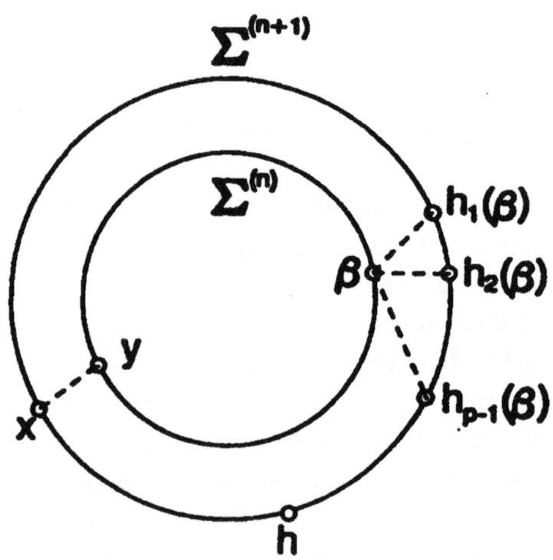

Fig. 4.1. Schematic representation of a Cayley tree used to derive renormalization recursions.

Since $\widehat{c}_h(s) = (p-1)\sum_{k\in\Sigma}\widehat{G}_{hk}(s)\widehat{c}_o(s)$, eq. (4.1) reads in the Laplace domain as

$$\frac{\widehat{M}(s)}{M_\infty} = \frac{h(p-1)N_\Sigma}{s^2 N}\left[1-(p-1)\sum_{k\in\Sigma}\widehat{G}_{hk}(s)\right] . \tag{4.2}$$

Indeed, by enforcing the symmetries, the term $S(s) = \sum_{k\in\Sigma}\widehat{G}_{hk}(s)$ is independent of h, for all $h \in \Sigma$.

The renormalization of the integral quantity $S^{(n)}(s)$ can be easily obtained by considering the schematic representation of Fig. 4.1, where the circles indicate the external shells at iteration n (inner) and $n+1$ (outer).

Let x be a generic site on $\Sigma^{(n+1)}$ connected at iteration $n+1$ with the site $y \in \Sigma^{(n)}$. Then,

$$S^{(n+1)} = \sum_{h\in\Sigma^{(n+1)}}\widehat{G}_{xh}^{(n+1)} = \widehat{G}_{xx}^{(n+1)} + \sum_{h\in\Sigma^{(n+1)}\neq x}\widehat{G}_{xh}^{(n+1)} . \tag{4.3}$$

But $\widehat{G}_{xx}^{(n+1)} = g + g\widehat{G}_{yx}^{(n+1)}$, $\widehat{G}_{xh} = g\widehat{G}_{yh}^{(n+1)}$, where $g = g(s) = (s+p)^{-1}$ is the Green function for a single site. Therefore,

$$S^{(n+1)} = g + g\sum_{h\in\Sigma^{(n+1)}}\widehat{G}_{yh}^{(n+1)} . \tag{4.4}$$

Let $\beta \in \Sigma^{(n)}$ and $h_i(\beta) \in \Sigma^{(n+1)}$ $(i = 1,..,p-1)$, the $p-1$ sites of $\Sigma^{(n+1)}$ connected to it. By applying eq. (1.3) it follows that

$$\widehat{G}_{yh}^{(n+1)} = \sum_{\beta\in\Sigma^{(n)}}\widehat{G}_{y\beta}^{(n)}\sum_{i=1}^{p-1}\widehat{G}_{h_i(\beta),h}^{(n+1)} . \tag{4.5}$$

By summing over all the sites $h \in \Sigma^{(n+1)}$, due to the symmetry of the structure it follows that

$$\sum_{h\in\Sigma^{(n+1)}}\widehat{G}_{yh}^{(n+1)} = (p-1)S^{(n)}S^{(n+1)} , \tag{4.6}$$

which substituted into eq. (4.4) yields

$$S^{(n+1)} = \frac{g}{1-(p-1)gS^{(n)}} . \tag{4.7}$$

The initial condition for this recursion comes from a structure with a central site connected to p neighbouring sites, and therefore is $S^{(1)} = g/(1-pg^2)$. Eq. (4.7) is linear fractional and can therefore be solved in closed form. For Cayley trees, mass saturation implies $S^{(n)}(0) = (p-1)^{-1}$.

As regards sorption properties, Cayley trees are highly singular structures since the ratio $N_\Sigma^{(n)}/N_n$ admits a non-vanishing limit for $n \to \infty$. The external perimeter and the entire structure therefore admit the same cardinality (dimensionality). By applying eq. (2.6) or eq. (2.7), $d_f = d_f^{\mathcal{B}}$, it readily follows that $\widehat{M}(s)/M_\infty \sim s^{-1}$, which implies that the structure almost immediately saturates towards a uniform concentration profile.

4.2 Ultrafractal Structures

The two-phase model (coupling) discussed in section 2.3 for macropore/micropore structures can be generalized to an arbitrarily large number of phases indicated as $q_1, q_2, \ldots$ This leads to the concept of *ultrafractals* and to a new phenomenology in diffusion on complex structures.

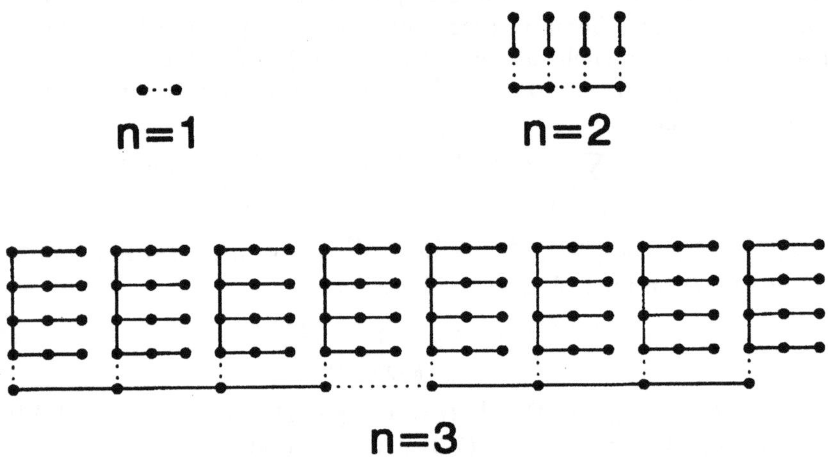

Fig. 4.2. Construction process of an ultrafractal structure starting from the one-dimensional chain.

The idea of ultrafractals can be introduced by considering the iterative construction scheme shown in Fig. 4.2, taking as an example the construction of ultrafractal structures generated from the one-dimensional line. At the first iteration, the phase $c = q_1$ is given by the two-point graph $\mathcal{G}^{(1)}$ of the one-dimensional line. At the second iteration, the q_1-phase develops in the corresponding four-point approximation $\mathcal{G}^{(2)}$, and a $\mathcal{G}^{(1)}$ lattice is attached to each node of $\mathcal{G}^{(2)}$ with transfer-rate constants k_1, k_2. At the next generation, the previous lattices develop respectively into $\mathcal{G}^{(3)}$ and $\mathcal{G}^{(2)}$ and to every point of $\mathcal{G}^{(2)}$ a new phase represented by the lattice $\mathcal{G}^{(1)}$ is attached with transfer-rate constants k_1, k_2, and the construction procedure continues.

This way of constructing ultrafractal lattices can be easily formalized. Let $\{\mathcal{G}^{(n)}\}$ be a family of graphs. The ultrafractal lattice family $\{\mathcal{U}^{(n)}\}$ generated by $\{\mathcal{G}^{(n)}\}$ is defined according to the recursion relations

$$\mathcal{U}^{(n+1)} = \mathcal{G}^{(n+1)} \leftarrow \mathcal{G}^{(n)} \leftarrow \cdot\cdot \leftarrow \mathcal{G}^{(1)} = \mathcal{G}^{(n+1)} \leftarrow \mathcal{U}^{(n)} \, , \qquad (4.8)$$

where each coupling is specified by the transfer rates k_1, k_2.

Let N_n be the number of sites of $\mathcal{G}^{(n)}$, $N_n = N_o \nu^n$, and $N_n^{\mathcal{U}}$ the number of sites of $\mathcal{U}^{(n)}$. From the construction process of ultrafractal structures, eq. (4.8), it follows that $N_{n+1}^{\mathcal{U}} = N_{n+1}(1 + N_n^{\mathcal{U}})$, i.e. $N_n^{\mathcal{U}} = \sum_{h=1}^{n} \pi_h$ where $\pi_h = \prod_{m=1}^{h} N_{n-m+1}$ For large n, assuming that $\nu > 1$ (ν is the number of copies of $\mathcal{G}^{(n)}$ necessary to generate $\mathcal{G}^{(n+1)}$ recursively, the leading term is π_n, and therefore

$$N_n^{\mathcal{U}} \sim \pi_n = N_o^n \nu^{\sum_{h=1}^{n} h} = N_o^n \nu^{n(n+1)/2} \,. \tag{4.9}$$

Eq. (4.9) implies that the number of sites of a ultrafractal structure grows faster than any power law with exponent n. This means that ultrafractal structures possess a non-finite dimension.

Let us now focus on diffusion in ultrafractal structures by regarding the ultrafractal graph as a model of a single phase. In this case, the transfer-rate coefficients are equal to the diffusion rate at iteration n, $k_1 = k_2 = \overline{D}^{(n)} = D/\Delta x_n^2$, where Δx_n is the distance between nearest neighbouring sites. We can take the dimensionless time $\theta = t\overline{D}^{(n)}$, so that the dimensionless diffusivity (hopping strength) becomes equal to 1.

From the construction rule, eq. (4.8), and from the analysis of two-phase coupling, eq. (2.9) with $k_1 = k_2 = \overline{D}_c = \overline{D}_q = 1$, the sorption Green functions $\widehat{G}_{i1}^{(\mathcal{U},n)}(s)$ for a ultrafractal lattice at iteration n are related to the sorption Green functions $\widehat{G}_{i1}^{(d,n)}(s)$ of $\mathcal{G}^{(n)}$ by the relations

$$\begin{aligned} \widehat{G}_{i1}^{(\mathcal{U},1)}(s) &= \widehat{G}_{i1}^{(d,1)}(s)\,, \\ \widehat{G}_{i1}^{(\mathcal{U},2)}(s) &= \widehat{G}_{i1}^{(d,2)}(s + (1 - \widehat{G}_{11}^{(d,1)}(s)))\,, \\ \widehat{G}_{i1}^{(\mathcal{U},3)}(s) &= \widehat{G}_{i1}^{(d,3)}(s + (1 - \widehat{G}_{11}^{(d,2)}(s + (1 - \widehat{G}_{11}^{(d,1)}(s)))))\,, \end{aligned} \tag{4.10}$$

and so on.

In order to analyze scaling properties in diffusion, it is convenient to consider the scaling of the first-order moments $m_i^{(\mathcal{U},n)}(1)$ at site i, $m_i^{(\mathcal{U},n)}(1) = -d\widehat{G}_{i1}^{(\mathcal{U},n)}(s)/ds|_{s=0}$ with the iteration order n.

Following the recursion process defining the sorption Green functions for ultrafractal structures, eq. 4.11), the moments $m_i^{(\mathcal{U},n)}(1)$ can be expressed as a function of the moments $m_i^{(n)}(1)$ for $\mathcal{G}^{(n)}$ as follows:

$$m_i^{(\mathcal{U},n)}(1) = m_i^{(n)}(1)\left(1 + \sum_{h=1}^{n-1} M_h^{(n)}.\right) \qquad \text{where} \qquad M_h^{(n)} = \prod_{j=1}^{h} m_1^{(n-j)}(1)\,. \tag{4.11}$$

From the moment analysis of diffusion on fractals (see [4]), it is known that the dimensionless moments $m_i^{(n)}(1)$ scale with the lengthscale as $m_i^{(n)}(1) \sim (L/\Delta x_n)^{\gamma_i^{(1)}}$, where $\gamma_1^{(1)} = d_f$ at the input node $I = 1$ and $\gamma_i^{(1)} = d_w$ for $i \neq 1$. The leading term for large n in eq. (4.11) is therefore given by

$$m_i^{(\mathcal{U},n)}(1) \sim m_i^{(n)}(1) M_{n-1}^{(n)} = m_i^{(n)}(1) \prod_{j=1}^{n-1} m_1^{(n-j)}(1) \sim \left(\frac{L}{\Delta x}\right)^{\gamma_i^{(1)}+(n-1)\gamma_1^{(1)}/2} . \tag{4.12}$$

The scaling exponent with respect to the lengtscale $L/\Delta x$ of first-order moment at the output site $m_2^{(\mathcal{U},n)}(1)$ therefore depends on the iteration n. This is a sign that ultrafractal structures are characterized by far more complex behavior in diffusion than regular fractals (for which this scaling exponent is independent of the iteration). To further ascertain the anomalous properties of ultrafractals, it is convenient to make use of the scaling of the first-order moment at the output site, as derived above, in order to obtain the behavior of the mean-square displacement $R^2(t)$ with time t. The first-order moment $m_2^{(\mathcal{U},n)}(1)$ is related to the effective diffusion coefficient D_e by the relation $m_2^{(\mathcal{U},n)}(1)(L/\Delta x)^{-2} \sim L^2 D/D_e$. Therefore,

$$D_e \sim \left(\frac{L}{\Delta x}\right)^{-[\gamma_2^{(1)}-2+(n-1)\gamma_1^{(1)}/2]} . \tag{4.13}$$

The order of iteration n can be expressed as $n \sim \log(L/\Delta x)/\log\lambda$ ($\lambda = 2$ for the one-dimensional line), and in the scaling analysis $L/\Delta x \sim R(t)$. This implies

$$D_e \sim R(t)^{-[d_w-2-d_f/2+\mu\log R(t)]} \sim R(t)^{-\mu\log R(t)} , \tag{4.14}$$

where $\mu = d_f/(2\log\lambda)$. Since $R^2(t) \sim D_e t$, from eq. (4.14) it follows that $R^2(t) \sim R^{-\mu\log R(t)} t$, and therefore

$$R(t) \sim t^{\zeta(t)} , \qquad \zeta(t) = 1/[\mu(\log t)^{1/2}] . \tag{4.15}$$

This result implies that the effective walk dimension of ultrafractal structures $d_{w,e}^{\mathcal{U}}(t)$ (strictly speaking, a walk dimension cannot be defined since the scaling of the mean-square displacement with time is not a power law) is time-dependent, and for large $t \gg 1$ behaves as $d_{w,e}^{\mathcal{U}}(t) \sim (\log t)^{1/2}$. Fig. 4.3 shows the behavior of $1 - \widehat{G}_{11}^{(\mathcal{U},n)}(s)$ vs s for an ultrafractal structure generated starting from the one-dimensional line ($d_f = 1$, $d_w = 2$). As can be observed, $1 - \widehat{G}_{11}^{(\mathcal{U},n)}(s)$ does not scale as a power law for small s, i.e. large times (Fig. 4.3, curves a)-c)), as the corresponding generating structure (one dimensional chain) does $1 - \widehat{G}_{11}^{(d)} \simeq s^{1/2}$ (Fig. 4.3, curve d)).

To sum up, ultrafractal structures can be regarded as a simple extension of the coupling model developed for hierarchical porous materials, but display a diffusional dynamics slower than in any regular fractal structure, as expressed by the scaling behavior eqs. (4.15).

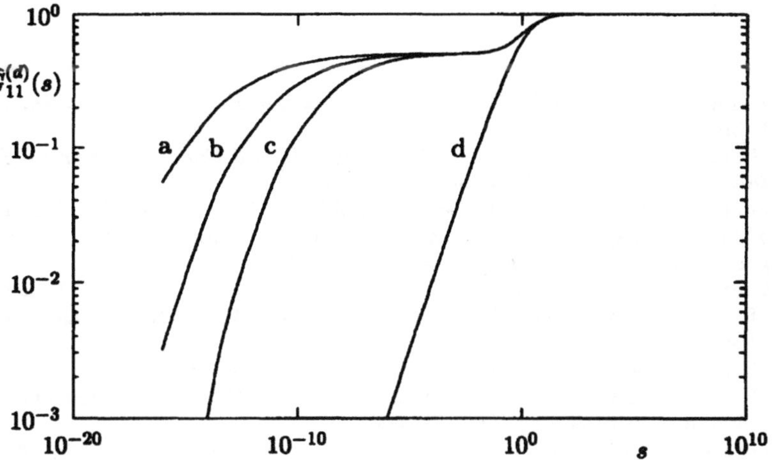

Fig. 4.3. $1-\widehat{G}_{11}^{(\mathcal{U},n)}(s)$ vs s for a ultrafractal structure generated from the 1-d chain. a) $n=10$, b) $n=9$, c) $n=8$. Curve d) is $1-\widehat{G}_{11}^{(d)}(s)$ for the one-dimensional chain

5. Concluding Remarks

In this article we have presented a brief overview of some engineering and mathematical implications of the study of transport processes by means of Green-function renormalization.

The analysis of the effectiveness factor developed for first-order reactions is substantially valid also for nonlinear elementary reactions, i.e. for reactions of the form $-r(c) = kc^m$.
This observation indicates that future numerical studies in transport/reaction schemes on fractals and disordered systems should be oriented primarily towards more complex reaction mechanisms involving more than one reacting species and multireaction pathways (e.g. parallel-competing reactions).

In this article we have also introduced the concept of ultrafractal structure. This class of lattices can hardly be regarded as a physical model of some realistic material (since, like Cayley trees, ultrafractals do not possess finite dimensions). Nevertheless, ultrafractal lattices represent a conceptually simple example of a structure that does not exhibit a power-law behavior in transport properties (as normal fractal structures do), and the analysis of the dynamical properties of these lattices may lead to other interesting and unexpected findings.

References

1. Avnir, A. (Ed.) (1990): *The Fractal Approach to Heterogeneous Chemistry*. John Wiley, New York.
2. Giona, M., Biardi, G. (Eds.) (1997): *Fractals and Chaos in Chemical Engineering*. World Scientific, Singapore, in press.
3. Giona, M., Schwalm, W.A., Schwalm, M.K., Adrover, A. (1996): Exact solution of linear transport equations in fractal media I - Renormalization analysis and general theory. *Chem. Engng. Sci.*, 51, 4717.
4. Giona, M., Schwalm, W.A., Schwalm, M.K., Adrover, A. (1996): Exact solution of linear transport equations in fractal media II - Diffusion and convection. *Chem. Engng. Sci.*, 51, 4731.
5. Giona, M., Adrover, A., Schwalm, W.A., Schwalm, M.K. (1996): Exact solution of linear transport equations in fractal media III - Adsorption and chemical reaction. *Chem. Engng. Sci.*, 51, 5065.
6. Giona, M., Schwalm, W.A., Adrover, A., Schwalm, M.K. (1996): Analysis of linear transport phenomena on fractals. *Chem. Eng. J.*, 64, 45.
7. Schwalm, W.A., Schwalm, M.K. (1992): Closed formulae for Green functions on fractal lattices. *Physica A*, 185, 195.
8. Schwalm, W.A., Schwalm, M.K. (1993): Explicit orbits for renormalization maps for Green functions on fractal lattices. *Phys. Rev. B*, 47, 7847.
9. de Archangelis, L., Redner, S., Coniglio, A. (1985): Anomalous voltage distribution of random resistor networks and a new model for the backbone at the percolation threshold. *Phys. Rev. B*, 31, 4725.
10. Aharony, A., Stauffer, D. (1993): *Introduction to Percolation Theory*. Taylor & Francis, London.
11. Froment, G.F., Bischoff, K. (1990): *Chemical Reaction Analysis and Design*. John Wiley, New York.
12. Ruthven, D.M. (1984): *Principles of Adsorption and Adsorption Processes*. John Wiley, New York.
13. Crank, J. (1975): *The Mathematics of Diffusion*. Claredon Press, Oxford.
14. Schwalm, W.A., Schwalm, M.K. (1988): Extension theory for lattice Green functions. *Phys. Rev. B*, 37, 9524.
15. Giona, M., Adrover, A., Schwalm, W.A., Schwalm, M.K. (1997): Solution of transport schemes on fractals by means of Green function renormalization - Application to integral quantities. *Fractals*, in press.
16. Domany, E., Alexander, S., Bensimon, D., Kadanoff, L. (1983): Solutions to the Schrödinger equation on some fractal lattices. *Phys. Rev. B*, 28, 3110.
17. Giona, M., Patierno, O. (June 25-27, 1997): Integral transforms of multifractal measures, this volume, page 19

Self-Similar Spatio-Temporal Structure of Material Filaments in Chaotic Flows

M.M. Alvarez[1], F.J. Muzzio[1]*, S.Cerbelli[1], and A. Adrover[2]

[1] Department of Chemical and Biochemical Engineering, Rutgers University, Piscataway, New Jersey 08855, USA

[2] Dipartimento di Ingegneria Chimica, Universita di Roma. Via Eudossiana 18, Roma 00184, ITALY

Abstract This paper focuses on the evolution of material filaments in chaotic flows. This evolution is numerically calculated for three cases of the sine flow corresponding to a mainly regular, a mainly chaotic, and a globally chaotic case without discernible islands. In all cases, the stretched filament has an extremely non-uniform spatial distribution, with densities spanning several orders of magnitude and being the largest in the neighborhood of hyperbolic periodic points. Such spatial non-uniformity is a *permanent* property of time-periodic flows. As expected, the length of the filament increases exponentially in time, but due to the spatial non-uniformities in filament density, its growth is much faster than predicted by the Lyapunov exponent. The stretched filaments are self-similar in time, as revealed by their spatial structure and by the frequency distribution of the logarithm of the striation thickness, which is described by a family of curves that have an invariant shape and that can be collapsed onto a single curve by means of a simple scaling.

1. Introduction.

The purpose of this paper is to communicate several new observations concerning the time evolution and the spatial properties of partially mixed structures generated by advection of passive tracers in chaotic flows. Experiments [1] [2], [3] have revealed that such structures are composed of thousands of thin parallel striations (Figure 2.1). More recently, [4] the presence of families of self-similar striation structures in stirred tanks operated in laminar regimes has been observed.

The spatio-temporal dynamics of such structures have never been characterized in detail. Specifically, the striation thickness distribution generated by chaotic flows has never been measured, either in real or in model flows. This is a topic of considerable importance; for example, the striation thickness distribution controls the relative amounts of products generated in reactive flows undergoing multiple simultaneous reactions [5], and also affects the rate of cross-linking reactions, physical properties, and quality of products generated by reactive polymerization processes.

Experimental characterization of mixture microstructure is impractical because chaotic flows create structures with a very wide distribution of length

* To whom all correspondence should be sent.

scales, most of which are well below the resolution achievable by image analysis instruments. Computations can be used to simulate the mixing process, but most previous studies have simulated mixing in chaotic flows by following point or vector tracers to determine their rate of spreading [6], [7], [8], and stretching [5], [9]. Such simulations do not preserve the continuity of macroscopic material lines as they evolve due to the flow, and are poorly suited to compute measures, such as the striation thickness distribution, that provide a direct quantification of microstructural properties of partially mixed systems. Long term evolution of material filaments has only been considered by Franjione and Ottino [10], who at the time concluded that the computational power required to simulate the exponential growth in the number of striations made such quantifications impractical.

In this paper, a streamlined algorithm is used to simulate the evolution of a macroscopic material filament (Figure 2.1) as it is stretched, advected, and folded. A two-dimensional time-periodic chaotic flow, the sine flow [11], is adopted here as a case study. The flow is defined in the unit square $0 \leq x \leq 1, 0 \leq y \leq 1$ by the two motions:

$$V_x = sin(2\Pi y), \quad V_y = 0; \qquad nT \leq t < (n+1/2)T \qquad (1a)$$
$$V_x = 0, \quad V_y = sin(2\Pi x); \qquad (n+1/2)T \leq t < (n+1)T \qquad (1b)$$

where V_x and V_y are the velocity components in the x and y directions, T is the flow period, n is the number of periods, and t is time.

This very simple flow is a good choice for an initial study because the velocity field is continuous and differentiable to any order, including at the boundaries. Imposition of periodic boundary conditions does not result in unphysical "cusps" in the advected filaments. In addition, in this periodic flow model the extension and location of the regular regions can be controlled by manipulation of the value of T. As it has been recently demonstrated [12], the spatial and temporal evolutions of stretching distributions in the sine flow are closely analogous to those observed for real three dimensional chaotic flow systems like the Kenics mixer. This analogy strongly suggests that the statistical properties of the striation thickness distribution in the sine flow could also be of general validity and that could be extended to include more realistic flows. The purpose of this paper is to give a brief account of progress-to-date focusing exclusively on the sine flow. Striation thickness distributions in other flows will be discussed in future publications.

2. Materials and Methods.

The algorithm for simulating evolution of the filament is quite simple. The filament is initially composed of only two points. Given the positions of the points at the beginning of a flow period, the algorithm finds the positions of the points a period later. The distances between consecutive points are then calculated; whenever two points move apart more than a certain distance d, new points are interpolated at the beginning of the period to minimize errors due to the bending of the filament. The algorithm is then iterated to simulate the effects of another period. Since the flow is piece-wise integrable, this computation can be performed in closed form, facilitating a fast and accurate simulation.

The self-similar temporal evolution of the iterated filament structure generated by this procedure (Figure 2.1,3.1,3.2) is made apparent by computing the striation thickness distribution and the spatial density of filament structures. Striation thicknesses are computed as follows: (i) the flow domain is covered with a set of uniformly-spaced horizontal lines; (ii) at the end of each period, the intersections of the filament with the lines are computed; (iii) at each intersection, the local angle of the filament relative to the horizontal coordinate is computed; (iv) striation thicknesses are found as the distances between consecutive intersections, corrected for the intersection angle, (v) the local filament density is computed by dividing the flow domain into equal-sized boxes and calculating the length of the filament contained in each box.

3. Results and Discussion.

Results of the filament evolution algorithm are shown in Figure 2.1 for $T = 0.8$ (a partially chaotic condition in which most of the flow domain is dominated by four large period-2 regular islands and many small ones). The filament was initially composed of two points located at $(0.5, 0.495)$ and $(0.5, 0.505)$. The value of $d = 0.01$ was adopted for this computation. The figure shows the resulting filament after 6 (Figure 2.1 a),8 (Figure 2.1 b), and 10 flow periods (Figure 2.1 c). After just a few periods, a complex structure emerges, composed of thousands of striations with a wide range of thicknesses and closely resembling the invariant manifolds of the central hyperbolic point in the flow. As expected from previous experimental work [3], [13], the structure generated by the mixing process displays strong symptoms of self-similarity. As time increases, the process preserves all features added to the mixture structure in earlier flow periods; i.e., every single fold and branch present in the structure at $n = 8$ is also present at $n = 10$. At each flow period, new features within and around previously existing ones are added. Results for $T = 1.2$, $n = 4$, 5, and 6 are shown in Figure 3.1 a-c, respectively. For this flow condition, the flow is mainly chaotic, and has four relatively small islands. As before, the filament grows very fast, quickly invading the entire chaotic region. The iterative growth process is readily observed and again generates a structure that is evidently self-similar in time. Finally, Figure 3.2 a-c shows the filament obtained for $T = 1.6$, $n = 3$, 4, and 5. Since this flow condition is globally chaotic and is devoid of any noticeable islands, the self-similar filament quickly invades the entire flow domain.

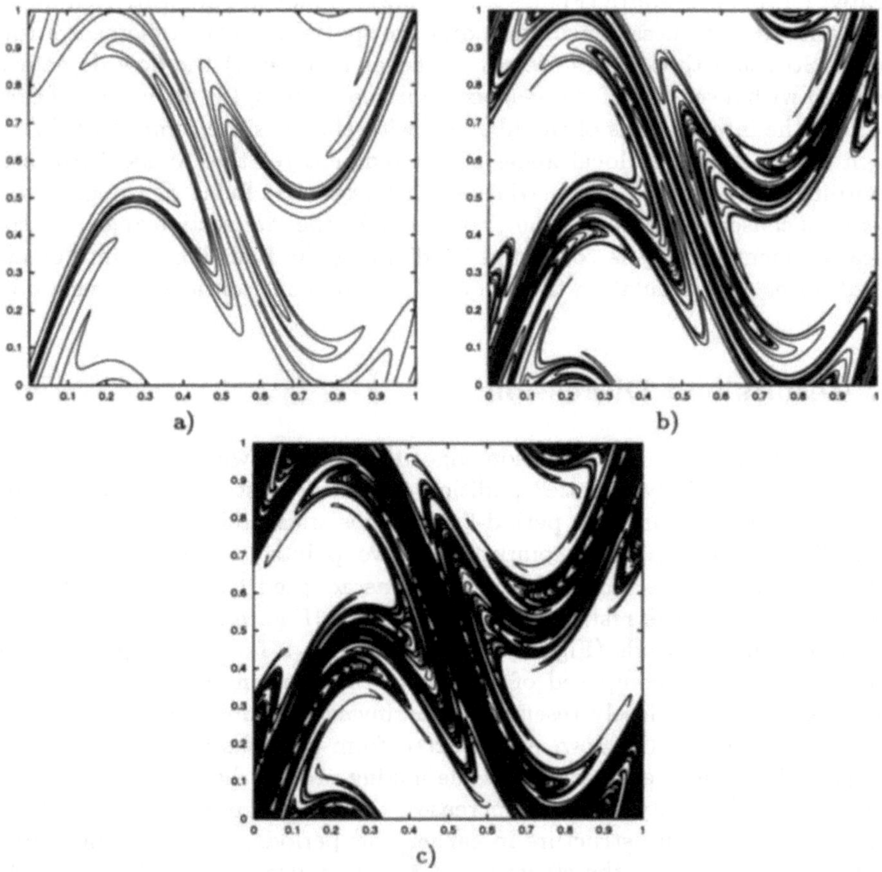

Fig. 2.1. Evolution of a material filament in the sine flow, $T = 0.8$, for (a) $n = 6$, (b) $n = 8$, and (c) $n = 10$.

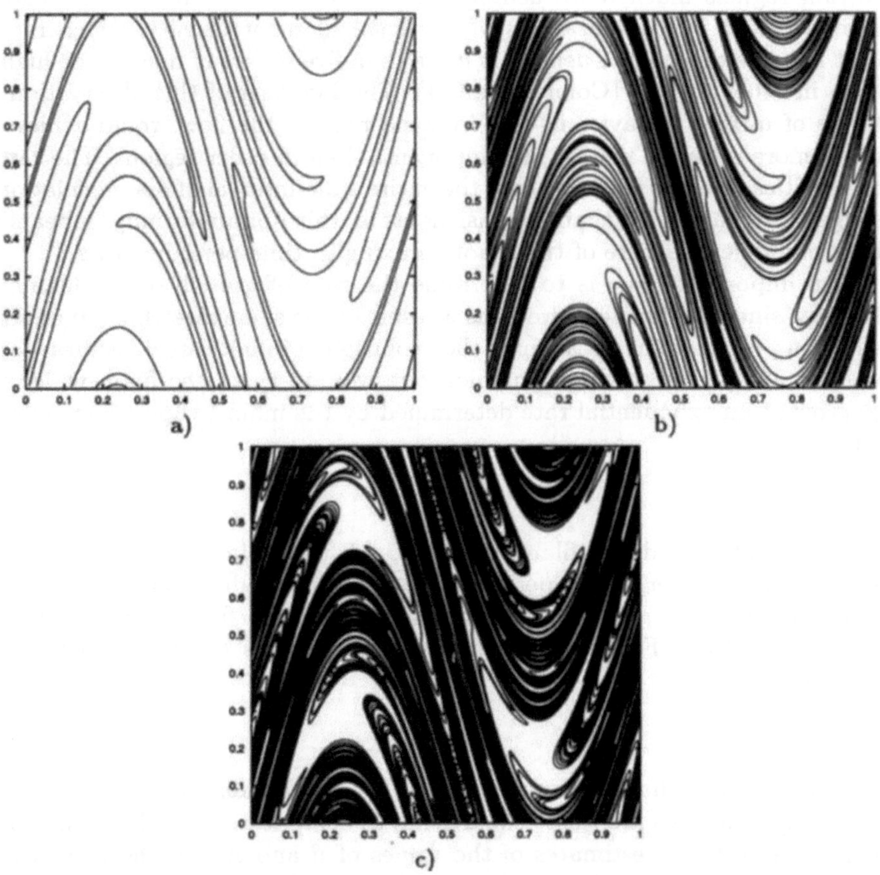

Fig. 3.1. Evolution of a material filament in the sine flow, $T = 1.2$, for (a) $n = 4$, (b) $n = 5$, and (c) $n = 6$.

In particular, it is interesting to notice the very wide distribution of striation thicknesses that is apparent in Figure 3.2. For such a globally chaotic flow condition, a Poincaré section would be entirely featureless, and would look like a random cloud of points. In fact, the (naive) conventional view is that a globally chaotic flow would eventually generate a completely homogeneous mixture. However, our simulations suggest that actual mixing behavior is not so simple. The figure clearly indicates that the stretched filament populates certain regions much more densely than others. This statement is readily demonstrated by computations of the local filament density, which reveal that in some regions the density is more than five orders of magnitude higher than in other regions (Color plate C.1). The meaning of this observation in terms of mixing behavior is that in some regions the flow would achieve a much more intimate mixing of components than in other regions. The range of densities remains constant over the entire time-interval of our calculations, suggesting that rather than a transient state, this non-uniformity appears to be a permanent feature of the chaotic mixing in time-periodic flows.

An important issue is to determine the rate of growth of the filament. Naively, since the filament grows as a result of the same stretching and reorientation processes that determine the evolution of infinitesimal vector tracers used to analyze stretching dynamics, one would expect the filament length to grow at an exponential rate determined by the main Lyapunov exponent, i.e.,

$$L_s = e^{\Lambda n} \tag{3.1}$$

where L_s is the predicted filament length, Λ is the main Lyapunov exponent, and n is the number of flow periods. However, this prediction is not accurate; while the filament growth is exponential, it is much faster than suggested by equation (3.1). For example, the number of points L_{np} that constitute a filament grows as

$$L_{np} = e^{\vartheta n}. \tag{3.2}$$

(Figure 3.3 a-c) shows a comparison of the predictions of equations (3.1) and (3.2). The lines are least−squares fits to computational values, and their slopes are accurate estimates of the values of ϑ and Λ. For the three cases considered here, the value of ϑ is significantly larger than the Lyapunov exponent (1.13 vs. 0.55 for $T = 0.8$, 1.87 vs. 1.37 for $T = 1.2$, and 2.40 vs. 1.92 for $T = 1.6$). This accelerated filament growth, which after just a few periods yields a filament that is several orders of magnitude longer than predicted by the Lyapunov exponent, is due to the above mentioned spatial non-uniformities in filament density. The Lyapunov exponent is a measure of the average rate of stretching in the flow, in which all locations within the chaotic region of the flow are assigned equal statistical density. However, the filament populates high-stretching regions of the flow much more densely than low stretching regions. It must be remarked that this phenomenon is

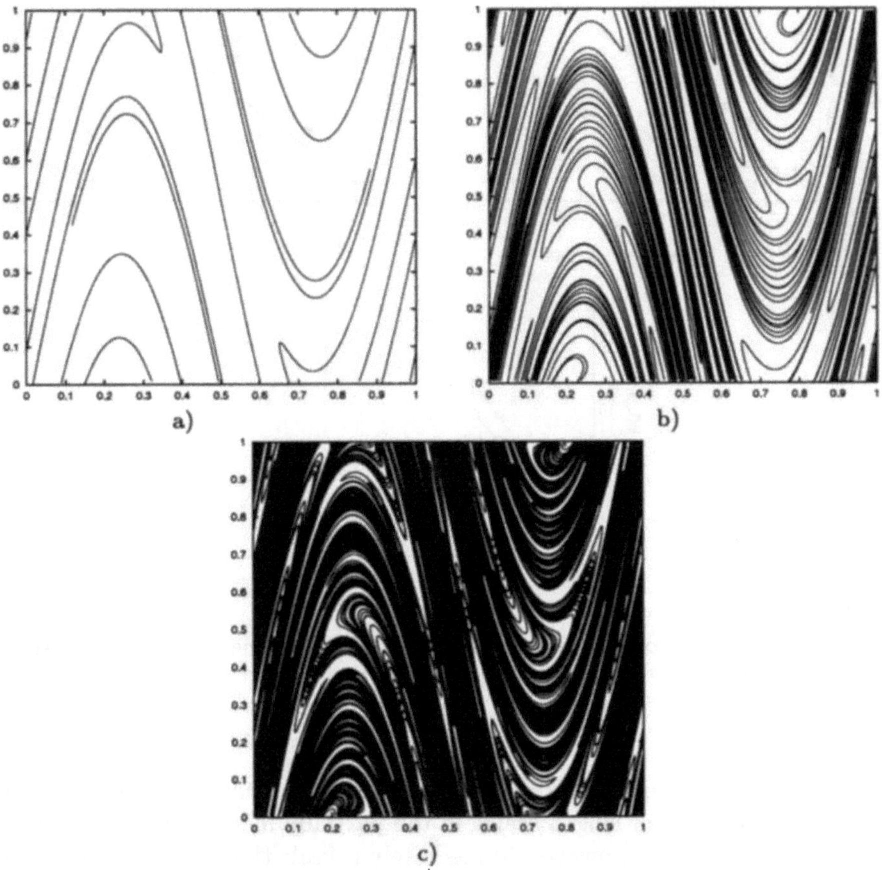

Fig. 3.2. Evolution of a material filament in the sine flow, $T = 1.6$, for (a) $n = 3$, (b) $n = 4$, and (c) $n = 5$.

independent of the initial location of the filament. After just a few periods of the flow, the structure of the filament becomes independent of the initial position, and the spatial structure generated by the process is always the same.

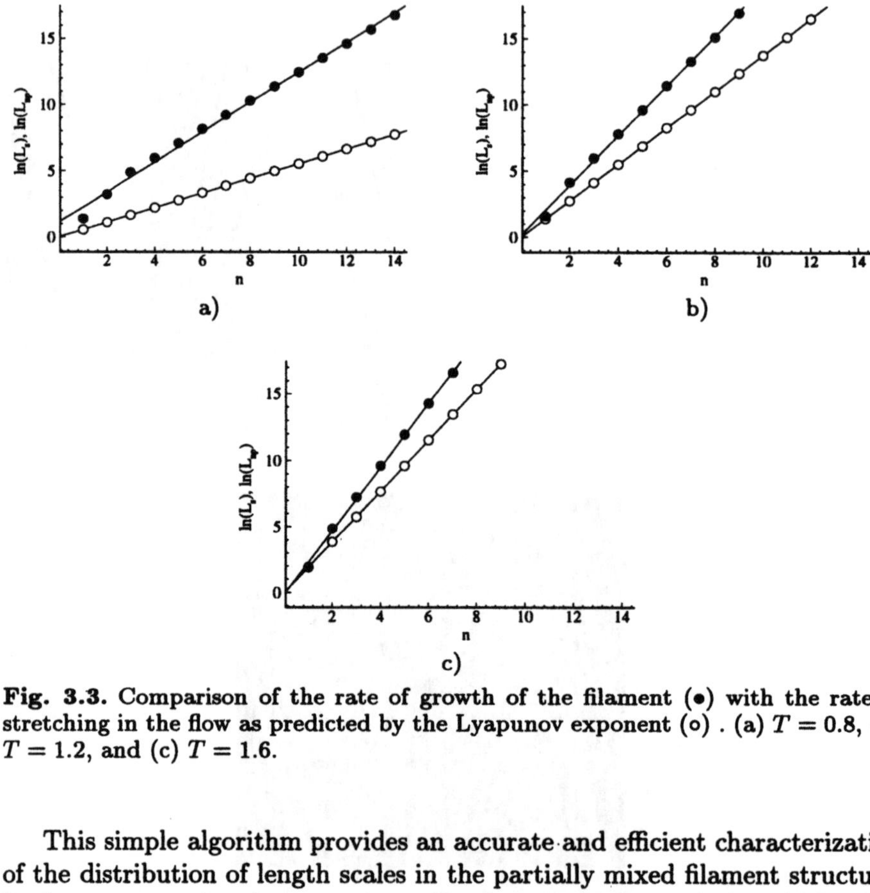

Fig. 3.3. Comparison of the rate of growth of the filament (•) with the rate of stretching in the flow as predicted by the Lyapunov exponent (∘) . (a) $T = 0.8$, (b) $T = 1.2$, and (c) $T = 1.6$.

This simple algorithm provides an accurate and efficient characterization of the distribution of length scales in the partially mixed filament structure. Such a computation, however, immediately reveals that striation thicknesses have values that span several orders of magnitude. For example, for $T = 1.6$, $n = 12$, thicknesses computed by the algorithm have values in the range from 10^{-5} to 10^{-15}. In order to compute a meaningful measure of statistical frequency over such a wide range of values, the structures are described in terms of the distribution of the logarithm of the striation thickness,

$$H(logs) = (1/N)dN(logs)/d(logs), \tag{3.3}$$

where s is a value of striation thickness, N is the total number of measurements, and $dN(logs)$ is the number of values of striation thickness on the interval between $logs$ and $logs + dlogs$. As shown in Figure 3.4 a) for the case corresponding to $T = 0.8$, $n = 9 - 14$, $H(logs)$ quickly evolves into a smooth curve that exhibits power law behavior with slope ≈ 1 for small s values (related to the use of the distribution of the logarithm of s) and exponential behavior for large s values. It is readily apparent in the figure that for $n \geq 12$, the curves all have the same shape. The meaning of this observation is that $H(logs)$ is asymptotically approaching a self-similar distribution of frequencies. The same behavior is observed for $T = 1.2$ for $n \geq 9$ (Figure 3.1 b) and for $T = 1.6$ for $n \geq 7$ (Figure 3.1 c). In all three cases, the curves become essentially identical to one another (except for a horizontal shift) after just a few periods that are needed to reach a "well-developed" structure.

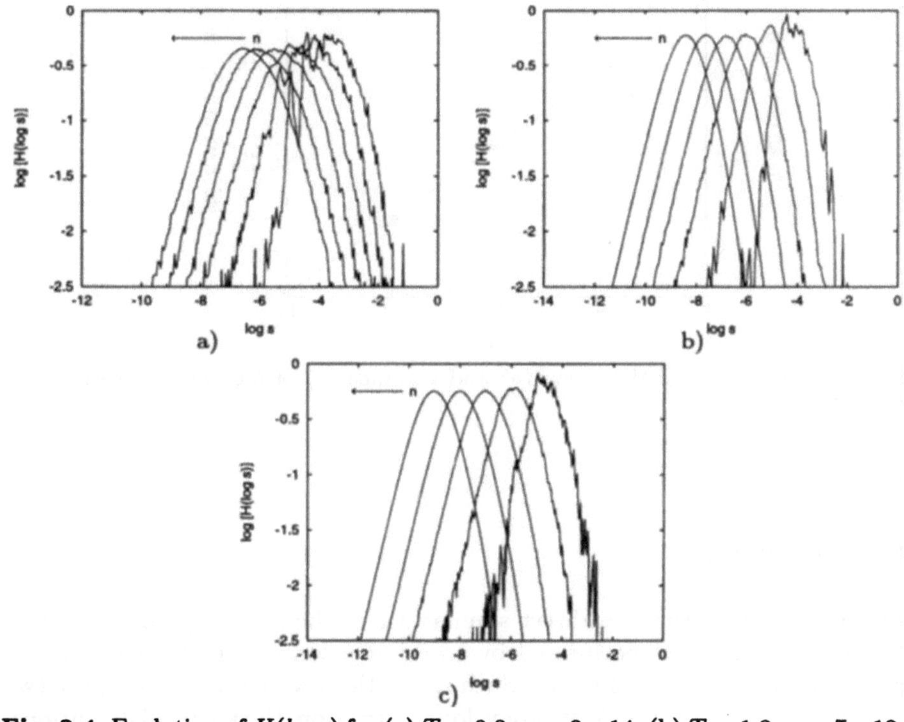

Fig. 3.4. Evolution of $H(logs)$ for (a) $T = 0.8$, $n = 9 - 14$; (b) $T = 1.2$, $n = 7 - 12$; and (c) $T = 1.6$, $n = 6 - 10$.

As shown in Figure 3.5, as time increases, $< logs >$ (the mean of $logs$), decreases linearly with time (as expected from the exponential growth of the filament length described in Figure 3.3), but its variance approaches a constant value. These results immediately suggest that the self-similarity of $H(logs)$ can be made explicit by applying a simple re-scaling that shifts the curves horizontally by redefining the independent variable in terms of a variable,

$$u = logs - < logs > . \tag{3.4}$$

As shown in (Figure 3.6 a-c), when re-scaled in this manner, the curves describing H(logs) for different periods asymptotically collapse onto a time-invariant curve that further demonstrates the self-similarity of the partially mixed structures created by the time-periodic flow.

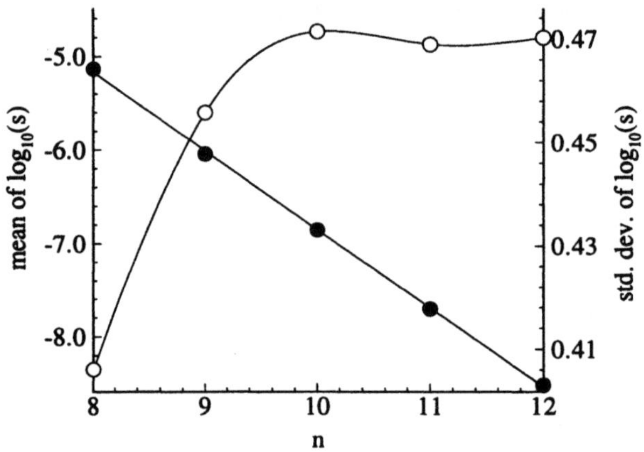

Fig. 3.5. Evolution of the mean (•) and variance (○) of *logs* for the case where $T = 1.2$.

4. Conclusions.

This brief paper discussed direct simulations of the mixing process in time-periodic chaotic flows performed by tracking a continuous filament as it is stretched and folded by the flow. Such simulations were used to compute (we believe for the first time) well-developed striation thickness distributions. Such distributions, which directly describe the microstructural topology of partially mixed systems, provide a physically meaningful quantification of the mixing process. Three main observations were apparent from the calculations presented here: (i) the spatial distribution of intermaterial area density, quantified as the local density of the stretched filament, is highly non-uniform; (ii) due to such non-uniformities, the filament length grows much more rapidly than predicted by the Lyapunov exponent; and (iii) the striation thickness distribution generated by the flow is self-similar and can be collapsed into an invariant curve through a very simple scaling transformation. All three

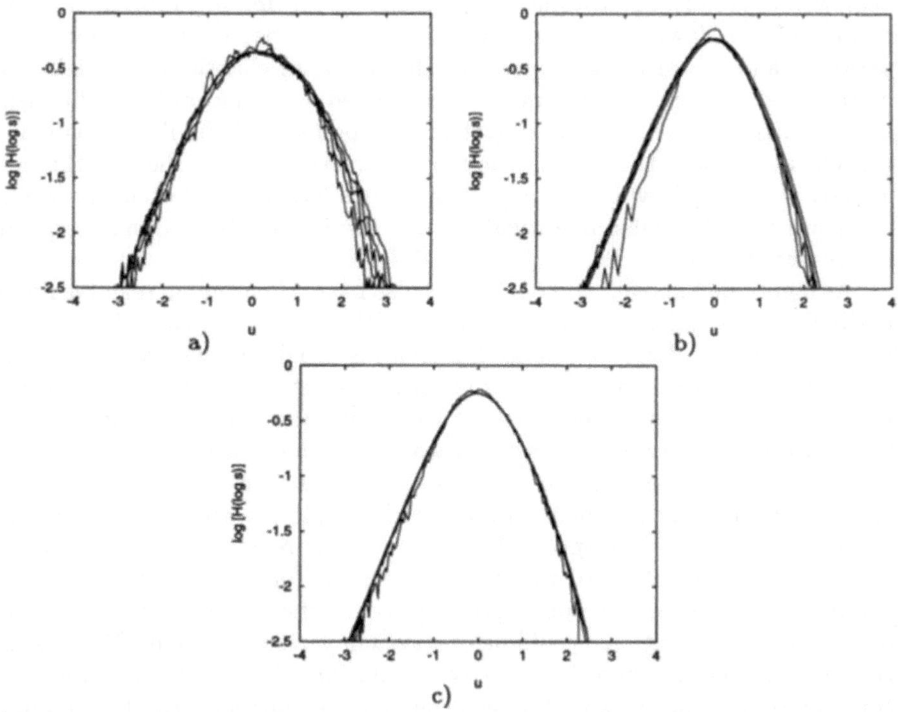

Fig. 3.6. $H(logs)$ collapses onto a time-invariant curve under a simple scaling. (a) $T = 0.8$, $n = 10 - 14$; (b) $T = 1.2$, $n = 8 - 12$; and (c) $T = 1.6$, $n = 7 - 10$.

conclusions have important implications for practical aspects of mixing, both for reactive and non-reactive flows.

Let us first consider non-reactive processes. The quality of an industrial mixture is often discussed, at least conceptually, in terms of an average striation thickness. As shown here, predictions of the average striation thickness based on the Lyapunov exponent would actually overestimate the average thickness by several orders of magnitude. Moreover, since the spatial distribution of the striation thickness is strongly non-uniform, mixing in some regions of a time-periodic chaotic flow is much more intimate than in other regions. Estimates of the overall mixture quality based on the average thickness would grossly overestimate the extent of the mixing process in large regions of the flow. A perhaps more interesting observation is that time-periodic (and, by direct analogy, spatially-periodic) chaotic flows might not be the most efficient mixing flows. The observed non-uniformities in filament density suggest that, from an industrial viewpoint, periodic flows wastefully concentrate most of their mixing action in narrow regions of the flow that are already well-mixed, while leaving other regions largely unaffected. Aperiodic flows are already under investigation and results will be communicated in future publications

As mentioned above, our conclusions also have a bearing on reactive flows. For high-viscosity reacting fluids (which are the most likely candidates for an industrial chaotic flow) molecular diffusivity is usually very small and rate of reaction depends directly on creation of inter-material surface between the reacting fluids. As a direct consequence of the observed non-uniformity in intermaterial density, such reactions would be much faster in regions that are densely populated by the filament. If the reaction is exothermic (as is the case with many polymerizations), such "explosive mixing" can explain the development of hot spots often observed in reactive extrusion, which can cause carbonization of polymers. Moreover, for systems with multiple competing reactions, this non-uniformity means that different reaction rates and product distributions would be observed in various locations. Under such conditions, accurate knowledge of the evolution of the striation thickness could be very useful. It is well-known that such polydisperse reactive laminar flows can be simulated as ensembles of striations undergoing diffusion, one-dimensional compression, and reaction. Our results show that once the well-developed striation thickness distribution is reached, it remains unchanged for all times (except for the mean thickness which decreases exponentially). Other statistical quantities of importance, such as spatial correlations of thicknesses of neighboring striations, can be readily derived from our simulations. Such an approach could greatly facilitate the development of simple and accurate models of reactive laminar flows.

Finally, it must be stressed that, from a computational viewpoint, simulating the development of self-similar striation thickness distributions is not a trivial exercise. Due to the extremely rapid growth of the filament, a huge amount of information is gathered and processed in order to compute $H(logs)$ for a long enough time. For example, for $T = 1.2$, $n = 12$, the filament is composed of more than $8X10^9$ points, and the calculation of $H(logs)$ is based on more than $51.1X10^7$ intersections. These computational demands would increase quite substantially for more realistic flows such as three-dimensional flows in tanks or ducts used in many industrial applications, because for such systems the striations are bounded by two-dimensional surfaces (rather than filament branches) and the velocity field is only known numerically. Results for such systems, which take a long time to compute, will be communicated in future publications.

References

1. Leong, C.W., Ottino, J.M. (1989): Experiments on mixing due to chaotic advection in a cavity. *Journal of Fluid Mechanics*, 209, 463.
2. Leong, C.W. (1990): Chaotic mixing of viscous fluids in time-periodic cavity flows. Ph.D. Dissertation, University of Massachusetts.
3. Swanson, P.D., Ottino, J.M. (1990): A comparative computational and experimental study of chaotic mixing of viscous fluids. *Journal of Fluid Mechanics*, 213, 227.
4. Lamberto, D.J., Muzzio, F.J., Swanson, P.D. (1996): Using time dependent RPM to enhance mixing in Stirred Vessels. *Chemical Engineering Science*, 51 (5), 733.
5. Muzzio, F.J., Ottino, J.M. (1990): Diffusion and reaction in a lamellar system: self-similarity with finite rates of reaction. *Phys. Rev. A*, 42, 5873.
6. Aref, H. (1984): Stirring by chaotic advection. *Journal of Fluid Mechanics*, 143, 1.
7. Khakhar, D.V. , Rising, H., Ottino, J.M. (1986): Analysis of chaotic mixing in two model systems. *Journal of Fluid Mechanics*, 172, 419.
8. Rom-Kedar, V., Leonard, A., Wiggins, S. (1991): An analytical study of transport, mixing and chaos in an unsteady vortical flow. *Journal of Fluid Mechanics*, 214, 347.
9. Beigie, D., Leonard, A., Wiggins, S. (1991): A global study of enhanced stretching and diffusion in chaotic tangles. *Physics of Fluids*, A 3, 1039.
10. Franjione, J.G., Ottino, J.M. (1987): Feasibility of numerical tracking of material lines and surfaces in flows. *Phys Fluids*, 12, 3641.
11. Liu, M., Muzzio, F.J., Peskin, R.L. (1994): Effects of manifolds and corner singularities on stretching in chaotic cavity flows. *Chaos, Solitons, and Fractals*, 4, 2145.
12. Hobbs, D.M., Alvarez, M.M., Muzzio, F.J. (in press) (to appear in 1997): Mixing in globally chaotic flows: a self-similar process. *Fractals*.
13. Muzzio, F.J., Swanson, P.D., Ottino, J.M. (1991): The statistics of stretching and stirring in chaotic flows. *Physics of Fluids A*, 3, 822.

Accessibility of a Catalyst's Fractal Surface to Diffusing and Reacting Molecules

M.-O. Coppens[1,2]

[1] Department of Chemical Engineering, Gilman Hall, University of California, Berkeley, CA 94720-1462, USA;

[2] Laboratorium voor Petrochemische Techniek, Universiteit Gent, Krijgslaan 281, B-9000 Ghent, BELGIUM

Abstract Many porous catalysts have a fractal internal surface on molecular scales. Similar to a natural coastline, the fractal roughness of the surface creates a cascade of fjords along the pores. In a heterogeneously catalyzed reaction, molecules diffuse through the pores and react on the surface. Because of the fractal cascades of fjords along the pores, some points on the catalyst's internal surface are more easily reached than other points. This influences the diffusion of molecules through the pore space and leads to a nonuniform reaction probability distribution over the surface. Gas diffusion within a cascade of fjords is mainly Knudsen diffusion, so that the results are fundamentally different when compared to the often studied molecular diffusion or analogous phenomena, like conduction. The obtained results are useful in the investigation of whether important diffusion limitations within the fjord cascades along a fractal catalyst surface exist, and, if so, on which subset of the surface the active phase may be concentrated without altering the effective reaction rate.

1. Introduction

This paper discusses a few aspects of the influence of a surface's fractal morphology on its accessibility to molecules which diffuse in the pore space close to it and on the rates of reactions catalyzed by it. This is an important problem, since many porous media, including heterogeneous catalysts, have a self-similar fractal internal surface on molecular scales [1]–[6]. Therefore, the surface area accessible to diffusing and adsorbing molecules is not a constant, but depends on the size of the molecules: the area S is a power law of the effective diameter δ of a molecule.

Molecules diffuse through the pores, the width of which is for many catalysts and sorbents of the order of 10 or a few tens of nanometers. For gases, this implies that Knudsen diffusion is often the dominating transport mechanism. On the catalyst surface, molecules can adsorb, possibly diffuse and react, after which the reaction products leave the surface.

The self-similarity of a surface leads to a cascade of "fjords" or indentations along it, comparable to the fjords along a rocky, natural coastline. The fjord size distribution is a power law, with cutoffs at a certain minimum value of atomic size, δ^s_{min}, and at a maximum value, δ^s_{max}, which could be of the order of the size of the pores through which a molecule moves. Among the points on the surface which can be reached by a diffusing molecule, it is clear that points sticking out into the pore space are more readily accessible than points hidden at the bottom of deep fjords: there exists an accessibility distribution along the surface. The question now is what the influence of the non-uniformity of the accessibility is on the diffusivity and on the effective reaction rates.

The first question has already been answered to some extent in previous papers, and only the main results will be cited in the next section. New is that some numbers will be given in order to show the reason for the diffusivity decrease with respect to porous media with a smooth surface. More attention will be paid to the second question, in Sec. 3.. Reaction rate or probability distributions along a surface were previously obtained from Monte–Carlo simulations, both for fractal and non-fractal surface morphologies. Such simulations and calculations have been performed by several research groups, but these studies have always been limited to *molecular diffusion* (Brownian motion) in the pore space along the surface [7]–[12]. While this may be so for molecular diffusion in liquids and for larger fjords, *Knudsen diffusion* is the main transport mechanism of gas molecules within the fjords in the vicinity of a surface which is fractal on molecular scales. This situation was not considered previously, and the results significantly differ from the case of molecular diffusion. Therefore, the analogy between catalysis and other phenomena, especially the conductivity of porous electrodes with a fractal surface [13]–[16], is very useful, but only valid for (mostly liquid phase) reactions with molecular diffusion through the fjords.

2. Diffusion in porous media with a fractal internal surface

Consider the "fjord tree" or cascade of fjords along a self-similar fractal surface. A molecule can enter a first fjord, and collide with its walls, or enter a smaller fjord which lies behind the first fjord. Then, it can hit the walls of this second fjord, or get into a third, even smaller one, and so on. When a molecule collides with the walls, it is adsorbed for a certain time, after which it either reacts or desorbs, and continues its trajectory. Ordinary molecular diffusion, in which the mean free path of the molecules is much smaller than the local pore size, is not notably affected by the surface morphology. On the other hand, within the fjords, the movement of a gas molecule generally is of the earlier mentioned Knudsen type: the molecule moves in a straight line from one point on the surface to the next, because the fjords are only a few times larger than the size of the molecule, and often much smaller than the mean free path in the bulk gas phase. If this is not the case, or if the molecules belong to a liquid, their movement is random, Brownian.

Nonlinear configurational effects could play a role as well. Indeed, the potential of the surface need not be a hard sphere potential, but could, for example, be a Lennard–Jones potential, which gradually drops along a direction perpendicular to the surface. This aspect is difficult to account for in analytical calculations, but could be included in Monte–Carlo simulations. Nevertheless, for a specific molecule, the problem can be simplified by assuming that the molecule cannot enter the outer Minkowski sausage of a certain width which is a function of the diameter of the molecule.

In [17, 18], a formula was derived for the ratio of the Knudsen diffusivity in a general porous medium with a statistically self-similar internal surface, $\mathcal{D}_K$, and the Knudsen diffusivity in a medium with the same pore network structure, but with a smooth internal surface, $\mathcal{D}_{K0}$. Hence, the fractality of the surface was treated as a perturbation. The ratio $\mathcal{D}_K/\mathcal{D}_{K0}$ depends on the smallest accessible fjord width, which, for a hard sphere potential, is the effective diameter δ of the molecule (or is directly proportional to it):

$$\frac{\mathcal{D}_K(\delta')}{\mathcal{D}_{K0}} = \frac{1}{1+\alpha\left[1-(\delta')^{\beta}\right]}, \tag{2.1}$$

in which $\delta' = \delta/\delta^s_{\max}$, and α and β are constants which can be calculated from the fractal adsorption dimension of the surface, D^s_{ads}, and a parameter p_0, the so-called return probability, *i.e.*, the probability for a molecule to leave a fjord it has entered already. Expressions for $\alpha(D^s_{\text{ads}}, p_0)$ and $\beta(D^s_{\text{ads}}, p_0)$ are given in [17, 18] (α should not be confused with the Hölder exponent, introduced in Section 3., to which it has no relation). This formula was derived from an evaluation of the ratio of the first-passage times of a molecule in the real medium with a fractal surface, and the non-perturbed medium, with a smooth surface. For not too low values of p_0 and δ', the following first-order approximation is useful, since it only depends on D^s_{ads}:

$$\frac{\mathcal{D}_K(\delta')}{\mathcal{D}_{K0}} = (\delta')^{D^s_{\text{ads}}-2}. \tag{2.2}$$

This equation can also be derived directly from a momentum transfer approach, and in fact merely expresses that the Knudsen diffusivity is inversely proportional to the accessible surface area. The latter is *not* true in general, because the non-uniformity of the accessibility distribution leads to a smaller effectively available surface area.

It is noteworthy that, although the probability p_e for a molecule to enter a fjord may be small, a relatively high probability $(1 - p_0)$ for the molecule to stay in the fjord and collide with its walls may result in a considerably enhanced residence time and, hence, a decrease in the diffusivity. The following example illustrates this point.

A surface segment is modeled by a Koch surface, formed by a generator which consists of a square with edge length L on which a cube with edge length $L/3$ is placed. This generator consists of 8 squares facing the pore space and 5 squares delimiting a cubic fjord with a square opening. To generate the surface by recursion, every square is replaced by a $b^3 = 3^3 = 27$ times reduced copy of the generator, *i.e.*, the base or contraction factor is $b = 3$. The second prefractal surface, for example, consists of $13^2 = 169$ squares with edge length $L/9$; there is 1 fjord of size $L/3$ and there are 13 fjords of size $L/9$ (Fig. 2.1). In general, there are 13^j fjords ($j = 0, \ldots, n-1$) of size $L/3^{j+1}$ in the nth prefractal.

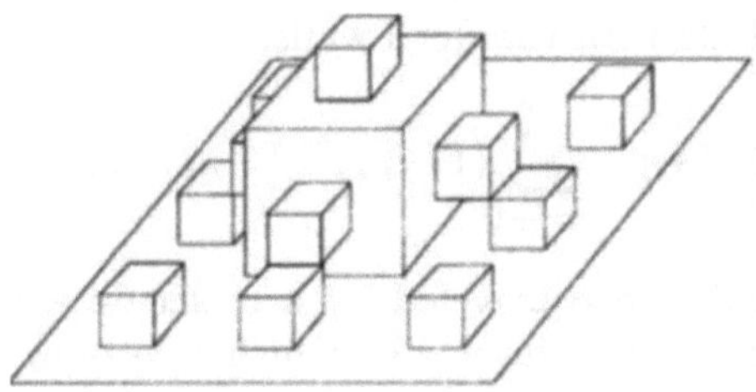

Fig. 2.1. Sketch of the second generation of the Koch surface.

The probability to enter a fjord is proportional to the area of its opening. If fjords down to the nth level are accessible to a molecule, but the $(n+1)$th level is not, then there is a probability $(1 - p_e)^n$ that a molecule will not enter any fjord in this segment. There is a probability p_e for the molecule to enter the largest fjord (level 1), a probability $(1-p_e)p_e$ to enter one of the 8 second largest ones (level 2), a probability $(1 - p_e)^{j-1}p_e$ to enter one of the 8^{j-1} fjords at level j which are directly accessible from the pore space. In this example, $p_e = 1/9$.

Once the molecule has entered a fjord of level j, it travels a distance l_j proportional to $L/3^j$. Then, it can either collide with this fjord's walls (probability $(1-p_e)^{n-j}$) or it can move into a smaller fjord of level k (probability $p_e(1 - p_e)^{k-j-1}$). In the latter case, the molecule moves another distance proportional to $L/3^k$, and can either collide with the walls or move into yet a smaller fjord, with the previously given probabilities. When a molecule has hit the pore surface in a fjord at level j, it can leave this fjord again and return into the pore space or into the larger fjord that carries it (probability p_O), or it can collide with the fjord's walls (probability $(1 - p_O)(1 - p_e)^{n-j}$) or it can enter a smaller fjord of level k (probability $(1-p_O)p_e(1-p_e)^{k-j-1}$). The probability to enter a fjord at a certain level j is always $(1 - p_e)$ times the probability to enter one of the larger fjords at a lower level $j - 1$. Every move through a fjord at level j adds a distance l_j proportional to $L/3^j$ (or, more exactly, $L/3^j - \delta$) to the distance the molecule has moved already. In this example, $p_O = 1/5$, being the ratio of the area of a fjord's opening to the area of its (unperturbed) walls [17, 18].

The average total length of the trajectory of a molecule through the fjord tree is:

$$s = p_e \sum_{j=0}^{n-1} (1-p_e)^j s_{j+1}, \tag{2.3}$$

in which s_j is the average distance a molecule travels from the moment it enters a fjord at level j until the moment it leaves it. The latter can be calculated, based on the previously given probabilities to enter fjords of a particular size or to collide with the fjord walls:

$$s_j = l_j + p_e \sum_{k=j+1}^{n} (1-p_e)^{k-j-1} s_k + s'_j. \tag{2.4}$$

The distance s'_j is the distance traveled by a molecule which has collided with the surface of the fjord at level j or which is leaving a fjord of level $k > j$ it has entered previously. The probability p_O to leave a fjord allows us to calculate s'_j:

$$s'_j = l_j + (1-p_O)\left[p_e \sum_{k=j+1}^{n} (1-p_e)^{k-j-1} s_k + s'_j\right]. \tag{2.5}$$

This equation can be solved, to yield an expression for s'_j:

$$s'_j = \frac{1}{p_O} l_j + \left(\frac{1}{p_O} - 1\right) p_e \sum_{k=j+1}^{n} (1-p_e)^{k-j-1} s_k, \tag{2.6}$$

which is substituted into Eq. (2.4):

$$s_j = \left(1 + \frac{1}{p_O}\right) l_j + \frac{p_e}{p_O} \sum_{k=j+1}^{n} (1-p_e)^{k-j-1} s_k. \tag{2.7}$$

This implicit equation for s_j is similar to the one which arises in the derivation of the Knudsen diffusivity for a statistically self-similar surface [17, 18]. The average path lengths s_j can be found recursively, starting with $n = j$: $s_n = (1 + 1/p_O)l_n$, $s_{n-1} = (1 + 1/p_O)l_{n-1} + (p_e/p_O)s_n$, $s_{n-2} = (1 + 1/p_O)l_{n-2} + (p_e/p_O)[(1-p_e)s_n + s_{n-1}]$, ... The calculated path lengths s_j, starting with a fjord of size b^{-j}, are substituted into Eq. (2.3) to obtain the total average path length s through the fjord tree.

However, these expressions do not yet teach which of the possible trajectories dominate the average time: some are short, while other trajectories are very long. Consider, *e.g.*, the 4th prefractal ($n = 4$) of the Koch surface, with $b = 3$, $p_e = 1/9$ and $p_O = 1/5$. Taking $l_j = b^{-j}$ (the proportionality factor does not influence the distribution anyway), it follows from Eq. (2.7) that the average path lengths: $s_4 = 2/27 \approx 0.074$, $s_3 = 2/9 + (5/9)s_4 \approx 0.263$, $s_2 = 2/3 + (5/9)(8s_4/9 + s_3) \approx 0.850$, $s_1 = 2 + (5/9)(64s_4/81 + 8s_3/9 + s_2) \approx 2.635$. Thus, from Eq. (2.3): $s \approx 0.406$.

Only a fraction $1 - (8/9)^4 \approx 0.376$ of the molecules will actually enter the fjord tree; the remaining, largest fraction, $f_0 \approx 0.624$, will collide with the external surface of the fractal segment, and the time within the fjord tree of this fraction is obviously zero. The fraction of molecules f_j entering fjords of size b^{-j} (level j) directly from within the pore space, are $f_4 \approx 0.078$, $f_3 \approx 0.088$, $f_2 \approx 0.099$ and $f_1 \approx 0.111$ respectively. All molecules entering the largest fjord (level 1, fraction $f_1 \approx 0.111$), will travel a path longer than the average path s, since the path length is at least $2/3 \approx 0.667$. Molecules entering a 2nd level fjord should make at least 2 additional collisions with the surface (probability $(4/5)^2 f_2 \approx 0.063$), or 1 additional collision and enter a fjord of the 3rd or 4th level and collide a few times with its walls, but the probability for this is smaller than 0.01. Nevertheless, once a molecule is in a fjord, the probability to go out after a collision with the walls is only $p_0 = 1/5$, which means that numerous collisions are highly probably, once the molecule has entered a fjord. In a similar way, molecules entering a 3rd level fjord from the pore space should make at least 10 collisions with the walls (probability $(4/5)^9 f_3 \approx 0.012$), or enter 4th level fjords on their surface, if less. The probability for the latter is only one ninth of the previous probability, however. Thirty-three collisions are needed for a molecule which has entered a 4th level fjord to reach a distance larger than the average distance s. The probability for this to happen, is negligible: $(4/5)^{32} f_4 \approx 6.2 \times 10^{-5}$. From these values, it follows that only about 19% of the molecules travel a distance larger than the average distance, and these molecules usually enter a large fjord and often collide many times with the surface, before leaving the fjord tree.

3. Diffusion toward and reaction on a fractal surface

3.1 Problem description

The diffusion of M components through a porous catalyst particle and their reaction on the catalyst surface is generally described by a set of continuity equations for the components, either for the complete catalyst particle which is treated as a pseudo-continuum, or for individual pores which constitute the pore network. A continuity equation for a component j in the steady state expresses the divergence of the molar flux of that component, N_j, as a result of its production by one or several reactions. For a pseudo-continuum, for example:

$$\nabla \cdot N_j = \rho_s R_j, \qquad j = 1, \ldots, M, \tag{3.1}$$

in which R_j is the production rate per unit catalyst mass for the jth component and ρ_s is the catalyst density. The equations for a single pore are very similar [19]. In these equations, expressions for the fluxes are substituted, typically the Stefan–Maxwell equations for multicomponent diffusion, possibly supplemented by a contribution due to transport by convective flow. The flux equations contain the binary diffusivities and, for gas reactions (mainly in mesoporous catalysts), also the Knudsen diffusivities. The influence of the fractal surface morphology on the Knudsen diffusivities was studied in the previous section.

If the internal surface is not uniformly accessible to the diffusing molecules, the reaction rates may also be non-uniformly distributed over the catalyst surface. The diffusivities which are substituted in the flux equations are those calculated in the *absence* of reaction. Strictly speaking, this is only permitted if the surface area is uniformly accessible. Under strongly diffusion limited conditions, the fact that some points on the surface may be more easily reached than other points can lead to concentration gradients within the tree or cascade of fjords in the neighborhood of the surface, and, therefore, to different effective reaction rates along the boundary of a pore cross-section. The intrinsic reaction rates in the continuity equations should then be substituted by effective reaction rates or, equivalently, multiplied by surface effectiveness factors, which account for these gradients. While the complete reaction probability distribution over the surface can be useful, the most important question for practical applications is whether the *surface effectiveness factors* η^s are close to 1 or not.

For a self-similar surface, the reaction rate distribution over the surface and the surface effectiveness factors can be evaluated in the Knudsen regime, without the need of Monte–Carlo simulations, such as those referred to in the Introduction for the case of molecular diffusion. Because the molecules move in a straight line from one point on the surface to another, instead of performing random walks, the local Knudsen diffusivity $\mathcal{D} = \mathcal{D}_K(u)$ is proportional to the local fjord size u, as opposed to molecular diffusion, where the diffusivity $\mathcal{D}$ is constant throughout the fjord tree. Therefore, the reaction probability distribution is expected to be entirely different in both cases.

In the simplest case, diffusion through the fjord tree is described by the Laplace equation, as is conduction through the electrolyte in which a porous electrode is put, but this assumes that the molecular diffusivity and the conductivity of the electrolyte in the fjords are constant. Since this is not so in the Knudsen regime, new formulas have to be derived.

3.2 Reaction probability distribution in the Knudsen regime

Consider a segment, of at least the size of the outer cutoff, of a catalyst's self-similar surface. Assume that the *intrinsic* activity is uniformly distributed over the segment, *i.e.*, the concentration of equally active sites in points of different accessibility is the same. This should not necessarily be so, but seems *a priori* reasonable, because of the self-similarity of the segment: even if the activity is linked to the fjord size, a fjord of a certain size could lie along the surface of a fjord of any larger size or could be immediately accessible from within the pore space.

In the steady state, the molar flux of a component j away from the surface equals the component's production by reactions at the surface:

$$R_j^s = -\mathcal{D}_j \nabla C_j, \tag{3.2}$$

in which R_j^s is the production rate per unit surface area. The diffusivity in the Knudsen regime, $\mathcal{D}_j = \mathcal{D}_{Kj}$, is proportional to the local fjord size, from which it is easily shown that the concentration profile in a fjord is independent of the fjord size. Assume indeed that this is so. Then, the concentration gradients ∇C_j are inversely proportional to the fjord size; multiplication with the local diffusivity yields scale invariant fluxes, and therefore scale invariant reaction rates, which is in agreement with the initial assumption that the concentration profile in a fjord is independent of its size. This can also be proven by comparing the possible trajectories of molecules through 2 similar fjords of different size. Since a trajectory consists of straight line segments from one point on the fjord's surface to another, these trajectories, including the number of hits which could possibly lead to reaction, are the same. Hence, the concentration profiles in both fjords are the same. For fjords in a fjord tree, in which a large fjord branches into smaller fjords which branch into still smaller ones, and so on, this holds if the fluxes at the openings of the fjords are also size independent and if the concentration profiles are not significantly altered by the openings in the surface, which lead to smaller fjords. That the first condition is met, follows directly from the self-similarity of the surface. The second condition, on the other hand, is not so obviously met, but is a very good approximation for molecular diffusion, as was verified by computer simulations for a situation that can be mapped on to the present problem when the reaction is instantaneous (extreme case) [20, 21]. Monte–Carlo simulations could be performed to avoid such approximations, but they probably give less insight.

The flux of a component j perpendicular to the surface of a single fjord is a function of the fluxes $N_0 = (N_{k0})_{k=1}^M$ at the fjord's opening:

$$N_j^{(1)} = g_j^{(1)}(N_0, x). \tag{3.3}$$

This function depends on the position x along the fjord surface and is linear in N_0 for first-order reactions:

$$N_j^{(1)} = h_j^{(1)}(x) N_{j0}, \tag{3.4}$$

but can be more complex for nonlinear reaction rate equations. The following is therefore correct for first-order reactions, but only approximately valid for other reactions. The surface effectiveness factor for one fjord is given by:

$$\eta_j^{(1)} = \frac{1}{N_{j0}S_1}\int_{S_1} N_j^{(1)}(x_{S_1})dS_1 = \frac{1}{S_1}\int_{S_1} h_j^{(1)}(x_{S_1})dS_1 = \langle h_j^{(1)}\rangle, \tag{3.5}$$

where the integration is performed over the complete accessible surface area of the fjord plus its base. For a strictly self-similar surface, S_1 is the area of the fractal generator.

For a general surface, the surface effectiveness factor is given by:

$$\eta_j^s = \frac{1}{S}\int_S h_j^s(x_S)dS. \tag{3.6}$$

For a strictly self-similar surface, the nth order prefractal has a surface area $S = S_n = (m/b^2)S_{n-1}$, and $h_j^s = h_j^{(n)}$, so that:

$$\eta_j^{(n)} = \frac{1}{S_1}\int_{S_1} \eta_j^{(n-1)}h_j^{(1)}(x_{S_1})dS_1 = \eta_j^{(1)}\eta_j^{(n-1)} = \left[\eta_j^{(1)}\right]^n. \tag{3.7}$$

This simple formula follows directly from the previous discussion: the fluxes at the surface of a prefractal are generated through a multiplicative cascade. The nth prefractal consists of m similar copies of the $(n-1)$th prefractal, lying along the boundary of the fractal generator, and, in each of these copies, the effectiveness factor is the same. In the more general case of a nonlinear reaction, $\eta_j^{(1)}$ should be treated as an operator, and the multiplication becomes an iterated transform. The result for the nth prefractal is calculated from the results for the previous prefractals by iteration.

This simple multiplicative cascade for the fluxes clearly does not exist in the case of molecular diffusion. The fluxes cannot be multiplied or transformed directly through the fractal generator, because the value of the operator h_j depends on the size of the fjord, and hence on the generation: concentration profiles in similar fjords of different sizes are not the same. For the same reason, the flux or reaction rate distribution along the surface is different for molecular diffusion and Knudsen diffusion. When the rates along the segments of the fractal generator are averaged over each segment, the multiplicative cascade in the Knudsen regime leads to a strict, multinomial reaction rate distribution along the surface. This distribution is easily derived as follows.

The fractal generator consists of m segments, which can be divided into k classes with different effectiveness factors: there are m_i segments with effectiveness factor $h_i \equiv h_j^{(1)}(i)$ $(i = 1, \ldots, k \leq m;\ \sum_i m_i = m)$. According to its position, every segment of the nth prefractal can be coded by a base k number with n digits. This code is determined by the number of times n_i $(\sum_i n_i = n)$ that a surface segment belonging to a certain class i has to be chosen in the iterative construction of the prefractal. Such a segment then belongs to class $n = n_1 n_2 \ldots n_k$. There are:

$$N^{(n)}(n) = \frac{n!}{n_1!n_2!\ldots n_k!} m_1^{n_1} m_2^{n_2} \ldots m_k^{n_k} \tag{3.8}$$

segments in this class. Along the surface of the limit fractal ($n \to \infty$), a multinomial multifractal measure is formed. Maximization of:

$$D^{(n)}(n) = \frac{1}{n} \log_b \left[N^{(n)}(n) \right] \tag{3.9}$$

leads to the following multifractal spectrum:

$$f[\alpha(n' = n/n)] = \lim_{n \to \infty} D^{(n)}(n) = -\sum_{i=1}^{k} n'_i \log_b n'_i + \sum_{i=1}^{k} n'_i \log_b m_i. \tag{3.10}$$

Using the notation $h \equiv h_j^{(n)}$ and $h_i \equiv h_j^{(1)}(i)$:

$$h(n) = h_1^{n_1} h_2^{n_2} \dots h_k^{n_k}, \tag{3.11}$$

so that the Hölder exponent α is given by:

$$\alpha = -\frac{1}{n} \log_b h(n) = -\sum_{i=1}^{k} n'_i \log_b h_i. \tag{3.12}$$

In this context, it should be added that the limit spectrum, Eq. (3.10), is approached very slowly and that the hull of the set of $D^{(n)}$-values can lie significantly below the $f(\alpha)$-curve for finite n. This is a general statement for multinomial measures (see also [22]), and a result of the bad convergence of Stirling's approximation for n!. For example, for a binomial measure with probabilities 0.6 and 0.4 (generated by $m = k = 2$, $h_1 = 0.6$, $h_2 = 0.4$ with the previous notations), the maximum of the spectrum of the 10th prefractal lies 22% off the maximum of the limit $f(\alpha)$-curve (1 in this example). Even for the 100th prefractal, the difference is still 3.7%. Note that the segments of this prefractal have a ridiculously small length of 7.9×10^{-31} times the length of the initiator: the limit $f(\alpha)$-curve will not be reached in a real situation. For statistically self-similar fractals, $b \to 1$, and $f(\alpha)$ will probably be reached even for a finite scaling regime, but this is stated without proof. These remarks are very important, because the scaling regime is often quite narrow, for example 1 to 2 decades for porous catalysts [1]–[6]. Results in the remainder of this section are discussed both for finite and for infinite n.

The properties of the multifractal measure are generated in a similar way to the well known special case where $m = k$. The reason for not normalizing the values of h, so that their sum is 1, is the physical interpretation of h: it is the value of the local effectiveness factor along the surface, and the average, defined by Eq. (3.6), is the global surface effectiveness factor η^s. The integral in this equation becomes a sum, which is readily evaluated by Newton's formula:

$$\eta^s = \frac{1}{m^n} \sum_n N^{(n)}(n) h(n) = \left(\frac{\sum_{i=1}^{k} m_i h_i}{\sum_{i=1}^{k} m_i} \right)^n = \left[\eta^{(1)} \right]^n. \tag{3.13}$$

The index referring to the component number j was dropped for convenience. Diffusion limitations in the fjord tree reduce the intrinsic reaction rate R_j^s to a lower effective reaction rate $R^e = \eta^s R^s$. This can also be expressed in another way. Although the surface has a fractal similarity dimension:

$$D_{\text{sim}}^s = \log_b m, \tag{3.14}$$

the molecules perceive it as if they would react on a surface which has the real intrinsic activity R_j^s, but a lower *effective dimension*, D^e, given by:

$$D^e = \log_b \left(\sum_{i=1}^{k} m_i h_i \right) = D_{\text{sim}}^s - \frac{\log \eta^s}{\log \delta'}, \tag{3.15}$$

where $\delta' = b^{-n}$ is the ratio of the size of the smallest accessible fjords (at level n, by definition) to the outer cutoff. The last formula is very general, but can be simplified for a strictly self-similar surface to:

$$D^e = D_{\text{sim}}^s + \log_b \eta^{(1)}. \tag{3.16}$$

Therefore, the effective dimension of a strictly self-similar surface can be calculated directly from the effectiveness factor of the fractal generator, $\eta^{(1)}$. This formula is very useful in checking whether diffusional limitations in the fractal fjord tree are important or not. For mesoporous catalysts, with a relatively narrow fractal scaling regime for their surface, it appears that $\eta^{(1)}$ is close to 1 in many practical situations, so that it is safe to use intrinsic reaction rates, R_j, instead of effective reaction rates, R_j^e, in the continuity equations, Eq. (3.1). An example is provided by [19]. For strongly diffusion limited reactions on a surface with a wide fractal scaling regime, the situation can be different.

When the moment generating function of the reaction probability distribution is defined by:

$$\tau(q) = \log_b \left(\sum_{i=1}^{k} m_i h_i^q \right), \tag{3.17}$$

the zeroth moment ($q = 0$) corresponds to the fractal dimension of the surface, $\tau(0) = D_{\text{sim}}^s$, and the first moment or average ($q = 1$) to the effective dimension of the surface, $\tau(1) = D^e$: the dimension of the support of the measure is D_{sim}^s, while the catalyst surface appears, on average, as a surface of dimension D^e. The moments $q \to +\infty$ and $q \to -\infty$ correspond to immediately accessible points sticking out into the pore space ($h = (h_{\max})^n$ where $h_{\max}$ is the largest of the h_i values), and points in the deepest fjords ($h = (h_{\min})^n$ where $h_{\min}$ is the smallest of the h_i values), respectively.

The following relations exist between a moment q, a class $n' = n/n$, and the Hölder exponent α:

$$\alpha(q) = -\frac{\sum_{i=1}^{k} m_i h_i^q \log_b h_i}{\sum_{i=1}^{k} m_i h_i^q}, \quad n_i'(q) = \frac{m_i h_i^q}{\sum_{i=1}^{k} m_i h_i^q}. \tag{3.18}$$

The multifractal spectrum, $f(\alpha)$, and the moment generating function, $\tau(q)$ are still linked by a Legendre transform. Table 3.1 summarizes the results for $q = -\infty$, 0, 1 and $+\infty$. The interpretation of this table is straightforward.

Table 3.1. Values for the moment generating function, $\tau(q)$, the Hölder exponent, α, the multifractal spectrum, $f(\alpha)$, and the (normalized) subset n' which corresponds to the previous values ($e_{\min}$ stands for a unit vector with k elements, with a 1 on the position corresponding to the smallest effectiveness factor, $h_{\min}$; analogous for $e_{\max}$).

q	$\tau(q)$	α	$f(\alpha)$	n'
$-\infty$	$\sim \log_b m_{\min} + q \log_b h_{\min}$	$\alpha_{\max} = -\log_b h_{\min}$	$\log_b m_{\min}$	$e_{\min}$
0	D^s_{sim}	$-\frac{1}{m}\sum_{i=1}^{k} m_i \log_b h_i$	D^s_{sim}	$\frac{\mathbf{m}}{m}$
1	$D^e = \log_b(m \cdot h)$	$-\frac{\sum_{i=1}^{k} m_i h_i \log_b h_i}{\sum_{i=1}^{k} m_i h_i}$	$\alpha + D^e$	$\left(\frac{m_i h_i}{m \cdot h}\right)_{i=1}^{k}$
$+\infty$	$\sim \log_b m_{\max} + q \log_b h_{\max}$	$\alpha_{\min} = -\log_b h_{\max}$	$\log_b m_{\max}$	$e_{\max}$

It is interesting to note that the activity of the surface is concentrated on the set $n(q = 1)$: even if all other segments were inactive, the activity of the catalyst would effectively be the same. Such an observation is useful for catalyst design purposes, because it means that it is not necessary to distribute the active phase uniformly over the complete surface, but that it can instead be concentrated over a particular subset of it, without decreasing the *effective* activity.

4. Conclusion

The molecular fractal roughness of the internal surface of many porous media has an important influence on the Knudsen diffusivity of gas molecules diffusing through these media. In the Knudsen regime, molecules move in straight lines from one point on the surface to another, as opposed to ordinary molecular diffusion. Knudsen diffusion occurs very often in mesoporous media, and, due to the small size of the fjords or indentations along a fractal surface, is also the diffusion mechanism for gas molecules within a fjord tree, *i.e.*, the series of smaller fjords upon larger fjords along the surface. The accessibility distribution over the surface is nonuniform, because some points can be more easily reached from within a pore than other points.

A formula for the Knudsen diffusivity was derived and applied previously. This formula shows the sometimes drastic decrease of the diffusivity in media with a fractal internal surface, when compared to smooth media. It has now also been shown that the molecular trajectories which lead to the decrease are a result of only a small fraction of all the possible trajectories. The path length of a molecule is often not or not significantly lengthened by the fractal roughness, but when the molecule enters one of the larger fjords, it can remain in it, or move into smaller ones and collide several times with the walls, before leaving the fjord tree. These relatively infrequent paths dominate the diffusion process.

Difficulties in reaching points on the surface of some fjords can also decrease the rate of a reaction on the surface, as compared to the situation where the surface is uniformly accessible. The Knudsen diffusion of gas molecules through a fjord tree leads to a multifractal reaction rate distribution which is different from the one when the molecules perform Brownian motions in the fjord tree. For a strictly self-similar fractal, the measure is multinomial. Often, there are no diffusion limitations in the fjord tree, but when there are, the surface acts as if it had an effective dimension which is lower than the true fractal dimension of the surface. A subset of the surface with a uniform reaction probability, and on which the *total* reaction rate is a maximum, has this effective dimension, so that catalytically active centers on the rest of the surface could be removed without changing the effective activity for that reaction. This is expected to be even so for statistically self-similar surfaces with a finite fractal scaling regime. For strictly self-surfaces, this result is only valid for the limit fractal.

Acknowledgement. A grant of "Aspirant" from the N.F.W.O. (Belgian National Fund for Scientific Research) and a Fellowship from the Belgian American Educational Foundation are gratefully acknowledged. The author is also very thankful to Professor Benoit B. Mandelbrot for the many discussions with him during work at Yale University and for his many valuable suggestions.

References

1. Avnir, D., Farin, D., Pfeifer, P. (1984). *Nature* **308**, 261.
2. Pfeifer, P., Avnir, D. (1983). *J. Chem. Phys.* **79**, 3558.
3. Pfeifer, P., Avnir, D. (1984). *J. Chem. Phys.* **80**, 4573.
4. Avnir, D. Farin, D., Pfeifer, P. (1983). *J. Chem. Phys.* **79**, 3566.
5. Coppens, M.-O. (1996). Ph. D. thesis, Gent University, Belgium.
6. Coppens, M.-O., Froment, G.F. (1995). *Chem. Engng Sci.* **50**, 1013.
7. Meakin, P. (1986). *Chem. Phys. Lett.* **123**, 428.
8. Gutfraind, R., Sheintuch, M., Avnir, D. (1990). *Chem. Phys. Lett.* **174**, 8.
9. Gutfraind, R., Sheintuch, M., Avnir, D. (1991). *J. Chem. Phys.* **95**, 6100.
10. Lee, C.-K., Liang, W.-H.,Lee, S.-L. (1994). *J. Chinese Chem. Soc.* **41**, 665.
11. Lee, C.-K., Lee, S.-L. (1995). *Surf. Sci.* **325**, 294.

12. Lee, C.-K., Lee, S.-L. (1995). *Surf. Sci.* **339**, 171.
13. Sapoval, B. (1991): *Fractals in Disordered Systems*. Eds. A. Bunde and S. Havlin (Springer, Heidelberg).
14. Sapoval, B. (1991). *C. R. Acad. Sc. Paris, Série II* **312**, 599.
15. Sapoval, B., Gutfraind, R., Meakin, P., Keddam, M., Takenouti, H. (1993). *Phys. Rev. E* **48**, 3333.
16. Sapoval, B. (1994). *Pour la Science* **198**, 30.
17. Coppens, M.-O., Froment, G.F. (1995). *Fractals* **3**, 807.
18. Coppens, M.-O., Froment, (1995): *Fractal Geometry and Analysis. The Mandelbrot Festschrift, Curaçao, Netherlands Antilles, 2–4 February 1995*, eds. C. J. G. Evertsz, H.-O. Peitgen and R. F. Voss (World Scientific, Singapore, 1996), 403.
19. Coppens, M.-O., Froment, G.F. (1994). *Chem. Engng Sci.* **49**, 4897.
20. Mandelbrot, B.B., Evertsz, C.J.G. (1990). *Nature* **348**, 143.
21. Evertsz, C.J.G., Mandelbrot, B.B. (1992). *J. Phys. A: Math. Gen.* **25**, 1781.
22. Mandelbrot, B.B. (1989). *Pure Appl. Geophys.* **131**, 5.

Lévy Distribution of Collisions for Knudsen Diffusion in Fractal Pores

S.B. Santra, B. Sapoval, and O. Haeberlé

Laboratoire de Physique de la Matière Condensée,
Ecole Polytechnique, 91128 Palaiseau Cedex, FRANCE

Abstract We show on a specific example a few anomalous properties of Knudsen diffusion in two dimensional irregular porous catalysts. The geometrical irregularity is modeled by prefractal shapes. The probability distribution of the number of collisions is found to obey a power law of the Lévy type with an exponent ≈ -2. A remarkable result is that the distribution does not change significantly between specular and random reflection in deterministic prefractals.

1. Introduction

The interaction of particles with irregular surfaces has many applications in chemical engineering. For example, many chemical and petrochemical processes use porous catalysts to increase reaction rates and to obtain higher selectivities toward the desired products. A better understanding of a catalyst's efficiency is then of great economical importance.

Since heterogeneous catalysis is a surface phenomena, the first effect of porosity is to increase the active area per unit mass. Moreover, porous catalysts have often very irregular surfaces down to the molecular scale. Geometrical irregularity may play an important role in confining reactants in the small scale structures of the catalyst. This effect has two consequences. Firstly it may increase the interaction with the surface. Secondly it may happen that two reacting molecules A and B are trapped in the same place increasing the reaction probability between A and B after activation by collision on the surface. The emergence of fractal geometry has provided an efficient tool to study the influence of strong geometrical irregularity on various physical or chemical processes. The pioneering work of Avnir and co-workers [1, 2, 3] has triggered several studies in order to improve the design of catalyst [4].

A frequent and important physical situation occurs when the pressure or the pore size are such that the reacting particles collide with the pore walls before particle-particle collisions occur. This is the so-called Knudsen diffusion regime that we discuss here [5, 6, 7]. In that case the mean free path is limited by the catalyst geometry. The so-called molecular diffusion regime in which the mean free path is much smaller than the pore size of the heterogeneous catalyst is discussed in [8, 9] .

Since Knudsen's original work several studies have been devoted to obtain general expressions for the macroscopic diffusivity and its possible dependence on the geometry [7, 10, 11, 12]. These results show that the macroscopic diffusion coefficient depends on the specific average of the free path between two successive collisions with the pore walls. There exists a dependence on geometry but the diffusivity D remains $D \approx cl$ where c is the molecular velocity and l is suitably averaged pore diameter. On the other hand it is known, at least empirically, that the geometrical structure of a porous catalyst may play an important role on a catalyst's efficiency.

The interaction of an atom with a surface is a complex phenomenon. When an atom or molecule hits a surface, it may stay adsorbed for some time and then desorbs exchanging momentum with the pore wall. For this reason the particle reflection is non specular and presents a random character which may depend on the temperature of the wall [13]. The random character of the trajectories may then come from the partial random character of individual collisions or from the micro-roughness of the prefractal surface. In this paper we present two dimensional computer simulations of these phenomena using a prefractal geometry to model the geometrical irregularity.

2. Numerical Method

We study numerically the ballistic trajectories of particles sent at a fixed angle of incidence ϕ with the horizontal line. Figure 2.1 shows three particle trajectories with specular reflection at 45° within a prefractal surface. This is the second generation of a quadratic Koch curve of dimension $\frac{3}{2}$. The source is inclined at a fixed angle $\phi = 45°$ with the horizontal.

In order to mimic the complex particle-wall interaction we introduce the random nature of the real collision by adding a random angle to the specular reflection angle. At each collision the angle of reflection θ_r is related to the angle of incidence θ_i by

$$\theta_r = \theta_i + \alpha \tag{2.1}$$

where α is a random angle uniformly distributed over a range $\pm\delta\theta$. The angle $\delta\theta$ is then a parameter which characterize the randomness of the particle-wall interaction. The system has been studied for different values of $\delta\theta$ ranging from $\delta\theta = 0°$ (purely specular reflection) to $\delta\theta = 16°$ in steps of 2° . In other words we select the direction of the momentum vector randomly within the range of disorder $\pm\delta\theta$ to describe qualitatively the effect of the real particle-wall interaction.

We study the "porous region" of the structure which is delimited by the two arrows in figure 2.1. Atoms are supposed to be point mass particles. A trajectory is described by the successive events made out of the collision positions and the successive angles of incidence and reflection. The surface is described by line segments. The point of collision is determined by solving the linear equations which describe the path of the atom and the prefractal surface. The path of the atom will intersect the prefractal surface at several

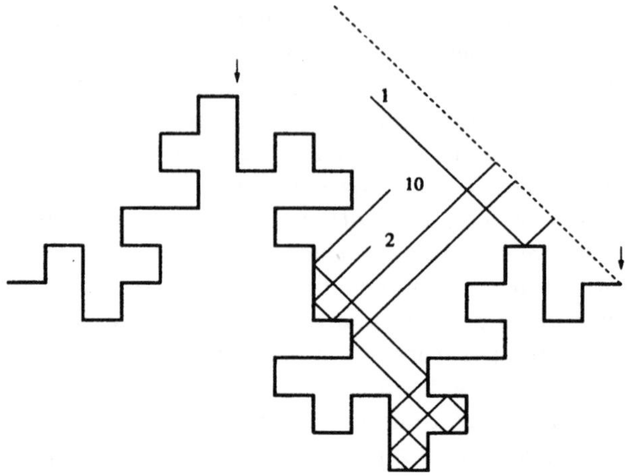

Fig. 2.1. A schematic representation of the trajectories of atoms at 45° angle of incidence on the prefractal surface at second generation. The thick line represents the prefractal surface and the dotted line represents the slit. Thin lines are the trajectories of the atoms and the numbers indicate the number of collisions made by the respective atomic rays in the case of specular reflection. The arrows represents the porous region under study.

different places. The real solution is the closest to the previous point of collision. The path after reflection is on the opposite side of the normal drawn at the point of collision.

We note that, for specular reflections $\frac{3}{15}$ of the particles are reflected after only one collision as can be seen by direct inspection. These particles are neglected in the following. The other particles undergo an even number of collisions and are finally retroreflected.

To avoid the effect of possible "numerical" ray splitting at the corners of the prefractal surface we calculate Δ, the distance of the point of collision from the nearest corner and the error Δ_x or Δ_y in the calculation of the point of collision. The error in the calculation of the point of collision will be either in the x coordinate or in the y coordinate since the prefractal surface is described by the horizontal and the vertical lines only. If Δ_x or Δ_y becomes greater than Δ at any point we stop computing for that trajectory.

3. Results and Discussion

Trajectories have been numerically studied for a source angle $\phi = 45°$. For a given system (given generation and given $\delta\theta$), we consider initial emission of 1500 equally spaced rays on the source and we note the number of collisions for each ray. In order to obtain the collision statistics we repeat the experiment 32 times and we discuss below the global statistical distribution of collisions.

In figure 3.1 we have plotted $\langle N \rangle$, the average number of collision per atom as a function of the randomness parameter $\delta\theta$ for second and third generations of the prefractal surface. System size is kept constant for all generations.

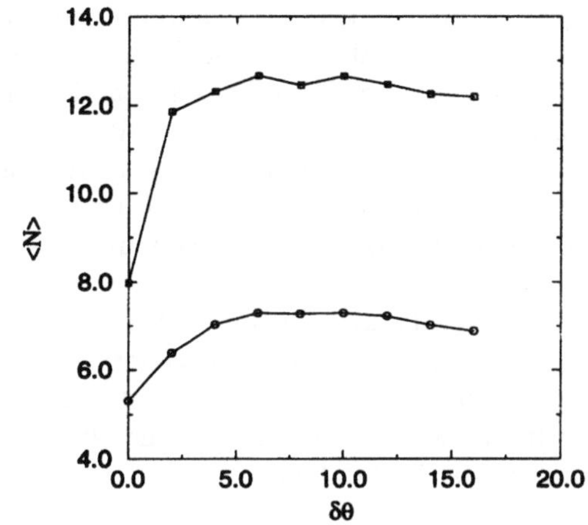

Fig. 3.1. A plot of the average number of collisions per atom $\langle N \rangle$ versus the randomness parameter $\delta\theta$ for second (○) and third (□) generation. The initial angle of incidence is 45°.

There are two effects. First, the number of collisions increases with the randomness parameter $\delta\theta$, but reaches quickly a saturation value which is essentially independent of the random character of the reflection. Secondly the saturation value of $\langle N \rangle$ at generation $\nu = 3$ is approximately two times larger than that of the generation $\nu = 2$.

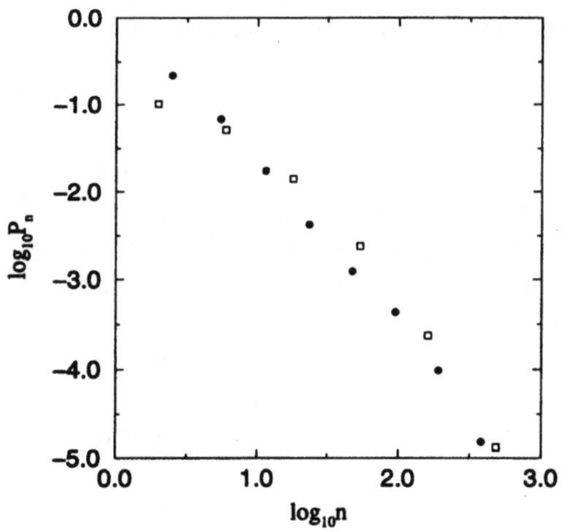

Fig. 3.2. A plot of the probability P_n of an atom to have n collisions versus $n(\bullet)$ at third generation of the prefractal surface for the randomness parameter $\delta\theta = 6°$. The squares represent the same probability distribution P_n of the number of collisions for specular reflections ($\delta\theta = 0°$) from another prefractal surface, shown in Fig. 4.1, at seventh generation.

In figure 3.2 we have plotted the distribution of the probability P_n for an atom to have n collisions (•) for third generation of the prefractal surface with a randomness parameter $\delta\theta = 6°$. One observes a power law with an exponent of the order of $-1.9 \approx -2$. So the distribution is of the Lévy type [14]. It should be noted that although working with the third generation prefractal only the power law extends over more than four orders of magnitude.

It is remarkable that the same type of distribution may be obtained for the number of collisions within another pre-fractal pore as in figure 4.1, in which we have shown specular reflection of three particles. For this particular geometry, the number of collisions for specular reflection can be easily found from a simple recursive relation, a trinomial multiplicative process, when the angle of incidence is 45° [16]. It is also possible to obtain an algebraic equation for the number of particles which make n collisions and consequently tests our numerical procedure.

In figure 3.2 we have plotted the probability distribution of the number of collisions P_n (□) for specular reflection within the seventh generation of the pre-fractal in figure 4.1. One observes the same type of Lévy distribution also with an exponent ≈ -2. The prefractal structure enhances the effect of trapping between parallel plates discussed by Lévitz [15].

4. Conclusion

In this preliminary work, we have shown that geometrical irregularity plays a dominant role in the number of collisions in Knudsen diffusion. The probability distribution of the number of collisions is found to be of the Lévy type. This has profound consequences. First, the majority of the particles have a small number of collisions. Second and most important, the average number of collisions is dominated by the few particles which undergo a large number of collisions. The first moment of the distribution can be written as

$$\langle N \rangle = \int n P_n dn \approx \int \frac{dn}{n} \tag{4.1}$$

If the maximum number of collisions increases with the fractal generation, which is true for the specular reflection in the pre-fractal in figure 4.1 with a 45° angle of incidence, the average number of collisions $\langle N \rangle$ will diverge logarithmically. This is of great practical significance in heterogeneous catalysis because it shows that the catalytic efficiency may be controlled by the very few molecules which undergo a very large number of collisions. In other words the catalytic activity may be largely disconnected from the macroscopic transport coefficient like diffusivity which depends on the behavior of the majority of the diffusing particles.

It should be emphasized also that the random character of the reflection process only plays a small role in the distribution of the number of collisions. In that sense there is little difference between random structures and

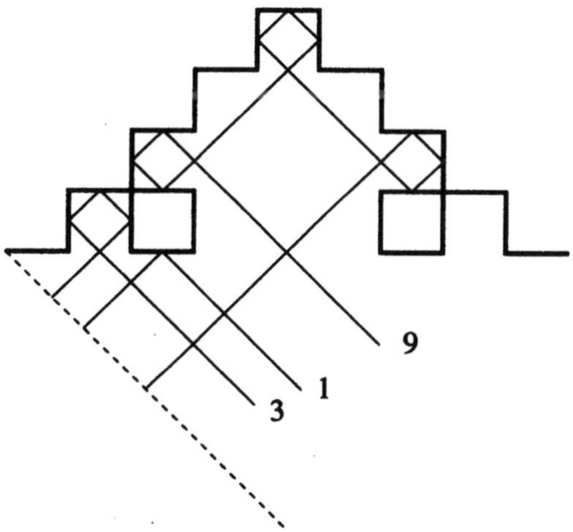

Fig. 4.1. A schematic representation of specular reflections of atoms at 45° angle of incidence on the prefractal surface at second generation of a quadratic Koch curve of fractal dimension $\frac{\log 5}{\log 3}$.

deterministic structures with the same type of geometrical irregularity. This confirms the role of fractal geometry as a tool to describe extreme geometrical disorder.

References

1. Avnir, D., Farin, D., Pfeifer, P. (1983). *J. Chem. Phys.*, **79**, 3566.
2. Avnir, D., Farin, D., Pfeifer, P. (1984). *Nature* **308**, 261, (1984).
3. Pfeifer, P., Avnir, D. (1983). *J. Chem. Phys.*, **79**, 3558.
4. Kopelmann, R. (1989): in *The Fractal approach to Heterogeneous Chemistry.* Ed D. Avnir (Wiley, New York), 295 and references therein.
5. Coppens, M.O., Froment, G.F. (1994). *Chem. Eng. Sc.*, **49**, 4897.
6. Coppens, M.O., Froment, G.F. (1995). *Chem. Eng. Sc.*, **50**, 1207.
7. Coppens, M.O., Froment, G.F. (1995). *Fractals*, **3**, 807.
8. Sapoval, B. (1996): in *Fractals and Disordered Systems.* 2nd edition ed by A. Bunde and S. Havlin (Springer Verlag), 232–261.
9. Pfeifer, P., Sapoval, B. (1995). *Mat. Res. Soc. Symp. Proc.*, **366**, 271.
10. Derjaguin, B. (1948). *C. R. Acad. Sci. URSS*, **7**, 623.
11. Lévitz, P. (1993). *Mat. Res. Soc. Symp. Proc*, **290**, 197.
12. Lévitz, P. (1993). *J. Phys. Chem.*, **97**, 3813.
13. Valleau, J.P., Diestler, D.J., Schoen, M., Hertzner, A.W., Riley, M.E. (1991). *J. Chem. Phys.*, **95**, 6194.
14. Klafter, J., Shlesinger, M.F., Zumofen, G. (February 1996). *Physics Today*, **49**, 33.
15. Lévitz, P. *Europhys. Lett.*, to be published.
16. Santra, S.B., Sapoval, B., Haeberlé, O. To be published.

NEW TOPICS

α-Stochastic Differential Equations and Option Pricing Model

L. Belkacem

INRIA - Groupe Fractales,
B.P. 105, Domaine de Voluceau,
78153 Le Chesnay Cedex, FRANCE
Lotfi.Belkacem@inria.fr

Abstract This paper develops a numerical model for option pricing under the hypothesis that the underlying asset price satisfies a stochastic differential equation driven by an α-stable Lévy motion. We feed this model with simulated European call option prices and use it to explain the skewed "smile effect" of the implied volatility.

1. Introduction

Option pricing models are usually based on stochastic calculus (martingale theory, stochastic integration, Itô's formula, theory of stochastic differential equations). Such models are continuously being developed, following the progress made in this area of probability theory.

Since 1973, date of the opening of the Chicago Board of Option Exchange (CBOE) and the exchange for trading standard option contracts, a huge amount of studies on option pricing models for different assets (interest rate, bonds, indexes, FX rate, ...) has been performed. Black and Scholes [3] and Merton [22] were the first to propose an approach to the foundation and search for a fair price which the holder should pay for the option. They developed also a theory that allows one to find an optimal hedging strategy which the seller of the option must entertain in order that the possible payments which are stated by the contract and depend on the random behavior of market prices be fulfilled with certainty. Cox, Ross and Rubinstein [6] introduced a simple binomial option pricing model. In this model, the stock prices changes follow a discrete binomial process. Black and Scholes formula is obtained as a limit case of this model when we divide the time interval $[0, T]$, where T is the expiration time of the option, into an infinite sub-periods of time.

More precisely, the Black and Scholes model assumes that the volatility and the interest rate are constant and stock prices are modeled by a geometric Brownian motion. However, these assumptions are incompatible with the market reality. Indeed, the empirical studies (see Latané and Rendleman [18]) exhibit systematic biases with respect to the Black and Scholes option pricing model due to a misspecification of the underlying stock's returns. These price distortions, well-known to practitioners, are usually documented in the empirical literature under the terminology of *smile effect*, where the so-called "smile" refers to the U-shaped pattern of implied volatilities across different strike prices.

A variety of option pricing models have appeared to improve on the Black and Scholes model. We cite in the Gaussian frame the stochastic volatility models (see Hull and White [13], Scott [33], Renault and Touzi [30]) which assume that the volatility of the stock return follows a specific stochastic process. Although these models have the potential to explain the smile effect of the implied volatility, the explicit calculation of option prices remains difficult.

It has become clear through the work of Mandelbrot [19, 20] and Fama [9] and corresponding empirical work by Mittnik and Rachev [23], Belkacem [1] and Belkacem et al. [2] that infinite variance stable distributions yield better fits of the stock price data than the Gaussian one. Mandelbrot and Taylor [21] and Clark [5] developed two competing models of stock returns based on subordinated processes. In the Mandelbrot-Taylor's model the resulting process is a Lévy stable motion while in the Clark's model, the process has finite variance. Rachev and Samorodnitsky [29] have generalized the binomial option pricing model from a discrete version of the Mandelbrot-Taylor model. Rachev and Ruschendorf [28] gave generalizations of the results for the binomial model in the case where the number of changes of stock prices is a random variable. These generalizations lead to richer classes of models that include models with stable distributions of index $0 < \alpha < 2$ and geometrically stable laws.

In this paper we develop a simple *Log-stable model* for option pricing under the hypothesis that the underlying asset price satisfies a *Stochastic Differential Equation* driven by an *α-stable motion* (α-SDE). This model is used to explain the skewed "smile effect" of the implied volatility.

The paper is organized as follows. Section 2 presents some definitions and properties of α-stable processes. In the third section, we study numerically the solution of an α-SDE using the Euler scheme. Finally, we present a numerical approach to calculate the price of European options when the underlying asset price is a solution of an SDE driven by an α-stable motion. This approach allows us to explain the skewed "smile effect" of the implied volatility.

2. α-stable stochastic processes

We recall in this section basic facts about the processes we will deal with in sections 3. and 4. To define stable processes, we will turn to the rich context of Lévy processes. For the following definitions and properties, we refer essentially to Feller [10] and Sato [32].

2.1 Definitions and properties

We assume as given a complete probability space $(\Omega, \mathcal{F}, P)$. In addition we are given a filtration $(\mathcal{F}_t)_{0\leq t\leq\infty}$. By a filtration we mean a family of σ-algebras $(\mathcal{F}_t)_{0\leq t\leq\infty}$ that is increasing ($\mathcal{F}_s \subset \mathcal{F}_t$ if $s \leq t$). The filtration $(\mathcal{F}_t)_{0\leq t\leq\infty}$ is right continuous.

Definition 2.1. *A stochastic process* $\{X(t) : t \in [0,\infty)\}$ *on* $\mathbb{R}^d$ *is called a* Lévy process *on* $\mathbb{R}^d$ *if the following five conditions are satisfied:*

1. $X(0) = 0$ *P-a.s.;*
2. *for any choice of* $n \geq 1$ *and* $0 \leq t_0 < t_1 ... < t_n$, $X(t_1)-X(t_0), ..., X(t_n)-X(t_{n-1})$ *are independent (independent increments property);*
3. *for any* $s \geq 0$ *and* $t \geq 0$, $X(s+t) - X(s) \overset{d}{=} X(t) - X(0)$ *(stationary increments property);*
4. *stochastically continuous, i.e.* $\forall t \geq 0$ *and* $\varepsilon > 0$, $\lim_{s\to t} P(|X(s)-X(t)| > \varepsilon) = 0$*;*
5. *there is* $\Omega_0 \in \mathcal{F}$ *with* $P(\Omega_0) = 1$ *such that, for every* $\omega \in \Omega_0$, $X(t,\omega)$ *is càdlàg (right continuous with left limits).*

Lévy processes can be identified through their characteristic functions.

Definition 2.2. *A stochastic process* $\{X(t) : t \in [0,\infty)\}$ *on* $\mathbb{R}^d$ *is called a* Lévy process *if its characteristic function is as follows:*

$$Ee^{i(\lambda, X(t))} = e^{-t\psi(\lambda)}$$

where

$$\psi(\lambda) = i(\lambda, \mu^0) + \frac{1}{2}Q(\lambda) + \int_{\mathbb{R}^d} \left(1 - e^{i(\lambda,x)} + i(\lambda,x)\mathbb{1}_{|x|\leq 1}\right) \nu(dx)\,; \qquad (2.1)$$

Q is a non-negative defined quadratic form (it represents the Brownian term) and $\nu(dx)$ is the Lévy measure of $X(t)$, i.e. a positive Radon measure on $\mathbb{R}^d - \{0\}$ satisfying

$$\int \left(1 \wedge |x|^2\right) \nu(dx) < \infty. \qquad (2.2)$$

This last representation is due to many authors: Finetti, Kolmogorov, Lévy et Kchintchine and is often called *Lévy-Kchintchine formula.*

Remark 2.1. Lévy processes include the Brownian motion, Poisson processes and α-stables motion [1]. Indeed,

- if $\mu^0 = 0$ et $Q = 0$, we obtain jump processes;
- if $\mu^0 = 0$ et $\nu = 0$, we obtain the Brownian motion;

[1] these processes are defined in subsection 2.3

- if $Q = 0$ and if there is a measure σ on the unit sphere S_d in $\mathbb{R}^d$ such that:

$$\nu(dx) = \frac{1}{r^{1+\alpha}} dr\sigma(ds) \tag{2.3}$$

where $r = |x|$ and $s = \frac{x}{|x|}$ belong to S_d, we obtain a α-stable motion.

Definition 2.3. $\{X(t) : t \in [0,\infty)\}$ *is called* stable process *on* $\mathbb{R}^d$, *if*

1. $\{X(t) : t \in [0,\infty)\}$ *is Lévy process on* $\mathbb{R}^d$ *;*
2. *for any* $a > 0$ *there are* $b > 0$ *and* $c \in \mathbb{R}^d$ *such that*

$$\left\{ X(at) : t \in [0,\infty)\right\} \stackrel{d}{=} \left\{ bX(t) + tc : t \in [0,\infty)\right\}. \tag{2.4}$$

$\{X(t) : t \in [0,\infty)\}$ *is called* strictly stable process *on* $\mathbb{R}^d$, *if it is a Lévy process such that for any* $a > 0$ *there are* $b > 0$ *satisfying*

$$\left\{ X(at) : t \in [0,\infty)\right\} \stackrel{d}{=} \left\{ bX(t) : t \in [0,\infty)\right\}. \tag{2.5}$$

The property (2.5) is referred to as *self-similar* property.

Properties

Property 2.1. if $\{X(t)\}$ is a stable process *non trivial* (i.e. for any $\gamma \in \mathbb{R}^d$, $X(t) \neq t\gamma$), then there exists a unique $\alpha \in (0,2]$ and there is $c \in \mathbb{R}^d$ such that, for any $a > 0$

$$\left\{ X(at) : t \in [0,\infty)\right\} \stackrel{d}{=} \left\{a^{\frac{1}{\alpha}} X(t) + tc : t \in [0,\infty)\right\}. \tag{2.6}$$

Property 2.2. if $\{X(t)\}$ is a strictly stable process *non zero* (i.e. $P\{ X(t) = 0 : t \in [0,\infty)\} \neq 1$), then there exists a unique $\alpha \in (0,2]$ such that, for any $a > 0$

$$\left\{ X(at) : t \in [0,\infty)\right\} \stackrel{d}{=} \left\{a^{\frac{1}{\alpha}} X(t) : t \in [0,\infty)\right\}. \tag{2.7}$$

2.2 Canonical forms

1. The characteristic function of an α-stable process, $\alpha \in (0,2)$, can be written in the following form:

$$Ee^{i(\lambda, X(t))} = \exp\left\{t\left[\int_{S_d}\int_0^{\infty}\left[e^{ir(s,x)} - 1 - i\frac{r(s,x)}{1+r^2}\right]\frac{dr}{r^{1+\alpha}}\sigma(ds) + i(\lambda,\mu^0)\right]\right\} \tag{2.8}$$

2. Let $0 < \alpha < 2$. If $\{X(t)\}$ is an α-stable process on $\mathbb{R}$, then there exists a finite measure Γ on S_d and $\mu^0 \in \mathbb{R}^d$ such that

$$Ee^{i(\lambda,X(t))} = \exp\left\{t\left[i(\lambda,\mu^0) - \int_{S_d} |(\lambda,s)|^\alpha \left(1 - i\ sign((\lambda,s))W(\alpha,s,\lambda)\right)\Gamma(ds)\right]\right\}, \tag{2.9}$$

with

$$\begin{aligned} W(\alpha,s,\lambda) &= \tan\frac{\pi\alpha}{2} && \text{if } \alpha \neq 1 \\ &= -\frac{2}{\pi}\log|(\lambda,s)| && \text{if } \alpha = 1 \end{aligned}$$

Remark 2.2. If $d = 1$, (2.8) can be written as follows

$$Ee^{i\lambda X(t)} = \exp\left\{t\left[i\mu\lambda - \gamma^\alpha|\lambda|^\alpha\left(1 - i\beta\ sign(\lambda)W(\alpha,\lambda)\right)\right]\right\}, \lambda \in \mathbb{R}; \tag{2.10}$$

where $\gamma > 0$, $\mu \in \mathbb{R}$, $\beta \in [-1,1]$ and $W(\alpha,\lambda)$ is given by .

$$\begin{aligned} W(\alpha,\lambda) &= \tan\frac{\pi\alpha}{2} && \text{if } \alpha \neq 1 \\ &= -\frac{2}{\pi}\log|\lambda| && \text{if } \alpha = 1 \end{aligned}$$

An α-stable process $\{X(t)\}$ satisfying (2.9), is called stable process with parameters $(\alpha,\beta,\mu,\gamma)$. Figures 2.1 to 2.3 display simulations of stable processes on $\mathbb{R}$ and their increments for different values of α.

2.3 Example: α-stable motion

A stochastic process $\{L_\alpha(t) : t \in [0,\infty)\}$ is called a (standard) α-stable motion if:

1. $L_\alpha(0) = 0$ *a.s.*
2. $L_\alpha(t)$ has independent and stationary increments.
3. $L_\alpha(t) - L_\alpha(s) \sim S_{\alpha,\beta}\left(0,(t-s)^{\frac{1}{\alpha}}\right)$ [2] for any $0 \leq s < t < \infty$.

Note that if $\alpha = 2$, $M_2(t) = \sqrt{2}B(t)$ where $\{B(t) : t \in [0,\infty)\}$ is the standard Brownian motion. α-stable motions are denoted $S\alpha S$ if $\beta = 0$. Symmetric (i.e. $\beta = \mu = 0$) α-stable motions are $\frac{1}{\alpha}$-self-similar, i.e. for any $c > 0$,

$$\left\{L_\alpha(ct) : t \in [0,\infty)\right\} \stackrel{d}{=} \left\{c^{\frac{1}{\alpha}}L_\alpha(t) : t \in [0,\infty)\right\}$$

Remark 2.3. α-stable Lévy motions are examples of semimartingales. For more details, see Protter [26] and Kwapien and Woyczynski [17].

---

[2] $S_{\alpha,\beta}(\mu,\gamma)$ denotes a stable law with parameters $(\alpha,\beta,\mu,\gamma)$.

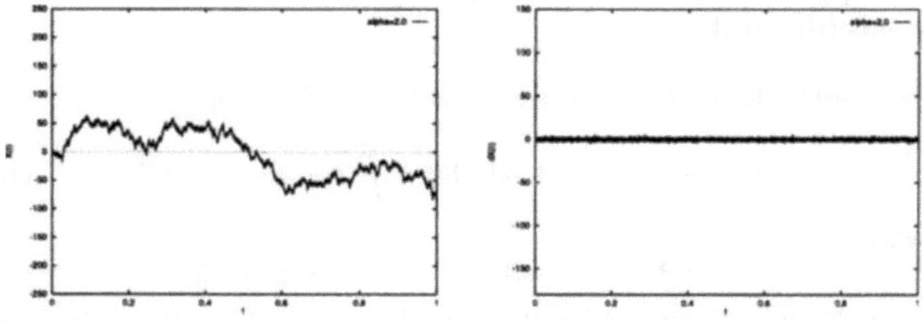

Fig. 2.1. Realization of a stable process with parameters (2,0,0,1) (left) and the corresponding increments (right).

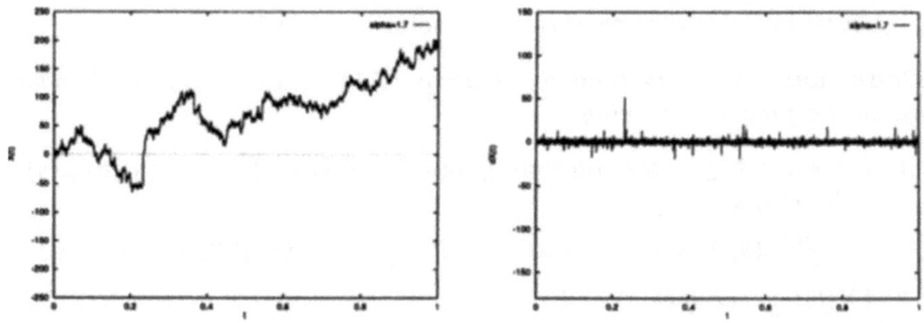

Fig. 2.2. Realization of a stable process with parameters (1.7,0,0,1) (left) and the corresponding increments (right).

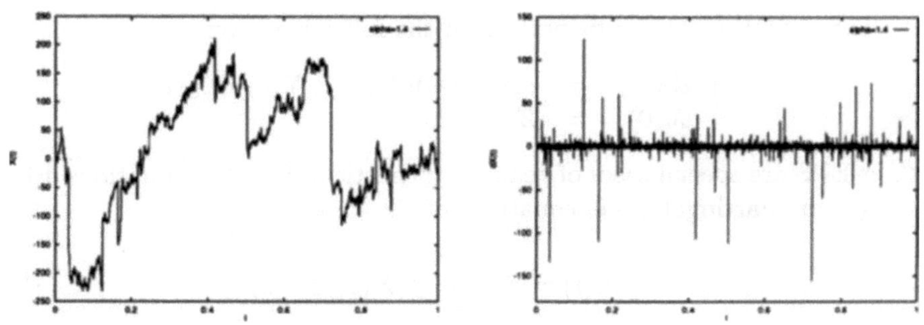

Fig. 2.3. Realization of a stable process with parameters (1.4,0,0,1) (left) and the corresponding increments (right).

3. Approximation of Diffusion process driven by α-stable motion

We consider Itô type equations expressed as the following :

$$X(t) = Z + \int_0^t \delta(s, X(s))ds + \int_0^t c(s, X(s))dL_\alpha(s). \tag{3.1}$$

where :

$$\delta : \mathbb{R}^+ \times \mathbb{R} \longrightarrow \mathbb{R}, \qquad c : \mathbb{R}^+ \times \mathbb{R} \longrightarrow \mathbb{R},$$

and Z is a $\mathcal{F}_0$-measurable stable random variable, $(L_\alpha(t))_{t\geq 0}$ is an $\mathcal{F}_t$-symmetric α-stable motion with $1 < \alpha < 2$.

By analogy with classical Itô differential equations, we call (3.1) a Stochastic Differential Equation with respect to α-stable motion (α-SDE). For details about the construction of the stochastic integral $\int_0^t c(s, X(s))dL_\alpha(s)$, see Rosinski and Woyczynski [31] and references therein.

Let us explain what we mean by a solution of (3.1) :

Definition 3.1. *A solution of equation (3.1) is a continuous $\mathcal{F}_t$ adapted stochastic process satisfying :*

1. *For every $t \geq 0$, the integrals $\int_0^t \mu(s, X(s))ds$ et $\int_0^t c(s, X(s))dL_\alpha(s)$ are well defined :*

   $$\int_0^t |\mu(s, X(s))|ds < +\infty \quad \text{and} \quad \int_0^t |c(s, X(s))|^\alpha ds < \infty \text{ P-a.s.}$$

2. *$\{X(t)\}_{t\geq 0}$ verifies (3.1) i.e. :*

   $$\forall t \geq 0 \quad P-a.s \quad X(t) = Z + \int_0^t \mu(s, X(s))ds + \int_0^t c(s, X(s))dL_\alpha(s).$$

Remarks

1. The α-SDE (3.1) can be written in the following way:

   $$\begin{cases} dX(t) &= \mu(t, X(s))dt + c(t, X(s))dL_\alpha(t) \\ X(0) &= Z \end{cases}$$

2. α-EDS are special cases of general stochastic differential equations driven by semimartingales, i.e. equations of the form

   $$X(t) = Z + \int_0^t f(X(s))dY(s), \tag{3.2}$$

   where $\{Y(t)\}$ is a semimartingale process.
   The existence and the uniqueness of the solution of (3.1) result from the use of some appropriate theorems concerning semimartingales (see Protter [26]).

3. The solution of (3.1) is generally not a α-stable process. A α-stable diffusion process can however be obtained in a special case, namely when the integrand in the stochastic integral is deterministic, i.e.

$$X(t) = Z + \int_0^t (a(s) + b(s)X(s))ds + \int_0^t c(s)dL_\alpha(s). \tag{3.3}$$

then, $X(t)$ may be expressed in the following form:

$$X(t) = f(t,0)Z + \int_0^t f(t,s)a(s)ds + \int_0^t f(t,s)c(s)dL_\alpha(s), \tag{3.4}$$

where $f(t,s) = \exp\left\{\int_s^t b(u)du\right\}$.

There exists many methods to simulate the solution of SDEs with respect to α-stable motion (see e.g. Pardoux and Talay [24]). We present here one of the simplest schemes, "the random Euler method"(see Janicki and Weron [16]). We adapt this numerical method to approximate the process $\{X(t) : t \in [0,T]\}$, solution of (3.1), by a discrete time process $\{X_{t_i}^\tau\}_{i=0}^I$, by fixing a regular mesh $\tau = \frac{T}{I}$ on the interval $[0,T]$ and considering the discrete time $t_i = i\tau$, $i = 1, ...I$.

The algorithm runs as follows:

- approximate the process $\{X(t) : t \in [0,T]\}$ which solves equation (3.1) by a discrete time process $\{X_{t_i}^\tau\}_{i=0}^I$ defined by

$$\begin{cases} X_{t_i}^\tau - X_{t_{i-1}}^\tau &= \mu(t_{i-1}, X_{t_{i-1}}^\tau)\tau + c(t_{i-1}, X_{t_{i-1}}^\tau)\Delta L_{\alpha,i}^\tau \\ X_0^\tau &= Z \end{cases}$$

  where

$$\Delta L_{\alpha,i}^\tau = L_{\alpha,t_i}^\tau - L_{\alpha,t_{i-1}}^\tau \sim S_{\alpha,\beta}\left(0, \tau^{\frac{1}{\alpha}}\right); \tag{3.5}$$

- replace random variables $\Delta L_{\alpha,i}^\tau$ with stable random samples $\{\Delta L_{\alpha,i}^\tau\}_{n=1}^N$ simulated by Chambers-Mallows-Stuck method [4].

Janicki, Podgorski et Weron [15] have shown, in the simple case of (3.1) where $c(t, X(t)) = c(t)$, that $\{X^\tau(t) : t \in [0,T]\}$ is uniformly converging in probability to the exact solution $\{X(t) : t \in [0,T]\}$ on $[0,T]$ when $\tau \to 0$. A more general result on the convergence of the Euler approximate solution of an SDE driven by a Lévy process was obtained by Protter and Talay [27] who have shown that the convergence rate depends on the tail of the Lévy measure.

4. "Stable" option pricing model

An *option* is a security which gives its holder the right to sell or to buy the underlying (for example shares, currency) within a stated period (American option), or at the end of the stated period (European option) at a predetermined price.

The options providing the right to buy are referred to as *call options.* If a contract gives the right to sell, it is called *put option.*

We consider a standard European call option which gives its holder the right to buy at a given time T (*expire date*) fixed in the future, a stock for the price K (*strike*) pointed out in the contract. If the random price of the stock is described by the process (S_t), then the gain of the holder at time T equals $f_T = \max(S_T - K, 0)$. (In case $S_T > K$, the holder buys the stock for the price K and sells it for the spot price S_T. He gains $S_T - K$. If $S_T < K$, it makes no sense to use the right given to him by the option of the contract.)

Let us start with the basic results of Black and Scholes [3] under the Gaussian hypothesis of the financial market behavior.

4.1 Black and Scholes option pricing model

Assume that the bank account S_t and the stock price $X(t)$ satisfy the following stochastic differential equation

$$
\begin{aligned}
dS(t) &= rS(t)dt, && (4.1)\\
dX(t) &= \mu X(t)dt + \sigma X(t)dB(t). && (4.2)
\end{aligned}
$$

where r denotes the riskless interest rate which is assumed to be constant, σ denotes the volatility coefficient which is assumed to be positive constant and $B(.)$ is the standard Brownian motion.

Equation (4.1) is an ordinary differential equation which allows a deterministic solution $S(t) = S(0)e^{rt}$, for every $t > 0$. Equation (4.2) is an SDE which has an explicit solution:

$$X(t) = X(0)\exp\left(\mu t + B(t) - \frac{\sigma^2}{2}t\right), \tag{4.3}$$

where $X(0)$ is the price observed at time 0. It results that the process $(\log(X(t)))$ is a Brownian motion.

Naturally, the price C_t of the call option at time t is given by the following expression:

$$C_t = E\left[e^{-r(T-t)}f(X_T)|\mathcal{F}_t\right], \tag{4.4}$$

where $f(x) = \max(x - K, 0)$ and T is the expire date of the option.

In such a model, one may obtain an explicit option price C_t at time t as function of t and $X(t)$. Finally we end up with

$$C_t = X_t F\left(h(X_t,t)\right) - Ke^{-r(T-t)} F\left(h(X_t,t) - \frac{\sigma}{\sqrt{T-t}}\right), \tag{4.5}$$

where

$$h(X_t,t) = \frac{\left[\log\frac{X_t}{K} + \left(r + \frac{\sigma^2}{2}t\right)\right]}{\sigma\sqrt{T-t}}, \tag{4.6}$$

and F denotes the standard normal distribution function:

$$F(t) = \frac{1}{\sqrt{2\pi}}\int_{-\infty}^{t} e^{-\frac{x^2}{2}}dx. \tag{4.7}$$

4.2 The model

As mentioned above, the Black and Scholes model assumes that instantaneous rate of the stock return follows a Gaussian law with mean μ and variance σ^2 for a unit of time. However, it has been observed that this assumption is incompatible with the market reality.

A variety of option pricing models (see e.g. Rachev and Ruschendorf [28], Rachev and Samorodnitsky [29], Hurst, Platen and Rachev [14]) have appeared to deal with the leptokurtic aspect of empirical distributions of stock returns and the instantaneous variations of financial markets. These models are developed under the hypothesis that the behavior of stock returns is described by a subordinate process such as the one suggested by Mandelbrot and Taylor [21] and Clark [5].

In the same context, we consider in this section a European call option pricing model where the underlying price process satisfies an α-SDE. Next we feed this model with simulated European call option prices and use it to explain the *smile effect* of the *implied volatilities*.

Let the bank account $S(t)$ and the stock price $X(t)$ satisfy the following SDE:

$$\begin{aligned} dS(t) &= rS(t)dt, & (4.8)\\ dX(t) &= \mu X(t)dt + \gamma X(t)dL_\alpha(t), & (4.9)\\ X(0) &= X_0. & (4.10) \end{aligned}$$

where r denotes the interest rate (for the sake of simplicity and to be able to compare this model with the Black-Scholes model, r is assumed to be constant); μ denotes the average rate of stock return; γ denotes the scale parameter of the rate of stock return (assumed to be positive and constant) and X_0 is a given constant.

The equation (4.9) can be written as follow:

$$X(t) = X_0 + \mu\int_0^t X(s)ds + \gamma\int_0^t X(s)dL_\alpha(s). \tag{4.11}$$

Remark 4.1. In the special case of $\alpha = 2$, $L_2(t) = \sqrt{2}B(t)$ and $\gamma = \frac{\sigma}{\sqrt{2}}$, we recover the well known Black and Scholes model.

Remark 4.2. The α-stable motion $L_\alpha(t)$ can be viewed as a subordinated process by substituting in the Brownian motion the physical time t with a random intrinsic time $\mathcal{T}$ that follows a totally skewed $\frac{\alpha}{2}$-stable process. More precisely:

If $B = \{B(t) \ : t \geq 0\}$ is a standard Brownian motion and $\mathcal{T} = \{\mathcal{T}(t) \ : t \geq 0$ $\mathcal{T} = \{\mathcal{T}(t) \ : t \geq 0\}$ **a totally skewed to the left** $\frac{\alpha}{2}$**-stable process, i.e.**

$$\mathcal{T}(t+s) - \mathcal{T}(t) \sim S_{\frac{\alpha}{2},1}\left(0; 2\cos\left(\frac{\pi\alpha}{4}\right)^{\frac{2}{\alpha}} s^{\frac{2}{\alpha}}\right) \tag{4.12}$$

for every s, $t \geq 0$ and $\alpha \in (0,2)$, then $L_\alpha = \{L_\alpha(t) = B(\mathcal{T}(t)) \ : t \geq 0\}$ is a standard symmetric α-stable motion. We thus recover the well known Mandelbrot-Taylor [21] hypotheses.

We consider now a standard European call option with the payment function $f_T = \max(X_T - K)$, where T is the expiration time, K is the striking price and the risky stock $X(t)$ is modeled by an α-EDS. A viable time t price of the option is given by

$$C_t = E^*\left[e^{-r(T-t)} f_T | \mathcal{F}_t\right], \tag{4.13}$$

where E^* denotes the expectation under risk neutral probability P^* also known as an equivalent martingale measure [3] (see Hurst-Platen-Rachev [14]). In other words, P^* is a probability measure equivalent to the objective probability measure P and such that the discounted traded asset price $X(t)e^{-rt}$ is a P^*-martingale. Hurst, Platen and Rachev [14] used the no-arbitrage concept derived in Harrison & Kreps [11] and Harrison & Pliska [12] to give, in the case of subordinate process, the form of the Radon-Nykodim density of a risk neutral probability P^* with respect to P.

In our framework, we use Monte-Carlo method of simulation to compute numerically option prices given by (4.13). We use for that, m realizations of N-random samples $\{X_T^{\mathcal{T}}\}_{n=1}^{N}$ where $\left\{X_{t_i}^{\mathcal{T}}\right\}_{i=1}^{I}$ is the discrete approximate solution of (4.9) such as described in section 4.

A relevant test to asses the quality of this numerical method is to compare call options prices obtained by Black-Scholes formula given by (4.5) with simulated prices in the special case $\alpha = 2$ and $\gamma = \frac{\sigma}{\sqrt{2}}$.

Consider for example call options with different striking prices and expiration time $T = 1$. The interest rate is chosen to be equal to 10% and

[3] Note that we are in a context of incomplete market. There are many equivalent martingale measures. The existence of such risk neutral probabilities are "essentially" equivalent to the no-arbitrage concept. For more details, see Dalang, Morton and Willinger [7] or Stricker [34].

volatility of the underlying asset σ is 20%; this correspond, when $\alpha = 2$, to a scale parameter $\gamma = 14.14\%$.

Table 4.1 presents expected option prices computed at time 0 based on 50 simulations of 2000 length-samples and over an interval of time subdivided into 1000 meshes, i.e. $m = 50$, $N = 2000$, and $I = 1000$. These results evidence the good agreement between the numerical model and the values obtained using Black-Scholes formula. In all cases the absolute approximation error is less than 1%.

X_0	K	$B\&S$	2-EDS	$\|\varepsilon\|(\%)$
50	55	4.09	4.06	0.7
60	68	4.11	4.10	0.2
80	85	7.91	7.93	0.2
90	100	6.94	6.92	0.3
100	112	10.17	10.26	0.9

Table 4.1. Comparison between Black and Scholes ($B\&S$) call option prices and 2-SDE simulated ones.

α	option price at $t = 0$
2.0	4.10
1.9	4.28
1.8	4.60
1.7	5.00
1.6	5.42
1.5	6.14
1.4	6.94
1.3	9.22

Table 4.2. Behavior of call option price with respect to the exponent α.

It is significant to study the behavior of the call option price as a function of the exponent α. This is equivalent to compare different kinds of modeling stock returns and their implications on the option prices. Let us consider for example a call option with a striking price $K = 55$ and expiration time $T = 1$, and let us assume also that $X_0 = 50$, $\gamma = 14.14\%$, $r = 10\%$, $\mu = 10\%$, $m = 50$, $N = 2000$ and $I = 1000$. Table 4.2 and figure 4.1 show that simulated price of the call option is a decreasing function of α in $(0, 2]$. This means that the $B\&S$ model corresponding to a 2-SDE model always under-valuate the option.

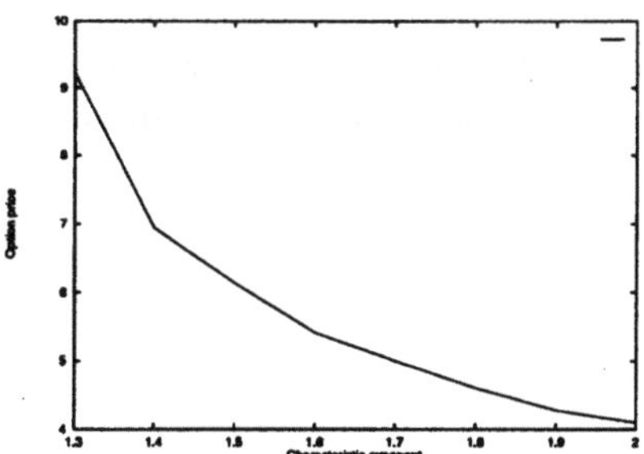

Fig. 4.1. Behavior of call option price with respect to the exponent α.

4.3 Implied volatility and smile effect

The *implied volatility* of a given European call option price is the volatility that equals the European call option price given by $B\&S$ model to that price. It is a volatility implicitly reflected in the option price (see Latané and Rendleman [18]). In the Black and Scholes model, if we know the option price, the stock price, the expiration time and the interest rate, we can extract from formula (4.5) the volatility of the stock return implied to the call option. The method of estimation of the implied volatility is as follows: we fix an initial value for σ, then we find by successive iterations the value of the parameter which minimizes the difference between actual option price and its theoretical price given by $B\&S$ formula. If the market prices for call options are consistent with $B\&S$ model, then all implied volatilities corresponding to various options written on the same asset should coincide with the volatility parameter σ of the underlying asset (independent of the striking prices). However, this is not the case in practice. One often observes that deep *in the money* and deep *out of the money* options have higher implied volatilities than *at the money* options. This phenomenon is called the *smile effect* of the implied volatilities (see Derman and Kani [8], Platen and Schweizer [25], Renault and Touzi [30]).

To estimate implied volatilities from our numerical model, we consider three kind of call options:

- option for a fictive stock with $\alpha = 1.6$, $\gamma = 15\%$, $\mu = r = 8\%$, $X_0 = 70$ and $T = 1$;
- option for *NYSE* index [4] with $\hat{\alpha} = 1.5$, $\hat{\mu}_{an} = 7.5\%$, $\hat{\gamma}_{an} = 18.3$, $r = 7\%$, $X_0 = 100$ and $T = 0.5$ (6 months);
- option for *CAC40* index [5] with $\hat{\alpha} = 1.7$, $\hat{\mu}_{an} = 7.61\%$, $\hat{\gamma}_{an} = 13.4$, $r = 8\%$, $X_0 = 100$ and $T = 0.25$ (3 months) then $T = 0.5$ (6 months).

Tables 4.3 to 4.6 show the simulated results for $m = 50$, $N = 2000$, et $I = 1000$, of option prices and implied volatilities for different call options with different striking prices.

The $B\&S$ implied volatility is U-shaped (smile volatility) as a function of the strike and attains its minimum for an *at-the-money* option. These properties of the smile volatility are consistent with market prices. Furthermore, on figures 4.2 to 4.5, we see that the implied volatilities display a skewness which frequently occurs in the market.

[4] The parameters μ et γ are estimated on daily data ($\hat{\mu}_q = 0.0205\%$, $\hat{\gamma}_q = 0.46\%$) and are often converted to year rates: $\mu_{an} = \mu_q \times 366$ and $\hat{\gamma}_{an} = \hat{\gamma}_q \times 250^{\frac{1}{\alpha}}$.

[5] $\hat{\mu}_q = 0.0208\%$ and $\hat{\gamma}_q = 0.52\%$

K	*optionprice*	implied σ(%)
60	17.58	29.06
65	13.98	27.46
68	12.06	26.89
70	10.88	26.62
73	9.28	26.36
75	8.33	26.30
77	7.48	26.33
80	6.37	26.51
84	5.17	26.89
88	4.25	27.52

Table 4.3. Option prices for a fictive stock, $\alpha = 1.6$, $T = 1$, $\gamma = 15\%$ et $r = 8\%$.

K	option price	implied σ (%)
80	26.85	49.56
85	22.14	42.81
90	17.87	38.05
95	14.34	35.79
100	11.28	34.21
105	8.95	33.98
110	7.03	33.88
115	5.62	34.40
118	4.96	34.90
120	4.64	35.51

Table 4.4. Option prices for NYSE index, $\alpha = 1.5$, $\gamma_{an} = 18.3\%$, $\mu_{an} = 7.5\%$, $r = 7\%$, $T = 0.5$.

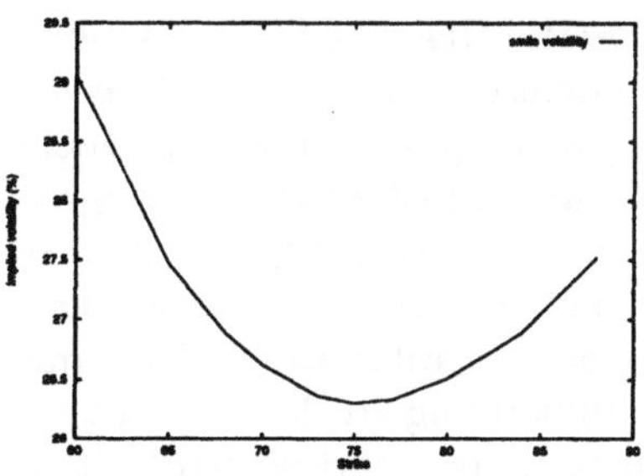

Fig. 4.2. Implied volatilities, $\alpha = 1.6$, $\gamma = 15\%$, $r = 8\%$, $T = 1$.

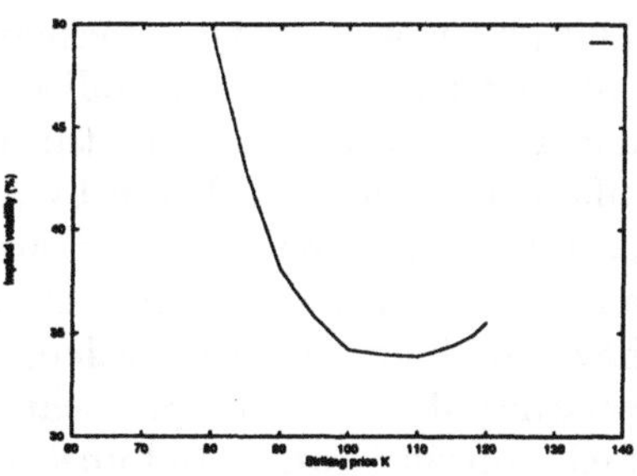

Fig. 4.3. Implied volatilities, $\alpha = 1.5$, $\gamma_{an} = 18.3\%$, $\mu_{an} = 7.5\%$, $r = 7\%$, expiration time 6 months ($T = 0.5$).

K	option price	implied σ (%)
85	17.12	27.15
90	12.55	23.82
95	8.34	21.19
100	5.06	20.50
105	2.76	20.24
108	1.92	20.73
110	1.49	21.01
113	1.06	21.83
115	0.93	23.05
118	0.63	23.39

Table 4.5. Option prices for CAC40 index, $\alpha = 1.7$, $\gamma_{an} = 13.4\%$, $\mu_{an} = 7.61\%$, $r = 8\%$, expiration time 3 months ($T = 0.25$) .

K	option price	implied σ (%)
85	19.39	26.43
90	15.05	23.69
95	11.25	22.41
100	8.10	21.89
105	5.60	21.62
110	3.76	21.63
113	3.03	22.11
115	2.63	22.45
118	2.18	23.19

Table 4.6. Option prices for CAC40 index, $\alpha = 1.7$, $\gamma_{an} = 13.4\%$, $\mu = 7.61\%$, $r = 8\%$, expiration time 6 months ($T = 0.5$) .

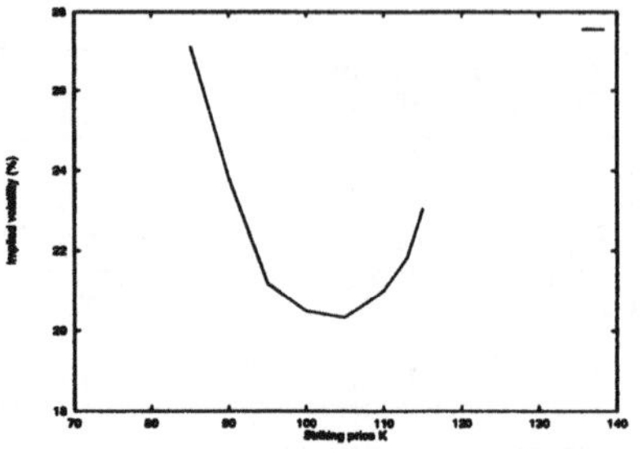

Fig. 4.4. Implied volatilities, $\alpha = 1.7$, $\gamma_{an} = 13.4\%$, $\mu_{an} = 7.61\%$, $r = 8\%$, expiration time 3 months ($T = 0.25$).

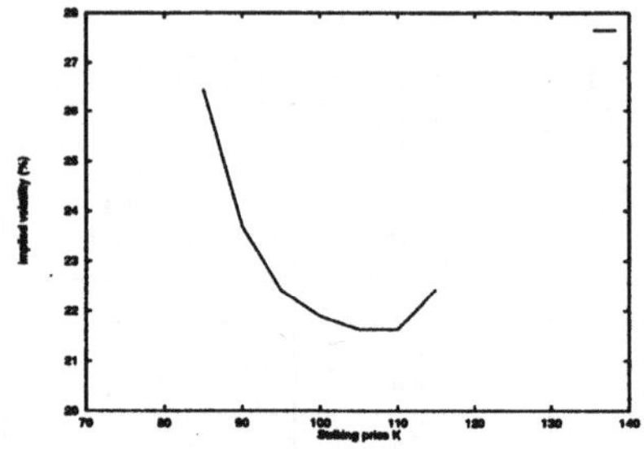

Fig. 4.5. Implied volatilities, $\alpha = 1.7$, $\gamma_{an} = 13.4\%$, $\mu = 7.61\%$, $r = 8\%$, expiration time 6 months ($T = 0.5$).

5. Conclusion

This paper provides an extension of the classical Black and Scholes model for option pricing. We have studied a simple numerical model for option pricing under the hypothesis that the underlying asset price satisfies a stochastic differential equation driven by an α-stable motion. We have investigated in particular the ability of the characteristic exponent α to explain the discrepancy between the *B&S* model prediction and the reality of the market. We have obtained numerical values of pricing and an estimation of the implied volatility due to a misspecification of the underlying stock's return that fits well empirical results obtained on the market. As a further work, it would be of value to prove analytically that stable processes produce automatically a smile effect. In addition, it would be interesting to compare smile effects produced by the stochastic volatility models and the ones produced by stable processes.

Acknowledgement. This work was supported by Crédit Lyonnais (France). The author thanks Jacques Lévy Véhel, Nizar Touzi and Christian Walter for helpful comments.

References

1. Belkacem, L. (1996): *Processus Stables et Applications en Finance : CAPM, Risque, Choix des Portefeuilles, Évaluation des Options dans un Marché α-stable.* PhD thesis, Université de Paris IX.
2. Belkacem, L., Lévy Véhel, J., Walter, C. (Janvier 1996): CAPM, Risk and Portfolio selection in "Stable" Markets. Rapport de recherche 2776, INRIA.
3. Black, F., Scholes, M. (1973): The pricing of options and corporate liabilities. *J. Political Economy*, 81, 637–659.
4. Chambers, J.M., Mallows, C., Stuck, B.W. (1976): A method simulating stable random variables. *J. Amer. Stat. Asso*, 71, 340–344.

5. Clark, P.K. (1973): A subordinated stochastic process model with finite variance for speculative prices. *Econometrica*, 41, 135–155.
6. Cox, J.C., Ross, S.A., Rubinstein, M. (1979): Option pricing: a simplified approach. *J. Financial Economics*, 7, 229–264.
7. Dalang, R.C., Morton, A., Willinger, W. (1990): Equivalent Martingale measures and no arbitrage in stochastic securities market models. *Stochastics and Stochastics Reports*, 29(2), 185–202.
8. Derman, E.H., Kani, I. (1994): Riding on smile. *Risk*, 7, 33–39.
9. Fama, E.F. (January 1965): The Behavior of Stock Market Prices. *Journal of Business*, 38.
10. Feller, W. (1971): *An Introduction to Probability Theory and its Applications*. Wiley, New York, 2nd edition.
11. Harrison, J.M., Kreps, D.M. (1979): Martingale and arbitrage in multiperiods securities markets. *J. Economic Theory*, 20, 381–408.
12. Harrison, J.M., Pliska, S.R. (1981): Martingales and stochastic integrals in theory of continuous trading. *Stochastic Processes App.*, 11, 215–260.
13. Hull, J., White, A. (1987): The pricing of options on assets with stochastic volatilities. *J. of Finance*, 42, 281–300.
14. Hurst, S.R., Platen, E., Rachev, S.T. (1995): Option pricing for asset returns driven by subordinated processes. Technical Report 281, Dep. of Statistics and Applied Probability, University of California, Santa Barbara, CA, 1995.
15. Janicki, A., Podgorski, K., Weron, A. (1991): Computer simulation of α-stable Ornstein-Uhlenbeck stochastic processes. In R. Karandikar S. Cambanis, J.K. Ghosh and P.K. Sen Eds., editors, *Stochastic Processes, A Festschrift in Honor of Gopinath Kallianpur*, Springer,New York, 1991.
16. Janicki, A., Weron, A. (1994): *Simulation and Chaotic Behavior of α-Stable Stochastic Processes*. Marcel Dekker, INC, New York.
17. Kwapien, S., Woyczynski, N.A. (1992): *Random series and stochastic integrals: single and multiple*. Boston.
18. Latané, H.A., Rendleman, R.J. (1976): Standard deviations of stock price ratios implied in option prices. *J. of Finance*, 31, 369–382.
19. Mandelbrot, B. (1963): New Methods in Statistical Economics. *Journal of Political Economy*, 56.
20. Mandelbrot, B. (1966): The Variation of Some Other Speculative Prices. *J. Business*, 39.
21. Mandelbrot, B.B., Taylor, H.M. (1967): On the distribution of stock price differences. *Operations Research*, 15, 1057–1062.
22. Merton, R. (1973): Theory of rational option pricing. *Bell J. Econ. Manag. Sc.*, 4, 141–183.
23. Mittnik, S., Rachev, S.T. (1993): Reply to Comments on Modeling Asset Returns with Alternative Stable Distributions and Some Extensions. *Econ. Rev.*, 12, 347–390.
24. Pardoux, E., Talay, D. (1985): Discretization and simulation of stochastic differential equations. *Applicandae Mathematicae*, 3, 23–47.
25. Platen, E., Schweizer, M. (1994): On smile and skewness. Statistics Research Report SSR 027-94, Center for Mathematics and its Applications, School of Mathematical Sciences, The Australian National University.
26. Protter, P. (1990): *Stochastic integration and differential equations: A new Approach*. New York.
27. Protter, P., Talay, D. (Juillet 1995): The Euler scheme for Lévy driven stochastic differential equations. Rapport de recherche 2621, INRIA.
28. Rachev, S.T., Ruschendorf, L. (1994): Models of option prices. *SIAM Theory of Prob. and its App.*, 39, 120–152.

29. Rachev, S.T., Samorodinsky, G. (1992): Option pricing formula for speculative prices modeled by subordinated stochastic processes. Technical report, Dep. of Statistics and Applied Probability, University of California, Santa Barbara, CA.
30. Renault, E., Touzi, N. (1996): Option hedging and implied volatilities in a stochastic volatility model, *Mathematical Finance*, 6(3), 279–302.
31. Rosinski, J., Woyczynski, W.A. (1986): On itô stochastic integration with respect to p-stable motion: inner clock, integrability of sample paths, double and multiple integrals. *The Annals of Probability*, 14, 271–286.
32. Sato, K. (1991): Self-similar processes with independent increments. *Probab. Th. Rel. Fields*, 89, 285–300.
33. Scott, L. (1987): Option pricing when the variance changes randomly: theory, estimation and application. *J. Financial Quant. Anal.*, 22, 419–438.
34. Striker, C. (1990): Arbitrage et lois de martingales. *Annales de l'I.H.P.*, 26, 451–460.

Modelling Contaminant Spread on the Ocean Surface and within Soils using fBm's: Two Civil Engineering Applications

P.S. Addison[1], B. Qu[1], A. Nisbet[2], and G. Pender[2]

[1] Department of Civil and Transportation Engineering, Napier University, Merchiston Campus, 10 Colinton Road, Edinburgh, EH10 5DT, Scotland, U.K.
[2] Department of Civil Engineering, Glasgow University, The Rankine Building, Oakfield Avenue, Glasgow, G12 8LT, Scotland, U.K.

Abstract A particle tracking method is outlined for modelling non-Fickian diffusion using fractional Brownian motions. Two environmental engineering applications of the model are presented, that is, the modelling of turbulent diffusion of spilled contaminants on the ocean surface and the sub/super-diffusive behaviour of contaminants in soils.

1. Introduction

The need for the accurate prediction of the transport of contaminants within the environment has led to a rapid escalation in the use of numerical models. The diffusive spreading of a contaminant cloud in the environment is often non-Fickian. Accurate modelling of the spread of contaminants requires a numerical method which can recreate non-Fickian diffusion. In the following, a particle tracking diffusion model is described which employs fractional Brownian motions (fBm's) to recreate non-Fickian diffusion. Application of the model to two typical environmental diffusion problems - ocean surface diffusion and diffusion within soil media - is discussed.

2. An fBm Particle Tracking Model

Traditional particle tracking models reproduce Fickian diffusion by simply releasing a large number of massless, marked particles into a known flow field. These are advected by the mean flow and diffused by taking random steps from a normal distribution. The standard deviation of the distribution is directly related to the diffusion properties of the flow under consideration [1]. The resulting cloud diffusion scales as,

$$\sigma_c = \sqrt{2Kt} \tag{2.1}$$

where σ_c is the standard deviation of the diffusion cloud, t is the time since release, and K is the diffusion coefficient.

A new particle tracking method has been proposed recently by Addison [2] and Addison and Bo [3] which generates marked particles in the flow field that follow fractional Brownian motions. The diffusion of such a particle cloud scales as

$$\sigma_c = (2K_f t)^H \tag{2.2}$$

where K_f is the fractal diffusion coefficient and H is the Hurst exponent $(0 < H < 1)$ [4]. The model is based on the fBm definition given by Mandelbrot and Van Ness [5],

$$B_H(t) - B_H(0) = \frac{1}{\Gamma(H+\frac{1}{2})}\left[\begin{array}{r} \int_{-\infty}^{0}\left[(t-t')^{H-\frac{1}{2}} - (-t')^{H-\frac{1}{2}}\right]R(t')dt' \\ -\int_0^t (t-t')^{H-\frac{1}{2}}R(t')dt' \end{array}\right] \tag{2.3}$$

Two discrete approximations to the above integral are currently being considered by the authors, these are as follows. The first discrete integral gives the fBm, $B_H(t_i)$, at the discrete time, t_i, as

$$B_H(t_i) = \frac{1}{\Gamma(H+\frac{1}{2})}\left[\begin{array}{r} \sum_{j=i-M}^{0}\left((i-j)^{H-\frac{1}{2}} - (-j)^{H-\frac{1}{2}}\right)R(t_j) \\ +\sum_{j=1}^{i}(i-j)^{H-\frac{1}{2}}R(t_j) \end{array}\right] \tag{2.4}$$

where Γ is the gamma function; M a finite memory; and $R(t)$ random numbers from a Gaussian distribution of known variance. The second integral gives the fBm step at each time, t_i, as,

$$B_H(t_i) - B_H(t_{i-1}) = \frac{1}{\Gamma(H+\frac{1}{2})}\left[\begin{array}{r} \sum_{j=i-M}^{i-1}\left((i-j)^{H-\frac{1}{2}} - (i-j-1)^{H-\frac{1}{2}}\right)R(t_j) \\ +\sum_{j=i-1}^{i}(i-j)^{H-\frac{1}{2}}R(t_j) \end{array}\right] \tag{2.5}$$

and the value of the fBm at time, t_i, is found by simply summing the increments produced by equation (2.5). Equations (2.4) and (2.5) are preferred by the authors over the standard spectral method of fBm generation [6] as, although slower, they allow both a limited memory to be defined and allow a restart of a model simply by adding more steps.

Figure 2.1 shows six fBm traces generated using the above discrete integrals and three Hurst exponents: 0.2 (anti-persistent or sub-diffusive), 0.5 (neutrally persistent or Fickian) and 0.8 (persistent or super-diffusive). Both the diffusion coefficient, K_f, and the time interval $(t_i - t_{i-1})$ are set to unity for simplicity, and two hundred time steps are shown for the fBm's. Although six traces are plotted in the figure, it can be seen that there are no noticeable differences in the $H = 0.2$ and 0.5 traces which essentially lie on top of each other. However, the $H = 0.8$ trace diverges noticeably through time. To investigate the statistical properties of the fBm traces through time many such paths were generated for particular values of H to simulate a diffusion cloud. This has been done for particle clouds generated using both discrete integrals given above (i.e. equations (2.4) and (2.5)). Figures 3.1 and 3.2 contain some

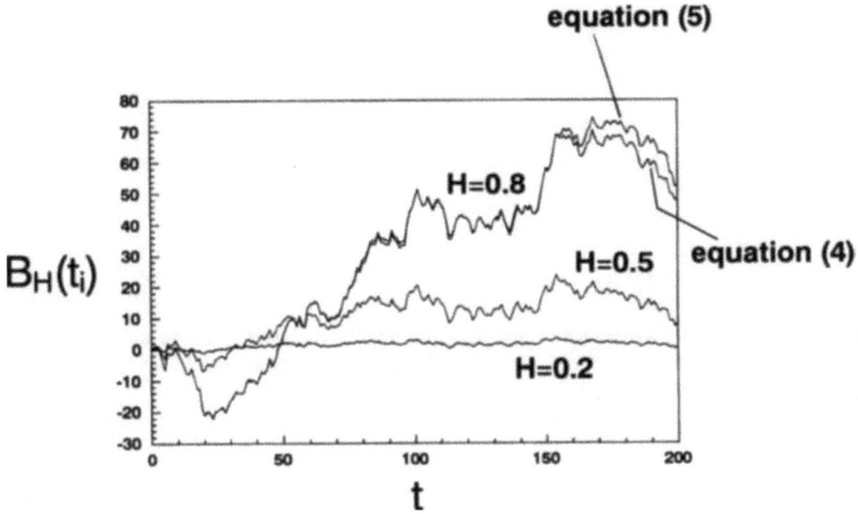

Fig. 2.1. Three fBm Traces Generated Using the Two Discrete Integral Approximations.

typical results of the tests of the two methods (equations (2.4) and (2.5)). Figure 3.1 plots $\sigma^{\frac{1}{H}}$ against time, which should be linear according to equation (2.2). As expected, increasing the number of particles produces a more consistent, i.e. linear, relationship in the plot. In figure 3.2, it can be seen that low memories (i.e. $M = N$ and $M = 2N$) result in an under-prediction of $\sigma^{\frac{1}{H}}$, and in addition a non-linear relationship. The larger memories produce more linear relationships, but tend to over-predict $\sigma^{\frac{1}{H}}$ by up to 10%. Variations in the memory, M, and the number of particles, N, have shown that both methods lead to accurate values of diffusion exponent, H, for the clouds with large M and N, however, both methods lead to an increase in the resulting diffusion coefficient (up to 10%). This increased diffusion coefficient (which we will call $K_{f(res)}$) which results from the simulation then has to be linearly scaled down within the model to the required diffusion coefficient K_f.

The particle tracking model superimposes the fBm tracks of a large number of marked particles (of the order of 10,000's in practice) onto a known velocity vector field of the fluid, (for more details, see for example Ozmidov [7]). The advective component is calculated at each time step for each particle using an interpolation scheme on the velocity field and added to the fBm diffusion [2] to give the total particle displacement at each time step. Simplistically, the advective-diffusive step, Δx, taken at each time step, Δt, is given by an equation of the form:

$$\Delta x = u\Delta t + \Delta B_H(t) \tag{2.6}$$

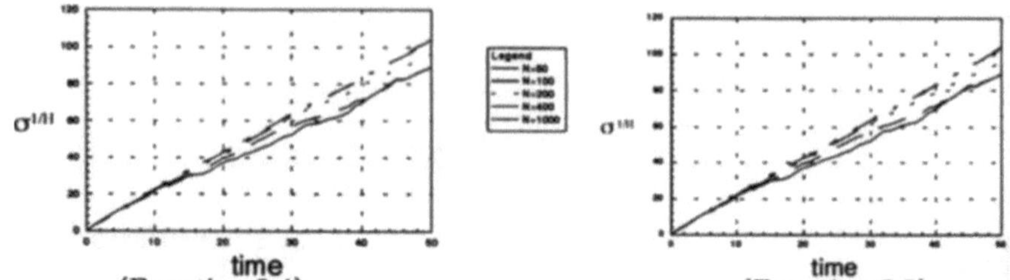

(Equation 2.4) (Equation 2.5)

Fig. 3.1. Typical Parametric Test Results for Simulated Clouds using Equations 2.4 and 2.5. Effect of Particle Number, N, on Cloud Standard Deviation through Time ($M = 400$; $H = 0.8$; $Kf = 1.0$)

where u is the advective velocity and $\Delta B_H(t)$ is the incremental step of the fBm, given by either equation (2.4) or equation (2.5). Similarly, Δy and Δz are computed from separate realisations of $B_H(t)$.

3. Applications of the fBm Particle Tracking Model

Two applications of the fBm particle tracking model are given below. Firstly, diffusion of a surface contaminant release is simulated in a coastal bay region. A comparison is made of two simulations with H set to 0.5 and then 0.75. In the second application, a brief comment is made on the possible application of the fBm particle tracking model to the prediction of contaminants within soils.

3.1 Ocean Surface Diffusion

It is known that the diffusion in coastal waters is super-diffusive [8]. In addition, recent surveys of ocean surface drifter trajectories have shown that they posses distinct fractal dimensions over a large range of scales [9, 10]. These surveys found the ocean surface drifter tracks to be persistent with Hurst exponents of the order of 0.79. Figures 3.3 and 3.4 contains brief results of the fBm particle tracking model with both regular Brownian motions and fBm's (using equation (2.4)) used to diffuse the particles. The numerical example models a coastal bay feature with a main external flow from North to South causing a recirculation pattern to form in the bay, (illustrated schematically in the figure). In both simulations, the contaminant cloud is released from grid point ($1500m, 1700m$), circled in the figures, at time, $t = 0$ hrs, and the spreading of the contaminant cloud monitored over a twelve hour period. A typical diffusion coefficient of $K_f = 0.01\ \mathrm{m^2s^{-1}}$ is used in both spatial directions. (Note that the units of K_f really depend upon H.)

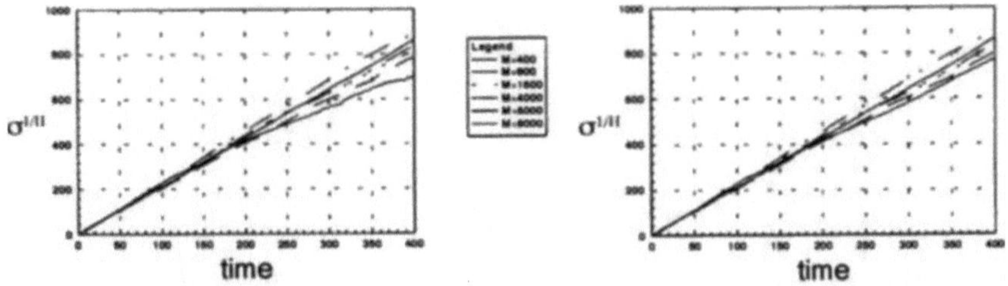

(Equation 2.4) (Equation 2.5)

Fig. 3.2. Typical Parametric Test Results for Simulated Clouds using Equations 2.4 and 2.5.The Effect of Memory, M, on Cloud Standard Deviation through Time ($N = 400; H = 0.8; Kf = 1.0$)

Comparison of the two figures reveals the significant differences that can occur in model predictions using fBm's over traditional Brownian motion models. The movement of the regular Brownian motion cloud ($H = 0.5$) is given in figure 3.3. The subsequent location of the cloud is given at two hours intervals since the time of release. The cloud is clearly advected around the bay circulation region, spreading due to both the action of diffusion, and the differential velocities experienced over the extent of the cloud. The enhanced spreading of the superdiffusive cloud ($H = 0.75$) in figure 3.4 produces quite a different spatial spreading of the cloud after the twelve hour simulation. As with the $H = 0.5$ cloud, it is advected around the bay, this time, however, some of the cloud has diffused into the North-South main flow and is transported onto the downstream shoreline - a region shown not to be affected in the traditional model!

The two simulations above have assumed that H is constant. In reality, the Hurst exponent may not be constant within the coastal zone [11]. However, of importance is the realisation that a simple change in the Hurst exponent has led to a fundamental change in the predicted spread of a contaminant release. The results have major implications for the prediction of environmental pollution.

3.2 Soil Contamination

Current techniques used to predict the spread of contaminants within soils usually attempt to produce numerical solutions to an advection-diffusion equation [12]. However, as with traditional particle tracking models, the method may only simulate Fickian diffusion. Recently it has become apparent that diffusion within soils may often be of a non-Fickian nature due to both the fractal structure [13, 14] and water content of the soil medium [15]. A porous medium with fractal dimension D_N leads to diffusion of the order

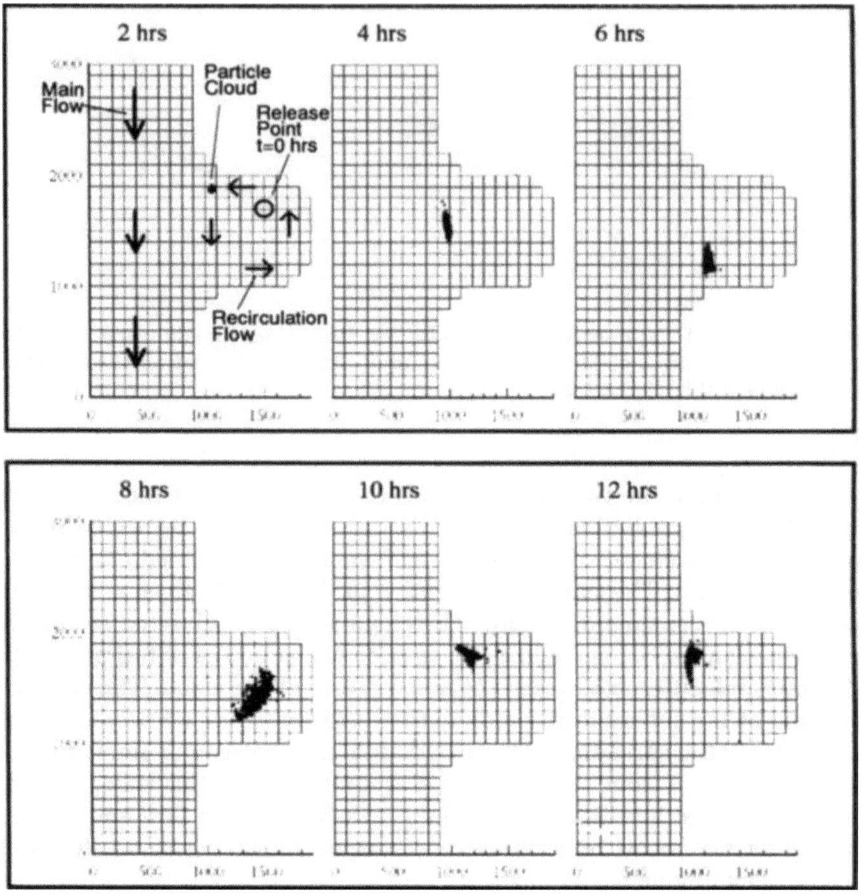

Fig. 3.3. Modelled Contaminant Release in a Coastal Bay - Regular and Fractional Brownian Motion Models. $H = 0.5$ (Time since release given at top of each grid.)

$$\sigma_c \propto t^{\frac{1}{2}\left[\frac{\bar{\bar{d}}}{D_N}\right]} \tag{3.1}$$

where $\bar{\bar{d}}$ is the fracton dimension, [16]. Equation (3.1) can be expressed in the form of equation (2.2), with $H = 0.5\left(\bar{\bar{d}}\,/D_N\right)$. In addition, D_N and $\bar{\bar{d}}$ can both be measured from thin sections of a soil sample [17]. Once the soil properties were known, numerical simulation of non-Fickian contaminant diffusion within the soil would then follow the same procedure as described above for the ocean surface diffusion model.

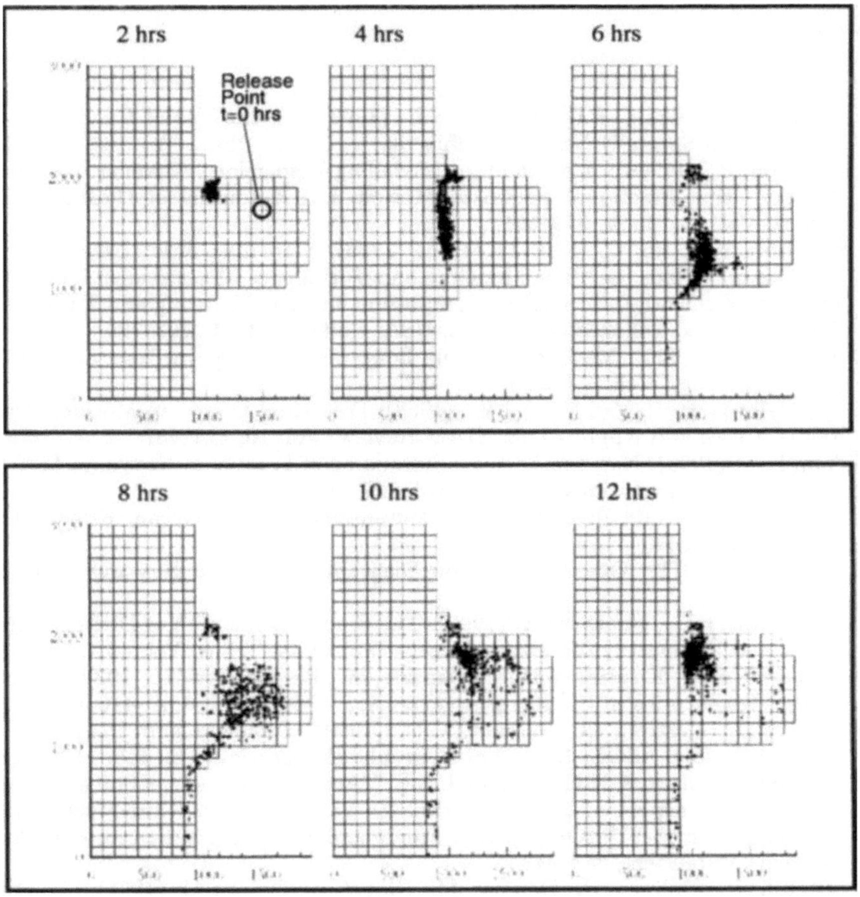

Fig. 3.4. Modelled Contaminant Release in a Coastal Bay - Regular and Fractional Brownian Motion Models. $H = 0.75$.

4. Concluding Remarks

The development of a general particle tracking diffusion model has been described which can recreate non-Fickian diffusion by using fractional Brownian motions. There are numerous possible applications for such a model, two of which have been described briefly above. It is a simple task to extend the model to simulate three-dimensional diffusive processes and, in addition, simulate super-diffusion with Hurst exponents above unity by employing accelerated fractional Brownian motions [9].

References

1. Hunter, J.R., Craig, P.D., Philips, H.E. (1993): On the Use of Random Walk Models with Spatially Variable Diffusivity. J. Comp. Phys., Vol. 106, 366–376.
2. Addison, P.S. (1996): A Method for Modelling Dispersion Dynamics in Coastal Waters Using Fractional Brownian Motion. IAHR Journal of Hydraulic Research, Vol. 34, No. 4, 549–561.
3. Addison, P.S., Bo, Q. (8-10 September 1996): Modelling Fractal Diffusion on the Ocean Surface. In 'Flow Modeling and Turbulence Measurements VI', Proceedings of the 6th International Symposium on Flow Modeling and Turbulence Measurements, Tallahassee, Florida, USA, Balkema, 1996, 703–709.
4. Hurst, H.E. (1956): Methods of Using Long-Term Storage in Reservoirs. Proc. Inst. Civil Eng., Vol. 5, Part 1, 519–577.
5. Mandelbrot, B.B., Van Ness, J.W. (1968): Fractional Brownian Motions, Fractional Noises and Applications. SIAM Review, Vol. 10, 422–437.
6. Yin, Z.-M.(1996): New Methods for Simulation of Fractional Brownian Motion. Journal of Computational Physics, Vol. 127, 66–72.
7. Ozmidov, R.V. (1990): Diffusion of Contaminants in the Ocean. Kluwer Academic Publishers.
8. Okubo A. (1971): Oceanic Diffusion Diagrams. Deep Sea Research, Vol. 18, 789–802.
9. Sanderson, B.G., Booth, D.A. (1991): The Fractal Dimension of Drifter Trajectories and Estimates of Horizontal Eddy-Diffusivity. Tellus, Vol. 43A, 334–349.
10. Osborne A.R., Kirwan, A.D., Provenzale, A., Bergamasco, L. (1989): Fractal drifter Trajectories in the Kuroshio extension. Tellus, Vol. 41A, 416–435.
11. List, E.J., Gartrell, G., Winant, C.D. (1990): Diffusion and Dispersion in Coastal Waters. ASCE, Journal of Hydraulic Engineering, Vol. 116, 1158–1179.
12. Jendele, L., Chan, A.H.C., Hellawell, E.E. (1995): Comparison of Software Modelling the Transport of Pollutants through Saturated Soil. Numerical Models in Geomechanics-NUMOG-V, G.N. Pande and S. Pietruszczak (editors), A.A. Balkema Publishers, 343–348.
13. Bartoli, F., Philippy, R., Doirisse, M., Niquet, S., Dubuit M. (1991): Structure and Self-Similarity in Silty and Sandy Soils: The Fractal Approach. Journal of Soil Science, Vol. 42, 167–185.
14. Young, I.M., Crawford, J.W. (1991): The Fractal Structure of Soil Aggregates: Its Measurement and Interpretation. Journal of Soil Science, Vol. 42, 187–192.
15. Toledo, P.G., Novy, R.A., Davis, H.T., Scriven, L.E. (1989): On the Transport Properties of Porous Media at Low Water Content. Proceedings of the International Workshop on Indirect Methods for Estimating the Hydraulic Properties of Unsaturated Soils, Riverside, California, Oct. 11-13, 97–114. Edited by M.Th. van Genuchten, F.J. Leij and L.J. Lund.
16. Havlin S., Ben-Avraham D. (1987): Diffusion in Disordered Media. Advances in Physics, Vol. 36, No. 6, 695–798.
17. Crawford, J.W., Ritz, K., Young, I.M. (1993): Quantification of Fungal Morphology, Gaseous Transport and Microbial Dynamics in Soil: An Integrated Framework Utilising Fractal Geometry. Geoderma, Vol. 56, 157–172.

The Retinal Vasculature: Towards an Understanding of the Formation Process

S. Kyriacos[1], F. Nekka[1], P. Vicco[2], and L. Cartilier[1]

[1] Faculté de Pharmacie, Université de Montréal
C.P. 6128, succ. centre-ville, Montréal, CANADA, H3C 3J7

[2] Université Libre de Bruxelles
1, rue Heger-Bordet, $B - 1000$ Bruxelles, BELGIQUE

Abstract Growth phenomena have been studied extensively in natural sciences. This interest has been renewed since the introduction of the fractal concept. In an attempt to understand the origin of irregular phenomena, several computer models and theoretical approaches have recently been developed. Studies using fractal analysis of the retinovasculature have proposed diffusion-limited aggregation (DLA) one of the most popular models to explain the formation of the retina. A deeper investigation of the physiological laws ruling the development of the retinovasculature has, however, revealed static and dynamic discrepancies with DLA, leading to rejection of the DLA model and reopening the debate. In light of comparison of the two formation processes and of the absence of a DLA structural characteristic in retinovasculature, we discuss the validity of some previously proposed models.

1. Introduction

Fractal geometry has been widely used to characterize irregular structures [1], i.e vascular structures [2, 3]. The degree of complexity is estimated by a fractal dimension (D) which may have a non-integer value. The network of blood vessels in the inner human retina has been demonstrated to be fractal [4]. Analysis of fundus images remaining a subjective process, several studies have been elaborated to develop an objective and noninvasive diagnostic technic [5, 6]. A more efficient box-counting method has been developed previously in our laboratory [7]. It was then applied to estimate the fractal dimension of the human retinovasculature RV to obtain one of the components used to support existing growth models that have been proposed so far. Our results of the estimated fractal dimension for the arteriovenous network were consistent with results reported in the literature that suggest Laplacian processes like diffusion-limited aggregation (DLA). However, deeper investigation relative to other components involved questioned the reliability of the proposed models. These models do not dynamically reproduce the physiological events responsible for RV development . Even more, from a static point of view, the RV does not share some of the characteristic features revealed in the geometry of DLA [8]. Understanding the formation mechanism of vascular networks could be useful in predicting the effects of many factors and the possible correlations with physiological and pathological events. This paper reopens the debate concerning the formation of retinal vessel patterns in the developing human eye.

2. Materials and Methods

2.1 Sources of Images and Image Processing

The retinal vessels of 5 subjectively normal humans were photographed with a Topcon $TRC - 50VT$ Camera. Photomontages were produced with photographs taken at a 30 deg angle (Fig. 2.1).

Fig. 2.1. Photomontage of the retina (15 images of 30 deg each)

The photographs were hand-traced onto paper at constant thickness. Arteries and veins comprising only the retinal circulation of each individual were analyzed separately and in combination for vessels larger than capillaries. Vascular networks were digitized on a two-dimensional 640 × 480 array of pixels. Pixel size on the retina was $38mm$, assuming an average optic disc diameter of $1.5mm$ in retinal photographs. Digitizing a structure consists of determining whether each pixel is occupied by any part of it. The tracings of vascular networks were digitized using a MTI digital imaging system (Dage-MTI Inc., Michigan City, IN, U.S.A.). The camera was a $MTI81$ with a Nikon $28 - mm$ macrolens set to $f/22$. Images were grabbed using Neotech Image Grabber software 2.1 (NeotechTM, U.K.). Grey level images were obtained on a graphic window of 640 × 480 pixels. Image processing was performed with Ultimage $TM/X1.41$ software (Graftek, Meudon-La-Foret, France) on a Quadra 900 Apple MacintoshTM II. To extract blood vessels, the images were thresholded to produce binary images. Thresholding consists of segmenting an image into two regions where the structure and background are characterized by two different intensities: pixels belonging to a given grey level interval are highlighted while others become black. An approximately same grey level interval was considered for all images. Then, the images were subjected to a L-skeleton command [9] (Fig. 2.2). Skeletons allow representation of the structural shape of an object by reducing it to a thickness of 1 pixel. L-skeleton uses masks, known as structuring elements, with a L-shape.

In fact, vessel width was ignored as this study was concerned with the spatial distribution of vessels.

2.2 Fractal Analysis

Fractal dimension of the binary images was estimated by the modified box-counting method, developed previously by Kyriacos et al. [7]. It is based on a modification of the classical computerized box-counting method [10] in which the counting method, the type of step sequence and range of length scales have been precisely defined in order to provide an accurate estimation of D. The counting is performed in a range defined by ϵ_{min} to ϵ_{max} . Hence, ϵ_{min} corresponds to a box size larger than the thickness of the structure - in our case, $\epsilon_{min} = 2$ - and ϵ_{max} is the one that gives a box number equal to 2. One important and specific feature of our method is that the step sequence is determined by an arithmetic subset, where repetitions of the same values of $N(\epsilon)$ as well as plateaus are avoided. A detailed description as well as its validation can be found in our previous paper [7].

3. Physiological Review

A review of the main events involved in RV development is necessary before presenting and discussing the results. The intrinsic vasculature of the adult

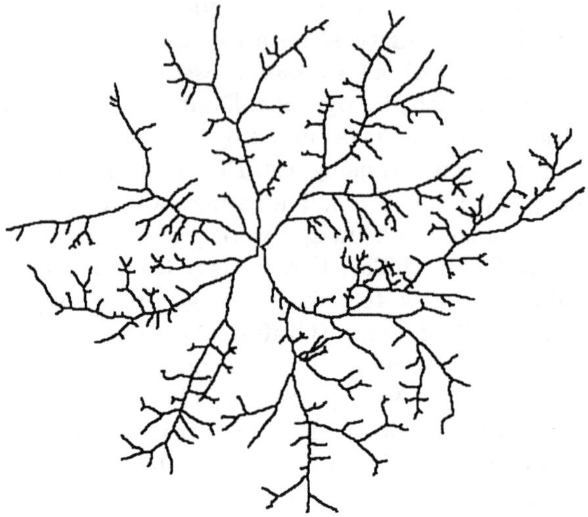

Fig. 2.2. Skeletonized image of the venous vascular network.

human retina is organized in two basic layers: a deep one (outer layer) in the inner nuclear layer and a more superficial one (inner layer) in the ganglion cellular layer and the nerve fiber layer [11] . The latter layer contains arterioles and venules whereas the former is comprised exclusively of capillaries. Each laminar plexus runs in a nearly planar direction and these are interconnected by vertical capillary segments. The distribution of these vascular laminae varies along the central-peripheral direction of the retina with vessels being absent in the fovea centralis and in the far peripheral retina rim adjacent to the ora serrata [11] (see Figs. 3.1 and 3.2).

Several studies [11, 12] were performed recently to understand the process leading to vascularization of the retina of animal and human embryos. Two distinct phenomena are currently described that are partly segregated in space and time: vasculogenesis and angiogenesis. The first, vasculogenesis, is the vascularization from precursor cells, called spindle cells. It is responsible for most of the inner layer vasculature of the retina [12]. Four stages can be distinguished:

- spreading of spindle cells over the inner retinal surface from the optic disc;
- formation of solid cords of endothelial cells; formation of a polygonal network;

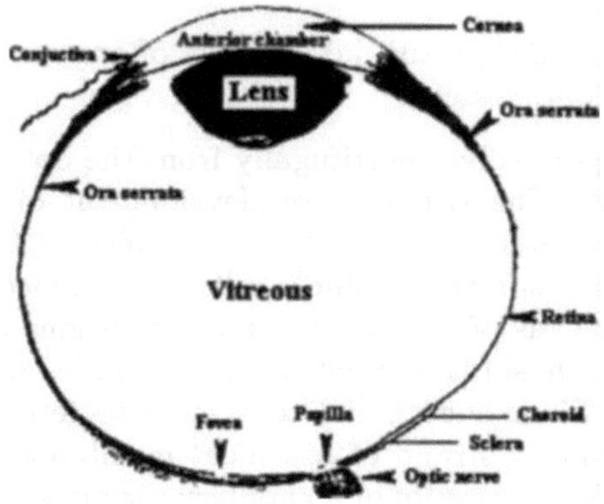

Fig. 3.1. Horizontal meridional section of the eye through the central fovea and optic nervehead. Adapted from ref. [45].

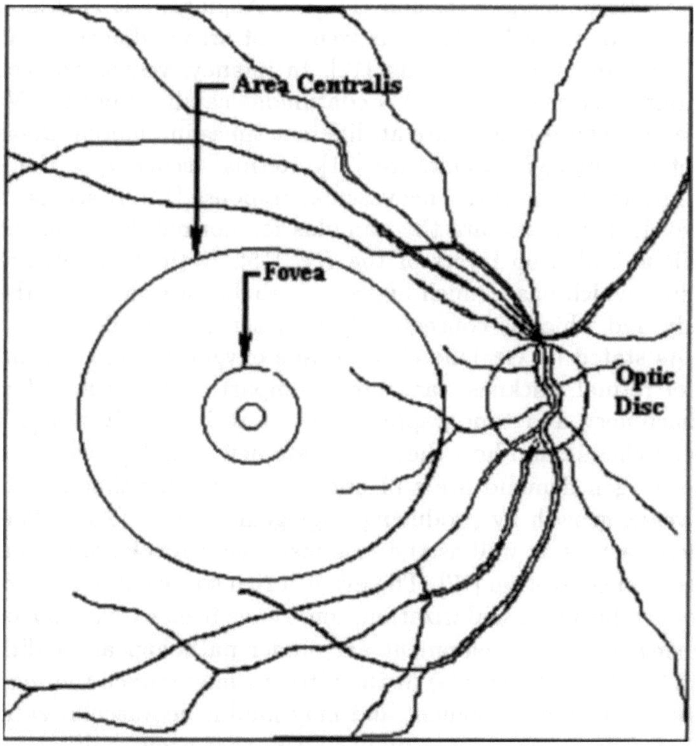

Fig. 3.2. Posterior retina. Adapted from ref. [45].

. permeation of the cords which then become patent to form a plexus of capillaries;
. maturation of the vascular plexus, i.e. clearing of excessive capillaries and selection of major blood vessels;

These different stages evolve centrifugally from the optic disc to the peripheral retina [11, 13]. The stimulus for development of intrinsic retinal vessels is still not completely elucidated. The pattern of formation is first determined by the earlier spread of spindle cells. Homogeneous narrow-mesh capillaries form. The plexus of immature capillaries begins afterward to differentiate (maturation phase) [12]. First, some capillaries narrow until they are no longer patent. Their endothelial cells seem to retract into adjacent capillaries, leaving behind a strand of basement membrane. This retraction process, still not elucidated, results in a capillary plexus with a larger mesh-size. Second, large vessels differentiate from the existing capillary plexus. It appears that around arteries, due to high tissue oxygen tension, the retraction of capillaries is particularly marked, creating the capillary-free zones. The second process is angiogenesis. It involves the vascularization via budding from existing vessels by classic angiogenic mechanisms and it is mainly controlled by tissue oxygen tension [11, 12]. Angiogenesis is responsible for vascularization of the outer layer as well as of areas adjacent to the foveal avascular zone and the ora serrata [12]. In infancy, vessels adjacent to the latter regions anastomose to form a continuous capillary border. Vasculogenesis and angiogenesis contribute at different times in different areas to form portions of the vitread vasculature [11]. Retinal ischemia is the generally accepted initiator of the entire neovascular response [14]. It seems that physiological levels of hypoxia are the stimulus for normal development of both the superficial and deep layers of the *RV* [15]. Retinal vessels are embedded in tissues which has a much higher metabolic rate than do other tissues in the body and which is consequently very sensitive to decreases in blood flow [16]. As stated by Gariano et al., tissue oxygen tension depends on the interplay of retinal thickness, anatomic compartmentalization of metabolic cellular machinery and topographic onset of retinal activities [11]. Which cells detect ischemia and how they stimulate new vessel growth so that they match the oxygen demand are still unclear. It appears that tissue hypoxia mediates vessel growth by producing angiogenic factors (*AF*). The production of these factors as well as the response they provoke are influenced by local oxygen concentration [17]. The eye is a rich source of many factors that promote or inhibit neovascularization; some have been extracted from retinal tissues during normal development and under pathological conditions such as hypoxia [14, 17, 18]; other(s) in the vitreous may induce the regression of blood vessels during development and may inhibit neovascularization in the adults [16]. It is becoming increasingly evident that neovascularization occurs when the delicate balance between growth stimulators and inhibitors is disturbed. The variability of responsiveness suggests that several factors may, in concert or independently, induce new vessel formation. The same mechanisms responsible for normal development may be involved in pathological neovascularization. The retina may contain two different growth factors, one with chemotactic/migrating properties and another with mitogenic activity [14, 16, 19]. It seems that endothelial cells can migrate up a concentration gradient of chemotactic factors [14]. There is also evidence that the secretion of an AF (vascular endothelial growth factor), which is an endothelial cell-specific mitogen and chemoattractant, is strongest in secreting cells that are at and just peripheral to leading vessels whereas more centrally its expression is diminished [15]. Although cystic spaces were thought to provide a structural framework in which vessels develop, recent studies do not support this function of these spaces [11]. Rather, glial structures may either mechanically support or direct vessel extension; astrocytes may produce fibronectin meshwork along which vascular precursor cells migrate [11] and secrete some AF [15].

4. Results and Discussion

Estimation of fractal dimension was one of the elements needed to draw conclusions on a possible growth mechanism. It is obvious that the fractal dimension alone is not a sufficient parameter to characterize a structure. In particular, the following points should be kept in mind when considering the relevance of the fractal dimension:

- Different growth processes can give rise to patterns with the same D.
- Patterns with different shapes may have the same D.
- Different measurement techniques of D may lead to different values of D.
- There are numerous sources of error in estimation of D (see text).

As shown in Table 4.1, no significant difference was found between fractal dimension of arterial (A) and venous (V) networks ($p = 0.78$, paired Student's t-test). However, this difference was significant ($p < 0.001$, paired Student's t-test) when the arteriovenous (AV) network was compared separately to each arterial and venous network. One can cautiously explain the lack of statistical difference found between A and V networks based on the fact that both networks originate from the same polygonal network.

However, one has to keep in mind that estimated fractal dimension is not an absolute measure but depends greatly on the experimental and methodological parameters involved. To list a few of them, we can mention some sources of variability in the results: capture of the image (image resolution, field angle of the fundus camera...), type of image (angiography, fundus photography...), its processing (differences in contrast leading to a loss of detail...), the method used to calculate D (box-counting, mass-radius, density-density correlation function method...), including the algorithm and range

Table 4.1. Estimated mean fractal dimension of the different networks

	Arterial network	Venous network	Arteriovenous network
Mean	1.537	1.532	1.664
Standard deviation	0.036	0.030	0.025

of ϵ (box sizes) used for the determination of D [7]. At this point, it must be remembered that natural structures have a fractal behavior only on a limited range of length scale in contrast to mathematical objects [20]. Some authors [21, 22, 23, 24, 25, 26, 27, 28, 29] have not specified the nature of the vascular network they explored, in other words, the range of vessel diameter they included in their analyses. The degree of detail included in the image is of great importance since the extent of fractal behavior is not known a priori. Depending on the type of image (fluorescein angiograms or fundus photographs...), the degree of detection of detail can vary. For example, in fluorescein angiograms, detection of single capillaries is possible only in a limited fashion, e.g. in areas of low capillary density or in regions of a single capillary layer [30]. Differences in field angles can also lead to similar problems [5]. Hence, all of these parameters must always be specified in order to reproduce the results as well as to compare absolute values of different studies and to interpret them. Studies performed on networks are summarized in Table 4.2.

The above mentioned factors partly explain the wide variations in results encountered in the literature. These variations can also be attributed to problems inherent to the methods used in estimating D. The studies of Masters et al. [21] confirm the above-mentioned influence of methods used to estimate D. Indeed, values obtained by the mass-radius technique were much lower than those obtained by the density-density (two-point) correlation function method. The former relationship does not seem to be adequate for structures exhibiting inhomogeneity such as the RV [32]. Others [6, 29] have used a box-counting method in their studies. This method is versatile, easy to implement, but suffers from a lack of precision. We have developed an efficient, modified box-counting method ($MBCM$) based on a new algorithm to objectively set parameters involved in the estimation of fractal dimension. The $MBCM$ allows application of the same process of calculation to the image using all the data, thus avoiding variations in D resulting from different processing. It is not therefore necessary, for example, to perform an interactive best fit estimation. The utility of measuring the dimension of a projection for a structure embedded in a d space onto a $d - m$ space has been questioned. For a set $F \subseteq R^2$, we denote its orthogonal projection onto L_θ by $Proj_\theta(F)$, where L_θ is the line through the origin of R^2 that makes an angle θ with the horizontal axis. We have the obvious relation:

Table 4.2. Summary of studies on retinovasculature: parameters and concluded mechanisms. A+V:whole network; A&V: each network separately; C: capillaries; N: normal; P: pathologic; BDR: background diabetic retinopathy; CRVO: central retina vein occlusion; * : from data of Toussaint D, Danis P. Arch Ophtalmol. 1970; 83: 281-285; (1) and (2): when considering two different slopes; (3): all data considered.

YEAR REF.	SAMPLE	NETWORK CONSIDERED	RANGE OF CALCUL	FUNDUS ANGLE
1989 [21]	3 N fundus photographs 2 P fundus photomontages 1 N angiography	Human A+V		30 deg &140 deg 60 deg 140deg
1990 [23]	15 N photomontages	Human A+V		small
1990 [31]	6 BDR angiographies collages	Human A&V	Vessels>C	wide field
1991 [25]	10 angiographies collages	Human A&V		
1992 [26]	3 indian-ink filling pictures from ref. [12] at different age 1 fundus photography	embryonic cat Human A+V		wide angle
1992 [27]	23 N angiograms 3 BDR angiograms 5 CRVO angiograms	Human A+V and A& V		50deg
1993 [24]	1 P fundus photography (evolution in time)	Human A+V		
1993 [6]	23 N angiograms	Human A+V and A&V	vessels> 40μ	60deg
1993 [28]	10 P photographs 14 N photographs	Human A+V		15deg
1994 [29]	10 N photomontages of 6 patients	Human A+V		30deg

Continued:

YEAR REF.	METHOD OF CALCULATION	VALUE OF D: A	V	A+V	C	CONCLUDED MECHANISM
1989 [21]	mass-radius density-density			1.72-1.82 1.64-1.71 1.69-1.88		DLA
1990 [23]	mass-radius density-density computerized			1.7		DLA
1990 [31]	mass-radius	1.63	1.71		1.82*	DLA
1991 [25]	mass-radius	1.6	1.7			Laplacian Processes
1992 [26]	density-density				1.825- 1.844 1.875	Invasion Percolation
1992 [27]		1.64 1.6	1.66 1.66	1.76 1.75 1.75		
1993 [24]	density-density validated			Initial state ≈ 1.67 New vessels ≈ 1.8 After regression ≈ 1.71		
1993 [6]	box-counting validated	1.48 1.19 1.64	1.49 1.19 1.66	1.63(1) 1.29(2) 1.76(3)		Laplacian Processes
1993 [28]	density-density validated			1.845 1.708		Invasion Percolation DLA
1994 [29]	box-counting			1.68-1.72		Laplacian process?

$dim_H(proj_\theta F) \leq min(dim_H F, 1)$, where dim_H is the Hausdorff dimension. The opposite inequality holds in the following manner: Let $F \subseteq R^2$ be a Borel set. a) If $dim_H F \leq 1$, then $dim_H(proj_\theta F) = dim_H F$ for almost all $\theta \in [0, \pi)$ b) If $dim_H F > 1$, then $dim_H(proj_\theta F) = 1$ for almost all $\theta \in [0, \pi)$ This projection theorem generalizes to higher dimensions. Hence, in general, projecting a $D \leq d - m$ dimensional fractal F embedded in $d-$dimensional space onto a $d - m$ dimensional surface results in a structure with the same fractal dimension, $D(proj(F)) = D$. For $D > d - m$, the projection fills the surface and $D(proj(F)) = d - m$ [33]. Some years ago, Masters et al. [4] introduced fractal analysis in the study of the RV by demonstrating the fractal nature of its structure. The retinal network has been extensively investigated since then. Fractal analysis has been proposed as a noninvasive diagnostic technique in ophthalmology [5]. However, fractal dimension seems more useful for a dynamic follow-up of pathological than for inter-comparison between patients. Again, this is due to the fact that estimated fractal dimension is not an absolute value. The different values found for D have led their corresponding authors to support different hypotheses concerning the mechanism of formation of retinal vessel patterns in the developing human eye [6, 21, 25, 26, 31, 34]. Based on the physical similarities of the structures and on D values of the same order, DLA was proposed as a possible mechanism of formation of the RV, involving diffusion of some AF [6, 21]. Fractal dimension of some DLA structures, estimated using $MBCM$, approaches the values found for the AV network i.e. $D = 1.6$. Nevertheless, as already mentionned, fractal dimension is not a sufficient parameter to characterize a structure and to conclude for a mechanism of its formation. Growth of many complex structures is governed by a diffusion-limited process under non-equilibrium conditions, that is, irreversible sticking of particles to form clusters. These processes, commonly called diffusion-limited growth, are governed by the Laplace equation. Witten and Sander [35] introduced DLA models in the early eighties. In these models, particles are added one at a time to a growing cluster or aggregate of particles, via random-walk trajectories originating from outside the region occupied by the cluster [36]. The growth probabilities, $\phi(x, y)$, at the surface of the growing cluster obey the Laplace equation:

$$\nabla^2 \phi = (\delta^2\ \phi(x, y)/\Delta\ x^2) + (\delta^2\ \phi(x, y)/\delta\ y^2) = 0 \tag{4.1}$$

with boundary conditions $\phi(x, y) = 0$ on the cluster and $\phi(x, y) = 1$ on a certain circle surrounding the growing object. According to the previous definition, the diffusing matter which reaches the cluster will aggregate and form part of it. Daxer [34] rejected the DLA hypothesis because of the fact that DLA structures are formed by the aggregation of diffusing particles: AF in this case. However, diffusion could involve particles diffusing to finally aggregate as well as particles diffusing to stimulate growth (aggregation). We have to mention contradictions reported in the literature concerning the mechanism of growth. Based on a D value for the whole network - statistically different from individual A and V networks - some authors [6, 21] suggest a DLA mechanism while others [31] have come to the same conclusion by analyzing the A and V networks separately. Moreover, the evolution of vascular vessels is not taken into consideration when proposing a model for its development: based on high D values obtained in cat embryonic retinal vessels [26] and in pathological conditions in humans, the percolation invasion model was proposed for vascular development while for lower values of D, the DLA model was suggested for the normal adult RV [36]. This raises the question about how progression occurs from one mechanism to the other: from percolation (in embryos) to DLA (in adults) to percolation (in pathological conditions). The subsequent Laplacian model proposed, the dielectric breakdown model (DBM), was initially developed to handle problems in potential theory to simulate growing clusters seeking to maximize a potential and subject to boundary constraints posed by the source of the diffusion and the growing cluster itself [37]. DBM has afterwards been used on a wide scale in physics to explain various fractal growth phenomena [38]. If by $\phi(x, y)$ we denote the potential, where (x, y) are the coordinates of location, function $\phi(x, y)$ satisfies the Laplace equation: $\nabla^2\phi(x, y) = 0$. The function ϕ of equation (4.1) could represent, for example, flow velocity or a nutrient concentration. Sites adjacent to the cluster grown so far are candidates for growth and have a probability of growth $p(x, y)$ proportional to some power μ, called the growth exponent of $\phi(x, y)$, that is :

$$p(x, y) \approx \phi^{\mu}(x, y) \tag{4.2}$$

Boundary constraints are defined as: $\phi(x, y) = 1$ for a certain circle surrounding the growing object $\phi(x, y) = 0$ in and on the edge of the cluster. The general shape of the pattern and their corresponding D values, obtained when parameter μ is set to 1 in equation (4.1), closely recall the results obtained from DLA [39]. Variation of the growth exponent μ generates a continuum of shapes for which D values vary in the whole interval $[1, d]$ where d is the dimension of the embedding space. This parameter μ allows the DBM to be quite versatile. This characteristic along with visual similarity could be tempting in favor of the DBM mechanism to explain the evolution of the vascular network. However, detailed study of the dynamic behavior of a growth structure is essential to detect the underlying formation mechanisms involved. DLA as well as DBM can not simulate dynamic growth of the retina because they do not take into account physiological events, especially regression phenomena. Retraction processes occur on the polygonal structure of capillaries, eliminating some capillaries and thereby forming a branched structure. Even if static description of the RV shows similar geometries and similar D values with DLA, Laplacian processes are growth irreversible ones that do not include the phenomenon of regression. The mature RV is the result of two distinct mechanisms, partly separated in time and space. In previous studies dealing with the involvement of Laplacian processes in the formation of the RV, it is not always clearly stated which step of the evolution of the retina these processes explain. In any case, dynamic evolution as well as a certain spatial preference ruled by vascular angiogenesis (areas adjacent to the foveal avascular zone and to the ora serrata) are neglected (see above). In general, analyzed images show mainly the gross vasculature resulting from the first mechanism (vasculogenesis). However, capillaries and anastomosis phenomena which occur in both stages of development are not considered: DLA structures have no loops [10]. The asymptotic behavior of DLA is still unclear. Some studies have supported a statistical mono-fractality of DLA structures [8]. In this case, the region which no longer grows, that is the frozen region, can be considered as representative of the asymptotic structure. A static comparison of this frozen region with the RV is thus allowed. Visual analysis of the RV shows some discrepancies with DLA structures. In the former, there is evidence of attraction in some areas (area centralis). This phenomenon is completely absent from DLA structures because of the screening effect: large branches hinder the growth of small ones, creating fjords between branches. Moreover, Arneodo et al. [8] reported the existence of screening angle distribution (distribution of angles between branches of successive generations) of preferential angle of $\Theta \approx 36$ deg. Starting from the optic disc, each bifurcation defines a ne w generation of vessels for which angles were estimated. Statistical data performed on our retinal images (for two successive generations) gives a mode of 70 deg for the arterio-venous network. Besides, other studies have suggested the non-self-similarity of plane DLA [40, 41]. Recently, simulation of DLA clusters (on-lattice, $N \approx 10^6$) showed at small scales a reasonably circular shape consisting of the frozen region and at larger scales a cross shape [42]. Fractal dimensions have been measured on such structures: until the shape is about circular, $D \approx 1.70$, while for larger sizes and non-circular shapes, this value apparently drops to about $D \approx 1.50$. This result raised a controversy about the effect of lattice anisotropy on the asymptotic behavior of the fractal dimension. This anisotropic behavior disappears if simulation takes place off-lattice. However, a controversy remains concerning the asymptotic behavior. Coniglio [41] proposed a multiscaling form which gives a different fractal dimensionality. It has been argued that this change in fractal dimension is a crossover effect due to the finite size of the cluster [43]. As mentioned above, this multiscaling behavior has been rejected [8, 44], suggesting rather fluctuations around one global dimension, even for very large clusters of the same size order as the one studied by Coniglio [41]. This controversy about the changes of D with size does not allow us to conclude whether a single value of D can characterize the whole structure. Finally, DLA is governed by a diffusional process defined by the Laplacian equation. However, at the molecular scale, one can suspect other intervening factors in the retina such as surface tension, capillary forces, topology of the environment, migration, convection, surface charge, hemodynamic factors... The existence of local gradient concentration of AF or oxygen is also to be considered to account for preferential growth on some sites rather than on others, separated by a short distance. In conclusion, Laplacian processes are not responsible - at least alone - for the formation of the RV as static and dynamic analyses of the models do not display characteristics similar to the existing RV. More complex phenomena seem to account for the formation of retinal vessels that involve mechanical and physico-chemical factors.

Acknowledgement. Mrs. Soula Kyriacos is the recipient of the Julien Braun Pharmacy Studentship from the Merck Frosst Company.

References

1. Mandelbrot, B. (1982): *The Fractal Geometry of Nature.* Freeman, San Francisco.
2. Vico, P.G., Cartilier, L.H. (1993): A new approach to the study of skin vascularization. *Plast & Reconstr Surg* 92, 463–468.
3. Nekka, F., Kyriacos, S., Kerrigan, C., Cartilier, L.H. (1996): A model of growing vascular structures. *Bull Math Biol* 58(3), 409–424.
4. Masters, B.R., Family, F., Platt, D.E. (1989): Fractal analysis of human retinal vessels. *Biophys J* 55 (Suppl): 575a.
5. Masters, B.R. (1990): Fractal analysis of human retinal blood vessels patterns: developmental and diagnostic aspects. In: Masters B R (ed) 1990 *Noninvasive Diagnostic Techniques in Ophthalmology.* Springer-Verlag, New York, 515–527.
6. Landini, G., Misson, G.P., Murray, P.I. (1993): Fractal analysis of the normal human retinal fluorescein angiogram. *Curr Eye Res* 12, 23–27.
7. Kyriacos, S., Buczkowski, S., Nekka, F., Cartilier, L.H. (1994): A modified box-counting method. *Fractals* 2, 321–324.
8. Arneodo, A,, Argoul, F., Bacry, E., Muzy, J.F., Tabard, M. (1992): Golden mean arithmetic in the fractal branching of diffusion- limited aggregates. *Phys Rev Lett* 68(23), 3456–3459.
9. Russ, J.C. (1990): *Computer-Assisted Microscopy-The Measurement and Analysis of Images.* Plenum Press, New York.
10. Feder, J. (1988): *Fractals.* Plenum Press, New York.
11. Gariano, R.F., Iruela-Arispe, M.L., Hendrickson, A.E. (1994): Vascular development in primate retina: comparison of laminar plexus formation in monkey and human. *Invest Ophthal Vis Sci* 65, 3442–3455.
12. Chan-Ling, T.L., Halasz, P., Stone, J. (1990): Development of retinal vasculature in the cat: processes and mechanisms. *Curr Eye Res* 9, 459–476.
13. Navaratnam, V. (1991): Organisation and reorganisation of blood vessels in embryonic development. *Eye* 5, 147–150.
14. Forrester, J.V., Shaffiee, A., Schroder, S., Knott, R., McIntosh, L. (1993): The role of growth factors in proliferative diabetic retinopathy. *Eye* 7, 276–287.
15. Stone, J,, Itin, A., Alon, T., Peer, J., Gnessin, H., Chan-Ling, T. et al. (1995): Development of retinal vasculature is mediated by hypoxia -induced vascular endothelial growth factor (VEGF) expression by neuroglia. *J Neurosci* 15(7), 4738–4747.

16. Sebag, J., Wallace McMeel, J. (1986): Diabetic retinopathy. Pathogenesis and the role of retina-derived growth factor in angiogenesis. *Survey Ophtalmol* 30(6), 377–384.
17. Boulton, M., Patel, B., Khaliq, A., Moriarty, P., Jarvis-Evans, J., McLeod, D. (1992): Modulators and milieu in preretinal neovascularisation. *Eye* 6, 560–565.
18. Penfold, P.L., Provis, J.L., Madigan, M.C., van Driel, D., Billson, F.A. (1990): Angiogenesis in normal human retinal development: the involvment of astrocytes and macrophages. *Graefe's Arch Clin Exp Ophtalmol* 228, 255–263.
19. McIntosh, L.C., Muckersie, L., Forrester, J.V. (1988): Retinal capillary endothelial cells prefer different substrates for growth and migration. *Tissue & Cell* 20(2), 193–209.
20. Caswell, W.E., Yorke, J.A. (1985): *Dimensions and Entropies in Chaotic Systems*. Springer-Verlag, New York.
21. Family, F., Masters, B.R., Platt, D.E. (1989): Fractal pattern formation in human retinal vessels. *Physica D* 38, 98–103.
22. Masters, B.R. (1989): Fractal analysis of human retinal vessels. *SPIE MedTech'89:Medical Imaging* 1357, 250–256.
23. Masters, B.R. (1990): Fractal patterns in the human retina and their physiological correlates. *SPIE Biostereometric Technology and Applications* 1380, 218–226.
24. Daxer, A. (1993): Characterisation of the neovascularisation process in diabetic retinopathy by means of fractal geometry: diagnostic implications. *Graefe's Arch Clin Exp Ophtalmol* 231, 681–686.
25. Kinoshita, M., Honda, Y. (1991): The fractal property of retinal vascular pattern. *Invest Ophthalmol Vis Sci* 32 (Suppl.), 1082.
26. Daxer, A. (1992): Fractals and retinal vessels. *Lancet* 339, 618.
27. Landini, G., Misson, G.P., Murray, P.I. (1992): Fractals and Ophtalmology. *Lancet* 339, 872.
28. Daxer, A. (1993): The fractal geometry of proliferative diabetic retinopathy: implications for the diagnosis and the process of retinal vasculogenesis. *Curr Eye Res* 12(12), 1103–1109.
29. Masters, B.R. (1994): Fractal analysis of normal human retinal blood vessels. *Fractals* 2(1), 103–110.
30. Schrder, S., Brab, M., Schmid-Schnbein, G.W., Reim, M., Schmid-Schnbein, H. (1990): Microvascular network topology of the human retinal vessels. *Fortschr Ophtalmol* 87, 52–58.
31. Mainsters, M.A. (1990): The fractal properties of retinal vessels: embryological and clinical implications. *Eye* 4, 235–241.
32. Vicsek, T. (1989): *Fractal Growth Phenomena*. World Scientific, Singapore.
33. Falconer, K. (1990): *Fractal Geometry. Mathematical Foundations and Applications*. John Wiley & Sons, New York.
34. Daxer, A. (1995): Mechanisms in retinal vasculogenesis: an analysis of the spatial branching site correlation. *Curr Eye Res* 14(4), 251–254.
35. Witten, T.A., Sander, L.M. (1983): Diffusion-limited aggregation. *Phys Rev* 27, 5686–5697.
36. Meakin, P., Tolman, S. (1990): Diffusion-limited aggregation. In: Fleishmann M, Tildesley D J, Ball R C (eds) 1990. *Fractals in the Natural Sciences*, Princeton University Press, Princeton, 133–148.
37. Batty, M. (1991): Cities as fractals: simulating growth and form. In: Crilly A J, Earnshaw R A, Jones H (eds) 1991. *Fractals and Chaos*, Springer-Verlag, New York, 43–69.
38. Kaandorp, J.A. (1994): *Fractal Modelling - Growth and Form in Biology*. Springer-Verlag, New York.

39. Niemeyer, L., Pietronero, L., Wiesmann, H.J. (1984): Fractal dimension of dielectric breakdown. *Phys Rev Lett* 52, 1033–1036.
40. Mandelbrot, B.B. (1992): Plane DLA is not self-similar; is it a fractal that becomes increasingly compact as it grows? *Physica A* 191, 95–107.
41. Coniglio, A. (1993): Is diffusion limited aggregation scale invariant? *Physica A* 200, 165–170.
42. Erzan, A., Pietronero, L., Vespignani, A. (1995): The fixed-scale transformation approach to fractal growth. *Rev Mod Phys* 67(3), 545–604.
43. Ossadnik, P. (1992): Branch order and ramification analysis of large diffusion-limited-aggregation clusters. *Phys Rev A* 45, 1058–1066.
44. Ossadnik, P. (1991): Multiscaling analysis of large-scale off-lattice DLA. *Physica A* 176, 454–462.
45. Sigelman, J., Ozanics, V. (1982): Retina In: Jakobiec F A (ed) *Occular Anatomy and Teratology* Harper & Row publishers, Philadelphia, 442–443.

COLOR PLATES

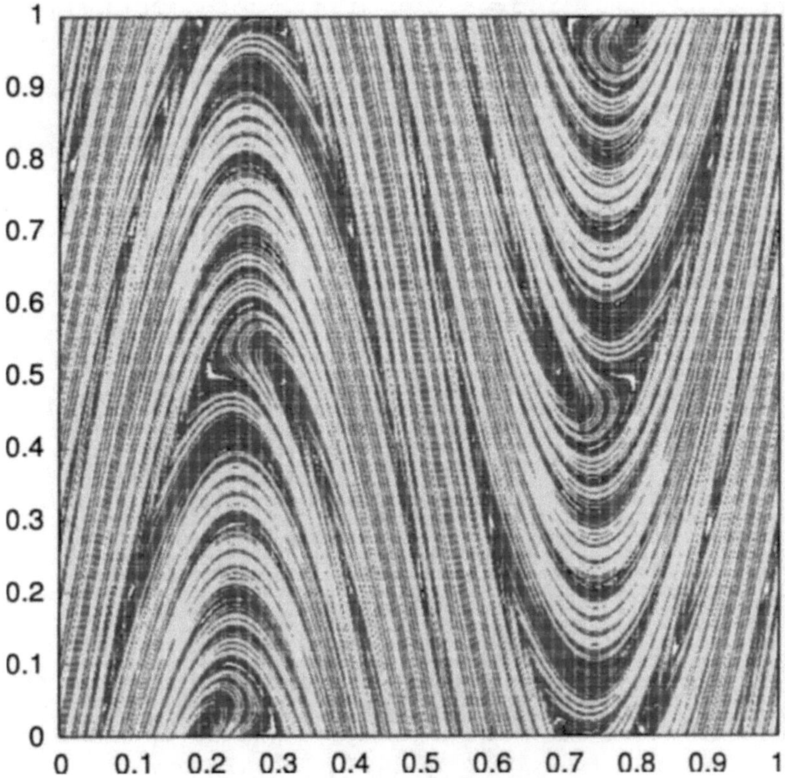

Plate C.1: Density of the material filament in different regions of the flow domain for the globally chaotic case ($T = 1.6$). Colors are assigned *ad hoc* to enhance visual representation of the spatial structure: red: $10^{-4} - 10^{-5}$; yellow: $10^{-5} - 10^{-6}$; green: $10^{-6} - 10^{-7}$; light blue: $10^{-7} - 10^{-8}$; dark blue: 10^{-8}–minimum > 0.

Plate C.2: Illustration of the method presented in 4.3

Plate C.3: Sum of two bidimensionnal attractors

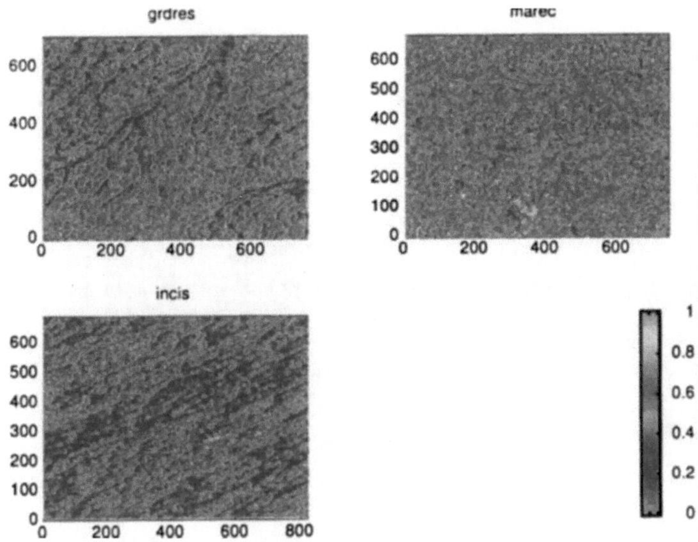

Plate C.4: Intensity maps in a normalized base 5 logarithmic scale

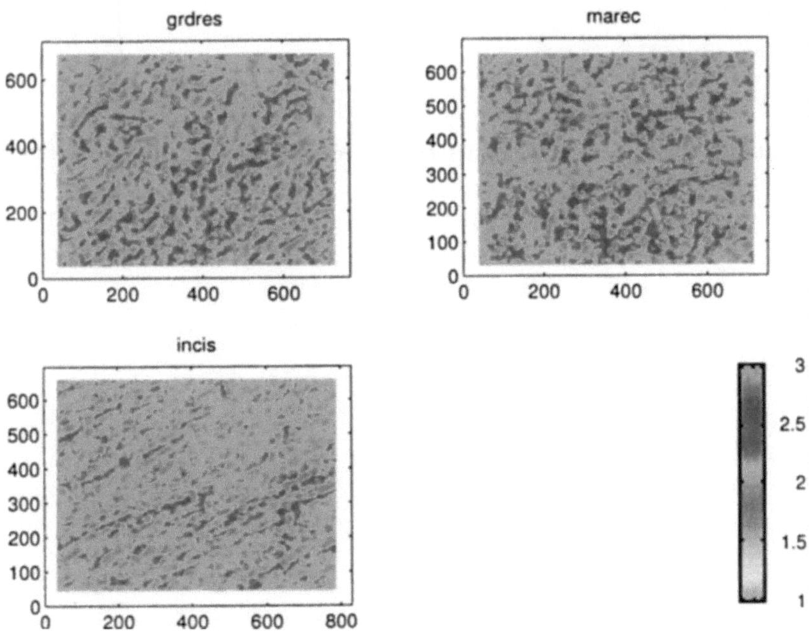

Plate C.5: Maps of singularities computed within a 73x73 pixels scope

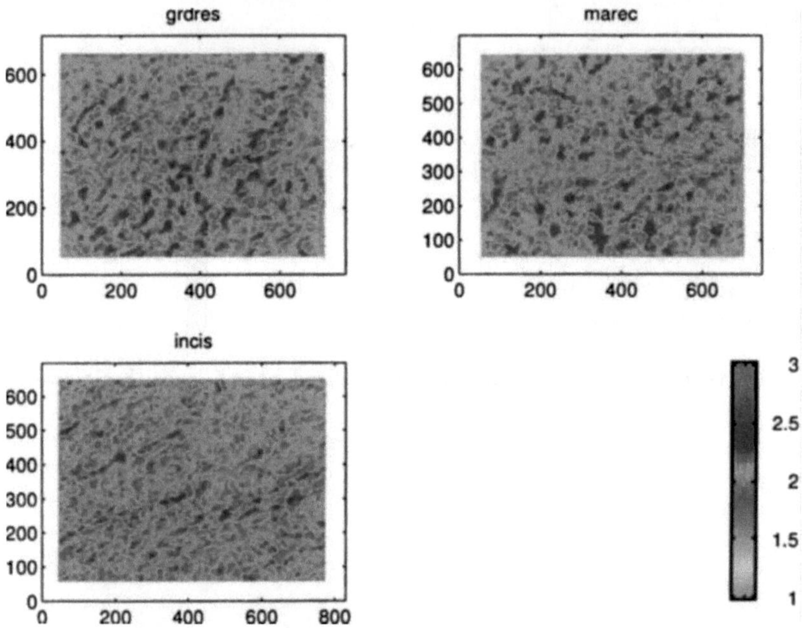

Plate C.6: Maps of singularities computed within a 101x101 pixels scope

MIX
Papier aus verantwortungsvollen Quellen
Paper from responsible sources
FSC® C105338

If you have any concerns about our products,
you can contact us on
ProductSafety@springernature.com

In case Publisher is established outside the EU,
the EU authorized representative is:
Springer Nature Customer Service Center GmbH
Europaplatz 3, 69115 Heidelberg, Germany

Printed by Libri Plureos GmbH
in Hamburg, Germany